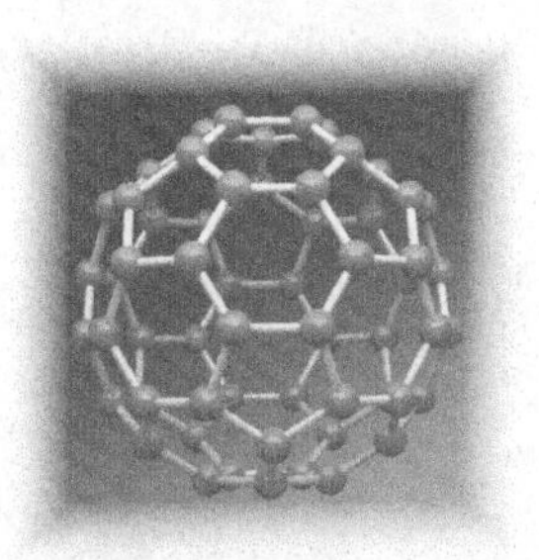

INTRODUCTION TO nanoscience and nanomaterials

INTRODUCTION TO nanoscience and nanomaterials

Dinesh C. Agrawal
Indian Institute of Technology
Kanpur, India

NEW JERSEY • LONDON • SINGAPORE • BEIJING • SHANGHAI • HONG KONG • TAIPEI • CHENNAI

Published by

World Scientific Publishing Co. Pte. Ltd.
5 Toh Tuck Link, Singapore 596224
USA office: 27 Warren Street, Suite 401-402, Hackensack, NJ 07601
UK office: 57 Shelton Street, Covent Garden, London WC2H 9HE

British Library Cataloguing-in-Publication Data
A catalogue record for this book is available from the British Library.

INTRODUCTION TO NANOSCIENCE AND NANOMATERIALS

ISBN 978-981-4397-97-1

Printed in Singapore by World Scientific Printers.

Preface

This book is meant as a textbook for the senior undergraduate and the first year graduate students from different science and engineering departments. Researchers in the area will also find the book useful as a reference material for topics which are outside their expertise. The text assumes only a basic level of competency in physics, chemistry and mathematics.

Apart from the first few chapters, which deal with the topics of surfaces and some relevant topics in physics, the rest of the book is mostly organized according to the dimensionality of the nanomaterials, *i.e.* zero, one and two dimensional nanomaterials followed by bulk nanocrystalline metals and the nanocomposites. The last two chapters are devoted to the topics on the molecules for nanotechnology and the self assembly. A significant amount of space is devoted to "soft matter" which is becoming increasingly important in the fields of nanomaterials and nanotechnology. The focus is on the basics rather than on providing a large volume of information. Sufficient number of solved examples and end of the chapter problems have been included.

Students in a course on nanoscience and nanomaterials are expected to have diverse backgrounds; it is therefore necessary to provide some amount of introductory matter in each chapter. It may be assigned for self study by the instructor if thought fit.

It is hoped that after going through a course based on the book, the student will have a good understanding of the field and will also be able to comprehend the current research literature.

Despite the author's best efforts, there are likely to be some errors. The readers' comments and suggestions to improve the book will be greatly appreciated.

Dinesh C Agrawal

Acknowledgments

The author wishes to acknowledge the support and help provided by various colleagues at the Indian Institute of Technology Kanpur in the writing of this book. In particular, I am grateful to Professors Satish Agarwal and Rajeev Gupta for going through chapters 3 and 5 respectively and for making valuable suggestions. The support from the University Grants Commission in the form of an Emeritus Fellowship is gratefully acknowledged. I am indebted to my family for the constant support and encouragement which has been largely instrumental in bringing the book to completion.

Dinesh C. Agrawal

Contents

Chapter 1

Introduction

A nanometer (nm) is one billionth (10^{-9}) of a meter. Nanoscience can be defined as the science of objects and phenomena occurring at the scale of 1 to 100 nm. The range of 1–100 nm was taken as the defining range by the US National Science and Technology Council in its report titled "National Nanotechnological Initiative: Leading to the Next Industrial Revolution", in February 2000 and a subsequent report in 2004. The committee noted that "Nanotechnology is the understanding and control of matter at dimensions of roughly 1 to 100 nanometers, where unique phenomena enable novel applications. ....At this level, the physical, chemical, and biological properties of materials differ in fundamental and valuable ways from the properties of individual atoms and molecules of bulk matter" [1]. It was also mentioned that "novel and differentiating properties and functions are developed at a critical length scale of matter typically under 100 nm". Actually, in most cases, this happens below 10 nm. On the other hand, some nanoscale phenomena extend beyond 100 nm. The range of 1 to 100 nm appears therefore to be appropriate for the purpose of defining the field as long as it is kept in mind that in some cases one may have to step across these boundaries.

Although the field of nanoscience and nanotechnology has consolidated and emerged as a distinct field only in the decades of eighties and nineties, the activities encompassed by the field have been carried on for decades and in fact for centuries prior to that. Some prominent examples are the use of noble metal particles to impart beautiful colors to the windows of the medieval cathedrals (dating back to fourth century CE), explanation of this effect by Gustav Mie in 1908, development of the ferrofluids in the 1960s, discovery of the magic numbers in the metal clusters in the 1970s, fabrication of the quantum wells in 1970s and many such developments including the development of photography and catalysis.

In 1959 Richard Feynman delivered his famous talk at Caltech titled "There is plenty of room at the bottom" [2]. In it Feynman visualized building of nanosized circuits for use in computers and using precisely positioned assembly of atoms to store vast amounts of information. The Feynman's talk was much ahead of its time — the field of nanoscience and nanotechnology can be said to have come into its own only in the 1990s. The development of newer characterization tools such as the scanning probe microscope and the atomic force microscope and the technique of electron beam lithography capable of making 10 nm structures in 1980s contributed much to the development of the field.

Another factor which provided a great impetus to the area was the need of the computer chip industry. The manufacture of integrated circuits used in the electronics industry is at present carried out by the so called "top down" method in which a thin film is first deposited and parts of it are then etched away to yield the desired pattern. The feature size on the integrated circuits that can be produced by this method has exponentially decreased over the years as predicted by Moore in 1965 [3, 4]. Table 1.1 illustrates this by showing how the line width and the number of transistors per chip in the microprocessors have changed over the years. The number of transistors per chip has increased from 2300 per chip in 1971 to 410 million in 2007 — an increase by a factor of 200,000! This is roughly in accordance with the prediction of Moore that the number of transistors that are placed on a chip should double approximately every two years. Continuing with this trend would mean that the transistor size would have to be scaled down to 9 nm by the year 2016 and to still lower size after that. Photolithography with excimer laser is currently used to fabricate integrated circuits with features below 45 nm and this trend is expected to continue to reach a feature size of about 10 nm. It is apparent that to sustain the growth of the massive semiconductor industry the feature size on the chips will have to be reduced to a few nanometers and new techniques will be needed to accomplish this.

The strategies that are going to be adapted for going below 10 nm are not yet clear. The bottom–up approach in which a structure is assembled from elementary building blocks using noncovalent interactions is of

Table 1.1. Microprocessor trends (reproduced with permission from IC Knowledge LLC)

Year	Product	Process type	Line width (nm)	Transis-tors (M)	Mask layers
1971	4004	PMOS	10K	0.0023	---
1972	8008	PMOS	10K	0.0035	---
1974	8080	NMOS	6,000	0.006	---
1976	8085	NMOS	3,000	0.0065	---
1978	8086	NMOS	3,000	0.029	---
1979	8088	NMOS	3,000	0.029	---
1982	80286	CMOS	1,500	0.134	12
1985	80386DX	CMOS	1,500	0.275	12
1989	80486DX	CMOS	1,000	1.2	12
1992	80486DX2	BiCMOS	800	1.2	20
1993	Pentium	BiCMOS	800	3.1	20
1994	80486DX4	BiCMOS	600	1.6	22
1995	Pentium Pro	BiCMOS	350	5.5	20
1997	Pentium II	CMOS	350	7.5	16
1998	Celeron	CMOS	250	19	19
1999	Pentium III	CMOS	180	28	21
2000	Pentium 4	CMOS	180	42	21
2001	Pentium 4	CMOS	130	55	23
2001	Itanium	CMOS	180	25	21
2002	Pentium 4	CMOS	130	55	26
2002	Itanium II	CMOS	130	220	26
2003	Pentium 4	CMOS	90	>55	29
2006	Core 2	CMOS	65	291	31
2007	Core 2 (Penryn)	CMOS	45	410	35

considerable interest. Bottom–up self–assembly is the way all the plant and biological structures are made. These are hierarchical structures which start out from a simple nanostructure and combine to form a higher level building block; these building blocks further assemble to give the next level of structure and so on. The connective tissue in the body is a hierarchical structure consisting of many levels of assembly. The beautiful colors and iridescence seen in the wings of butterflies and the feathers of many birds such as a peacock arise due to specific nanostructures created by nature using the bottom–up self assembly. Nanotechnologists derive inspiration form these natural nanostructures to design useful nanomaterials. Such materials are called biomimetic materials.

The zero dimensional nanostructures and nanomaterials are defined as those in which all the three dimensions are less than 100 nm. Similarly in the one and two dimensional nanostructures, two and one dimensions are < 100 nm, respectively. Typical examples of the zero, one and two dimensional nanostructures are semiconductor quantum dots, carbon nanotubes and quantum well structures respectively.

The bulk nanostructured materials are those materials which are synthesized in bulk quantities (*i.e.* > mm^3) and have structural features which are controllable at the nanoscale. Examples are nanocrystalline metals and composites of clay or carbon nanotubes with polymers or other matrices.

In the nanomaterials, the surface to volume ratio is very high and increases rapidly as the relevant dimension decreases. For a 3 nm diameter particle approximately 90 % of the atoms reside on the surface. The result is that the surfaces play an increasingly important role in determining the properties of nanomaterials. In the next chapter (Chapter 2), we first discuss the structure and properties of surfaces and illustrate this with some examples of their application to nanostructures. As an illustration, the use of nanostructuring of surfaces to render them superhydrophobic or superhydrophilic is discussed.

Apart from the issue of fabrication of nanomaterials, another issue which has to be faced is the new phenomena which arise when the size is scaled down to such levels. For example in the integrated circuits the quantum tunneling becomes important and the conventional circuits may no longer work. The changes in the properties of materials with size in the range of 1 to 100 nm is a key issue in Nanoscience. Such changes in properties begin to occur when the size approaches a characteristic *scaling length* which is usually different for each property or phenomenon. Many of these changes, *e.g.* those in the electronic and optical properties, arise due to quantum effects. A review of some topics in physics and a discussion of the various scaling lengths is given in Chapter 3 to help understand such changes in properties.

The first example of the zero dimensional nanomaterials, the semiconductor quantum dots, is discussed in Chapter 4. The nanostructures in which the carriers (electrons and holes) are confined within nanometer sized regions in all three dimensions are called quantum dots. This confinement leads to effects such as a change in the band gap and Coulomb blockade. These effects together with the principles involved in the fabrication of self assembled quantum dots is described in Chapter 4.

One manifestation of the change in the band gap of the semiconductor quantum dots with size is the appearance of distinct size dependent colors in the suspensions of these particles. The suspensions of the nanoparticles of some metals also show bright colors; however, the source of this is a different phenomenon called the surface plasmon resonance. This is discussed in Chapter 5.

Several characteristic length scales in the nanometer range exist in magnetism. An important length scale for the magnetic phenomena at nanoscale is the domain size in ferromagnetic crystals. Typically the domain size is in the range 10–200 nm. One of the important effects that is observed as a result of this is the phenomenon of *superparamagnetism* in nanometer sized ferromagnetic crystals. The phenomenon of giant magnetic resistance arises due to the spin of the electrons and is observed in a stack of alternating, nanometer thick, magnetic and nonmagnetic layers. The ability to control and manipulate the spins of electrons, has given rise to the whole new field of spin based electronics or *spintronics* where the electron carries not only the charge but also information in the form of its spin. In Chapter 6, some of these topics are discussed. The chapter ends with a brief discussion of spintronics and the applications of superparamagnetic and other materials.

Colloids can be defined as systems which are microscopically heterogeneous and in which one component has at least one of its dimensions in the range between 1 nm and 1 μm. A model colloidal system is a suspension of particles having diameters ranging from 1 nm to 1 μm in a liquid. Such suspensions are of interest in nanotechnology in the preparation of nanoparticles, their modification and their assembly to yield nanomaterials and nanodevices. In order for the particles to remain suspended (*i.e.* for the dispersion to be stable) for any useful length of time despite the van der Waal's attraction, the presence of a repulsive force between them is essential. Such repulsive interaction is produced either by the presence of electrical charges or by the adsorption of some polymers on the particle surface. The stabilization of a dispersion by these techniques is called *electrostatic stabilization* and *steric stabilization* respectively. In chapter 7 these interactions are first discussed. Secondly, the example of colloidal crystals is used to illustrate the use of nanoparticles in the fabrication of nanostructures.

The carbon nanostructures — graphenes, fullerenes and carbon nanotubes — have remarkable properties making them useful for a large number of applications. In chapter 8 the structure, preparation, properties and some potential applications of these nanostructures are discussed.

The layer structure of graphite is conducive for the growth of fullerenes and one dimensional carbon nanotubes. By folding into tubes or spheres the dangling bonds at the edges of a graphene sheet are eliminated making the structure more stable. This holds true for other layered structures also. The chalcogenides of metals such as tungsten or molybdenum are well known inorganic layered structures. Indeed the first noncarbon nanotubes to be observed consisted of MoS_2 and WS_2.

After the discovery of carbon nanotubes, efforts were intensified to prepare nanotubes of other materials. Although, the layered inorganic compounds were the first ones to be explored because of the similarity with graphite, at present synthesis strategies are available to produce one dimensional nanostructures from almost any material. Nanowires of materials other than carbon are dealt in Chapter 9.

The two dimensional (2D) nanostructures are defined as structures in which two dimensions (say along the x and y axes) are large while the third dimension along the z axis is less than 100 nm. Preparation of the

2D nanostructures is carried out by the techniques of thin film deposition. Thin film deposition is a key process in nanotechnology. In addition to the direct formation of the 2D nanostructures, thin film deposition is used for the preparation of core–shell particles, hollow shells, functionalization of particles and surfaces, and other miscellaneous structures. In chapter 10 the various deposition techniques and their applications in 2-D nanostructures are discussed and one important example, multilayers for mechanical engineering applications, is discussed in some detail.

The bulk nanostructured materials such as nanocrystalline metals and ceramics and nanocomposites have the potential to offer materials with extremely high strengths and other mechanical properties. The polycrystalline metals show a deviation from the expected trends in properties as the grain size goes below a few nanometers. Thus the hardness and the strength of a polycrystalline metal increase with decreasing grain size as (grain size)$^{-1/2}$ according to the well known Hall-Petch relation. However, there is a change in behavior when the grain size approaches a few nanometers. Such changes in the properties of nanocrystalline solids are discussed in this Chapter 11. In Appendix II, an introduction to the mechanical properties of solids is provided.

A composite is a material made from two chemically distinct materials which maintain their identity and have a distinct interface between them in the finished structure. The mechanical properties of a composite are a strong function of the properties of the reinforcement. From this point of view, the carbon nanotubes (CNT) with a Young's modulus of 600 to 1000 GPa, strength ranging from 35 to 110 GPa and a density as low as 1.3 gcm^{-3} can be said to be the ultimate reinforcement material. The polymer–CNT composites are of special interest because of their low density for applications such as aerospace and automobile where weight saving is a very important concern.

In addition to the CNT, several other nanofillers have been used to prepare nanocomposites. Most important amongst these are the composites of polymer with exfoliated clay particles. In Chapter 12, after discussing some basic ideas about composites, the polymer–CNT and the polymer–clay nanocomposites are discussed.

Soft materials such as polymers have come to occupy a special place in nanoscience and nanomaterials. In Chapter 13 some important classes of molecules widely used in soft matter are briefly discussed. These are polymers and biopolymers, surfactants and dendrimers. These are also of relevance to the topic of self assembly discussed in the following chapter.

As mentioned earlier, one of the approaches being investigated to assemble nanosized structures is the self assembly process. In this process the aim is to assemble a structure starting from some building blocks. In contrast to the strong covalent or other primary bonds that are used in the synthesis of a molecule, it is the much weaker secondary bonds that are used in the self assembly process. The weak bonds permit self correction of any defects during the assembly process because such bonds can be easily broken and remade to create a structure with the lowest free energy. The self assembly process thus operates under the "thermodynamic control". In Chapter 14, the various secondary interactions that are made use of in the self assembly are first described. This is followed by a discussion of some important examples.

Chapter 2

Surfaces

2.1 Introduction

The surface of an object becomes increasingly important as its size decreases. Generally the surface effects extend to a depth of few atoms below the surface. A plot of the fraction of atoms on the surface of a cubic crystal with a lattice parameter of 0.2 nm and assuming that the surface layer extends to a depth of 4 atoms from the surface is shown in Fig. 2.1. It is seen that the number of atoms on the surface increases rapidly for particle size $d < 10$ nm and reaches ~ 90% for $d = 3$ nm. This shows that a majority of atoms in a nanoparticle reside on its surface. The surface atoms differ from the bulk atoms in many ways. The coordination number of a surface atom is smaller than that of a bulk atom *i.e.* a surface atom has unsatisfied bonds which make it chemically very active. Therefore, the chemical activity of a material increases drastically as the particle size reaches a few nm. A dramatic manifestation of this is the auto ignition of fine metal powders in air. Because of a smaller coordination number, the surface atoms experience an unbalanced force which is relieved by a readjustment in their positions by *"surface relaxation"* and *"surface reconstruction"*. The surfaces also have many types of defects. Because of these factors, in addition to the chemical activity mentioned above, the other properties (electronic, magnetic, mechanical) of the surfaces may be quite different from those of the bulk. In the macroscopic objects, the fraction of the surface atoms is small and so their influence is weak. However, in the nanosized objects, which are nearly all surface, these properties can be quite different for the object as a whole.

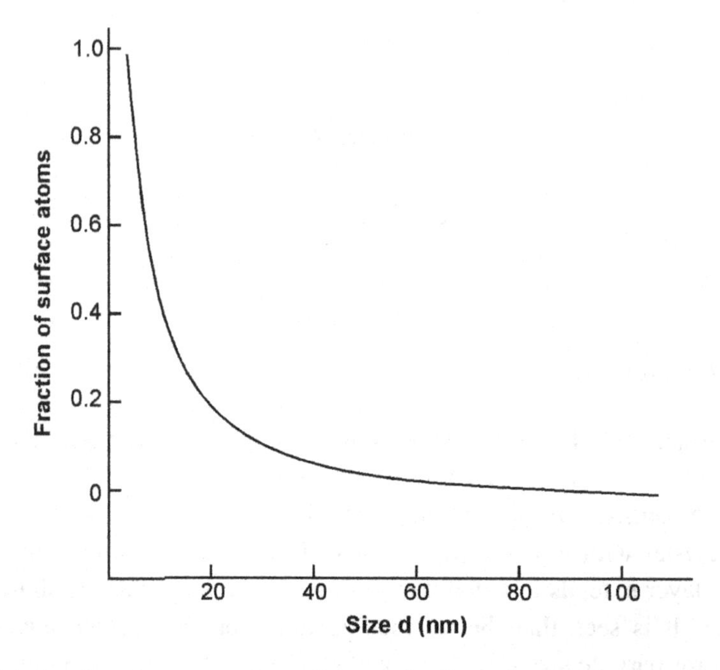

Fig. 2.1. Fraction of atoms on the surface *vs.* size for a simple cubic crystal having a lattice parameter of 0.2 nm and assuming that the surface layer extends to 4 atoms below the surface.

The surfaces play a crucial role in such phenomena as the shape of the nanoclusters, synthesis of nanoparticles, catalysis, sintering, *etc.* Additionally, many of the important processes such as nucleation, growth, dissolution and precipitation, chemical reactions, adsorption, surface modification, self assembly occur at the interface between different phases (solid, liquid and gas). Synthesis, processing, self assembly and properties of nanoscale objects are therefore very sensitive to parameters such as specific surface area, surface structure, surface curvature, and surface energy. These quantities play an increasingly dominant role as the size of an object decreases to below submicron levels. In this chapter, we first discuss these surface properties and illustrate them with some examples of their application to nanostructures. This is followed by a description of the properties of the solid–liquid interface. As an illustration, the use of nanostructuring of surfaces to render them superhydrophobic or superhydrophilic is discussed.

2.2 The solid–vapor interface

A surface is in fact an interface between two phases. What we commonly call the surface of a solid is actually the interface between the solid and the surrounding air. We can similarly have different interfaces such as solid–liquid, liquid–gas or solid–solid and liquid–liquid. In this section the surface refers to the interface between a solid and the air (or vacuum). The solid–liquid interface is dealt with later in this chapter.

2.2.1 *Surface energy*

The formation of bonds between two atoms in a solid is accompanied by a decrease in the energy. As the atoms on the surface have fewer bonds than the atoms in the bulk (Fig. 2.2), the energy of the surface atoms is higher. The extra energy associated with the surface atoms is called surface energy.

A simple estimate of surface energy, γ, can be made by multiplying the number of bonds broken per unit area, N_b, by one half of a bond energy ε

$$\gamma = N_b\left(\frac{1}{2}\varepsilon\right)$$

$$= \frac{1}{2} n_a . n_b \, \varepsilon \tag{2.1}$$

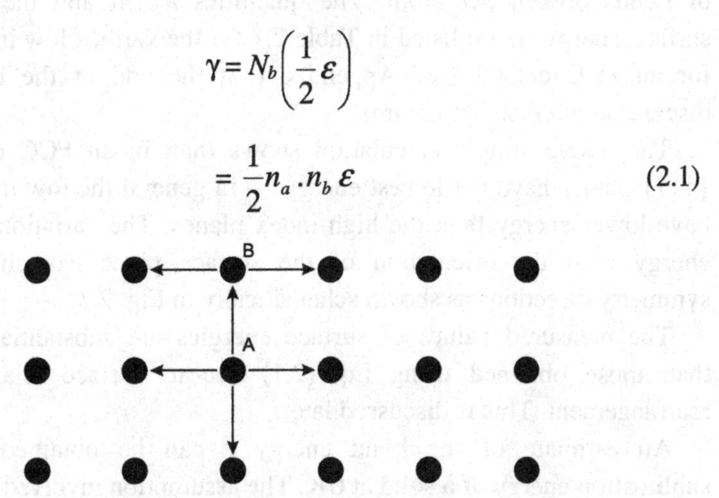

Fig. 2.2. The atoms on the surface of a solid have fewer bonds as compared to the atoms inside the solid as illustrated for a two dimensional solid here. Thus the atom A has four bonds as compared to the atom B which has only three bonds.

Table 2.1. A simple calculation of the surface energy of the planes of an FCC crystal in terms of bond energy ε.

Plane	n_a	n_b	γ
(100)	$\frac{2}{a^2}$	4	$\frac{4\varepsilon}{a^2}$
(110)	$\frac{2}{\sqrt{2}a^2}$	5	$\frac{5\varepsilon}{\sqrt{2}a^2}$
(111)	$\frac{4}{\sqrt{3}a^2}$	3	$\frac{2\sqrt{3}\varepsilon}{a^2}$

where n_a = number of atoms per unit area on the surface and n_b = number of bonds broken per atom. The quantities n_a, n_b and the calculated surface energy, γ, are listed in Table 2.1 for the various low index planes for an FCC crystal (see Appendix 1 at the end of the book for a discussion of crystal structure).

The above simple calculation shows than in an FCC crystal, the {111} planes have the lowest energy — in general the low index planes have lower energy than the high index planes. The variation in surface energy with the orientation of the surface plane has cusps in the symmetry directions as shown schematically in Fig. 2.3.

The measured values of surface energies are substantially smaller than those obtained using Eq. (2.1) due to surface relaxation and rearrangement. This is discussed later.

An estimate of the bond energy ε can be obtained from the sublimation energy of a solid at 0 K. The assumption involved is that the

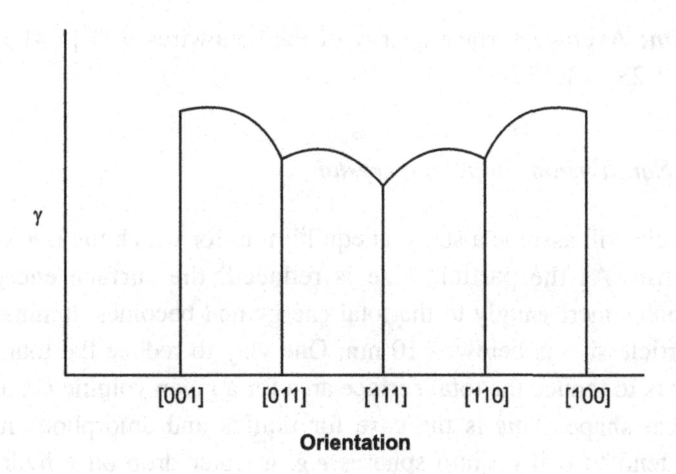

Fig. 2.3. Schematic of the variation of surface energy with orientation of the crystal planes for a FCC crystal; the plot has cusps in the symmetry directions.

cohesive energy of the solid is equal to its sublimation energy at 0 K and that only the nearest neighbor bonds contribute to the cohesive energy. Then, if Z is the coordination number of atoms in the solid and N the Avogadro's number, the sublimation energy can be written as

$$\Delta H_{sub} = \frac{1}{2} ZN\varepsilon$$

so that,

$$\varepsilon = 2\Delta H_{sub}/ZN \tag{2.2}$$

An empirical relation between the cohesive energy of a metal and its melting point, T_M, can also be used. The relation is as follow

$$E_{cohesive} = T_M/(-3.5 \pm 0.3) \tag{2.3}$$

Example 2.1. *During the VLS (vapor–liquid–solid) growth of silicon nanowires (see Chapter 9), the nanowires grow as prisms with nearly equal numbers of {211} and {110} faces at the growth temperature of 1000°C. Estimate the surface energy of the nanowire from the following data: surface energy of the {111} Si = 1.23 Jm^{-2}, ratio of the density of broken bonds on the {211} and {110} faces with respect to the {111} face = 1.41 and 1.22 respectively.*

Solution: Average surface energy of the nanowires = ½ [1.41 × 1.23 + 1.22 × 1.23] = 1.62 Jm^{-2}. □

2.2.2 *Equilibrium shape of a crystal*

A particle will assume a shape at equilibrium for which the free energy is minimum. As the particle size is reduced, the surface energy term contributes increasingly to the total energy and becomes dominant when the particle size is below ~ 10 nm. One way to reduce the total surface energy is to reduce the total surface area for a given volume *i.e.* assume a spherical shape. This is the case for liquids and amorphous materials which tend to ball up into spheres *e.g.* a water drop on a hydrophobic surface. The liquids and amorphous solids are characterized by isotropic surface energy *i.e.* the surface energy is independent of the direction of the orientation of the surface. However, in a crystal it is not so. Each crystal plane is characterized by a different value of the surface energy. To make a crystal assume a nearly spherical shape, many high energy planes may have to be exposed (Fig. 2.4) and the reduction in the total surface energy due to the reduced area may not be enough to compensate for the increase in surface energy due to the presence of the high energy planes on the surface. The crystal therefore tends to assume an equilibrium shape in which the total surface energy is minimized by preferably exposing the low energy planes while keeping the total

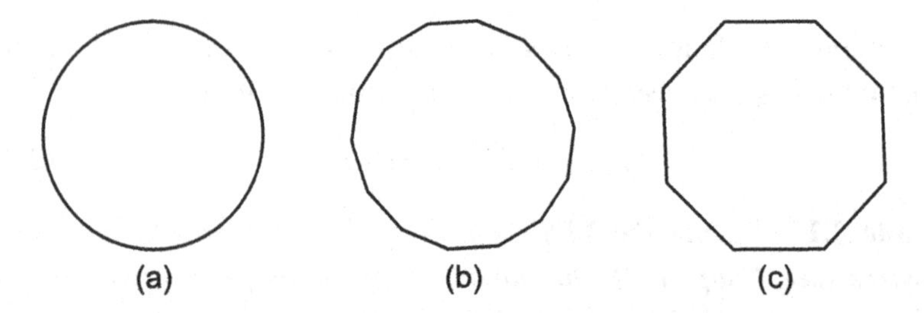

Fig. 2.4. (a) Particle of a liquid or an amorphous solid assumes a spherical shape to minimize the total surface area (b) A crystal can be made to have a nearly spherical shape but this will not be a low energy configuration as this exposes a large number of high energy facets (c) A crystal can better minimize its surface energy by having mostly low index low surface energy facets while simultaneously minimizing the total surface.

surface area to a minimum. As an example, in an FCC crystal, the {111} planes (see Appendix I) have the lowest energy. A shape consisting entirely of {111} planes is an octahedron with a square base and four {111} planes forming a pyramid on each side (Fig. 2.5 a). However, this shape has a high surface area. One way to make the shape closer to spherical is to truncate the corners of the octahedron which exposes the {100} planes, with higher energy than the {111} planes. As the depth of the cut increases, the total surface area of the octahedron decreases but the area of the exposed {100} planes increases. An optimum depth of cut is to be chosen to minimize the total surface energy. This optimization can be done by using a construction called Wulff construction which minimizes the surface energy of a crystal at a fixed volume.

Wulff [5, 6] developed a simple construction which minimizes the surface energy of a crystal at a fixed volume. The Wulff theorem states that "for an equilibrium crystal shape there exists a point in the interior such that its perpendicular distance, h_i, from the ith face is proportional to the surface energy of that face". In the Wulff construction, a set of vectors is drawn from a common point in directions which are normal to the various crystal planes and the length of each vector is proportional to the surface energy of the corresponding plane. Planes are drawn at the end of the vectors and normal to the vectors. A

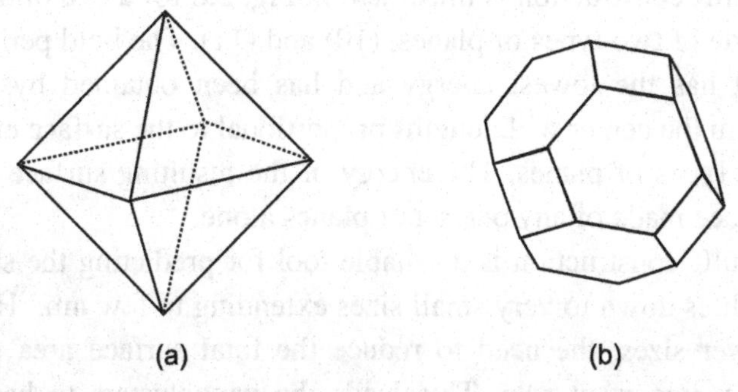

Fig. 2.5. (a) An FCC crystal with only the {111} planes making the surface; the surface to volume ratio is high and so the total surface energy is high even though the {111} planes have the lowest energy. (b) the surface energy can be reduced by truncating the corners of the octahedron; this exposes some high energy {100} planes but reduces the total surface area.

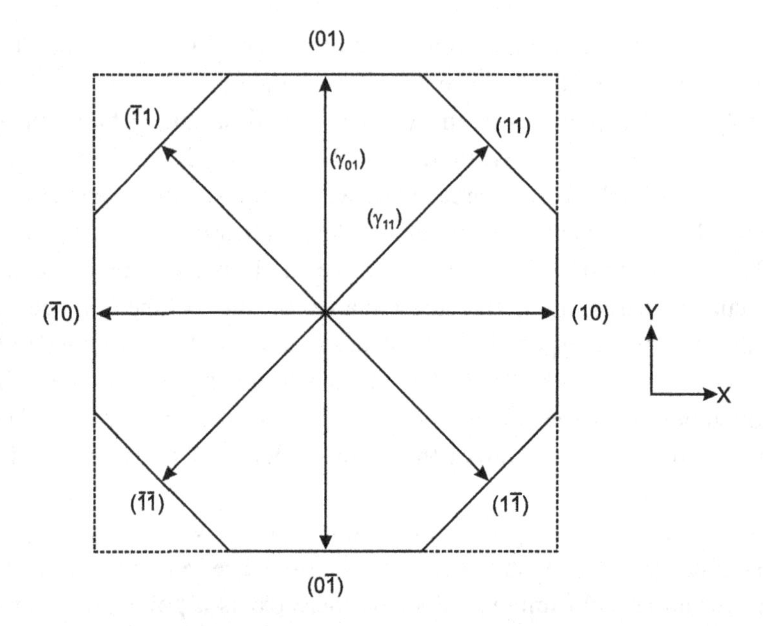

Fig. 2.6. Wulff construction for obtaining the lowest surface energy shape for a two dimensional crystal made of (10) and (11) planes.

geometrical figure formed from such planes, with no planes intersecting other planes is the required surface. An analytical proof of the theorem is given by Benson and Patterson [7].

The Wulff construction is illustrated in Fig. 2.6 for a two dimensional crystal made of two types of planes, (10) and (11). The bold perimeter of the crystal has the lowest energy and has been obtained by drawing vectors from the center with lengths proportional to the surface energies of the two types of planes. The energy of the resulting surface is lower than a surface made of any one set of planes alone.

The Wulff construction is a reliable tool for predicting the shapes of the crystallites down to very small sizes extending to few nm. However, at still lower sizes, the need to reduce the total surface area plays an increasingly important role. This leads the nanoclusters to have other stable shapes which are more spherical. The structure of nanoclusters is discussed in the next section.

Very often the larger crystals also do not have the Wulff shapes. This can happen because of several reasons. If the kinetics of a process involving the crystal (*e.g.* preparation of nanoparticles by wet chemical methods) is very rapid, then there may not be enough time for the equilibrium Wulff shape to be attained. Because of this reason, most of the nanoparticles are spherical. A second reason is that as the temperature increases, the difference in the surface energy for different orientations decreases; the cusps in Fig. 2.3 become blunt and the corresponding facets shrink. Facets finally disappear at a temperature called the *roughening transition temperature* T_R, producing a rough and rounded surface. An example is the cylindrical silicon crystals produced by the Czochralski method.

2.2.3 *Structure of nanoclusters*

A good example of the effect of surface energy on the shape of crystals can be found in the structure of nanoclusters. Aggregates of atoms containing 2 to 10^6 atoms are called nanoclusters. Considering a cluster to be a sphere with a volume equal to the volume of the total number of atoms in it (the so-called Spherical Cluster Approximation, see problem 6), the diameters of the clusters with 125 and 10^6 atoms come out to be 2 nm and 40 nm, respectively.

The nanoclusters are usually grown from the vapor phase or by using the colloidal route. The essence of the vapor phase process is the production of the vapor of the material and then condensing it rapidly to form the clusters. The vapors can be generated by directly heating the bulk material thermally or by hitting it by a laser beam or an ion beam. The hot vapors are then cooled by bringing them in contact with cold inert gas or by first mixing them with high pressure inert gas and then expanding the mixture to get adiabatic cooling.

The kinetics of cluster formation is very important in determining the structure of the cluster. Often the time for the formation of the cluster is not sufficient (see the example below) for the cluster to rearrange itself in the most stable morphology so that clusters with metastable structures are usually obtained.

Example 2.2. *Clusters of Au are growing from the vapor at 1250 K and 10 mbar pressure. Assuming that all the atoms which strike the cluster are incorporated in it, find the time required for the growth of a cluster containing 1000 atoms.*

Solution. From kinetic theory, the number of gold atoms hitting the cluster per second is given by

$$\phi = pA_{eff}/(2\pi mkT)^{1/2}$$

Here p is the pressure of the vapor and $A_{eff} = 4\pi R^2$, where R is the radius of the gold atom. Taking the radius of gold atom as 0.144 nm and the mass of a gold atom as 3.272×10^{-25} Kg and using 1 bar = 100 kPa, $k = 1.38 \times 10^{-23}$JK^{-1}, we get

$$\varphi = \frac{10\times10^{-3}\times100\times1000\times4\times\pi\times(0.144\times10^{-9})^2}{\sqrt{2\times\pi\times3.272\times10^{-25}\times1.38\times10^{-23}\times1250}}$$

$$= 1.38\times10^6 \; atoms \; s^{-1}$$

The time taken to grow a cluster of 1000 atoms is then

$$= \frac{1000}{1.38\times10^6} = 0.72 \text{ ms.}$$

□

As mentioned earlier the Wulff construction is a reliable tool for predicting the shapes of the crystallites down to very small sizes extending to few nm. For the FCC metals, the shape predicted by the Wulff construction is a truncated octahedron. In the truncated octahedron, the crystalline nature is preserved; the shape is, however, far from spherical. There is thus scope for further reduction in energy by devising shapes which are more nearly spherical. Some such shapes, described below, have been predicted and have also been observed experimentally. In these shapes, the lattice has to be distorted to minimize the total surface area. This introduces some strain energy. If d is the size of the cluster, then the total surface energy would be expected to increase with increasing surface area (*i.e.* as d^2) while the total strain energy would increase with the total volume (*i.e.* as d^3). Thus below a critical value of d, the strain energy can be more than compensated by a reduction in the surface energy due to more spherical shape. Two such shapes are described below.

2.2.3.1 *Icosahedron (Ih)*

An icosahedron was proposed by Mackay in 1962. An icosahedron can be thought of as consisting of 20 FCC tetrahedra packed together. As the tetrahedra do not pack closely, they have to be distorted to fill the gaps. If the structure is imagined to consist of shells of atoms arranged around a central atom, then the required distortion causes compression of the radial (intershell) bonds and expansion of the intrashell bonds producing a highly strained structure (Fig. 2.7).

2.2.3.2 *Decahedron (Dh)*

A decahedron consists of two pyramids pointing in opposite directions with a common pentagonal base (Fig. 2.8). The facets of the decahedron are close packed (111) planes. The joining together of the (111) facets necessitates straining of the structure. However, the strains are lower than those in an icosahedron.

To make the surface more spherical, Ino [8] proposed truncating the decahedron at the five corners of the pentagon to reveal five rectangular (100) facets (Fig. 2.8 b). The high energy due to the exposed (100)

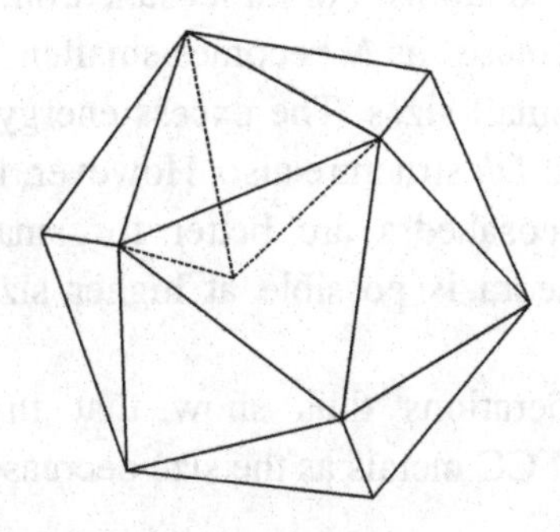

Fig. 2.7. An icosahedron; a tetrahedron unit is also shown.

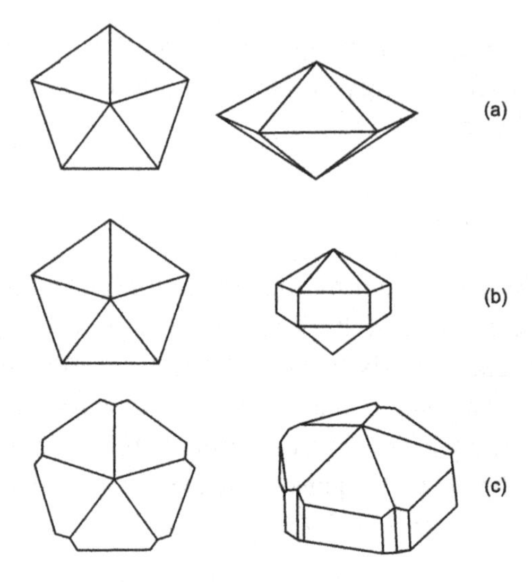

Fig. 2.8. Top and side views of (a) Decahedron (b) Ino's decahedron (c) Marks' decahedron.

facets can be reduced by a further truncation so as to create (111) like reentrant faces as proposed by Mark (Fig. 2.8 c) [9].

To compare the stability of different shaped clusters consisting of identical number of atoms, a parameter $\Delta(N)$ is defined as follows

$$\Delta(N) = (E_b(N) - N\varepsilon_{coh})/N^{2/3}$$

where $E_b(N)$ is the binding energy of the cluster, ε_{coh} is the cohesive energy per particle in the bulk solid; Δ is thus the excess energy divided by the number of surface atoms. For an icosahedron, Δ is proportional to $N^{1/3}$ and so rapidly decreases as N becomes smaller. Thus the icosahedra should be favored at small sizes. The excess energy parameter Δ varies as $N^{1/3}$ for the truncated *Dh* structure also. However, because of the lower surface energy, the icosahedra are better for small clusters while a change over to decahedra is possible at higher sizes because of their lower strain.

The above considerations thus show that the sequence of the structures expected in FCC metals as the size decreases are as follows

FCC → truncated octahedron → decahedron → icosahedron

This sequence of structures is observed experimentally in many cases as well as predicted theoretically from other models for the rare gas and FCC metal clusters. Experimentally, the structure determination of clusters is usually carried out by mass spectroscopy and by electron microscopy and electron diffraction. The mass spectroscopy data agree with the smaller clusters having icosahedral or decahedral shapes. In the mass spectroscopy experiments, the clusters formed by the vapor phase method are ionized by UV radiation and then analyzed in a mass spectrometer. Pronounced peaks are obtained for clusters containing a certain number of atoms. These high intensity mass spectral peaks correspond to clusters of high relative stability. This stability can be understood by considering the shapes such as icosahedra and decahedra to consist of shells of atoms arranged around a central atom. The number of atoms in the nth shell is given by $10n^2 + 2$ so that for a cluster composed of k completed shells, the total number of atoms is

$$N(k) = 1 + \sum_{n=1}^{k} (10n^2 + 2)$$

On expanding this becomes

$$N(k) = \frac{1}{3}(10k^3 + 15k^2 + 11k + 3)$$

Thus the total number of atoms in clusters with 1,2,3, *etc.* shells around the central atom are 13, 55, 147, *etc.* These numbers are called magic numbers and such clusters with a complete, regular outer geometry are designated as full-shell, or '*magic number*' clusters. The stability of clusters with magic number of atoms (relative to their neighbors) is because the complete geometric shells have low surface energies. Mass spectroscopy and other experiments show that the Ar clusters have polyicosahedral structure composed of 13 atom interpenetrating icosahedra for $20 < N < 50$ while for $50 < N < 750$, the multishell Mackay icosahedra are formed. A transition to FCC or a mixture of close packed structures is noted for $N > 750$. Theoretically, the Lennard–Jones potential is the most common model for inert gas clusters. This model also shows that the icosahedron motif is dominant at small sizes (N< 150). For large N (>1600), Marks decahedra are lower in energy than icosahedra. A crossover to the FCC structure is predicted for $N > 10^4$.

Gold clusters produced in inert gas aggregation were identified as FCC for N > 400. For smaller clusters, theoretical models are used to arrive at the possible structure. Planar structures are predicted and have been confirmed experimentally for $N \leq 7$ [10]. A reason advanced for this is that in planar clusters, the d electrons contribute more to bonding than in nonplanar structures.

A number of structures such as decahedron, truncated octahedron, icosahedron and amorphous structures have been observed in passivated gold particles of a few nm diameter. Here, one has to be careful while comparing the results with the theory because the gold clusters are capped by a layer of thiols as soon as they are produced to prevent their aggregation.

The structures arrived at from geometrical considerations have been experimentally observed for the clusters of rare gases and metals like Ag and Au. However, in the alkali metals the geometrical criterion is not found useful. Here, another model, called "*electronic shell closing*" leads to better agreement with experiments. In this model, the cluster is considered to be a superatom and the valence electrons are delocalized and move in a background of positive charge. The model works for small clusters with total number of atoms, N < 2000. For N > 2000, the magic numbers are found to agree with those derived from the geometric considerations. This is interpreted as a transition from a liquid to a solid cluster, because the melting point of a cluster decreases as the cluster size decreases. The effect of the "cluster temperature" and its relation to the geometric and electronic shell effects is yet to be fully understood.

2.2.4 *Surface reconstruction and surface relaxation*

The surface of a solid has a surface energy which is mainly due to the surface atoms having missing bonds on one side, as discussed in Sec. 2.2.1. However, to view the surface of a solid as having the same relative arrangement of atoms as the atoms in the bulk would not be correct. The surface atoms adjust their positions by undergoing small movements so as to reduce the surface energy. The various types of these atom movements are as follows.

2.2.4.1 *Surface relaxation*

Consider the (100) surface of a crystalline solid having a simple cubic structure. An atom at the surface has bonds to four atoms in the (100) plane and one bond to the atom below. Due to one less bond, the attraction exerted by the atom below is not balanced and the surface atom experiences an inward pull. This reduces the spacing between the surface plane and the plane below (Fig. 2.9). Such decrease in spacing may extend to few planes below the surface. This is termed surface relaxation.

2.2.4.2 *Surface reconstruction*

The lateral distances between the surface atoms may also change to reduce the surface energy. Some solids (*e.g.* Au, Ir, Pt, W, Si) show such a change in the lateral spacing. This is called surface reconstruction or surface restructuring. Two examples of surface reconstruction in FCC metals are shown in Fig. 2.10.

As a result of the surface reconstruction, the unit cell vectors for the surface change from (a_1, a_2) to (b_1, b_2). If the relation between the two

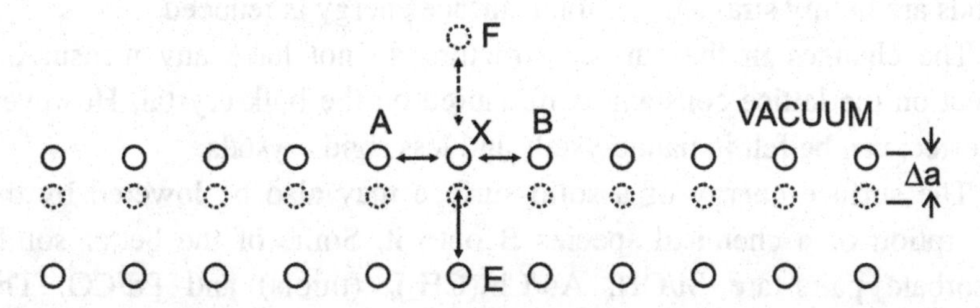

Fig. 2.9. An atom X at the surface is bound to atoms A, B and C, D (not shown) in the plane of the surface and to atom E below the plane but not to the atom F above the plane which is missing; this causes a net downward force leading to a decrease in the spacing of the planes near the surface by an amount Δa as shown by the dotted circles.

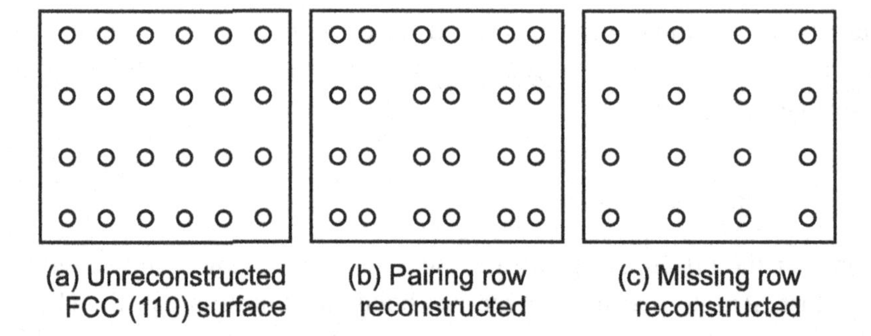

Fig. 2.10. Two types of surface reconstruction of the (110) surface of FCC metals; both types of reconstructions lead to a doubling of lattice spacing along [001] direction.

is such that,

$$\overline{b_1} = n.\overline{a_1} \text{ and } \overline{b_2} = m.\overline{a_2}$$

where n and m are integers, then the structure of the *(h k l)* plane of a crystal A after reconstruction is represented as

$$A(h\,k\,l)\;(n\;x\;m) \tag{2.4}$$

Another well known example is the reconstruction of the (100) surface of silicon by joining together of the bonds from the neighboring surface atoms (Fig. 2.11). The atoms pair up to form rows of dimers which reduces the density of dangling bonds by half. Even though such bonds are highly strained, the total surface energy is reduced.

The changes in the surface structure do not have any measurable effect on the lattice constant as measured on the bulk crystal. However, its effect can be felt in nanocrystals and less rigid crystals.

The surface energy of a solid surface may also be lowered by the adsorption of a chemical species B onto it. Some of the better solid–adsorbate pairs are Si/OH, $Au/Ch_3(CH_2)_n$ (thiols) and Pd/CO. The adsorbed molecules very often form an overlayer, called a superlattice, on the solid. As the number of adsorbate molecules are generally lower than the number of atoms on the surface of the solid, the unit cell of the

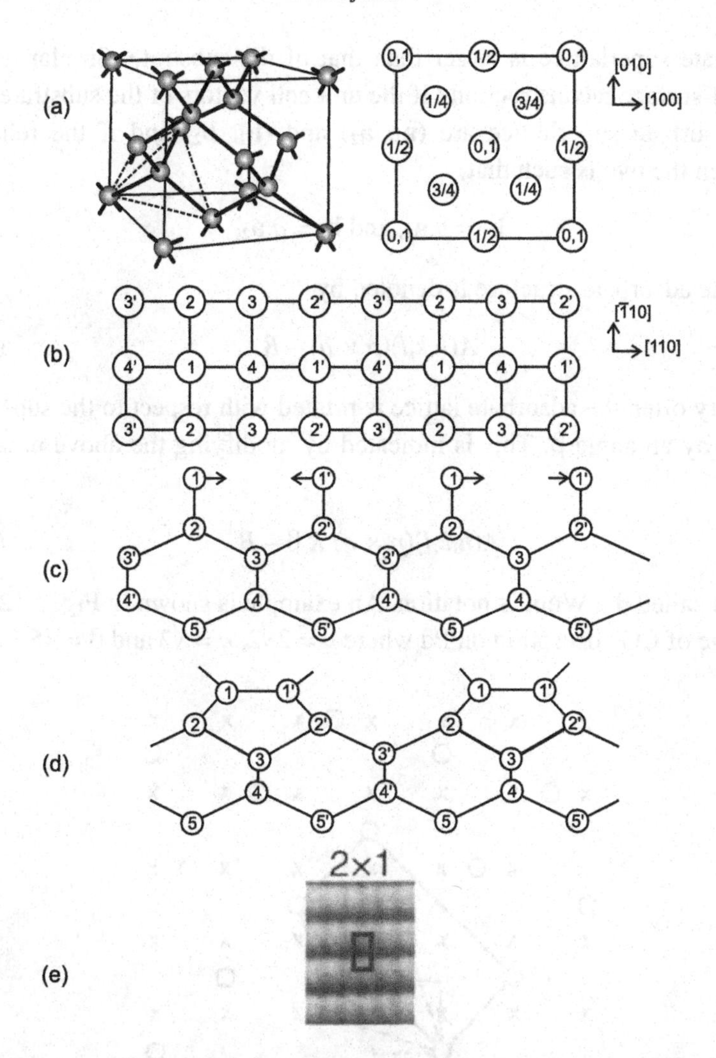

Fig. 2.11. (a) Unit cell of Si showing the arrangement of atoms in a unit cell (left) and top projection of the (001) face (right); the numbers indicate the depth of the atom from the top, taking the height of the unit cell to be unity (b) top view showing the layers of atoms with 1, 1' being the atoms in the top layer, 2 and 2' in the second layer and so on (c) side view of (b); (d) the 1-1' atoms move towards each other to form dimers giving the (2 x1) reconstructed surface (there are small movements of atoms in the lower layers also which have been omitted for the sake of clarity) (e) STM image of 2 x1 reconstructed Si surface with the corresponding unit cell outlined (reprinted from [11] copyright (1997) with permission from Elsevier).

adsorbate superlattice is larger than that of the substrate. Similar to the case of surface reconstruction , if the unit cell vectors of the substrate and the adsorbate superlattice are ($\mathbf{a}_1$, $\mathbf{a}_2$) and ($\mathbf{b}_1$, $\mathbf{b}_2$) and if the relation between the two is such that,

$$\mathbf{b}_1 = p.\mathbf{a}_1, \text{ and } \mathbf{b}_2 = q.\mathbf{a}_2,$$

then the adsorbate structure is denoted by

$$A(h,k,l)(p \times q) - B \tag{2.5}$$

Very often the adsorbate lattice is rotated with respect to the substrate lattice by an angle β. This is indicated by modifying the above notation to

$$A(h,k,l)(p \times q)\, R\,\beta - B \tag{2.6}$$

This is called the Wood's notation. An example is shown in Fig. 2.12 for the case of CO adsorption on Pd where $p = 2\sqrt{2}$, $q = \sqrt{2}$ and $\beta = 45^0$.

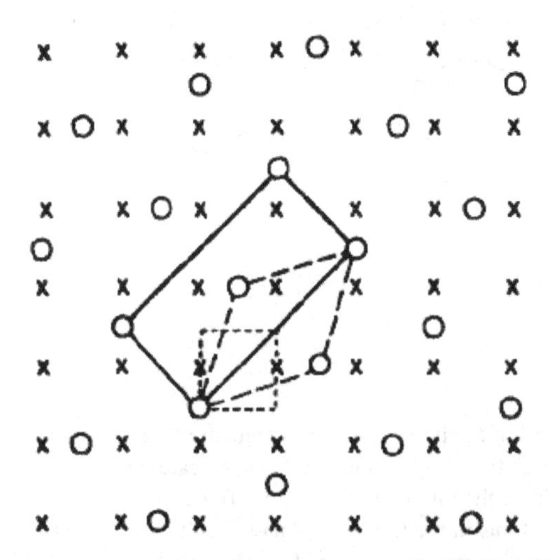

Fig. 2.12. Superlattice of adsorbed CO on Pd; the crosses are the Pd atoms, the circles are the CO molecules, the dotted line is the lattice cell of Pd while the dashed line is the unit cell of the CO superlattice. The structure is denoted by Pd(100)c(2√ × √2)R 45° – CO. (reprinted from [12] copyright (1997) with permission from Elsevier).

2.2.4.3 *Reconstructed surfaces as templates for nanostructures*

Nanopattern formation on a substrate by surface reconstruction can be used to grow nanostructures having a uniform size and spacing. An example of the exploitation of reconstructed surfaces in fabricating nanostructures is the use of the (111) surface of gold to fabricate a self assembled array of Co nanodots. The (111) surface of Au reconstructs, on annealing, into a $(23 \times \sqrt{3})$ reconstructed unit cell. The surface consists of domains in which the stacking is alternately FCC and HCP (ABAB...). At the boundary of the domains there is a row of raised atoms to accommodate the boundary mismatch (which is equivalent to a surface dislocation). These atoms form 0.15 A° high ridges along the $\langle 1\bar{1}2 \rangle$; these ridges separate the FCC regions from the BCC regions. The ridges form a zigzag pattern to relieve the strain better. The result is a herringbone structure.

When Co is deposited on this surface, the Co islands nucleate at the domain boundaries where two ridges meet. These sites provide the nucleation points for the Co islands. Uniform distribution of these sites leads to the formation of uniform sized Co islands forming a regular pattern on the surface. This technique for the formation of nanostructures has been exploited in several cases [13].

2.2.5 *Surface defects and crystal defects*

The surface of a crystal almost invariably has various defects. These defects play an important role during the deposition of films and in the formation of nanostructures by the self assembly of the deposited film.

2.2.5.1 *Defects on the vicinal surfaces*

Consider the surface produced by cutting a crystal at a small angle to a low index plane $(h\ k\ l)$. The surface thus generated is called *vicinal* to $(h\ k\ l)$. To accommodate the misorientation θ of the surface with respect to the $(h\ k\ l)$ plane, steps are produced on the surface. The height of each step, h, is given by $\sin\theta = (h/d)$ where d is the average distance between the steps. Figure 2.13 (a) illustrates the different features of a vicinal

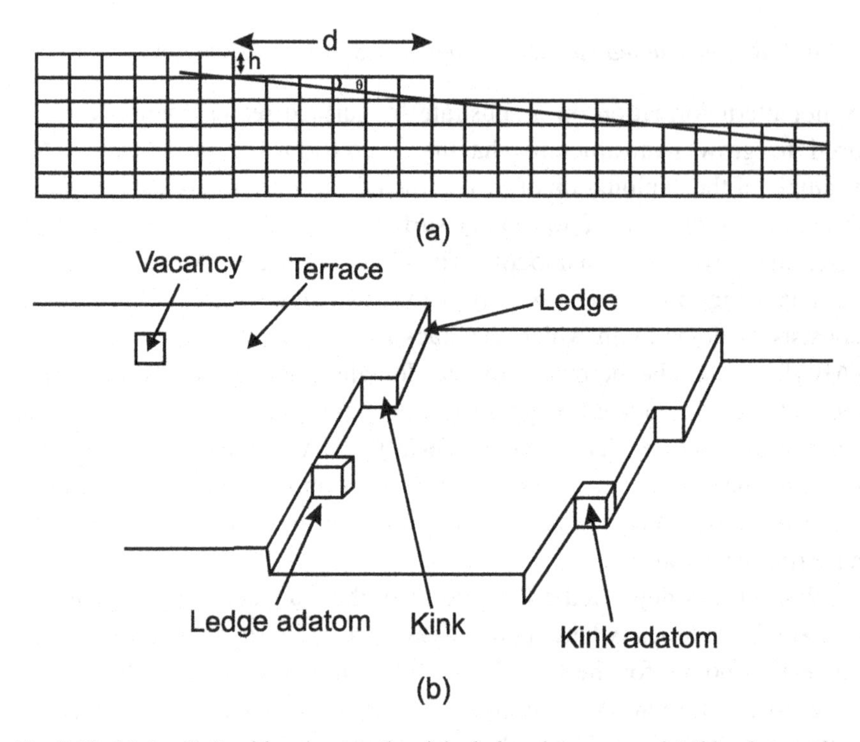

Fig. 2.13. (a) the tilt θ with respect to the vicinal plane is accommodated by the creation of steps. (b) Defects on a vicinal surface.

surface. If there is a component of tilt at right angle to the inclination θ, then kinks are also produced to accommodate this additional tilt (Fig. 2.13 (b)). The presence of ledges on a vicinal surface causes the roughening transition temperature to be lowered to nearly half the melting point. The surface features of the vicinal surfaces can be exploited for the fabrication of a variety of nano- structures. Thus the surface vicinal to W(110) has been used to fabricate parallel nanostripes of Fe. In this case, the surface normal of W deviated from the [110] direction by 1.4° which caused the formation of ~ 9 nm wide steps along the [001] direction. Iron was deposited by evaporation in amounts sufficient to form less than a monolyer. After annealing at 800 K, continuous nanostripes, one atom thick, were formed. Such nanostructures are useful to understand the various interactions which control the properties of magnetic nanostructures [14].

2.2.5.2 *Crystal defects*

A *vacancy* is a lattice site from which an atom is missing. Vacancies are thermodynamically stable defects because their presence increases the entropy. They are present in the bulk as well as on the surface.

Another kind of defect present in a crystal is *dislocation.* It is a line defect. A dislocation in general will have two components to it — an *edge* component and a *screw* component. A pure *edge dislocation* is the edge of an extra half plane of atoms in the crystal (Fig. 2.14 a). If a path ABCDE is traversed around the edge dislocation line, it fails to close even though equal number of atomic steps have been traversed along the top and the bottom rows. The vector **b** required to close the circuit is called the Burger's vector. It is perpendicular to the dislocation line in case of an edge dislocation.

In a pure *screw dislocation*, the dislocation line effectively results in the crystal planes forming a spiral around it (Fig. 2.14 b). A step is formed where the dislocation line emerges on the surface. In the case of the screw dislocation, the Burger's vector is parallel to the dislocation

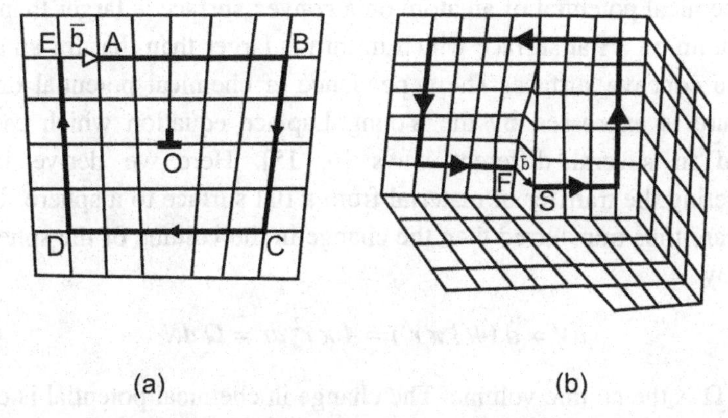

Fig. 2.14. (a) An edge dislocation is the edge of the extra half plane of atoms *i.e.* the line going through the plane of the paper at point O; ABCDE, the Burger's circuit fails to close when a dislocation is enclosed by it (b) a screw dislocation; is produced by a vertical shear by **b** as shown.

line. Note that a dislocation line always ends at the surface unless it forms a junction with another dislocation inside the crystal.

Dislocations are important in processes such as diffusion, adsorption, nucleation, *etc.* During the growth of a nanocrystal, a screw dislocation on the surface provides a ready made site from which the growth can start without the need for nucleation.

2.2.6 *Effect of surface curvature*

The surface energy values given in the handbooks are for flat surfaces. There is a small but significant effect of the curvature of a surface on its surface energy and, consequently, on the chemical potential of an atom on the surface. As the particle size decreases and the curvature increases, this effect becomes important. Dependence of the chemical potential on the curvature also leads to changes in many other important quantities with curvature. Some of these are discussed below. The discussion below is valid for any type of interface although for clarity the solid-vapor or the liquid–vapor interfaces are considered as examples.

2.2.6.1 *Chemical potential at the curved surfaces*

The chemical potential of an atom on a convex surface is larger than that of an atom on a flat surface which in turn is larger than that for an atom under a concave surface. The dependence of chemical potential on the curvature is expressed by the Young–Laplace equation which can be derived in several different ways [6, 15]. Here we derive it by considering the transfer of material from a flat surface to a sphere. If dN atoms are thus transferred then the change in the volume of the sphere is given by

$$dV = d\,(4/3\,\pi\,r^3) = 4\,\pi\,r^2\,dr = \Omega\,dN \qquad (2.7)$$

where Ω is the atomic volume. The change in chemical potential is equal to the change in surface energy per atom on the sphere

$$\Delta\mu = \mu_C - \mu_\infty = \frac{\gamma dA}{dN} = \gamma . \frac{8\pi r dr . \Omega}{dV} = 2\gamma \frac{\Omega}{r} \tag{2.8}$$

If the surface is not spherical but has arbitrary curvature, then this equation becomes

$$\Delta\mu = \gamma\Omega\left(\frac{1}{r_1} + \frac{1}{r_2}\right) \tag{2.9}$$

where r_1 and r_2 are the two radii of curvature of the surface. This equation is called the *Young–Laplace* equation. The radius of curvature for a flat surface is infinite; it is positive for a convex surface and negative for a concave surface. Equation (2.9), therefore, shows that the chemical potential of an atom on convex, flat and concave surfaces has the following relative values

$$\mu_{\text{convex}} > \mu_{\text{flat}} > \mu_{\text{concave}}$$

2.2.6.2 *Pressure under a curved surface*

The existence of a curvature on a surface causes a pressure to exist on the material under the surface. This can be better understood by considering the case of liquids. As a liquid drop tries to minimize its surface area to reduce the surface energy, the molecules inside the drop are compressed. Thus the pressure in the liquid inside the drop is higher than the pressure of the outside gaseous phase. In case of a bubble, the liquid is on the outside and is under a lower pressure than the air inside the bubble. The relation between the surface curvature and the pressure can be derived in the following manner.

We are familiar with the dependence of the free energy on pressure

$$G = G_0 + RT \ \ell n \, P$$

or, equivalently, for the chemical potential

$$\mu = \mu_o + kT \ \ell n \, P$$

(where μ_o is the chemical potential under a standard state and k is the Boltzmann's constant).

Let μ_∞ and μ_c denote the chemical potential of an atom on a flat surface and on a curved surface. Then from Eq. (2.8),

$$\mu_\infty = \mu_0 + kT \, \ell n \, P_\infty$$

and

$$\mu_c = \mu_0 + kT \, \ell n \, P_c$$

where P_∞ and P_c denote the pressure experienced by the atoms at a flat surface and at a curved surface respectively. Taking the difference,

$$\mu_c - \mu_\infty = \Delta\mu = kT \, \ell n \frac{P_C}{P_\infty}$$

But from Eq. (2.7) this is also given by

$$\Delta\mu = \gamma\Omega\left(\frac{1}{r_1} + \frac{1}{r_2}\right) \tag{2.10}$$

So that

$$\ell n\left(\frac{P_C}{P_\infty}\right) \approx \frac{P_C - P_\infty}{P_\infty} \approx \frac{\Delta P}{P_\infty} = \frac{\gamma\Omega}{kT}\left(\frac{1}{r_1} + \frac{1}{r_2}\right) = \gamma\frac{V_m}{RT}\left(\frac{1}{r_1} + \frac{1}{r_2}\right) \tag{2.11}$$

where R is the gas constant and V_m is the molar volume.

or,

$$\Delta P = \gamma\left(\frac{1}{r_1} + \frac{1}{r_2}\right) \tag{2.12}$$

For a spherical surface of radius r

$$\Delta P = \frac{2\gamma}{r}$$

Here $\Delta P = P_c - P_\infty$ is the difference in the pressures under a curved and a flat surface and ideal gas law $PV_m = RT$ has been used. Equation (2.12) is called the *Kelvin equation.* It can be derived independently by considering the force balance on an element of a curved surface [15]. As the radius of curvature of a drop of liquid is positive, the pressure inside the drop is larger than that of the outside atmosphere while in case of a bubble, for which the liquid surface is concave and so has a negative curvature, the pressure in the liquid is lower than that inside the bubble. The same holds true if instead of a liquid drop and the bubble we have a solid sphere and a pore inside a solid respectively.

Some familiar consequences of the change of pressure with curvature are the rise of a liquid in a capillary and the condensation of vapor in a capillary at temperatures above the dew point.

2.2.6.3 *Vapor pressure over a curved surface*

Equation (2.10) also holds true for the vapor phase over a drop of liquid or over a solid particle. If instead of the chemical potential of an atom of the surface we consider the chemical potential of an atom of the liquid or the solid in the vapor phase, then the result is again Eq. (2.12) but the pressure in this case is the vapor pressure over the drop. Thus the vapor pressure over a convex surface is larger and over a concave surface is smaller than that over a flat surface, respectively. This difference in the vapor pressure leads to coarsening of larger particles at the cost of the smaller particles by vapor phase transport as described later. It is also an important mechanism of coarsening during the sintering of particles.

2.2.6.4 *Vacancy concentration under a curved surface*

All crystalline solids have an equilibrium concentration of vacancies at any temperature. As the solid under a convex surface is under compression and that under a concave surface is under tension, the work to create a vacancy in the two cases is larger and smaller respectively than that for a flat surface. As a result, the vacancy concentration is smaller under a convex surface and large under a concave surface than under a flat surface. The dependence of the vacancy concentration on curvature is given by the following relation [16].

$$\ell n \frac{C_C}{C_\infty} = -\frac{\gamma\Omega}{kT}\left(\frac{1}{r_1} + \frac{1}{r_2}\right) \tag{2.13}$$

Here C_C and C_∞ are the vacancy concentrations under a curved surface and a flat surface respectively. According to the above relation, the concentration of the vacancies is largest under a concave surface, smaller under a flat surface and still smaller under a convex surface. This dependence of vacancy concentration on curvature is important in sintering as described later.

2.2.6.5 *Solubility of curved surfaces*

If lumps of salt with rough surfaces are placed in water, soon they will have smooth surfaces. This is because the sharp edges and corners with high positive curvature have more solubility than the regions which are flat or are concave. The dependence of the solubility on curvature is given by the following relation:

$$\ell n \frac{C_C}{C_\infty} = \frac{\gamma\Omega}{kT}\left(\frac{1}{r_1} + \frac{1}{r_2}\right)$$

Here C_C and C_∞ are the solubility of a curved surface and a flat surface respectively. Thus sharp corners and edges with a small positive value of r would have a high solubility. This dependence of solubility on curvature leads to the phenomenon of Oswald ripening (described below) and is important in liquid phase sintering [17].

Example 2.3. *Two spherical particles of 300 nm diameter of a pure element form a neck of radius 50 nm at the point of contact at 600 C.*

(a) *Calculate the difference in the chemical potential between the neck and the surface of the particle.*
(b) *What is the difference in the vapor pressure of the element at the two locations if the vapor pressure over a flat solid surface is 10^{-3} torr?*

Given: atomic volume Ω of the element = 3×10^{-29} m^3, surface tension of the solid = 1.4 Nm^{-1}, k = 1.38×10^{-23} $J K^{-1}$.

Solution. (a) Assume the surfaces to be spherical. The difference in chemical potentials between a curved surface with radius r and a flat surface is given by

$$\mu_c - \mu_\infty = \Delta\mu = 2\gamma\frac{\Omega}{r}$$

Hence

$$\mu_{neck} - \mu_o = 2\gamma\frac{\Omega}{r_{neck}}$$

and

$$\mu_{surface} - \mu_o = 2\gamma\frac{\Omega}{r_{surface}}$$

Therefore

$$\mu_{neck} - \mu_{surface} = 2\gamma\Omega\left[\frac{1}{r_{neck}} - \frac{1}{r_{surface}}\right]$$

$$= \frac{2\times1.4\times3\times10^{-29}}{10^{-9}}\left[-\frac{1}{50} - \frac{1}{150}\right]$$

$$= -2.24 \times 10^{-21} \text{ J}$$

(b) The vapor pressures over a flat surface P_∞ and a curved surface, P_c are related by Eq. (2.11) as follows

$$\frac{P_C}{P_\infty} = 1 + \frac{2\gamma\Omega}{rkT}$$

So,
$$\frac{P_{neck}}{P_\infty} - \frac{P_{surf}}{P_\infty} = \frac{2\gamma\Omega}{kT}\left[\frac{1}{r_{neck}} - \frac{1}{r_{suf}}\right]$$

$$= \frac{2\times1.4\times3\times10^{-29}}{1.38\times10^{-23}\times300\times10^{-9}}\left[-\frac{1}{50} - \frac{1}{150}\right]$$

$$= -0.54$$

Hence, $(P_{neck} - P_{surf}) = -0.54 \times P_\infty = -0.54 \times 10^{-3}$ torr. □

2.2.7 *Examples and applications*

The following examples illustrate the effect of curvature on various phenomena.

2.2.7.1 *Melting point of nanoparticles*

The melting point of a solid decreases drastically as its size approaches a few nm. An example is shown later for gold nanoparticles in Fig. 3.1 (a). The lowering of the melting point can be calculated by taking into account the change in pressure due to the curvature of the surface. Because of the curvature of the nanoparticle, it experiences an excess pressure ($= 2\gamma/r$) given by the Laplace equation. This adds to the chemical potential of the solid nanoparticles and results in a lowering of the melting point (Appendix 2.1).

2.2.7.2 *Coarsening of particles — the Ostwald ripening*

Consider a case in which particles of different sizes are deposited on a flat substrate (Fig. 2.15). This can occur during the process of thin film deposition. As the vapor pressure of the solid species is higher over smaller particles due their smaller radii, there is a transport of these atoms from above the smaller particles to the coarser particles due to the gradient in the vapor pressure. To maintain the smaller vapor pressure over the larger particles, these excess atoms get deposited on the larger particles while atoms from the smaller particle would continue to go into

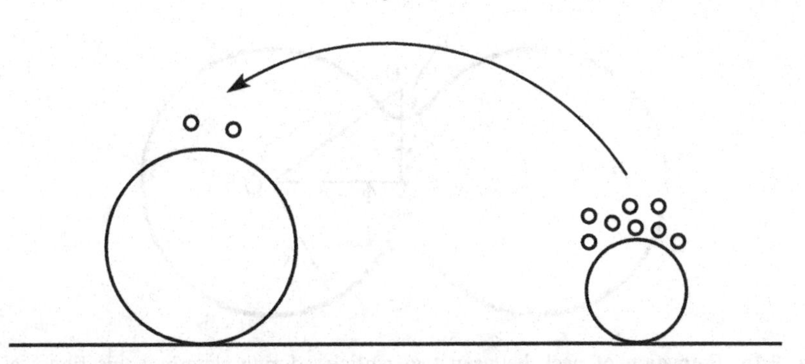

Fig. 2.15. Two particles of different radii on a substrate are shown. The concentration of the atom species of the particles in the vapor phase over the smaller particle is higher due to its smaller radius. Atoms are transported through the vapor phase from the smaller particle to the larger particle.

the vapor phase. Consequently the larger particles would coarsen while the smaller particles would eventually disappear. This process is called Ostwald ripening [18]. The Ostwald ripening also occurs in case of particles suspended in a liquid. In a suspension of particles the saturation solubility of a smaller particle is larger than that of a larger particle because of its smaller radius as discussed above. So if the concentration of solute in the liquid is between these two limits, then the smaller particles dissolve and the solute, after being transported through the liquid phase deposits on the larger particle. The smaller particles eventually disappear and the larger particles increase in size.

2.2.7.3 *Sintering of solids*

A metal can be given a desired shape by melting it and casting in a suitably shaped mould or by deforming the metal. Both of these options are not available with ceramics (such as alumina, Al_2O_3) because they have a very high melting point and are difficult to deform. Ceramics are therefore shaped by compacting their powders and then heating this compact at a temperature about 0.6 to 0.7 of the melting point. During the heating step the phenomenon of sintering occurs due to which the density of the compact increases and it becomes mechanically strong.

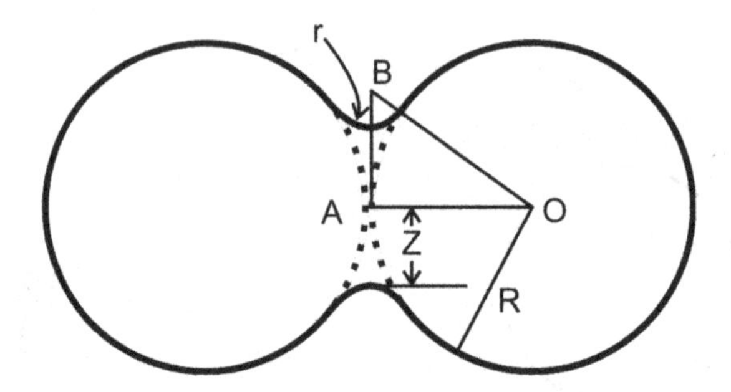

Fig. 2.16. Formation of neck between two particles during sintering; this figure also illustrates the drawing together of two particles due to a small amount of liquid trapped between them.

First a neck forms at the point of contact between two particles, the boundary between the particles now becoming the grain boundary (Fig. 2.16). Several mechanisms operate during the sintering process. One of the important mechanisms is the transport of vacancies from the grain boundary to the neck. This is driven by the difference in the concentration of vacancies at the two sites. The grain boundary can be assumed to be nearly flat while the neck has two radii of curvature, r and z, as shown in Fig. 2.16; r is much smaller than z and is also negative. The concentration of the vacancies at the neck is, therefore, much larger than that at the grain boundary. The net effect is a diffusion of atoms from the grain boundary to the neck. This fills the pores space around the neck and also leads to shrinkage due to a reduction in the distance between the centers of the two particles, leading to sintering.

2.2.7.4 *Condensation of vapor in a capillary: agglomeration of nanoparticles due to pendular forces*

We saw above that the vapor pressure over a curved surface depends on the curvature. A familiar consequence of this effect is the condensation of vapor in a capillary or a fine pore. This is illustrated for a capillary in Fig. 2.17. On a flat surface, the condensation of liquid occurs only when the vapor pressure exceeds the saturation pressure. However, in a capillary, the liquid condenses and forms a concave meniscus. As the

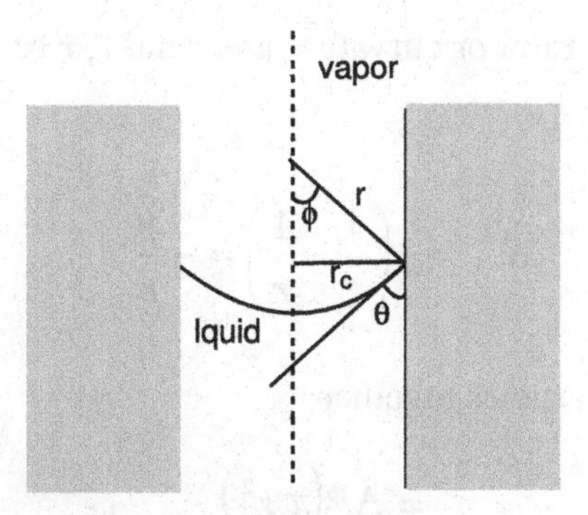

Fig. 2.17. Vapor condensation in a capillary. The equilibrium vapor pressure over the concave liquid surface is smaller than the vapor pressure out side the capillary, leading to condensation; the radius of curvature of the meniscus is $r = r_c/\cos\theta$ where r_c is the capillary radius and θ is the contact angle.

equilibrium vapor pressure over the concave meniscus is lower than in the outside air, the vapor condenses in the capillary *i.e.* the effective dew point in the capillary is higher. This effect also causes moisture to condense in the pores of a body and near the contact point between two fine particles even at temperatures above the Dew point (see Example 2.4). The contact region between the nanoparticles behaves as a capillary, leading to the condensation of the moisture. The miniscule amount of the condensed water forms a bridge between the particles, holding them together with a force known as the pendular force [19] causing them to agglomerate.

Example 2.4. *Show that two spherical particles in contact are pulled together by a force equal to $2\pi\gamma r$ where r is the radius of each particle and γ is the surface tension of the liquid which condenses between the particles.*

Solution. The liquid condensed between the particles has an excess (negative) pressure due to its curvature given by

$$\Delta P = \gamma\left(\frac{1}{r_1} + \frac{1}{r_2}\right)$$

For the drop, the two radii or curvature are z and r, r being much smaller than z (Fig. 2.16).

Therefore $$\Delta P = \gamma\left(\frac{1}{z} - \frac{1}{r}\right) \approx -\frac{\gamma}{r}$$

Force attracting the particles together

$$= \Delta P\left(\pi z^2\right)$$

Now from triangle OAB, $OB^2 = OA^2 + AB^2$

or $$(R + r)^2 = R^2 + (z + r)^2$$

$$R^2 + r^2 + 2Rr = R^2 + z^2 + r^2 + 2zr$$

or

$$2Rr = z^2 + 2zr \approx z^2$$

Hence the attractive force between the particles

$$= \Delta P\left(\pi z^2\right) = 2\pi Rr.\frac{\gamma}{r} = 2\pi\gamma R$$ □

2.3 The solid–liquid interface

2.3.1 *The surface tension*

In case of the liquids, the surface energy per unit area (Jm^{-2}) is also called "*surface tension*" (units $Nm^{-1} = Jm^{-2}$). The term brings out the fact that the surface of a liquid experiences a contractile force which tends to shrink the surface. The liquids are able to respond readily to this force and assume a shape which minimizes the total surface energy. The surface tensions of liquids are typically 20 – 80 $mN.m^{-1}$. A good estimate of the surface tension can be obtained from the heat of vaporization of

Table 2.2: Surface tensions of some liquids at 25°C

Liquid	Temp. °C	γ (mNm^{-1})
Water	25	71.99
Methanol	25	22.07
Ethanol	25	21.97
Isopropanol	20	21.7
1-propanol	25	23.32
1-butanol	25	24.93
2-butanol	25	22.54
Acetone	25	23.46
Glycerol	20	63
Mercury	15	487
Acetic acid	20	27.6
Diethyl ether	20	17
n Hexane	20	18.4
n octane	20	21.8

liquids. The surface tension of some liquids is listed in Table 2.2.

The surface tension of a liquid can be drastically lowered by the addition of small amounts of a suitable *surface active agent (i.e. a surfactant)*. The surfactants and their action will be described in Chapter 13.

2.3.2 *The contact angle*

Figure 2.18 compares a drop of mercury and a drop of water on a solid substrate such as glass. Mercury has a lesser tendency to spread or wet the glass surface.

One way to measure this spreading tendency is by the contact angle θ defined in Fig 2.19. Note that the contact angle is measured inside the liquid. A lower contact angle implies a greater tendency to spread on the solid or equivalently, more wetting of the solid by the liquid. The contact angle is essentially determined by the attraction of the molecules of the liquid towards the surface (adsorption force) and the attraction of the

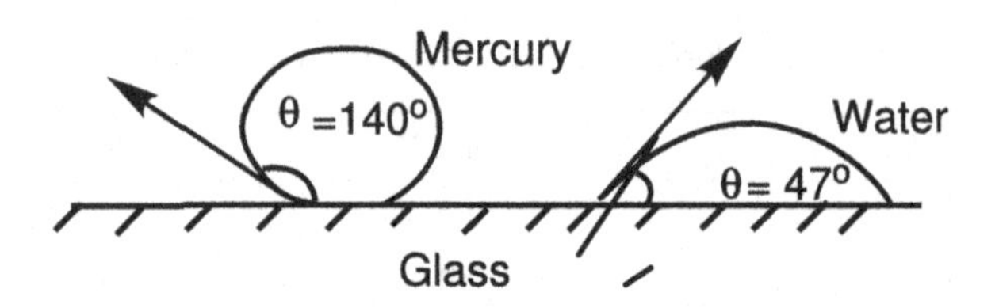

Fig. 2.18. A drop of mercury makes a larger contact angle ($\theta = 140^0$) with glass than a drop of water ($\theta = 47^0$).

liquid molecules towards one another (cohesive force). A liquid will not wet the surface if the cohesive force is much stronger than the adhesive force.

A relation between the contact angle θ and the surface energies γ_{LV}, γ_{SL} and γ_{SV} for the liquid–vapor, solid–liquid and solid–vapor interfaces respectively, can be derived by using the equivalence of surface energy and surface tension and taking a balance of forces at a point on the line of contact (Fig 2.19).

In addition to the surface tension forces, another force due to the line tension of the liquid–solid–vapor contact circle needs also to be considered in nanotechnology because the magnitude of this force is not negligible when the drop size becomes small. This line tension force is equal to τ/r and acts inwards on the liquid–solid–vapor contact circle (Fig. 2.19). Here τ is the line tension of the contact circle and r is the radius of the contact circle. Considering the equilibrium of the forces in the horizontal direction (i.e. in the direction of γ_{LS}) we get

$$\gamma_{LS} + \gamma_{LV}\cos\theta + \frac{\tau}{r} - \gamma_{SV} = 0$$

or

$$\gamma_{LV}\cos\theta = \gamma_{SV} - \gamma_{LS} - \frac{\tau}{r} \tag{2.13}$$

or

$$\theta = \cos^{-1}\left[\frac{\gamma_{SV} - \gamma_{LS} - \frac{\tau}{r}}{\gamma_{LV}}\right]$$

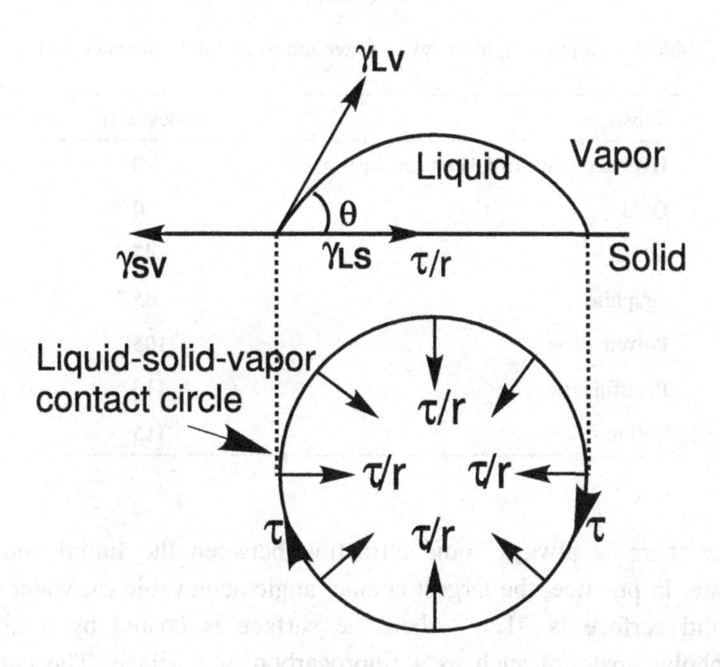

Fig. 2.19. At the equilibrium contact angle the three surface tensions and the force due to the line tension are balanced; the line tension force is τ/r and acts inwards on the liquid–solid–vapor contact circle as shown in the lower figure.

When the drop size is macroscopic (*i.e.* > 100 nm), the term τ/r is very small and is neglected so that Eq. (2.13) becomes

$$\gamma_{LV} \cos\theta = \gamma_{SV} - \gamma_{LS} \qquad (2.14)$$

Equation (2.14) is called the *Young's equation* [20]. When $\theta = 0$, *i.e.* $\gamma_{SV} = \gamma_{LS} + \gamma_{LV}$, then the surface of the solid can be completely covered by the liquid without any gain or loss of energy. If $\gamma_{SV} > \gamma_{LS} + \gamma_{LV}$, then there is a net reduction in energy by complete coverage of the solid surface by the liquid. The liquid spreads completely on the solid replacing γ_{SV} by $(\gamma_{LS} + \gamma_{LV})$. For this case the contact angle is infinite and Eq. (2.14), which is for the equilibrium case, is no longer valid.

The contact angle for various systems varies between 0° and 180°. A value of θ = 180° for a liquid drop on a smooth surface is impossible

Table 2.3. Contact angles between water and some solid substrates [21]

Substrate	θ (degrees)
Hydrated silica (SiOH on surface)	0
Gold	0
Silica	47
Graphite	85.7
Polyethylene	105
Paraffin wax	110
Teflon	115

because there is always some attraction between the liquid and the substrate. In practice, the largest contact angle achievable for water on a flat solid surface is ~120° when the surface is coated by a highly hydrophobic material such as a fluorocarbon or a silane. The contact angles between water and some solids are given in Table 2.3.

A useful application of a finite equilibrium contact angle between two solids is the deposition of nano islands of one solid on another *e.g.* nickel on silicon. A thin film of nickel is first deposited on SiO_2 coated Si by evaporation or sputtering. The film is in a metastable equilibrium. On heating the film to a temperature where the diffusion of nickel is significant, the film breaks into islands which assume the equilibrium contact angle with the substrate surface (Fig. 2.20). In an example discussed later, Lau *et al.* [22] have first prepared metal islands in this manner and then grown carbon nanotubes on the metal islands to prepare superhydrophobic surfaces.

2.3.3 *The contact angle hysteresis*

The values of the contact angles given in Table 2.3 are measured on flat, chemically homogeneous surfaces; the actually measured contact angle

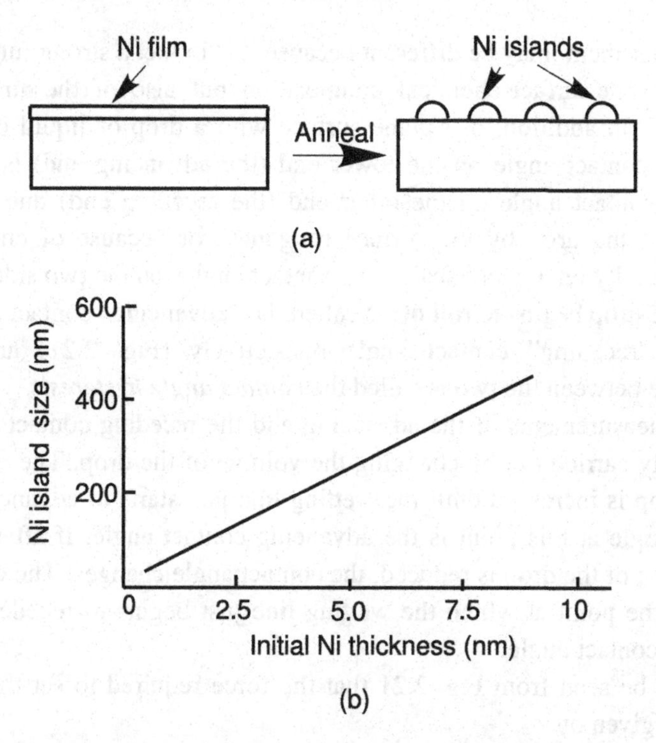

Fig. 2.20. (a) Layer of Ni deposited by sputtering on a 50 nm layer of SiO_2 on Si breaks into islands after annealing at 750 °C in 20 Torr of H_2 for 15 min; (b) The average Ni island size after annealing versus initial Ni layer thickness [23].

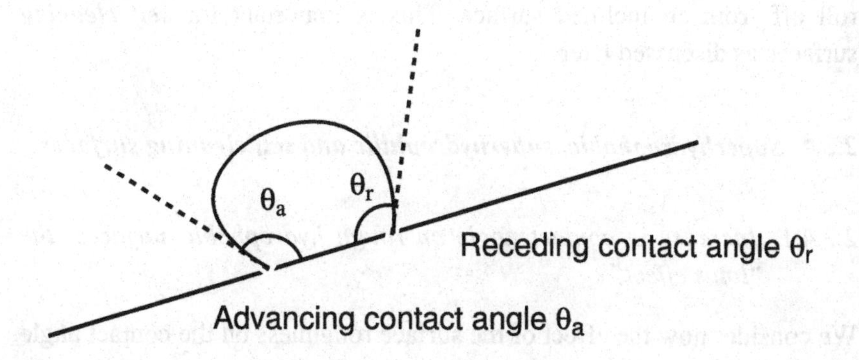

Fig. 2.21. The advancing and the receding contact angles for a drop on an inclined plane.

in an experiment may be different because it is in fact a strong function not only of the surface chemical composition but also of the surface roughness. In addition, if a plane surface with a drop of liquid on it is tilted, its contact angle on the lower end (the advancing end) is larger than the contact angle on the other end (the receding end) due to the pinning of the drop by the surface roughness or because of chemical inhomogeneity on a micro scale. The contact angles on the two sides, just before the drop begins to roll off is called the "advancing" contact angle and the "receding" contact angle respectively (Fig. 2.21) and the difference between the two is called the *contact angle hysteresis.*

The measurements of the advancing and the receding contact angles are usually carried out by changing the volume of the drop. The volume of the drop is increased until the wetting line just starts to advance. The contact angle at this point is the advancing contact angle. If afterwards the volume of the drop is reduced, the contact angle changes. The contact angle at the point at which the wetting line just begins to recede is the receding contact angle.

It can be seen from Fig. 2.21 that the force required to set the drop rolling is given by

$$F = \gamma_{LV} (\cos \theta_r - \cos \theta_a)$$

Thus smaller the contact angle hysteresis, easier it is for the drop to roll off from an inclined surface. This is important for *self cleaning* surfaces as discussed later.

2.3.4 *Superhydrophobic, superhydrophilic and self cleaning surfaces*

2.3.4.1 *Increase in contact angle on rough hydrophobic surfaces: the "lotus effect"*

We consider now the effect of the surface roughness on the contact angle in view of its importance for the *superhydrophobic* and *self cleaning* surfaces. A surface is called superhydrophobic or superoleophobic if the contact angle on it for water or oil is greater than 150°, respectively. On the other hand, if the contact angle for water or oil is close to zero, then the surface is called superhydrophilic or superoleophilic respectively.

On a flat surface, the largest value of the contact angle that can be achieved in practice is about 120° by coating the surface with highly hydrophobic materials like fluorocarbons and silicones. However, it is possible to achieve contact angles higher than 170° by suitably engineering the surface roughness provided the surface is slightly hydrophobic (with intrinsic contact angle > 90°) to start with. The inspiration for this has been provided by some spectacular examples in nature (Fig. 2.22). The surfaces of the leaves of some plants such as lotus and the wings of some insects exhibit superhydrophobicity. Due to this superhydrophobicity, a water drop does not wet the surface and easily rolls off, taking with it any dirt on the surface, making the surface self cleaning. This effect is called the *"lotus effect"* [24]. Scanning electron microscopy studies of the lotus leaf have shown that it has a surface roughness on two length scales. There are pillar like structures on the surface called *micropapillae* which are 5–9 μm high. The surface of the micropapillae, in turn, is nanostructured at a scale of about 120 nm (Fig. 2.22 c). The whole surface is coated with hydrophobic wax crystals which makes the surface intrinsically hydrophobic ($\theta > 90°$). The two scale surface roughness further enhances the hyrophobicity of the leaf, making it *superhydrophobic* with a contact angle of more than 150°. This, together with a low contact angle hysteresis, leads to the superhydrophobic as well as the self cleaning properties of the lotus leaf.

2.3.4.2 *The Wenzel and the Cassie–Baxter equations*

Though a complete understanding of the "lotus effect", particularly the role of the hierarchical roughness, is still lacking, a large volume of literature now exists in which the surfaces have been purposely imparted roughness on different scales to successfully mimic this effect. The effect of roughness is usually expressed in terms of two simple to use equations which apply to different configurations of a liquid drop on the surface. Two of the possible configurations that a drop of liquid, say water, can assume on a rough surface are depicted in Fig. 2.23. In Fig 2.23 (a) the

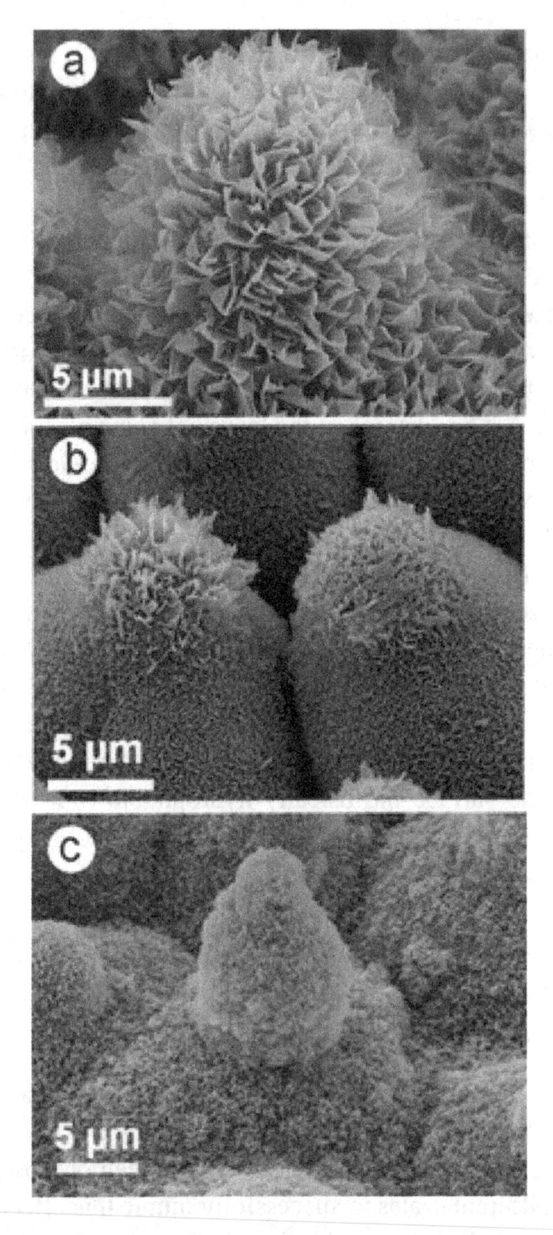

Fig. 2.22. SEM micrographs of superhydrophobic leaves with two-level hierarchical sculptures (a) Euphorbia myrsinites (b) Colocasia esculenta and (c) Nelumbo nucifera (lotus) (reprinted with permission from [24] copyright (2009) American Chemical Society).

liquid follows the surface topography while in Fig. 2.23 (b), the liquid drop rests on asperities on the surface, enclosing air beneath; in the latter case a so called "composite" interface is formed consisting of both the liquid–solid and liquid–air interfaces.

In the first case, Fig. 2.23 (a), where the liquid enters between the asperities and follows the surface profile, the area in contact with the liquid is increased by a *roughness factor r* given by

$$r = \frac{\text{actual roughness enhanced surface area}}{\text{projected surface area}}$$

Equation (2.14), the Young's equation, was therefore modified by Wenzel [25] as follows

$$\gamma_{LV} \cos\theta^* = r(\gamma_{SV} - \gamma_{LS})$$

where θ^* is the effective increased contact angle due to the surface roughness. Combining this with Eq. (2.14) gives the Wenzel relation for the apparent contact angle θ^*

$$\cos\theta^* = r \cos\theta \quad (2.15)$$

For a hydrophobic surface for which $\theta > 90^\circ$, the contact angle is, therefore, further enhanced due to the surface roughness while a hydrophilic surface ($\theta < 90^\circ$) becomes more wetting due to the roughness.

In the second case, Fig. 2.23 (b) the liquid rests partly on the solid surface and partly on the air trapped below it. For this case, Cassie and Baxter [26] derived the following relation

$$\cos\theta^* = f_1 \cos\theta_1 + f_2 \cos\theta_2$$

Here f_1 and f_2 are the area fractions of the solid and the vapor in contact with the liquid, as defined below, and θ_1, θ_2 are the contact angles of the liquid with the solid and the vapor respectively.

$$f_1 = \frac{\text{area in contact with the liquid}}{\text{projected surface area}}$$

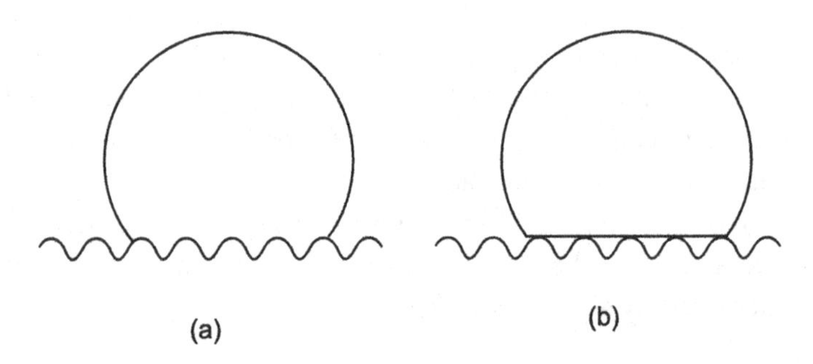

Fig. 2.23. Possible configurations of a liquid drop on a rough surface (a) the liquid follows the surface contour (b) the liquid drop rests on asperities with air trapped below, thus forming a composite interface with the solid and the air.

$$f_2 = \frac{\text{area in contact with the air trapped beneath the liquid}}{\text{projected surface area}}$$

Taking $\theta_1 = \theta$ (the contact angle for the liquid–solid interface), θ_2 (the contact angle for the liquid–air interface) $=180^0$ and using $f_1 + f_2 = 1$, one gets,

$$\cos\theta^* = f_1 \cos\theta - f_2 = f_1(1 + \cos\theta) - 1 \tag{2.16}$$

From Eq. 2.15 and Eq. 2.16 the value of the contact angle for the rough surfaces can be calculated as illustrated by the example below.

Example 2.5.

(a) *A surface is patterned so as to have square pillars s μm on the side and h μm high placed d μm apart in a square array (Fig. 2.24). Derive expressions for r, f_1 and f_2 in terms of s, h and d.*

(b) *If the contact angle for the flat surface is measured to be 114^0, find the apparent contact angle for the patterned surface according to the Wenzel and the Cassie–Baxter equations. For the Cassie-Baxter equation, assume that*

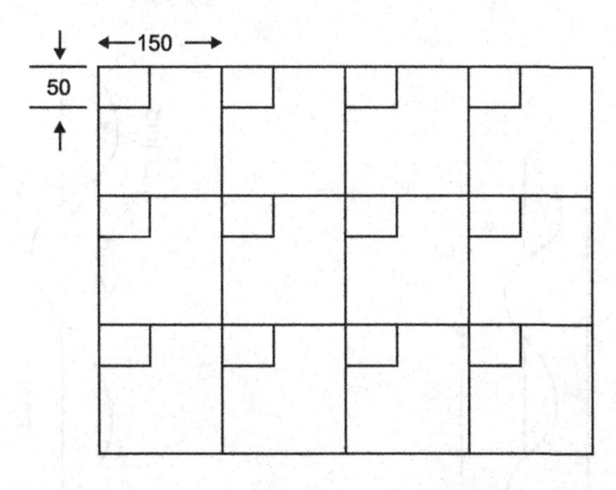

Fig. 2.24. Pattern of square pillars for Example 2.5.

the liquid covers the top surfaces of the pillars completely. Given: $s = 50\ \mu m$, $h = 10\ \mu m$ and $d = 150\ \mu m$.

Solution. The roughness factor for the given geometry is

r = (surface area of the top of the pillars and the remaining substrate + surface area of the sides of the pillars) / projected area

$= [d^2 + (4s \times h)]/d^2 = 1 + 4sh/d^2$

Given: $d = 150\ \mu m$, $s = 50\ \mu m$, $h = 10\ \mu m$

Hence $r = 1 + 4 \times 50 \times 10/150 \times 150 = 1.0889$

Thus for the Wenzel case, $\cos\theta^* = 1.0889 \cos 114$ or $\theta^* = 116°$

For the Cassie Baxter case: $f_1 = s^2/d^2 = 0.111$; $f_2 = 1 - f_1 = 0.889$

Hence $\cos\theta^* = 0.111 \cos 114 - 0.889$ or $\theta^* = 159.6°$ □

2.3.4.3 *Superhydrophilic surfaces: Enhanced hydrophilicity due to surface roughness*

Surfaces can also be made superhydrophilic by engineering the surface roughness. As seen above for the Wenzel case the change in the contact angle due to surface roughness is given by Eq. (2.15)

$$\cos\theta^* = r\cos\theta \quad (2.15)$$

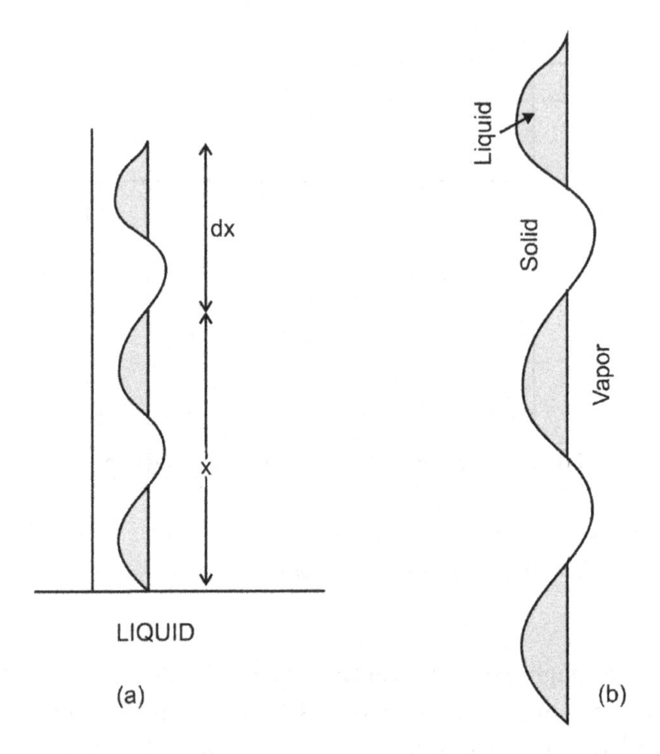

Fig. 2.25. (a) Spread of liquid on a superhydrophilic rough surface in contact with a reservoir of the liquid (b) magnified view, showing the liquid in the grooves of the surface roughness [27].

According to this relation, as $r > 1$, the contact angle for a hydrophilic surface ($\theta < 90°$) is further reduced in the Wenzel case due to the surface roughness making the surface more hydrophilic. If the intrinsic contact angle θ is only a little less than 90°, then this reduction in the contact angle $\Delta\theta = (\theta^* - \theta)$ is small. But if θ is less than a critical value, $\theta < \theta_c$, it has been shown [27] that the contact angle reduces to a value close to zero, so that the liquid drop spreads on the surface making the surface *superhydrophilic*. This can be seen by considering the energy change due to spreading of a liquid on a rough surface [28]. In Fig. 2.25, a rough surface is in contact with a liquid reservoir. The liquid is in the process of spreading on the substrate and has already spread to a length x. We consider the energy change per unit width of the surface when x increases by a small amount dx.

Let r be the surface roughness factor as defined earlier and ϕ be the fraction of the total surface area which remains dry. The fraction of the total area which is wetted is then $(1-\phi)$. The total surface area under the incremental length is rdx and the area of the liquid–vapor interface in the incremental length dx is the projected area of the total liquid–solid contact area and is, therefore, given by

$$\text{Liquid–vapor interface area} = (1-\phi)\, rdx/r = (1-\phi)dx$$

The required energy change, therefore, is given by

dE = Energy after spreading of the liquid by the length dx – energy before spreading

$$= [\gamma_{SL}(1-\phi)\, rdx + \gamma_{SV}(\phi rdx) + \gamma_{LV}(1-\phi)dx] - \gamma_{SV}(rdx)$$
$$= (\gamma_{SL}-\gamma_{SV})(1-\phi)rdx + \gamma_{LV}(1-\phi)dx$$
$$= (1-\phi)dx[r(\gamma_{SL}-\gamma_{SV}) + \gamma_{LV}]$$

Substituting, from Young's equation, $\gamma_{LV}\cos\theta = \gamma_{SV}-\gamma_{SL}$, one gets

$$dE = (1-\phi)dx[-r.\gamma_{LV}\cos\theta + \gamma_{LV}] = (1-\phi)\gamma_{LV}dx(1-r\cos\theta)$$

For the liquid to spread, the energy should decrease and so dE should be negative. At the critical point $dE = 0$ and so

$$1-r\cos\theta_c = 0 \text{ or } \cos\theta_c = 1/r \qquad (2.17)$$

At this point the Wenzel contact angle is given by $\cos\theta^* = r\cos\theta_c = 1$ or $\theta^* = 0$ *i.e.* surface has an effective zero contact angle and is superhydrophilic if the intrinsic contact angle of the surface is

$$\theta \leq \theta_c$$

The result (2.17) shows that the critical contact angle increases as the surface roughness increases *i.e.* it is possible to make a surface superhydrophilic even when it's intrinsic contact angle is only slightly less than 90° by sufficiently increasing the value of r.

While the Wenzel and the Cassie-Baxter equations provide a useful framework to study the wetting properties of rough surfaces, it must be pointed out that the measured values of the contact angels do not agree in many cases with the predictions of these equations. The topic is of continuing research interest and many articles have appeared attempting to modify these equations [29-31] as the rough surfaces are potentially technologically useful in many applications such as chemical sensors, catalyst supports, biomedical applications, antireflective coatings in addition to the superhydrophobic and superhydrophilic surfaces. These relations provide a working guideline to design such surfaces. Two examples of the engineered superhydrophobic and superhydrophilic surfaces are provided below.

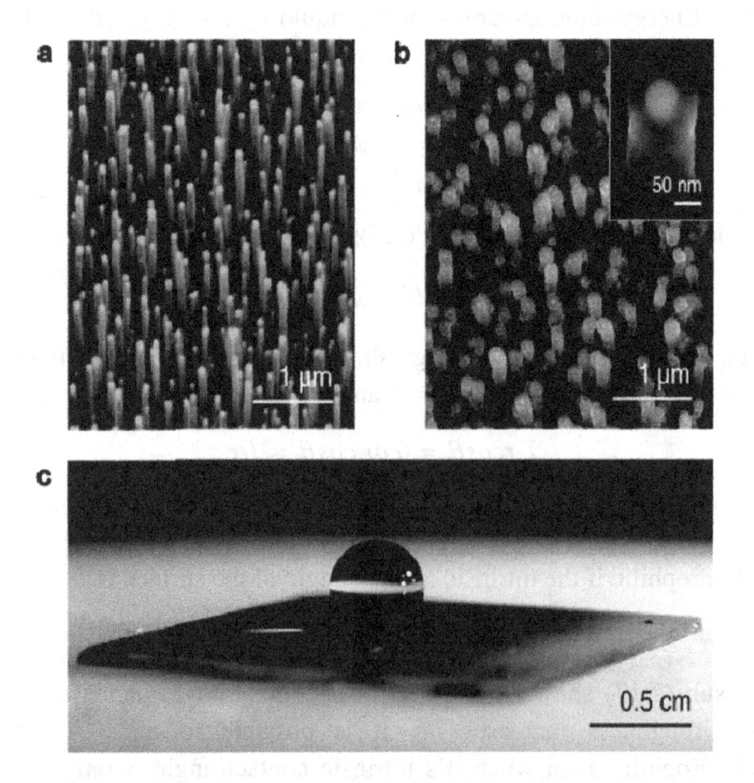

Fig. 2.26. SEM image of carbon nanotube forests (a) As grown nanotube forest with nanotube diameter 20 μm and height 2 μm (b) the nanotube forest after coating with PTFE to impart hydrophobicity to the carbon nanotube surface (c) a water droplet suspended on the nanotube forest – the drop is nearly spherical showing a contact angle approaching 180° (reprinted with permission from [22] copyright (2003) American Chemical Society).

2.3.4.4 *Examples of superhydrophobic and superhydrophilic surfaces by nanostructuring*

The first example is from the work of Lau *et al.* [22]. They prepared parallel arrays of carbon nanotubes on silicon surfaces patterned with nickel catalyst islands which form spontaneously when a thin film of nickel (5 nm) is sintered. The island formation occurs because of the finite equilibrium contact angle of nickel with the silicon surface. The size of the nickel islands and their spacing can be controlled by controlling the thickness of the nickel film, a thinner film breaking into smaller and more closely spaced islands. The nanotubes grow on the nickel islands. These are then coated by PTFE to make their surface hydrophobic. While the contact angle for a smooth surface of PTFE is 108°, the contact angle of 170° (advancing) and 160° (receding) are measured on the PTFE coated nanotube forest (Fig. 2.26).

In the second example Liu and He [32] prepared hierarchically structured coatings with surface roughness on micro as well as nanoscale by a hydrothermal treatment of soda lime glass surfaces. Two types of surface morphologies were obtained depending on the temperature and time of the hydrothermal treatment: (a) surfaces with flower like features made of nanoflakes (Fig. 2.27 a) and (b) surfaces with urchin like features made of nanowires (Fig. 2.27 b). The as prepared surfaces were superhydrophilic with a water contact angle of 0°.

After a hydrophobic treatment of the surfaces by a silane, the surfaces became superhydrophobic with a contact angle of 160° and a sliding angle of 1°. Thus depending on the initial contact angle of the surface, the roughness can make the surface either superhydrophilic or superhydrophobic.

2.3.4.5 *Titania coated superhydrophilic and self cleaning surfaces*

In the above, we have discussed the generation of superhydrophobic and superhydrophilic surfaces, both of which can be self cleaning, by engineering the surface roughness. However, it should be pointed out that at present a more common route to generate superhydrophilic and

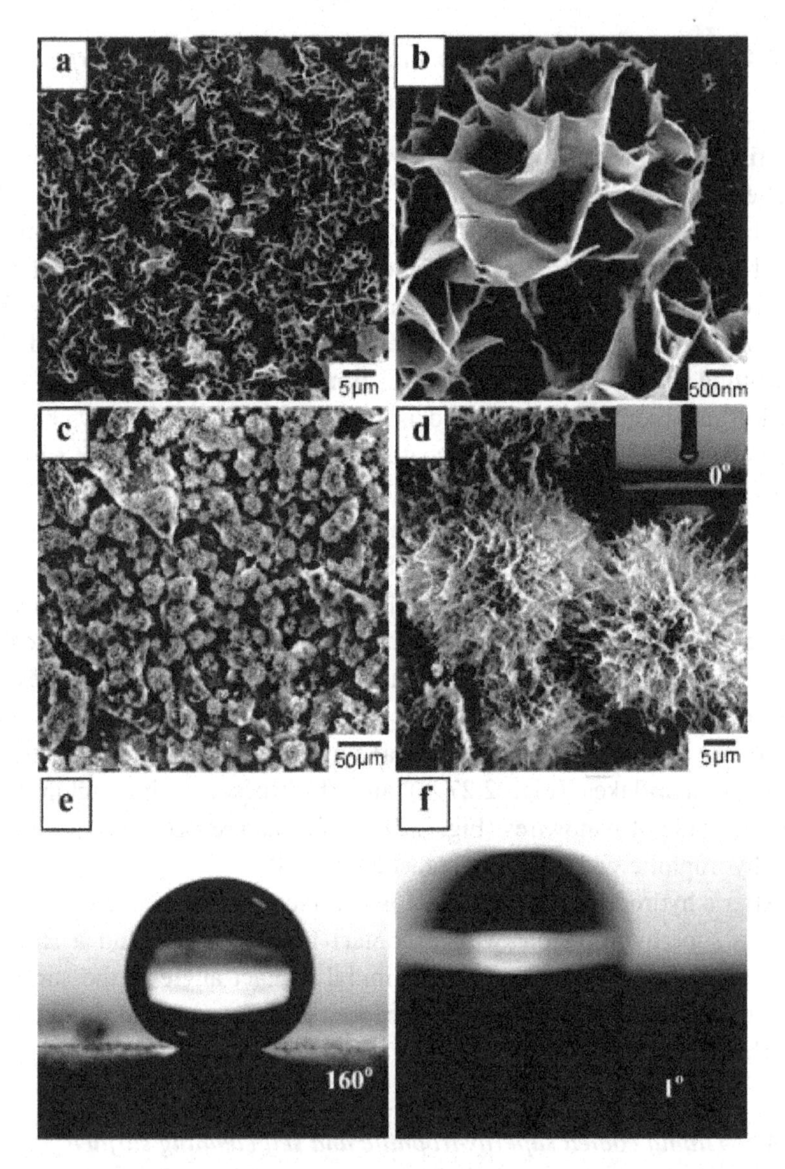

Fig. 2.27. (a) and (c) Examples of two coatings with different surface roughness prepared on soda lime glass by a hydrothermal treatment; (b) and (d) enlarged views of (a) and (c); as the glass surface is intrinsically hydrophilic, the surfaces are superhydrophilic with a contact angle of 0°; (e) a water drop on (c) after the hydrophobic treatment; the contact angle is now 160° and (f) the drop rolls off at a small sliding angle of 1° making the surface self cleaning (reprinted with permission from [32] copyright (2009) American Chemical Society).

and self cleaning surfaces is actually by coating the surface with an oxide semiconductor — the surface can be flat and the roughness is not the critical factor here. Such surfaces are now being commercially produced, a major application being self cleaning glass for use in windows. The most commonly used oxide semiconductor for such surfaces is titania, TiO_2, in anatase phase [33].

Appendix 2.1 : Melting point of nanoparticles

The lower melting point of the nanoparticles can be explained by taking into account the change in vapor pressure due to curvature [34, 35]. It is assumed that the nanoparticle is made of a pure element and is spherical, isotropic and homogeneous. Its chemical potential (free energy per mole) is expressed as a series expansion about the melting temperature and pressure of the bulk, T_0 and P_0, respectively as follows

$$\mu(T, P) = \mu(T_0, P_0) + \frac{\delta\mu}{\delta T}(T - T_0) + \frac{\delta\mu}{\delta P}(P - P_0) + \ldots\ldots \quad \text{(A2.1.1)}$$

The Gibbs–Duhem relation is

$$-VdP + SdT + n\mu = 0$$

where S is the entropy and n the number of moles. Using this relation, we can write,

$$\frac{\delta\mu}{\delta T} = -\frac{S}{n} \quad \text{and} \quad \frac{\delta\mu}{\delta P} = \frac{V}{n} = \frac{1}{\rho} \quad \text{(A2.1.2)}$$

In the bulk phase, at the melting point, the chemical potentials are the same in the solid and the liquid

$$\mu_S(T_0, P_0) = \mu_L(T_0, P_0) \quad \text{(A2.1.3)}$$

One can expand $\mu_S(T, P)$, $\mu_L(T, P)$ as in (A2.1.1) and use (A2.1.2) and neglect the higher order terms to get the difference between the two as

$$\mu_L(T,P) - \mu_S(T,P) = \mu_L(T_0,P_0) - \mu_S(T_0,P_0) + \frac{\delta\mu_L}{\delta T}(T - T_0)$$
$$- \frac{\delta\mu_S}{\delta T}(T - T_0) + \frac{\delta\mu_L}{\delta P_L}(P_L - P_0) - \frac{\delta\mu_S}{\delta P_S}(P_S - P_0)$$

(A2.1.4)

If T and P are taken to be the quantities corresponding to the melting of the nanoparticle, then

$$\mu_S(T, P) = \mu_L(T, P) \qquad \text{(A2.1.5)}$$

Also, at the melting point of the bulk, $\Delta H = nL = T_0\,\Delta S = T_0\,(S_L - S_S)$ where L is the latent heat of fusion

so that
$$L = \frac{S_L - S_S}{n} T_0 \qquad \text{(A2.1.6)}$$

Using (A2.1.2), (A2.1.3), (A2.1.5) and (A2.1.6) and writing $\Theta = T/T_0$, Eq. (A2.1.4) can be written as

$$0 = L(1-\Theta) + \frac{1}{\rho_L}(P_L - P_0) - \frac{1}{\rho_S}(P_S - P_0)$$

Here P_L and P_S are the pressures experienced by the nanoparticle in the liquid and the solid state while P_0 is the pressure experienced by the bulk. Because of the curvature of the nanoparticle, it experiences an excess pressure (= $2\gamma/R$) given by the Laplace equation. The above equation then is

$$0 = L\,(1-\Theta) + \frac{1}{\rho_L}\frac{2\gamma_L}{R_L} - \frac{1}{\rho_S}\frac{2\gamma_S}{R_S} \qquad \text{(A2.1.7)}$$

Also, as the mass of the nanoparticle is unchanged on melting, we have

$$\frac{4}{3}\pi R_S^3\,\rho_S = \frac{4}{3}\pi R_L^3\,\rho_L \quad \text{or} \quad R_L = \left(\frac{\rho_S}{\rho_L}\right)^{\frac{1}{3}} R_S$$

Then from (A2.1.7)

$$1-\Theta = \frac{2}{LR_S\rho_S}\left[\gamma_S - \gamma_L\left(\frac{\rho_S}{\rho_L}\right)^{\frac{2}{3}}\right] = \frac{A}{R_S}$$

Putting
$$A = \frac{2}{LR\rho_S}[\gamma_S - \gamma_L(\frac{\rho_S}{\rho_L})^{\frac{2}{3}}]$$

one gets
$$T = T_0(1 - \frac{A}{R_S}) \quad \text{(A2.1.8)}$$

Thus the melting point of the nanoparticle decreases as $1/R$. In Fig. 3.1 (see the next chapter) the solid line is calculated using the above model with the second order terms also included. The result is in good agreement with the experiments.

Problems

(For the first few problems, refer to Appendix I on crystal structure)

(1) Show, using a figure, the planes (111), (110) and (100) in the unit cell of a cubic crystal.

(2) Show, using a figure, that the set of {111} planes in the cubic system forms an octahedron.

(3) Copper has FCC structure with a lattice parameter of 0.361 nm
   (c) Find the number of atoms per unit area in the (111), (110) and the (100) planes of copper. Which is the most closely packed plane?
   (d) Find the number of atoms per unit length in the (100), (110) and the (111) directions of copper. Which is the most closely packed direction?

(4) For a simple cubic crystal with lattice parameter a , show that the fraction of atoms on the surface of a cube with edge length = na is given by

$$f = \frac{6n^2 + 2}{(n+1)^3}$$

(5) An octahedral shape is made by the (111) planes of an FCC crystal with n atoms in each edge of the octahedron.
   (a) What is the total number of atoms in the octahedron?
   (b) Show that the total number of atoms on the surface of the octahedron is $4n^2 - 8n + 6$.
   (c) For the smallest octahedron possible, what are the total number of atoms in the octahedron and atoms on the surface?
   (d) If the octahedral shape above is truncated so that the top m layers (counting the apex atom as one layer) are removed, then show that the total number of atoms in the truncated octahedron is $n(2n^2+1)/3 - 2m^3 - 3m^2 - m$.
   (e) Find the total number of atoms on the surface of the truncated octahedron.

(f) The surface energies of the (111) and the (100) planes of an FCC metal are 4.0 and 4.2 mJm^{-2} respectively. Find the value of m in terms of n in accordance with the Wulff criterion.

(6) In the Spherical Cluster Approximation, a cluster is modeled as a sphere with the volume of the cluster assumed to be equal to the number of atoms N in the cluster multiplied with the volume of one atom. This is an oversimplification as the hard spheres do not fill the space exactly, but nevertheless provides a good approximation, especially for somewhat larger clusters. Derive the following relations, using this model

(a) Radius of the cluster, R_c in terms of the radius of an atom, R_a

$$R_c = N^{1/3} R_a.$$

(b) Surface area of the cluster, S_c, in terms of the surface area of an atom, S_a

$$S_c = N^{2/3}\, S_a$$

(c) Number of surface atoms, N_s (assume that this is equal to the surface area of cluster divided by the cross sectional area of an atom, an assumption which is more valid for large clusters)

$$N_s = 4N^{2/3}.$$

(d) Fraction of atoms on the surface

$$f_s = 4N^{-1/3}$$

(7) Calculate the surface energy of (100) and (110) planes of NaCl given that the equilibrium interionic distance in NaCl is 0.283 nm. Use the following expression for the potential energy *vs* distance r between the ions to calculate the bond energy.

$$V = -\frac{Z_1 Z_2 e^2}{4\pi\, \varepsilon_0\, r} + \frac{B}{r^n}$$

($Z_1 = Z_2 = 1$ for NaCl; ε_0 = permittivity of vacuum = 8.854×10^{-12} fm^{-1})

[Hint: First obtain B by recognizing that V is a minimum at r_o. The bond energy is the value of V at the minimum].

(8) If the (10) and (11) planes for the two dimensional crystal in Fig. 2.6 have surface energies of 1.2 and 1.0 mJ m^{-1}, find the lowest

energy shape using the Wulff construction. Calculate the difference in the total surface energy between this shape and the shapes made of (i) {10} planes only (ii) (11) planes only.

(9) Show that the equilibrium ratio of the dimensions of a tetragonal crystal with surface energies γ_a and γ_c for the prismatic and the square faces respectively, is given by $a/c = \gamma_a/\gamma_c$.

(10) The surface tension of alumina is estimated to be 9×10^{-5} J cm^{-2}; for a liquid it is 1.72×10^{-4} J cm^{-2}. Under the same conditions, the interfacial tension (liquid – alumina) is 2.3×10^{-4} J cm^{-2}. What is the contact angle? Does the liquid wet the alumina?

(11) A silicon surface is patterned to have cylindrical pillars of diameter *d* and height *h* placed a distance *s* apart in a square array. The surface is slightly oxidized so as to have a contact angle of 50°.

(a) Find the roughness factor r and the fractions f_1 and f_2 of the Cassie Baxter equation in terms of *d*, *h* and *s*.

(b) Find the apparent contact angle of the surface according to the Wenzel and Cassie-Baxter equation under the following conditions (i) as oxidized silicon (ii) surface coated with PTFE to give a contact angle of 114°.

(c) What is the critical contact angle to make the surface superhydrophilic if *d* = 50 μm, *h* = 10 μm and *s* = 150 μm (Fig. 2.28).

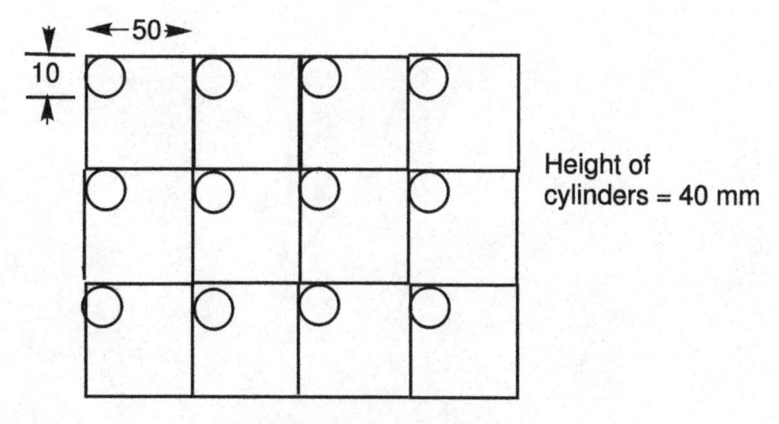

Fig. 2.28. Patterned silicon surface

(12) On the square pillars formed in Example 2.5 (Fig. 2.24), a fine scale roughness is further created by forming finer square pillars of side s_2, height h_2 and center to center spacing d_2. Calculate the apparent contact angle for this two scale roughness using the Wenzel and the Cassie–Baxter equations (Hint: First calculate the apparent contact angle for the fine pillars. Then use this angle as the true angle for the coarse scale roughness – see [36]).

Chapter 3

Zero Dimensional Nanostructures I Review of Some Topics in Physics

3.1 Introduction

Zero dimensional nanostructures are defined as those in which all the three dimensions are less than 100 nm. Similarly, the one and two dimensional nanostructures are those in which two and one dimensions are < 100 nm, respectively. In this and the next few chapters we confine ourselves to the zero dimensional nanostructures. Several terms are used to further distinguish the different types of such structures. *Nanoparticle* is a broad term which includes all zero dimensional objects with size < 100 nm. Nanoparticles which are single crystals are called *nanocrystals*. A *quantum dot* is a nanoparticle displaying quantum effects. Aggregates of atoms containing 2 to 10^6 atoms are called nanoclusters [37].

A striking characteristic of nanostructures is the changes in their properties with size. Such changes in properties begin to occur when the size approaches a characteristic scaling length which is usually different for each property or phenomenon. Many of these changes, *e.g.* those in the electronic and optical properties, arise due to quantum effects. A review of some topics in physics and a discussion of the various scaling lengths are given below to help understand such changes in properties. Specific zero dimensional nanostructures such as the semiconductor quantum dots, the metal nanoparticles, the magnetic nanostructures, the colloidal particles and colloidal crystals are covered in the following chapters.

3.2 Size dependence of properties in the nanometer range

Consider a bulk solid such as gold. It is metallic in nature, has high electrical and thermal conductivity, a lustrous yellow color and other properties such as ductility and malleability. These characteristic properties of gold arise as a result of the association of a large number of atoms – to associate such properties to an individual gold atom is meaningless. The question naturally arises: how many atoms of gold have to come together before the assemblage can be termed a metal? Or, if the number of atoms in a large chunk of gold is progressively reduced, at what size it ceases to be a metal?

It turns out that if the number of atoms in a body *i.e.* its dimension is progressively reduced, a stage comes when its intrinsic properties (properties which are independent of size or mass in the macroscopic regime) begin to change with size, the character of the bulk eventually being lost when the size reaches a low enough value. Some examples of this dependence of properties on size are shown in Fig. 3.1. It can be seen that the point at which the size dependence starts depends on the property being considered. Thus the melting point of gold begins to decrease rapidly below 7.5 nm; on the other hand a rapid drop in the value of the band gap of CdS begins when the size is reduced below about 40 nm. Generally a physical phenomenon responsible for a certain property has associated with it a characteristic length. When the dimensions of the solid approach this length, the corresponding physical phenomenon is influenced and the property undergoes a change. Although the length scales in physics can range from very large (light year) to very small (femto meter), our concern here is with the characteristic length scales in the nanometer range. Examples of some of these characteristic lengths are given later.

Some of the changes in properties with size such as the change in the melting point, the pressure induced phase transition, can be explained without resorting to quantum mechanics. As an example, an expression for the lowering of melting point of nanoparticles from thermodynamic considerations has been derived in Chapter 2. Changes in many of the other properties can be similarly accounted for by using

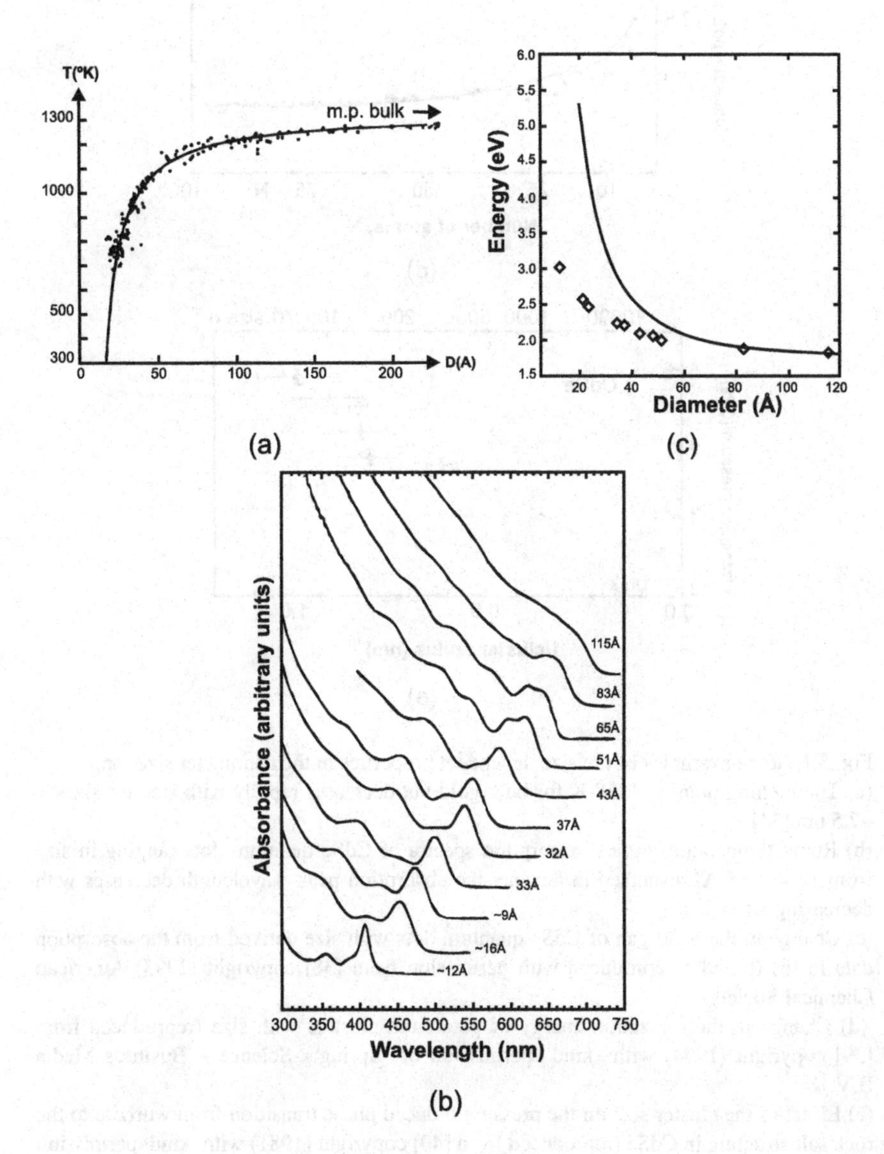

Fig. 3.1

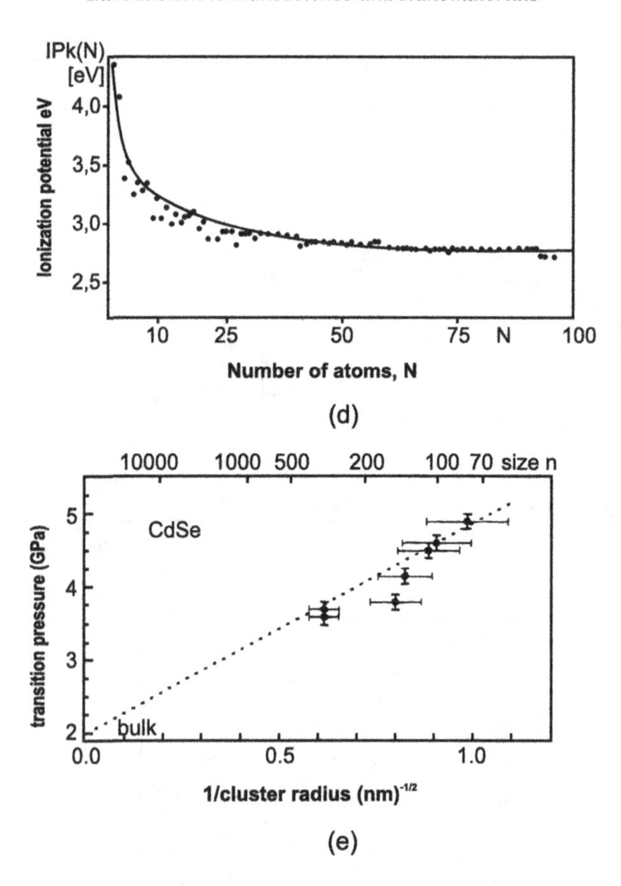

Fig. 3.1. Some examples of the size dependent properties in the nanometer size range:
(a) The melting point is 1338 K for bulk gold but decreases rapidly with size for sizes < ~7.5 nm [34].
(b) Room temperature optical absorption spectra of CdSe quantum dots ranging in size from 12 to 115 A° dispersed in hexane; the absorption peak wavelength decreases with decreasing size.
(c) Change in the band gap of CdSe quantum dots with size derived from the absorption data in (b) (b and c reproduced with permission from [38] copyright (1993) American Chemical Society).
(d) Change in the ionization energy of potassium clusters with size (reproduced from [39] copyright (1994) with kind permission of Springer Science + Business Media B.V.).
(e) Effect of the cluster size on the pressure induced phase transition from wurtzite to the rock salt structure in CdSe (reproduced from [40] copyright (1981) with kind permission of Springer Science + Business Media B.V.).

thermodynamics [35, 41]. However, changes in properties such as the band gap, the photoluminescence, the optical absorption *etc.* can only be understood on the basis of the quantum mechanics. In the following we review some ideas regarding the band structure and density of states in solids, particularly the semiconductors. These ideas form the basis for understanding phenomena such as the changes in the band gap and other electronic and optical properties with size. These, together with the formation of self assembled quantum dots, a topic of practical interest, are treated in the next chapter. A detailed quantum mechanical treatment of the nanostructures can be found elsewhere [42-44].

3.3 Review of some topics in physics

3.3.1 *Particle and wave: The Schrodinger equation*

Several experimental observations in the early twentieth century led scientists to the realization that the entities such as electrons or photons behave both as particles as well as waves. Thus, light was thought of as having a wave nature because of the easily observed phenomena of interference and diffraction but the explanation of the photoelectric effect by Einstein in 1905 showed that light also has a particle nature, the particle being called a photon. On the other hand the early picture of electrons was as particles but the diffraction of a beam of electrons by a nickel crystal surface, demonstrated by Davison and Germer in 1927, showed the wave nature of electrons. Thus the concept of wave-particle was well established by experimental results. In 1924 de Broglie gave the relation between the momentum p of an entity when it is viewed as a particle and its wavelength λ when it is viewed as a wave

$$\lambda = \frac{h}{p}$$

where h = Planck's constant (= 6.62×10^{-34} J.s).

The other relation which connects the wave and the particle is the Planck's relation (1901) between the energy E of a particle and its frequency ν

$$E = h\nu$$

Example 3.1. *Find the wavelengths and frequencies of waves corresponding to (a) electrons (b) photons having energies of 0.1 eV, 1 eV and 5 eV.*

Solution. 1 eV = 1.602×10^{-19} Joules; h = 6.62×10^{-34} Joules.second
Energy $E = h\nu$
If E is in eV, then E(eV) $\times$ 1.602 x $10^{-19} = 6.62 \times 10^{-34}\ \nu$
or the frequency ν = E (eV) $\times 2.42 \times 10^{14}$ Hz
which gives the frequencies for both the electrons and the photons. Thus
$E = 0.1$ eV $\nu = 24.2 \times 10^{12}$ Hz = 24.2 THz
$E = 1$ eV $\nu = 242 \times 10^{12}$ Hz = 242 THz
and $E = 5$ eV $\nu = 1210 \times 10^{12}$ Hz = 1210 THz

For photons, the relation between the frequency, wavelength and the velocity is
$\nu = c/\lambda$ where c = velocity of light = 2.998×10^{8} m.s^{-1}
Accordingly, we get wavelengths of 12.39 μm, 1.239 μm and 0.248 μm for the photons of energies 0.1 eV, 1 eV and 5 eV respectively.
For electrons we have to use the following relations
$E = p^2/2m$ and $\lambda = h/p$ where p is the momentum of the particle and m is the rest mass of the electrons= 9.109×10^{-31} kg
Using these relations we get $\lambda = 1.225 / \sqrt{E\,(eV)}$ nm
This gives wavelengths of 3.87 nm, 1.225 nm and 0.5478 nm for electrons of energies 0.1 eV, 1 eV and 5 eV respectively. □

In Table 3.1 the frequencies, the wavelengths and the energies of the photons in the electromagnetic spectrum are tabulated. The wavelengths of the electrons with the same energies as the photons are also given.

The relations so far described give some information about the wave nature of a particle. However, to completely describe a wave one has to express it as a function giving its dependence on space and time. In 1926 Schrodinger proposed such a function in the form of the solutions to a second order linear differential equation called the Schrodinger equation. Here we will use only the time independent form of this equation given below which is applicable when the total energy of a particle, *e.g.* an electron, is constant

$$[-\frac{\hbar^2}{2m}\nabla^2 + V(r)]\,\psi(r) = E\,.\psi(r) \qquad (3.1)$$

Here ∇^2 is the operator $\frac{\delta^2}{\delta x^2}+\frac{\delta^2}{\delta y^2}+\frac{\delta^2}{\delta z^2}$, V is the potential energy which is a function of position and time (but is assumed to be independent of time here), $i = \sqrt{-1}$ and $\hbar = h/2\pi$.

All the information about an object such as an electron, an atom, a molecule or a solid can, in principle, be obtained by solving the

Table 3.1. Frequency, wavelength and energy of photons in the electromagnetic spectrum; the wavelengths of the electrons with the same energies are also given in the last column

Radiation	Frequency (Hz)	Wave length	Energy	λ Electron (nm)
Radiowave	$10^7 - 10^9$	30–0.3 m	0.0413- 4.13 μeV	
Microwave	$3x10^8 - 3x10^{11}$	1 m –1 mm	1.24 – 1239 μeV	
Far IR	$3x10^{11} - 3x10^{13}$	1 mm –10 μm	1.239 –123.9 meV	34.81 – 3.481
Mid IR	$3x10^{13} - 1.2x10^{14}$	10 μm –2.5 μm	0.1239 –0.4956 eV	3.481 –1.74
Red	$4x10^{14} - 4.84x10^{14}$	750–620 nm	1.652 –1.999 eV	0.953 –0.866
Orange	$4.84x10^{14} - 5.08x10^{14}$	620 –590.5 nm	1.999 –2.098 eV	0.867 –0.846
Yellow	$5.08x10^{14} - 5.26 x 10^{14}$	590.5 –570 nm	2.098 –2.172 eV	0.846 –0.831
Green	$5.26x10^{14} - 6.06 x 10^{14}$	570 –95 nm	2.172 –2.503 eV	0.831 –0.775
Cyan	$6.06 x 10^{14} - 6.30 x 10^{14}$	495 –76 nm	2.503 –2.602 eV	0.775 –0.760
Blue	$6.30 x 10^{14} - 6.68 x 10^{14}$	476 –449 nm	2.602 –2.759 eV	0.760 –0.738
Violet	$6.68 x 10^{14} - 7.89 x 10^{14}$	449 –380 nm	2.759 –3.259 eV	0.738 –0.679
Near UV	$7.5 x 10^{14} - 1 x 10^{14}$	400 –300 nm	3.0975 –4.13 eV	0.695 –0.603
Mid UV	$1 x 10^{15} - 1.5 x 10^{15}$	300 –200 nm	4.13 –6.195 eV	0.603 –0.492
Far UV	$1.5 x 10^{15} - 2.46 x 10^{15}$	200 –122 nm	6.195 –10.156 eV	0.492 –0.385
X rays	$3 x 10^{16} - 3 x 10^{19}$	10 –0.01 nm	120 eV –120 keV	0.112 –0.0035
γ rays	$> 10^{18}$	< 0.3 nm	>100 keV	< 0.003875

Schrodinger equation for it. However, in practice it is possible to solve this equation only for a few simple cases such as (i) a free particle (V(r) = 0) (ii) particle in an infinite square potential well (iii) a harmonic oscillator and (iv) a hydrogen atom or a hydrogenic (hydrogen like) atom with any nucleus but only one electron. For molecules and solids, approximate methods have to be used to solve the Schrodinger equation. For solids, a simple approach is to consider the valence electrons in the solid to be free to move and not bound to any ions. This assumption is quite valid for metals. In a metal, the inner core electrons are bound to the individual metal atoms but the electrons in the outer shells (the valence shells) are "delocalized" *i.e.* not bound to any atom but free to roam throughout the solid. As an example, in aluminum, the inner core electrons are in the shells 1s, 2s and 2p and the delocalized electrons are in the outer (valence) shells 3s and 3p. In the semiconductors and insulators at low temperatures (~ 0^0 K), there are no free electrons. However, even for them, the results obtained using the free electron assumption are not too much off from those obtained using more sophisticated models if instead of the true mass of the carrier (electron or hole) an "effective mass" is used. In the following, we first consider the case of hydrogen atom, and then use the free electron model to derive some properties of semiconductors.

3.3.1.1 *The hydrogen atom*

The case of the hydrogen atom is of interest because the effects of impurities and *excitons* in semiconductors are modeled on the lines of the hydrogen atom. In the hydrogen atom the electron experiences a Coulombic potential V(r) ~ 1/r due to the proton. The solution of the Schrodinger equation for this case leads to the following allowed values of the energy of the electron

$$E_n = -\frac{m_r e^4}{2\,(4\pi\varepsilon_o)^2\hbar^2}\cdot\frac{1}{n^2} = -\frac{13.6}{n^2}\ \text{eV} \tag{3.2}$$

where m_r = reduced proton –electron mass = $\frac{m_e m_p}{m_e + m_p} \approx m_e$ where m_e and m_p are the masses of the electron and the proton respectively.

3.3.1.2 *A free particle*

Another simple case for which the Schrodinger equation can be solved exactly is that of a free particle. "Free" here means that there is no field present so that the potential energy of the particle is constant – it can be taken as zero for convenience, *i.e.* V(r) = 0. The problem is of special interest for solids as we shall see shortly.

The time independent Schrodinger equation for the free particle, *i.e.* with V(r) = 0 becomes

$$\left[-\frac{\hbar^2}{2m}\nabla^2 + V(r)\right]\psi(\mathrm{r}) = \mathrm{E}\,\psi(\mathrm{r}) \tag{3.3}$$

A solution of this equation, in one dimension, is

$$\psi(x) = A\sin kx + B\cos kx \tag{3.4}$$

where
$$k = \left(\frac{2mE}{\hbar^2}\right)^{1/2}$$

The function ψ(x) describes a wave with momentum and energy given by $p = \hbar k$ and $E = p^2/2m$.

3.3.1.3 *Particle in an infinite potential square well*

The case of an electron in an infinite square potential well is relevant to solids in which the electrons are confined within the solid and face a large energy barrier at the surface of the solid. The situation can be described as the electron moving along the x axis between two walls, which are separated by l_x and represent infinite potentials which the electron can not cross (Fig. 3.2). The wave function therefore must vanish outside the walls, and at each wall itself as it is a continuous function *i.e.* $\psi = 0$ at $x = 0$ and $\psi = 0$ at $x = l_x$. These are additional conditions, known as boundary conditions. This particular set of conditions is known as fixed boundary conditions. The solution can be obtained by substituting the fixed boundary conditions in Eq. (3.4).

Using $\psi = 0$ at $x = 0$, we get $B = 0$ so that the solution of the wave equation for this case is

$$\psi = A \sin \frac{2\pi}{\lambda} x$$

Using $\psi = 0$ at $x = l_x$, we get

$$\sin \frac{2\pi}{\lambda} l_x = 0 \quad \text{or} \quad \frac{2\pi}{\lambda} l_x = n\pi \quad \text{which gives } \lambda = \frac{2l_x}{n}$$

with n = 1, 2, 3, 4, ... The allowed values of the wave vector k are, therefore,

$$k = \frac{2\pi}{\lambda} = \frac{n\pi}{l_x}$$

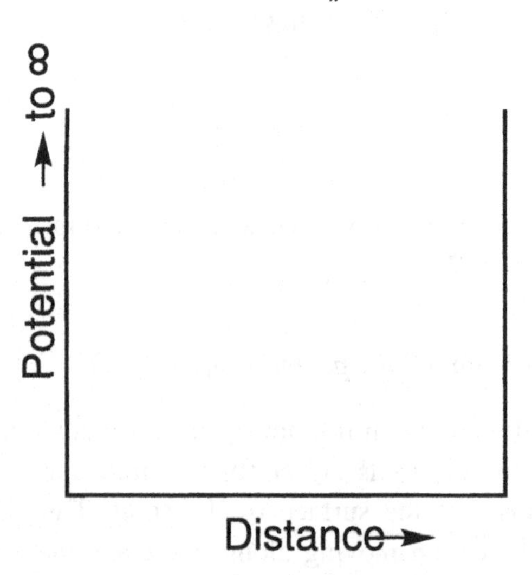

Fig. 3.2. Schematic representation of a one dimensional square well with boundaries of infinite potential.

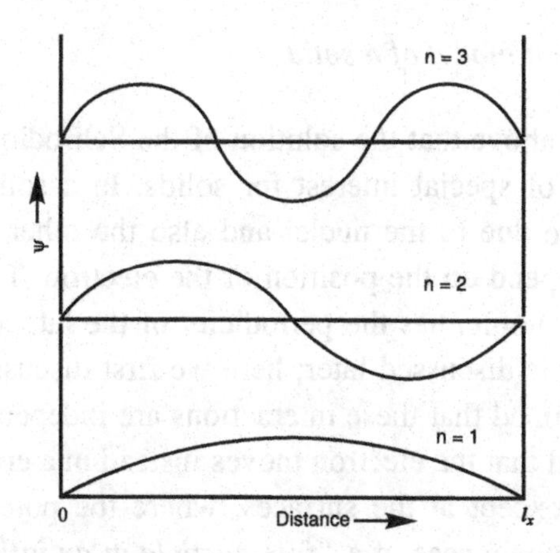

Fig. 3.3. The solutions of the wave function for a free particle in an infinite square potential well; only those wave lengths for which λ = 2l_x/n are allowed.

and the wave function is

$$\psi = A \sin \frac{n\pi}{l_x} x$$

The allowed values of energy of the electron for this case are

$$E_n = \frac{\hbar^2 k^2}{2m} = \frac{n^2 h^2}{8ml_x^2} \text{ ; n = 1, 2, ...}$$

Thus the electron can exist only with discrete energies. If l_x is of macroscopic dimensions, then the difference between the successive energy levels is very small and the energy levels form a quasicontinuous spectrum known as an "energy band".

The situation is similar to the vibration of a string fixed between two end points. The solutions are standing waves with nodes at each wall *i.e.* not all the wave lengths are possible but only those for which the length l_x is an integral multiple of half wave length, *i.e.* $l_x = n\lambda/2$ as shown in Fig. 3.3.

3.3.2 *Free electron model of a solid*

It was mentioned above that the solution of the Schrodinger equation for a free particle is of special interest for solids. In a solid, the electrons experience a force due to the nuclei and also the other electrons; these forces actually depend on the position of the electron. The potential due to the ions, for example, has the periodicity of the lattice. The case of a periodic potential is discussed later; here we first discuss a simpler case in which it is assumed that these interactions are independent of position of the electron and that the electron moves instead in a constant potential, $V(r)$ = constant, except at the surfaces, where the potential is infinite. This is the well known case of a *"free particle in an infinite square well potential"*. Thus in the solid, the electrons are assumed to move freely, with infinite potential barriers at the surfaces of the solid which does not allow them to escape. This model is quite close to the actual situation in case of metals; for the other solids also the conclusions arrived at by using this free electron model remain valid to a large extent if the mass m_o of the electrons is replaced by an effective mass m^* which empirically contains the corrections for the electron-crystal and the electron-electron interaction.

Consider a solid with dimensions l_x, l_y and l_z along the three coordinate axes. In the case of the free particle in an infinite square potential well, the fixed boundary conditions were used and the solutions were standing waves. To study the behavior of electrons in solids it is necessary to impose the *periodic boundary conditions, i.e.*

$$\psi(x, y, z) = \psi(x + l_x, y, z) = \psi(x, y + l_y, z) = \psi(x, y, z + l_z)$$

This is equivalent to imposing the condition that the wave function has the same value at both the boundaries of the solids in any direction as illustrated in Fig. 3.4.

Applying the periodic boundary condition, first in one dimension

$$A sin\frac{2\pi}{\lambda}x = A\sin\frac{2\pi}{\lambda}(x+l_x)$$

or
$$\frac{2\pi}{\lambda}(x+l_x) = \frac{2\pi}{\lambda}x \pm 2n\pi$$

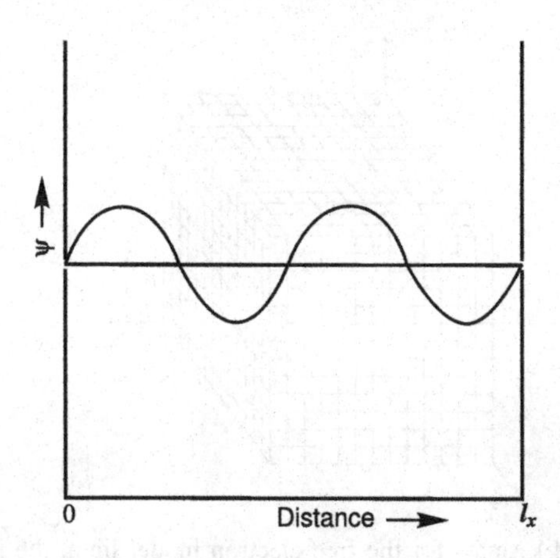

Fig. 3.4. In the periodic boundary conditions, the component of the wave function along any axis is identical at two points separated by the dimension of the solid in that direction *e.g.* $\psi(0) = \psi(l_x)$.

or $\frac{2\pi}{\lambda} l_x = \pm 2n\pi$ which gives $\lambda = \frac{l_x}{n}$ with n =0, ± 1, ± 2, ± 3,..

from which the allowed values of k_x are

$$k_x = \frac{2\pi}{\lambda} = \frac{2\pi}{l_x} n_x \text{ with } n_x = 0, \pm 1, \pm 2, ..$$

Similarly for the other two directions

$$k_y = \frac{2\pi}{\lambda} = \frac{2\pi}{l_y} n_y \text{ with } n_y = 0, \pm 1, \pm 2, ..$$

and

$$k_z = \frac{2\pi}{\lambda} = \frac{2\pi}{l_z} n_z \text{ with } n_z = 0, \pm 1, \pm 2, ..$$

The allowed values of energy of the electron for this case are

$$E_n = \frac{\hbar^2 k^2}{2m} \text{ with } k^2 = k_x^2 + k_y^2 + k_z^2$$

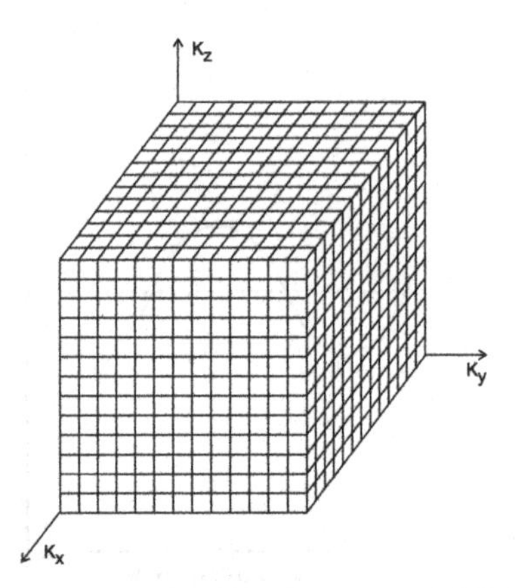

Fig. 3.5. The allowed k states for the free electron model lie at the intersections of the lines drawn with spacings $2\pi/l_x$, $2\pi/l_y$, $2\pi/l_z$ along the k_x, k_y, k_z axes respectively.

Each set of numbers n_x, n_y, n_z corresponds to a state of the electron with wave vector and energy given by the above relations. The allowed values of k are uniformly distributed in the k space with separation $\Delta k_x = 2\pi/l_x$, $\Delta k_y = 2\pi/l_y$ *and* $\Delta k_z = 2\pi/l_z$ (Fig. 3.5). If each atom in the solid contributes one free electron, then there would be N free electrons in a solid with N atoms. Each energy state *i.e.* each point in the k space, can be occupied by two electrons with opposite spins according to the Pauli exclusion principle. All the states lie within a sphere with a radius equal to the wave number for the highest energy electrons. This radius is

Fig. 3.6. (a) The closely spaced ("quasicontinuous") energy levels according to the free electron theory; the energy levels up to the Fermi energy are occupied in the absence of the electric field (b) when an electric field is applied, some electrons acquire higher velocity and move to higher energy levels while the others lose speed giving a redistribution of the occupied states as shown.

denoted as k_f and the corresponding energy is called the Fermi energy E_f. At O° K, all the states up to the Fermi energy are occupied by electrons while the states above the Fermi energy are empty.

The above description is able to explain the high electrical conductivity of metals. Figure 3.6 shows the energy levels in an *E vs. K* plot. The energy levels up to the Fermi level are filled with electrons and the states above the Fermi level are empty. Note that for every positive *k*, there is a negative *k* also *i.e.* for every electron moving in one direction, there is another electron moving with the same speed but in the opposite direction; thus there is no net flow of electrons. When an electric field is applied, electrons moving in one direction (towards the positive end of the field) get accelerated and acquire an extra velocity while the electrons moving in the opposite direction are slowed down. For the former to acquire additional kinetic energy in the field, there must be empty states available just above the Fermi energy which is the case for the metals. This results in electrical conduction as there is a net drift of the electrons in one direction.

3.3.3 *Electron in a periodic potential: the occurrence of the band gap*

The free electron model of a solid discussed above shows that the energy levels are quasi continuously distributed in a band. There are empty energy states available just above the highest filled energy level and the electrons can move up to these empty levels if an electric field is applied or a small energy is imparted by some other means. This explains well the high electrical conductivity of metals but not the behavior of the semiconductors or insulators. In the semiconductors and insulators, there is a large gap in the energy between the highest filled state and the next available empty state. This gap in energy is called the "band gap". This is shown schematically in Fig. 3.7.

The presence of band gaps is revealed if in the above free electron model the constant potential is replaced by a periodic potential. This is more realistic as the dominant potential experienced by the charge carriers in the solid is due to the ions. Figure 3.8 shows schematically

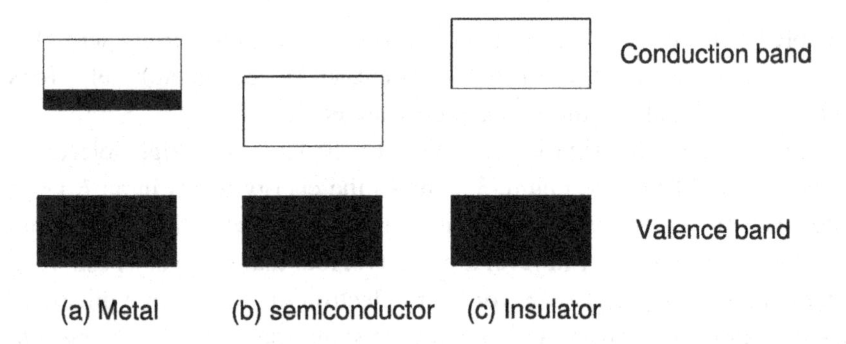

Fig. 3.7. Band positions in metals, semiconductors and insulators. (a) In metals, there exists a partially filled band, called the conduction band above the fully filled valence band (b) in pure (intrinsic) semiconductors like Si at 0° K, the conduction band is empty and is separated from the valence band by a band gap (c) in insulators, as in semiconductors, the valence band is full and is separated from the empty conduction band by a band gap; however, the band gap is much larger than that for semiconductors.

this periodic potential. The potential is lowest when the electron is near an ion core.

The effect of the other electrons is assumed to be to provide an average constant background potential. When this problem of nearly free electron is solved using the quantum mechanical perturbation theory, it is found that there are discontinuities in the energy values at certain values of the wave vector, *k*. For simplicity, we assume the solid to be one dimensional with a lattice spacing equal to '*a*'. These discontinuities are then found to occur at the values of *k* given by

$$k = \frac{n\pi}{a} \quad \text{with} \quad n = \pm 1, \pm 2, \pm 3, \ldots$$

The result is shown schematically in Fig. 3.9. It is seen that there is a gap in the energy at certain *k* values, the gap being equal to the band gap for the respective energy bands.

The *E vs. k* plot shown in Fig. 3.9(a) is called the extended zone representation. A different representation called the repeated zone representation is obtained if it is recognized that the wave functions are periodic in the *k* space. This implies that the segments in the *E vs. k* plot

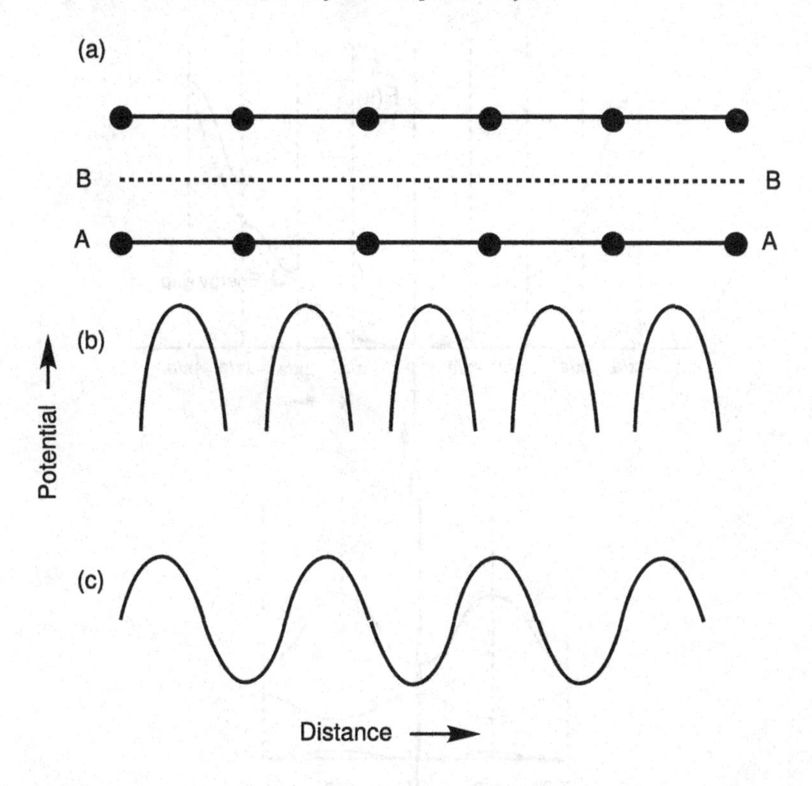

Fig. 3.8. The potential in a crystalline solid (a) AA is a line along a row of atoms and BB is a line between two rows of adjacent atoms (b) potential along the line AA (c) potential along the line BB.

in Fig. 3.9(a) can be translated by multiples of $\pm\ 2\pi/a$. This leads to the repeated zone representation shown in Fig. 3.9(b). This is simplified by further recognizing that the portions of the plot for $k > \pi/a$ and $k < -\pi/a$ are just the repetition of the plot between $-\ \pi/a$ and $+\ \pi/a$ and so are redundant. The resultant plot after removing these redundant portions is shown in Fig. 3.9(c). This is called the reduced zone representation and is most commonly used.

The sketch on the right in Fig. 3.9(c) depicts the energy range in which the various bands extend and the band gaps.

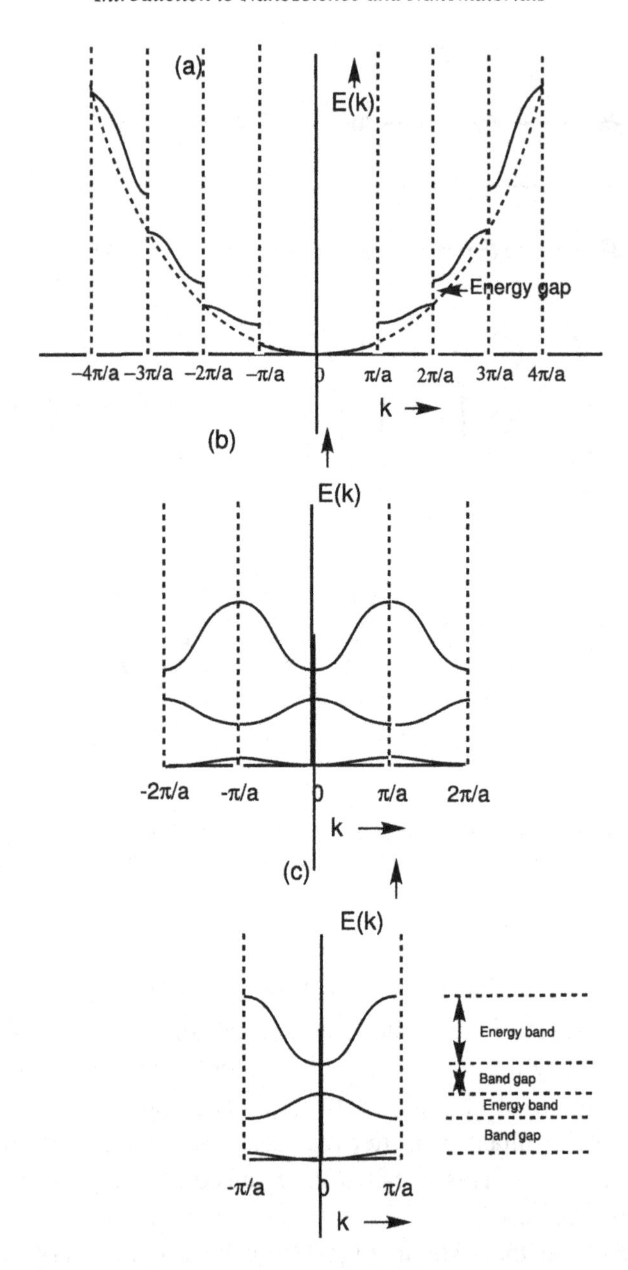

Fig. 3.9. The E *vs.* k plot for a solid showing that there are gaps in energy at certain energy values (a) extended zone representation (b) repeated zone representation (c) the reduced zone representation.

3.3.4 *Reciprocal lattice and the Brillouin zone*

3.3.4.1 *The reciprocal lattice*

In the above discussion, we have used the *E vs. k* plots. As *k* is a vector, we need to specify the direction with respect to the crystal in every such plot. The crystal exists in the real space and has a lattice defined in the real space. We can imagine a *k* space with axes k_x, k_y, k_z with reference to which any wave vector *k* can be described. As k ($= 2\pi/\lambda$) has dimension of the reciprocal of length, the k space is also called the reciprocal space. The question we wish to answer is, how the reciprocal space corresponding to the real crystal lattice can be constructed. A reciprocal space which is useful for representing the magnitudes and directions of wave vectors with respect to a crystal lattice is constructed as follows. For illustration, we chose a two dimensional crystal having a square lattice with parameter *a*. The reciprocal lattice is constructed according to the following rules

(i) Pick a point as the origin.
(ii) Draw normals to the different sets of crystal planes from this point.
(iii) Mark the points on each of these normals at a distance from the origin which is 2π times the reciprocal of the interplanar spacing of the respective set of planes.

The points thus obtained constitute the reciprocal lattice. The reciprocal space lattice is a set of imaginary points constructed in such a way that each point represents a set of crystal planes. In the reciprocal lattice, the direction of a vector from one reciprocal lattice point (which can be taken to be the origin) to another coincides with the direction of a normal to the real space planes represented by the latter point and the distance between the points is equal to the reciprocal of the real interplanar distance. For our two dimensional lattice of Fig. 3.10 (a) the reciprocal lattice is shown in Fig. 3.10(b). In Fig. 3.10(a), the (1,0) planes have an interplanar distance of *a* and their normal is horizontal. The corresponding reciprocal lattice point is therefore at a distance of $2\pi/a$ from the origin in the horizontal direction. Similarly, the points

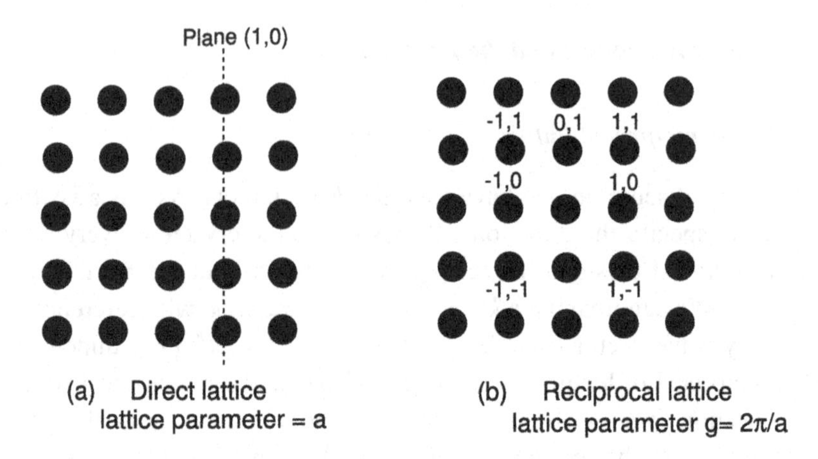

Fig. 3.10. (a) A two dimensional direct lattice (b) the corresponding reciprocal lattice.

corresponding to all the sets of planes can be constructed. The planes corresponding to the points in the reciprocal lattice are marked by their Miller indices. The unit cell of the reciprocal lattice has a parameter equal to $2\pi/a$.

It should be noted that in the reciprocal lattice the points are determined by the interplanar spacing between the planes in the real lattice. Any other arbitrary point has no meaning.

The same steps can be used to construct the reciprocal lattice for three dimensions. The reciprocal lattice for a simple cubic structure is simple cubic. However, the reciprocal lattice for a BCC crystal is FCC and vice versa; this is so because the primitive axes have to be used in constructing the reciprocal lattice. For further details, the reader should refer to elementary texts on solid state physics or materials science.

3.3.4.2 *The Brillouin zones*

While discussing the development of band gaps in semiconductors above, we saw that there are discontinuities in the values of the energies of the electron at certain values of k. In three dimensions, the end points of these wave vectors lie on a surface. This surface forms the boundary of what is called a Brillouin zone. The first discontinuity occurs at the boundary of the first Brillouin zone, the second discontinuity at the boundary of the second Brillouin zone and so on. It is usual to present the energy *vs.* k data with reference to the position in a Brillouin zone. Such data is useful in understanding the electrical and optical properties of a material. Later in the book we shall see how such data explains the presence of semiconducting or insulating behavior in different carbon nanostructures. It would therefore be useful to discuss briefly how the Brillouin zones of any crystal are constructed.

The Brillouin zones for a given crystal can be constructed graphically from its reciprocal lattice by following the steps below. The steps are illustrated in Fig. 3.11 with reference to a two dimensional lattice.

(i) Chose any point in the reciprocal lattice as origin and draw a line connecting this point to one of its nearest neighbors. Now draw a perpendicular bisector of this line. This bisector represents a plane as it is perpendicular to a reciprocal lattice vector. It is called a Bragg plane.

(ii) Similarly draw Bragg planes by joining the origin to the other nearest neighbors.

(iii) The volume (area in the case of a two dimensional lattice in Fig. 3.10) enclosed by these Bragg planes is the first Brillouin zone. There are no Bragg planes inside the first Brillouin zone.

(iv) Now draw the Bragg planes corresponding to the second nearest neighbors. The differently hatched area in Fig. 3.11(b) between the second set of Bragg planes and the first set of Bragg planes defines the second Brillouin zone. There is only one Bragg plane between any point within this zone and the origin.

The third and higher Brillouin zones can be obtained in a similar manner. However, in most cases it is the first Brillouin zone which is of interest. The planes constituting the Brillouin zone are called Bragg planes because at these planes the Bragg condition for diffraction is satisfied [45]

$$\lambda = 2d \sin\theta$$

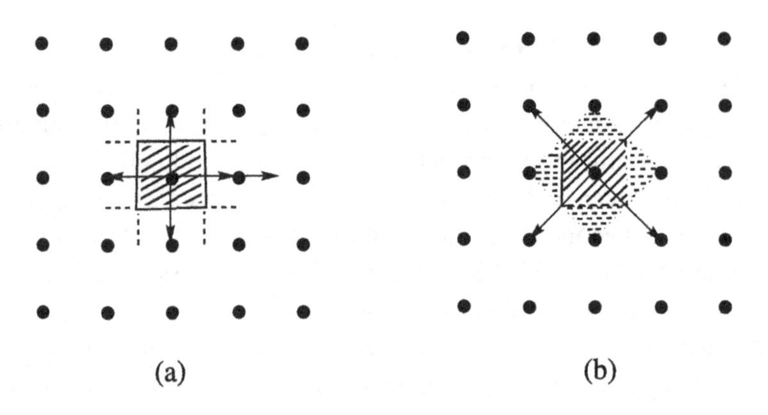

Fig. 3.11. Brillouin zones for a two dimensional lattice. (a) The first Brillouin zone is obtained by drawing bisectors of the reciprocal lattice vectors to the nearest neighbors. It is shown as a hatched area in the center and any point inside it contains no Bragg plane. (b) The second Brillouin zone is the differently hatched area surrounding the first Brillouin zone. Its outer boundary is defined by the bisectors to the lattice vectors joining the origin to the second nearest neighbors. There is only one Bragg plane between any point in this zone and the origin. The areas of the first and the second Brillouin zones are the same.

Here λ is the wavelength of the electron approaching the Bragg plane, θ is the incident angle and d is the interplanar spacing of the set of planes represented by the Bragg plane. An electron is diffracted at the boundary of the Brillouin zone *i.e.* it can not acquire energy which is just higher than that corresponding to its incident point in the Brillouin zone, meaning that the Brillouin zone represents the edge of the band gap.

3.3.5 *The band structures of semiconductors*

The electrical, optical and other properties of a material depend strongly on how the energy of a delocalized electron in it varies with the wave vector, *k*. It is usual to present such energy *vs.* *k* plots with reference to the location of *k* in the Brillouin zone. As an example, Fig. 3.12 shows the Brillouin zone of III-V and II-VI semiconducting compounds which all have the same cubic zinc-blende structure. The high symmetry points and the high symmetry lines in the Brillouin zone are marked by Greek or Roman letters.

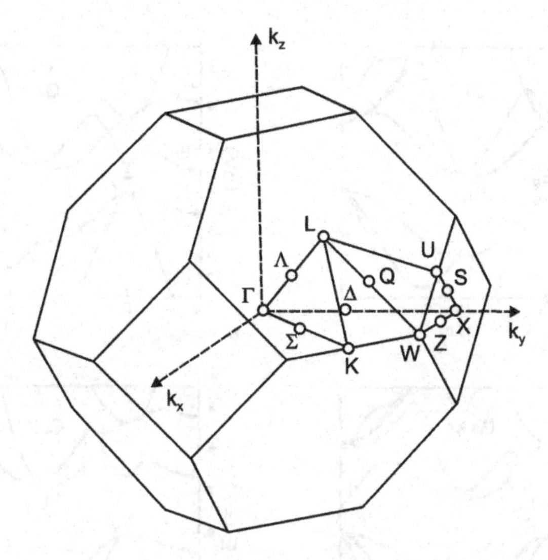

Fig. 3.12. Brillouin zone for the cubic zinc-blende structure; the high symmetry points and the high symmetry lines in the Brillouin zone are marked by Greek or Roman letters.

Figure 3.13 shows the band structures of some semiconductors. When the top of the valence band and the bottom of the conduction band are positioned one above the other, at the same value of *k*, the semiconductor is called a *direct band gap* semiconductor. Thus GaAs, InP, and InAs are direct band gap semiconductors. This is illustrated in Fig 3.14 in which the plot for InP is reproduced from Fig. 3.13. It can be seen that the maximum in the valence band at Γ_{15} and the minimum in the conduction band at Γ_1 are at the same value of *k*. Same is the case for GaAs and InAs (Fig. 3.13).In these semiconductors an electron in the valence band at point Γ_{15} can get excited to point Γ_1 in the conduction band with no change in the wave vector *k*.

On the other hand, in Si, Ge and GaP the maximum of the valence band and the minimum of the conduction band occur at different values of *k*. These semiconductors are called the *indirect band gap* semiconductors. Here, the absorption or emission of a phonon is required for the photon emission or absorption as shown schematically in Fig. 3.14(b).

Light absorption is a convenient technique to determine the band gap of semiconductors and also to find if the semiconductor has a direct band

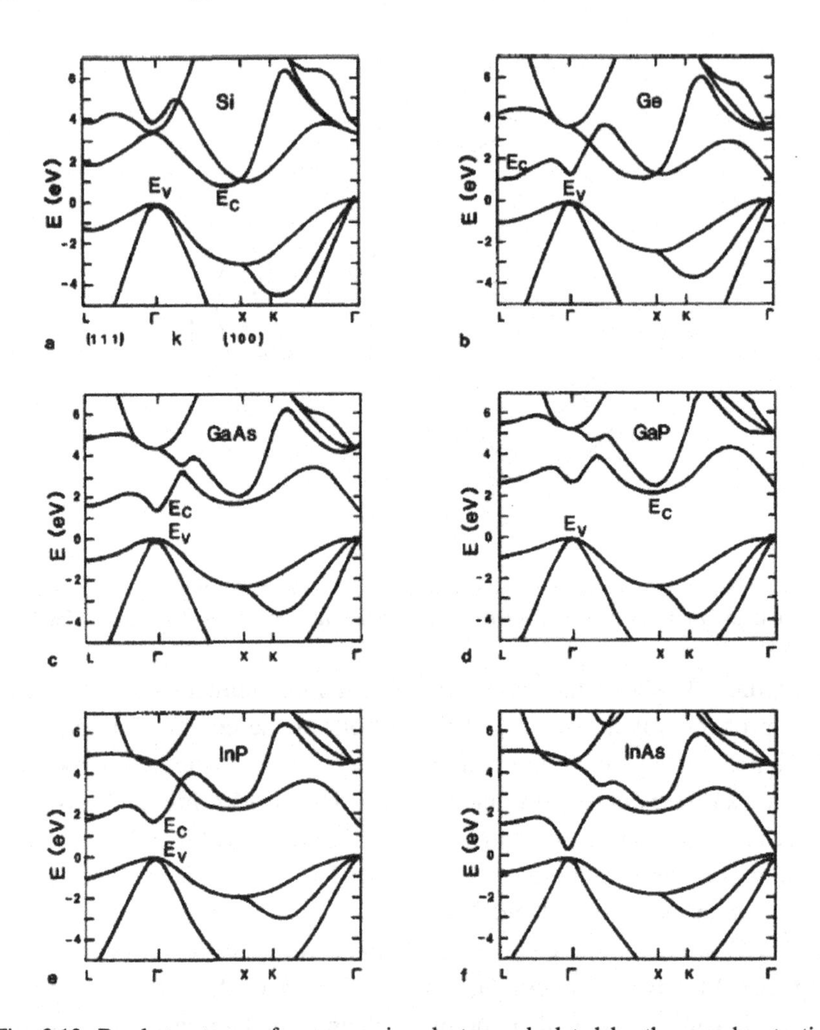

Fig. 3.13. Band structures of some semiconductors calculated by the pseudopotential method [46]. Each figure is an energy *vs.* wave vector plot along different directions in the Brillouin zone as shown in Fig. 3.12 : from Γ to L along Λ, from Γ to X along Δ, from Γ to K along Σ and along the line joining the points X and K. The highest filled band at absolute zero is marked as the valence band E_V and the lowest empty band is marked as the conduction band Ec. The band gap in each case is defined as the region between Ev and Ec where no band appears.

gap or an indirect band gap. When the light of energy $\geq E_g$ is incident on a semiconductor, it can excite an electron from the valence band to

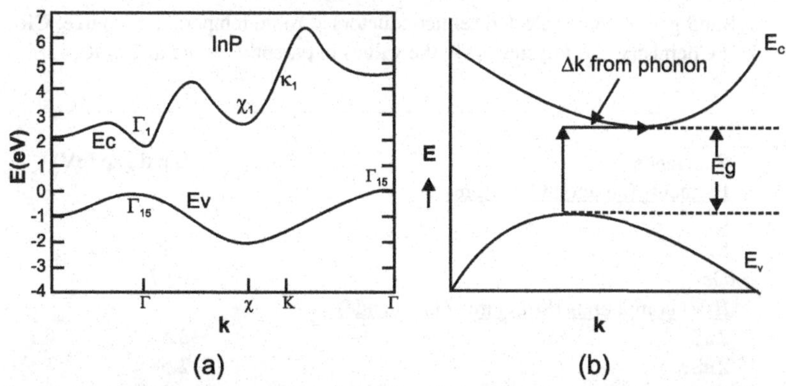

Fig. 3.14 (a) Band structure of InP from Fig. 3.13; only the curves corresponding to the conduction band and the valence band are shown. The maximum in the valence band at Γ_{15} and the minimum in the conduction band at Γ_1 are at the same value of k, indicating that this is a direct band semiconductor. (b) In an indirect band semiconductor, the maximum in the valence band and the minimum in the conduction band are not at the same k; consequently assistance of a phonon of suitable momentum is needed to cause the transition from the valence band to the conduction band.

the conduction band and get absorbed in the process. Therefore, if the frequency of the incident light is scanned from the lower side, there is a sudden increase in absorption as the frequency approaches an energy corresponding to the band gap of the semiconductor. In practice, in such an experiment, usually the transmittance T, defined as $T = I/I_0$, where I_0 is the incident intensity and I the transmitted intensity, is measured as a function of the wavelength of the incident light. From this the absorption coefficient α can be determined using standard relations [47]. The absorption coefficient near the absorption edge (*i.e.* the wavelength where there is a sudden increase in the absorption) is related to the band gap as follows

For the direct band semiconductor : $(\alpha h\nu)^2 = A\,(h\nu - E_g)$

For the indirect band semiconductor : $(\alpha h\nu)^{1/2} = B\,(h\nu - E_g)$

Here h is the Planck's constant, ν is the frequency of the incident light and E_g is the band gap. A plot of $(\alpha h\nu)^{1/2}$ *vs.* $h\nu$ or $(\alpha h\nu)^2$ *vs.* $h\nu$, depending on the type of semiconductor, should therefore yield a straight line with an intercept equal to the band gap E_g on the energy axis. Figure 3.15 shows such a plot for a zinc nitride film. The straight line fit is

Table 3.2. Band gap of some selected semiconductors at room temperature (source: CRC Handbook of Chemistry and Physics [48] ; the values in parentheses are at 286 K [49]).

Substance	Band gap (eV)
Elements (Diamond structure)	
C	5.7
Si	1.12
Ge	0.67
II-VI compounds (Structure: Zinc blende)	
ZnS	3.54
ZnSe	2.58
ZnTe	2.26
CdTe	1.44
II-VI compounds (Wurtzite structure)	
ZnO	3.2
ZnS	3.67
CdS	2.42
CdSe	1.74
CdTe	1.5
III-V compounds (Zinc blende structure)	
AlP	2.45
AlAs	2.16
AlSb	1.6
GaP	2.24
GaAs	1.43
GaSb	0.67
ZnP	1.27
InAs	0.36
InSb	0.163
III-V compounds (Wurtzite structure)	
AlN	6.02
GaN	3.34
InN	2.0
IV-VI compounds (NaCl structure, Halite)	
PbS	(0.37)
PbSe	(0.26)
PbTe	(0.25)
V-VI compound	
Bi_2Te_3 (trigonal)	(0.15)
II-V Compound	
Cd_3As_2 (tetragonal)	(0.13)
CdSb (orthorhombic)	(0.48)

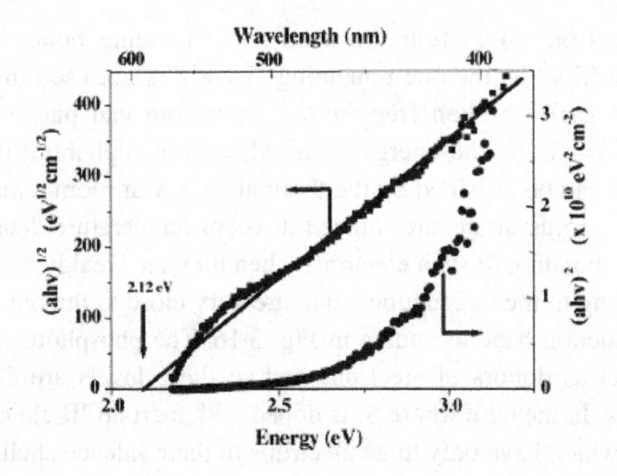

Fig. 3.15. $(\alpha h\nu)^{1/2}$ and $(\alpha h\nu)^2$ vs photon energy plots for zinc nitride films (reprinted from [50] copyright(2006) by permission of the American Physical Society).

obtained for $(\alpha h\nu)^{1/2}$ *vs.* hν only with an intercept of 2.12 eV. It is, therefore, concluded that the material is an indirect band gap semiconductor with a band gap of 2.12 eV. The values of the band gaps for some well known semiconductors are given in Table 3.2.

3.3.6 *Energy levels within the band gap – shallow traps and the deep traps*

We saw that in a semiconductor, the valence band and the conduction band are separated by a band gap in which there are no energy states. However, it is possible under certain conditions for energy levels to exist within the band gap as described below.

3.3.6.1 *Shallow traps in doped semiconductors*

Silicon, a well known semiconductor, is a group IV element. It has four electrons in its valence shell which form covalent bonds with four other silicon atoms in its crystal structure. If a group V element such as P, Sb or As is added to Si, its atoms go to the lattice sites of Si. Out of the five valence electrons of P, four are used up in forming bonds with the surrounding Si while the one remaining electron is not used in bonding and can be easily broken free from P atom and can pass on to the conduction band. As the energy required for accomplishing this is so small that it can be provided by the thermal energy at room temperature, all the phosphorus atoms are ionized at room temperature. The energy levels corresponding to such electrons when they are weakly bound to P, before moving to the conduction band, are very close to the bottom edge of the conduction band as shown in Fig. 3.16. The phosphorus atoms in this case act as donors of electrons and so these levels are called the donor levels. In the case where Si is doped with a group III element such as B or Al which have only three electrons in their valence shells, one of the bonds from the dopant atom to the surrounding Si atoms lacks an electron and remains unsatisfied. An electron from the valence band (*i.e.* from another Si atom) can migrate to this hole leaving behind a hole in the valence band, if the energy difference is small enough to be provided by the thermal energy at the room temperature. The dopant atom thus acts as an "acceptor" of electrons; due to the small activation energy, nearly all of the acceptor atoms are ionized at room temperature. The energy levels corresponding to an electron attached to the acceptor

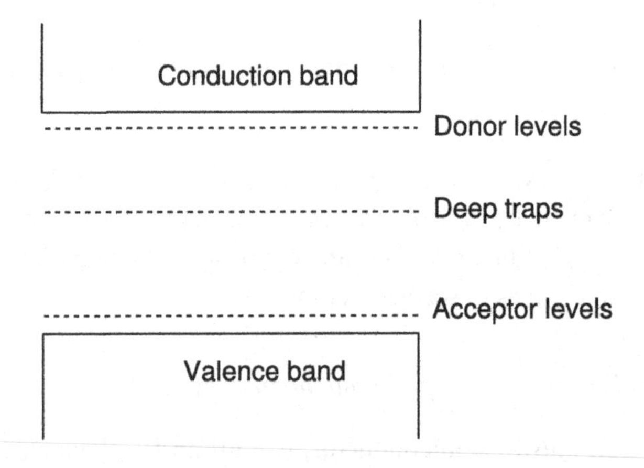

Fig. 3.16. The positions of donor levels, the acceptor levels and the deep traps in the energy diagram of a semiconductor.

atom are called "acceptor states" and lie a little above the top of the valence band edge as the electrons from the valence band are promoted in energy to the acceptor level . These acceptor levels are also shown in Fig. 3.16.

3.3.6.2 *Deep level traps*

The donor levels and the acceptor levels described above are known as "shallow traps" as they lie close to the band edges. A semiconductor can also have electron states lying close to the middle of the band gap. These are called deep level states. These deep levels are produced, for example, due to broken bonds or displacement of atoms from their equilibrium positions due to a strain. Presence of transition metals in Si generally leads to the generation of deep level traps.

There also exist *surface states* in a semiconductor which are states lying within the semiconductor band gap. These result from the incomplete covalent bonds at the surface. At the surface the periodicity of the bulk is broken, so that the periodic boundary conditions are no longer valid. The solution of the Schrodinger equation in this case gives wave functions which decay exponentially both into the bulk and into the vacuum and are thus located at the surface and are called the surface states. The surface states can act as donor states or acceptor states.

3.3.7 *Characteristic lengths in the nanometer range*

In Section 3.1 it was pointed out that a physical phenomenon responsible for a certain property has associated with it a characteristic length scale and when the dimensions of the solid approach this length, the corresponding physical phenomenon is influenced and the property undergoes a change. Examples of some of these characteristic lengths are given below.

3.3.7.1 *de Broglie Wavelength*

The de Broglie wavelength for an electron with a momentum p is given by

$$\lambda = \frac{h}{p} = \frac{h}{m^* v}$$

where m^* is the effective mass and v is the velocity of the electron. The effective mass m^* is different from m_o, the mass of electron in vacuum. It is defined by the relation

$$\frac{1}{\hbar^2}\frac{d^2E}{dk^2} = \frac{1}{m^*} \tag{3.5}$$

where $\frac{d^2E}{dk^2}$ is the curvature of the E *vs.* k plot for the electron.

The use of m^* in place of m_o allows one to treat the electrons in the solid as free electrons (*i.e.* moving in a zero field), without worrying about the interactions of the electron with the other electrons or the lattice – in effect these interactions are taken care of by using m^* instead of m_o. The effective mass is quite small for semiconductors *e.g.* 0.067 m_o for GaAs and 0.014 m_o for InSb while for the metals, in which the electrons naturally behave as free electrons, it is close to m_o. The de Broglie wavelength is therefore much larger for the semiconductors. Consequently, the size effects are observed at much larger sizes in semiconductors than in metals.

3.3.7.2 *Mean free path of electrons*

For the properties such as electrical conductivity, photoluminescence, *etc.*, the characteristic lengths associated with the electrons in the solid need to be considered. For the electrical conductivity, the characteristic length can be the mean free path of the electron. As the electron moves in a crystal, it is scattered by interactions with crystal imperfections such as lattice vibrations (phonons), impurities, defects, *etc.* The scattering event is inelastic *i.e.* it causes a change in the momentum and energy of the system. The average distance between two scattering events is called the *mean free path* ℓ and the average time between two successive collisions is called the relaxation time. For a macroscopic solid, the dimension of the solid is much larger than ℓ so that the electron suffers multiple scattering as it travels; this motion of the electron is termed *diffusive transport.* This mode is characterized by a high electrical resistivity. If, however, the dimension of the solid is less than the mean free path, the electrons can travel through it without any scattering except from the surfaces. The electrons in this case move by *ballistic transport* leading to a high electrical conductivity as in the carbon nanotubes. Thus the mean free path is the characteristic length for the conduction to change from the diffusive transport to the ballistic transport.

3.3.7.3 *Exciton and Bohr exciton radius*

When an electron from the valence band of a semiconductor or an insulator is excited to the conduction band (say by a photon having an energy greater than the band gap of the material), a hole is left behind in the valence band. In most cases, the electron will move away and the two exist separately for a short time until an electron recombines with the hole. In some cases the electron and the hole stay together for a longer time (~ milliseconds) as a pair due to the Coulombic attraction between them and have a natural physical separation which depends on the material. This (bound) electron-hole pair is called an exciton. Many properties of the exciton can be obtained by modeling it as a Bohr atom. Since the excitons play an important role in the semiconductor nanostructures, it is useful to discuss them in some detail.

We first recapitulate the Bohr model to derive the parameters for the exciton. In the Bohr atom the electron revolves around the nucleus in an orbit of radius r. The electrostatic force between the electron and the proton is balanced by the centripetal force

$$\frac{e^2}{4\pi\varepsilon_0 r^2} = \frac{m_r v^2}{r} \tag{3.6}$$

Here, ε_0 is the permittivity of vacuum, r is the radius of the orbit and m_r is the reduced mass of the electron and the proton given by

$$m_r = \frac{m_e m_p}{m_e + m_p} \approx m_e \tag{3.7}$$

m_e and m_p being the masses of the electron and the proton respectively. Also, the total energy of the electron is the sum of its kinetic and the potential energies;

$$E = \frac{1}{2} m_e v^2 - \frac{e^2}{4\pi\varepsilon_0 r} = -\frac{1}{2}\frac{e^2}{4\pi\varepsilon_0 r}, \tag{3.8}$$

(substituting the value of v from Eq. 3.6)

The quantization of energy is obtained by requiring that the angular momentum be an integral multiple of $\hbar$; this condition has been reinterpreted by de Broglie as a standing wave condition that the circumference of the electron orbit be equal to an integral multiple of the half wavelength of the electron;

$$2\pi r = n\lambda/2 \tag{3.9}$$

The kinetic energy of the electron can also be equated to hv/λ in the wave formulation *i.e.*

$$\tfrac{1}{2} m_e v^2 = hv/\lambda \tag{3.10}$$

Combining Eqs. (3.6), (3.9) and (3.10), one gets

The Bohr radius $r = \frac{4\pi\varepsilon_0 \hbar^2}{m_e e^2}$, for the first orbit, n =1 (3.11)

$$= 0.0529 \text{ nm}$$

and the allowed electron energies

$$E_n = -\frac{m_r e^4}{2(4\pi\varepsilon_0)^2 \hbar^2}\frac{1}{n^2} = -\frac{13.6}{n^2} eV \tag{3.12}$$

A representation of the exciton is shown in Fig. 3.17. It is shown as a bound electron-hole pair orbiting around its center of mass. This is analogous to the Bohr model of an atom. The energy levels and the radii of the exciton can be obtained by proceeding in a manner similar to that done above for the Bohr model by making the following changes:

(i) The reduced mass is now effective reduced mass of the electron and the hole, given by

$$\mu_r = \frac{m_e^* m_h^*}{m_e^* + m_h^*}$$

where m_e^* and m_h^* are the effective mass of the electron and the hole respectively.

(ii) The permittivity of the solid, ε, is to be used in place of ε_0, the permittivity of vacuum. This is given by $\varepsilon = \varepsilon_r\varepsilon_0$ where ε_r is the dielectric constant of the solid.

With these changes in (3.11) and (3.12), the expressions for the energy and the radius of the exciton are

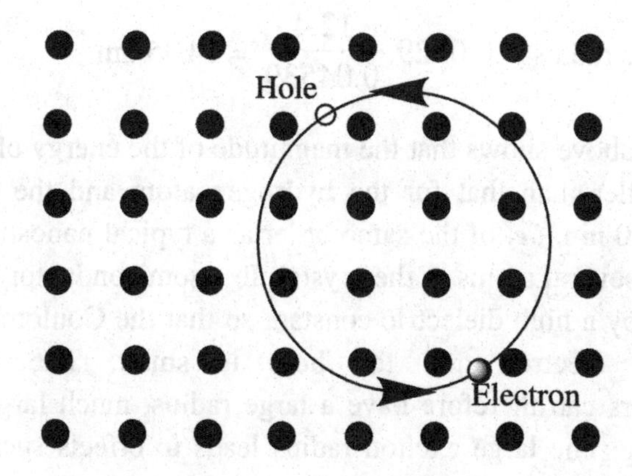

Fig. 3.17. Representation of an exciton as a bound electron-hole pair orbiting around its center of mass.

$$E_n = -\frac{\mu_r e^4}{2(4\pi\varepsilon_r\varepsilon_0)^2\hbar^2}.\frac{1}{n^2} = -\frac{\mu_r/m_e}{\varepsilon_r^2}.\frac{13.6}{n^2} = -\frac{\mu_r}{m_e\varepsilon_r^2}.\frac{13.6}{n^2}\ \text{eV}; \tag{3.13}$$

And $$r_{exciton} = \frac{4\pi\varepsilon_0\varepsilon_r\hbar^2}{\mu_r e^2} = 0.0529\frac{\varepsilon_r}{\mu_r/m_e}\ \text{nm} \tag{3.14}$$

Example 3.2. *Calculate the ground state energy ($n = 1$) and the radius of the exciton for the semiconductor InP using the following data: electron effective mass, $m_e^* = 0.073\ m_e$; hole effective mass, $m_h^* = 0.12\ m_e$; dielectric constant of InP = 12.4.*

Solution. The effective reduced mass of the electron-hole pair *is*

$$\mu_r = \frac{m_e^* m_h^*}{m_e^* + m_h^*} = 0.04539\ m_e$$

Using Eqs. (3.13) and (3.14), we get, for $n = 1$

$$\text{Ground state energy} = \frac{0.04539}{(12.4)^2}.13.6 = 0.004015\ \text{eV} = 4.015\ \text{meV}$$

$$\text{Ground state radius} = 0.0529.\frac{12.4}{0.04539} = 14.45\ \text{nm}$$ □

The example above shows that the magnitude of the energy of an exciton is much smaller than that for the hydrogen atom and the size of the exciton is ~ 30 nm, *i.e.* of the same order as a typical nanostructure, and extends over several atoms in the crystal. In a semiconductor the charges are screened by a high dielectric constant so that the Coulomb attraction between the electron and the hole is small. The exciton in semiconductors can therefore have a large radius, much larger than the lattice spacing. The large exciton radius leads to effects such as higher energy of the excitons in the confined space of a nanostructure, as discussed later. The energy levels of the exciton and the hydrogen atom are compared in Fig. 3.18.

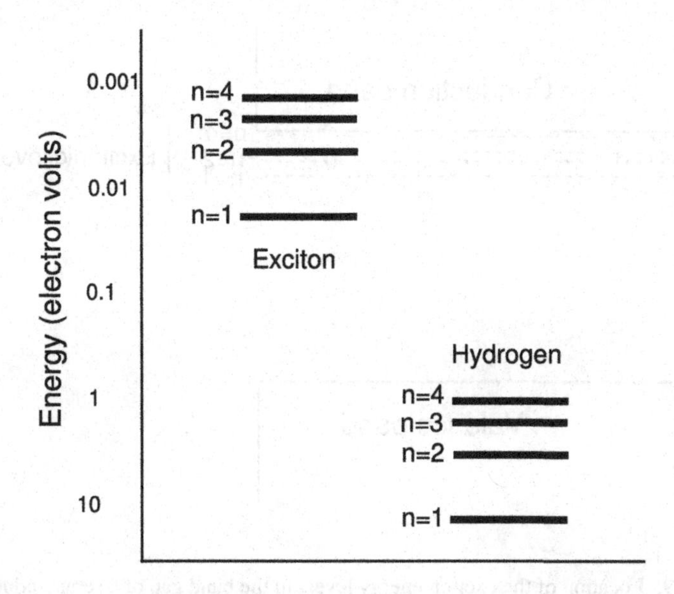

Fig. 3.18. A comparison of the energy levels of an electron in a hydrogen atom and an exciton.

The electron and the hole in an exciton are bound by the coulombic force. An energy, equal to this binding energy is therefore necessary to separate them. The process of photo excitation of an electron from the valence band of a semiconductor to its conduction band by a photon can be imagined to consist of two steps: creation of an exciton and then breaking up of the exciton into unbound electron and hole. The exciton energy levels, therefore, lie within the band gap, just below the edge of the conduction band as shown in Fig. 3.19.

The exciton energy levels can be probed at very low temperatures where $kT \ll$ exciton binding energy. Fig. 3.20 shows the exciton absorption spectrum of GaAs at 1.2 K. At such a low temperature, the first three energy levels of the exciton are clearly resolved.

Because of the large separation between the hole and the electron, there is difficulty in their wave functions overlapping so that the life time of an exciton *i.e.* the time before which the electron and the hole recombine, can be large (life times of up to several milliseconds have been observed in Cu_2O).

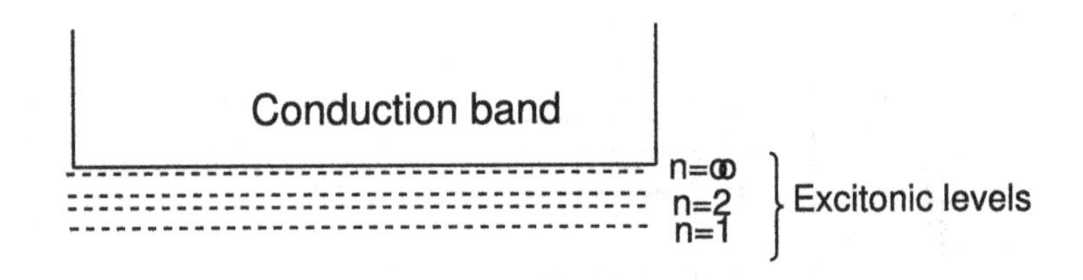

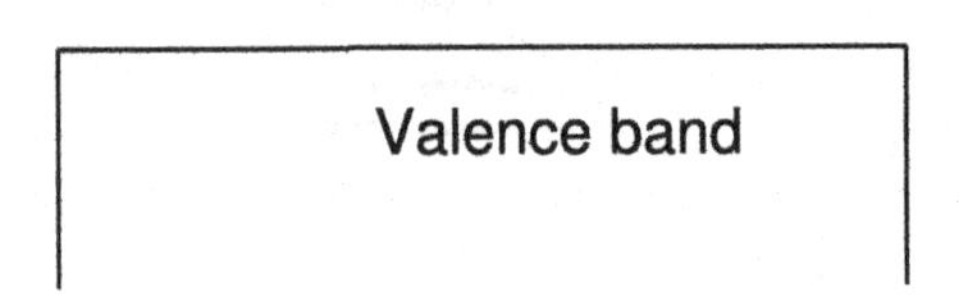

Fig. 3.19. Location of the exciton energy levels in the band gap of a semiconductor0.

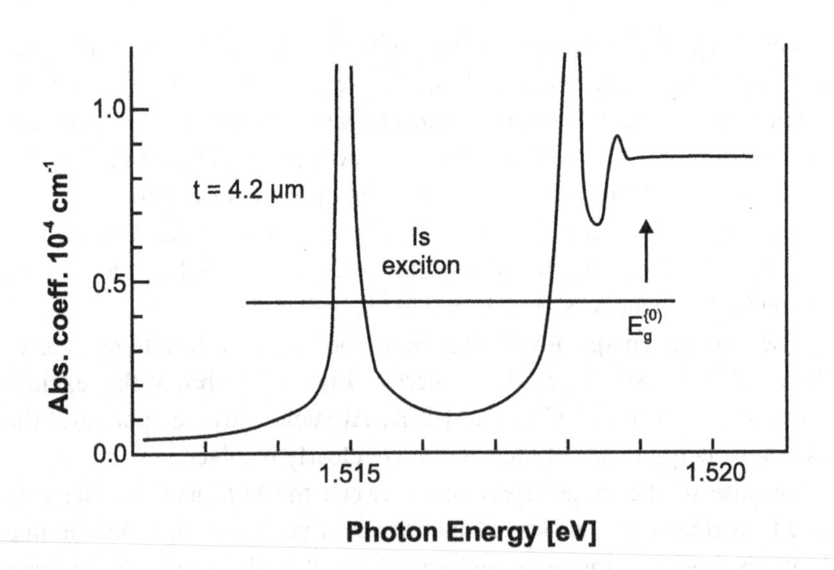

Fig. 3.20. Absorption spectrum from excitons in GaAs sample of thickness t=4.2μm at 1.2 K (reprinted from [51] copyright (1985) with permission from Elsevier).

Exciton radius is often used as the length scale to define a quantum dot. When the radius of the particle R is smaller than the Bohr exciton radius, the exciton can not achieve the desired separation; an energy called the confinement energy is needed to confine the exciton within the quantum dot. In fact, the confinement energy becomes appreciable for particle radius smaller than about twice the exciton radius ($R \leq 2\ r_{exciton}$) At this size the confinement energy is much larger than the Coulombic attraction energy of the exciton and the particles are treated as quantum dots. Thus for CdS, the excitonic Bohr radius is 29 A^o so that the particles of CdS smaller than about 50 A^o should behave as quantum dots.

3.3.7.4 *Other length scales*

The length scales discussed above relate to the electronic and photonic properties. There can be other characteristic length scales for these properties *e.g.* screening length for the screening of the potential due to the ionized dopant atoms in semiconductors by free carriers. In case of the charged particles in liquid suspension, a characteristic length is the double layer thickness which is discussed in detail in the chapter on colloids. In materials science, the size of the smallest crystal which is free of defects such as a vacancy or a dislocation is of concern since many properties of crystalline solids depend on the concentration of such defects.

Noting that the equilibrium concentration of vacancies (in atom fraction) is given by $exp(-Q/kT)$, it can be easily shown that the size d below which a crystallite can not contain a vacancy is [52]

$$d_c = 2[\{3exp(Q/kT)/4\pi n_v\}]^{1/3}$$

where n_v is the number of atoms per unit volume. This critical size comes out to be in the nanometer range for most crystal.

The mechanical properties of crystals are controlled by dislocations. It is almost impossible to create a dislocation free crystal but if the size is sufficiently reduced then the dislocations will be pinned by the grain boundaries and the interior of the grain would be essentially dislocation free (see Appendix II). Taking the stress required to move a dislocation to be $\tau = 2\alpha Gb/\lambda$, (where G is the shear modulus of the crystal, b is the Burgers vector of the dislocation, α is a constant (0.5 - 1.5) depending upon the type of dislocation), the grain size at which the dislocation will be pinned at both ends by the grain boundary of the crystallite comes out to be $d_c = 2\alpha Gb/\tau_s$ where τ_s is the theoretical strength. Here again the critical size is in the nanometer range.

It should be noted that even though reducing the grain size to below critical size can create a perfect, defect free crystal, the fraction of the surface atoms becomes so large, that other effects may intervene, preventing the realization of the properties expected from a perfect crystal [52]. As an example, the hardness increases at first according to the Hall-Petch relation, as the crystal size is decreased but then decreases as the grain size is reduced below a critical value, which is in the nanometer range. This is discussed in Chapter 11.

3.3.8 *The Density of States function for low dimensional systems*

The electrical, transport, optical and many other properties of solids are related to a function called the "density of states (DOS)" function. The DOS function $D(E)$ at a given value E of energy is defined such that $D(E)\Delta E$ is equal to the number of states in the interval from E to $E + \Delta E$. In the case of the solid with periodic boundary conditions as discussed above, each allowed value of $k(k_x, k_y, k_z)$ occupies a volume in k space

given by $\frac{2\pi}{l_x}.\frac{2\pi}{l_y}.\frac{2\pi}{l_z} = \frac{8\pi^3}{V}$ where V is the volume of the solid. The number of energy states lying between k and $k+dk$ is denoted as $D(k)\,dk$. It is obtained by dividing $4\pi k^2 dk$, the volume of the spherical shell of radius k and thickness dk by the volume occupied by each value of k in the k space. This gives

$$D(k)\Delta k = \frac{4\pi k^2 dk.V}{(2\pi)^3} = \frac{Vk^2 dk}{2\pi^2}$$

The number of states per unit volume is obtained by dividing this by V. Furthermore, since each state can be occupied by two electrons of opposite spin, the number of electron states per unit volume becomes

$$D_{3d}(k)dk = \frac{k^2 dk}{\pi^2}$$

so that
$$D_{3d}(E)dE = D_{3d}(k)dk = \frac{k^2 dk}{\pi^2} \tag{3.15}$$

As $E = \frac{\hbar^2 k^2}{2m}$, we can write

$$dE = \frac{\hbar^2 k dk}{m} \text{ or } \frac{dE}{dk} = \frac{\hbar^2 k}{m} \tag{3.16}$$

Substituting this in Eq. (3.15) we get

$$D_{3d}(E) = \frac{k^2}{\pi^2 \, dE/dk} = \frac{k^2}{\pi^2}\frac{m}{\hbar^2 k} = \frac{km}{\pi^2\hbar^2} = \frac{\sqrt{2mE}}{\hbar}\frac{m}{\pi^2\hbar^2}$$

or
$$D_{3d}(E) = \frac{\sqrt{2}}{\hbar^3}\frac{m^{\frac{3}{2}}}{\pi^2}\sqrt{E} \tag{3.17}$$

The density of states function for a three dimensional solid therefore is proportional to $\sqrt{E}$. For a bulk solid the values of l_x, l_y, l_z are large and $\Delta k = 1/l_x$ *etc.* are very small such that the points in the k space, and hence, the energy levels are quasi continuously distributed.

As was mentioned toward the beginning of this chapter, many new phenomena are observed when at least one of the dimensions of a solid is reduced to lengths comparable to appropriate scaling length. Depending on how many dimensions are small enough, the structures are given different names. These are quantum well, quantum wire and quantum dot for one, two and three dimensions being small respectively. The confinement of electrons in the nanometer regions results in a drastic change in the density of states function. This in turn leads to many changes in properties, particularly the transport and the optical properties, as well as the appearance of some new phenomena. In this chapter we proceed now to derive the density of states function for the three cases mentioned above.

3.3.8.1 *The quantum well*

A structure is called a quantum well when only one of the dimensions is in the nanometer range, the other two dimensions being large *e.g.* a thin sheet with thickness in the nm range. In this case the motion of the electrons is confined in the thickness (z) direction while the electron is free to move in the plane of the sheet *i.e.* in the x and y directions. A practical way to confine the electrons within the sheet in the z direction is to provide high enough potential barriers at the free surfaces of the sheet. This is done, for example, by sandwiching a thin film of a semiconductor such as GaAs (E_g=1.4 eV), between the layers of a higher band gap semiconductor such as $Al_xGa_{1-x}As$ (a band gap of $E_g \approx 2.0$ eV at x = 0.3). This results in a barrier of ~0.4 eV for the electrons and ~0.2 eV for the holes (Fig. 3.21) in the GaAs layer.

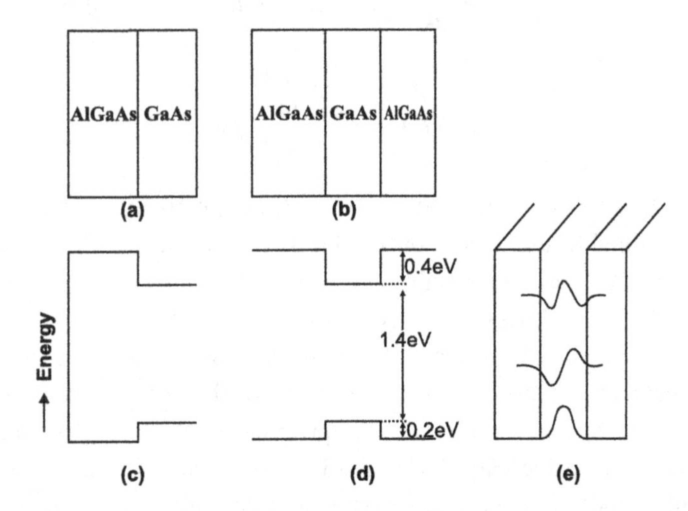

Fig. 3.21. (a) A "heterojunction" is formed at the junction of two different semiconductors, in this case AlGaAs and GaAs. (b) A "quantum well" is formed by sandwiching a layer of a low bandgap semiconductor (GaAs) between the layers of a higher bandgap semiconductor (AlGaAs) (c) the band structure of the heterojunction and (d) of the quantum well; in a quantum well barriers for the electrons and the holes are produced (in this case 0.4 eV for the electrons and 0.2 eV for the holes respectively). (e) Due to the finite barrier height, the wave function penetrates into the surrounding layers; also the number of the confined states is now finite in contrast to the case for an infinite potential well.

As the motion of the electrons is unrestricted in the x-y plane, *i.e.* it is like a free particle as far as the x-y plane is concerned, the allowed values of the components of the wave vector k_x and k_y are the same as obtained in the case of a free particle as discussed earlier, *i.e.*

$$k_x = \frac{2\pi}{l_x} n_x \qquad n_x = 0, \pm 1, \pm 2, , \pm 3, \pm 4, \ldots$$

and $$k_y = \frac{2\pi}{l_y} n_y \qquad n_y = 0, \pm 1, \pm 2, , \pm 3, \pm 4, \ldots$$

The situation is different in the z direction. Here the motion of the carriers is restricted by the potential barriers at both the surfaces. If the potential barriers are assumed to be infinite for simplicity, then the problem reduces to the well known case of a particle in a box with walls of infinite potential. As the wave function of the particles has to be zero outside the box, it must also be zero at the two walls to be continuous inside the box. With this boundary condition, the solutions of the Schrodinger equation are sine waves such that the width of the box is an integral multiple of the half wave length *i.e.*

$$\lambda = \frac{2l_z}{n_z} \text{ with } n_z = 1,2,3,\ldots$$

or $$k_z = \frac{n_z \pi}{l_z}$$

with the allowed values of the energy being,

$$E = \frac{n_z^2 h^2}{8m^* l_z^2} = \frac{\hbar^2 k_z^2}{2m^*} \text{ with } n_z = 1,2,3,\ldots \qquad (3.18)$$

The total energy of the electron in the potential well is, therefore,

$$E = \frac{\hbar^2}{2m_e^*}(k_x^2 + k_y^2) + \frac{\hbar^2\pi^2}{2m_e^* l_z^2} n^2 \text{ ; } n=1,2,3.. \quad (3.19)$$

Considering first only the second term, we can show graphically the energy levels in the quantum well due only to the motion in the z direction ($k_x = k_y = 0$) by Fig. 3.22 (a).

Figure 3.22 (a) shows that the minimum energy of the electron (the zero point energy) is not zero but a finite value E_1. This is the direct consequence of the confinement of the electrons in the z direction. This energy due to quantum confinement is a fraction of an electron volt, comparable to kT at room temperature. Low temperature measurements are therefore needed to observe the confinement effects easily. In the semiconductors the quantum confinement effects can be observed even at room temperature because, the confinement energy, being inversely proportional to the effective mass, is large as compared to the metals. Combining the parabolic dependence of energy on k_x, k_y with the discrete energy levels due to k_z, one gets the E *vs.* k plots as shown in Fig. 3.22 (b). For each value of E_n, there are energy sub-bands shown by the parabola.

The points in the k space are separated by π/l_z along the z direction. The thinner the solid in the z direction, the larger is this spacing. On the other hand, the spacing between the points in the k_x-k_y plane is $2\pi/l_x$ or $2\pi/l_y$ which is very small as l_x and l_y are macroscopic dimensions. Thus the states in the k space lie on planes parallel to the k_x–k_y plane with relatively large separation, π/l_z along the k_z direction. The states are distributed quasicontinuously in each plane. In any such plane, the number of states within a ring bound by radii k and $k+dk$ is obtained by dividing the area of the ring by the area $(2\pi/l_x).(2\pi/l_y)$ corresponding to each allowed state in this plane *i.e.*

$$n_{2d}(k)dk = \frac{2\pi k dk}{\frac{2\pi}{l_x}\frac{2\pi}{l_y}} = \frac{Akdk}{2\pi}$$

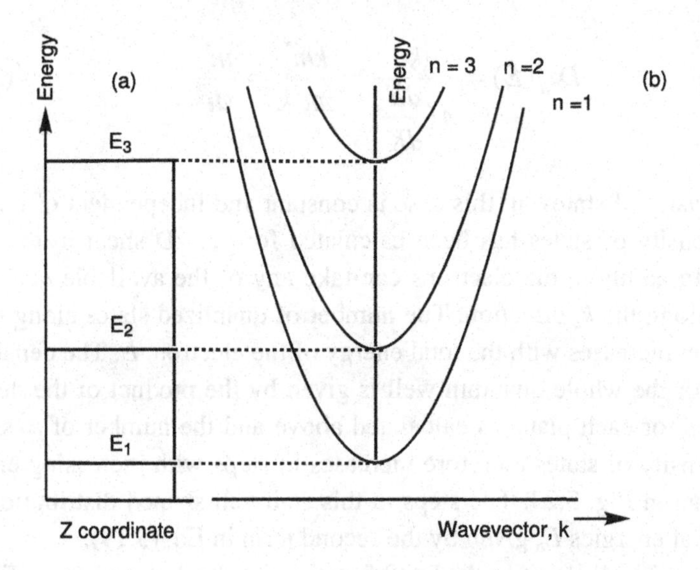

Fig. 3.22 (a) The energy levels in the quantum well due only to the motion in the z direction (b) the parabolas give the E *vs.* k relation in the x, y directions and constitute energy subbands for the quantum well.

where $A = l_x l_y$ is the planar area of the quantum well. The number of states per unit area, considering that each state can be occupied by two electrons, is

$$D_{2d}(k) = \frac{k}{\pi}$$

To find the density of states in energy, we note that the number of states between energy E and $E + dE$ is equal to that between k and $k + dk$.

$$D_{2d}(E)dE = D_{2d}(k)\, dk$$

As $E = \dfrac{\hbar^2 k^2}{2m_e^*}$, we have $dE = \dfrac{\hbar^2 k dk}{m_e^*}$ or $\dfrac{dE}{dk} = \dfrac{\hbar^2 k}{m_e^*}$

So,
$$D_{2d}(E) = \frac{k}{\pi(\frac{dE}{dK})} = \frac{km_e^*}{\pi\hbar^2 k} = \frac{m_e^*}{\pi\hbar^2} \tag{3.20}$$

The density of states in this case is constant and independent of energy. This density of states has been calculated for the 2D sheet in the k_x–k_y plane. In addition, the electrons can take any of the available quantized states along the k_z direction. The number of quantized states along the k_z direction increases with the total energy of the electron, E. The density of states for the whole quantum well is given by the product of the density of states for each plane as calculated above and the number of k_z states. This density of states therefore increases in steps with increasing energy as shown in Fig. 3.23. The steps in this stairwell shaped distribution are located at energies E_n given by the second term in Eq. (3.19).

The stairwell shape of the DOS function for the 2–d case is confirmed by the optical measurements [53]. Figure 3.24 shows the absorption

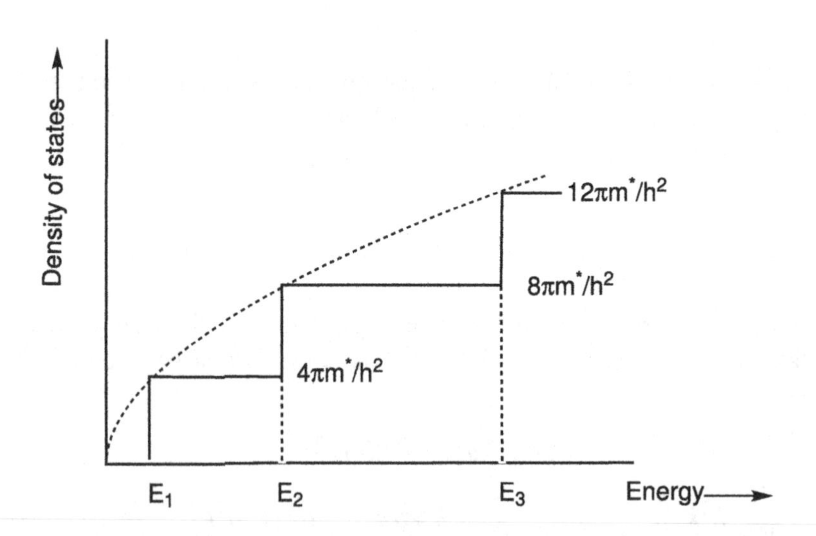

Fig. 3.23. The density of states function for a quantum well; the dotted curve shows the density of states for the bulk case.

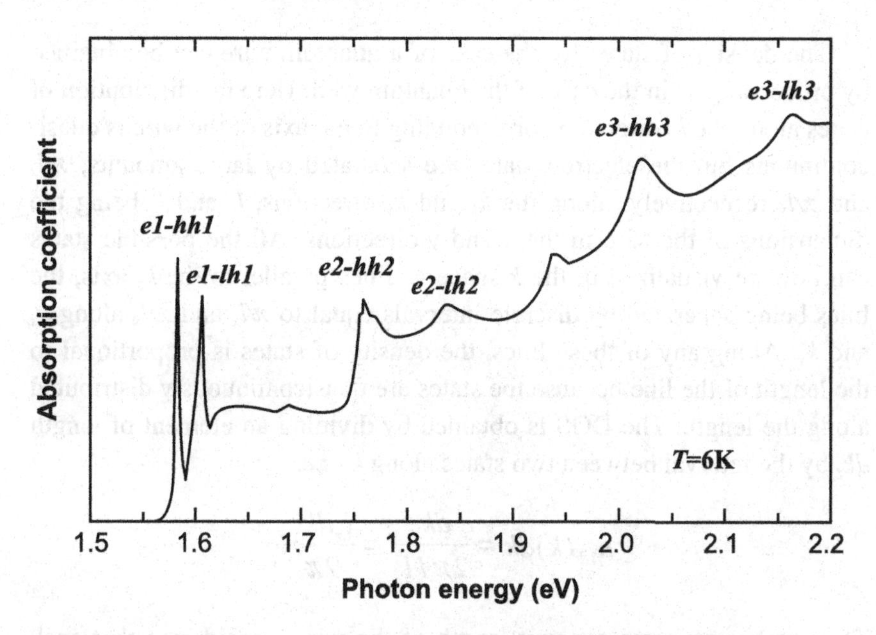

Fig. 3.24. Absorption spectrum of a GaAs/AlAs quantum well with quantum wells of width 7.6 nm; the transitions corresponding the peaks are marked (reprinted from [53] by permission of the publisher (Taylor and Francis Ltd. (http://www.tandf.co.uk/journals).

spectrum of a GaAs/AlAs quantum well of width 7.6 nm. The spectrum in general follows the steps of the DOS function. Peaks at the beginning of each step are due to excitons.

3.3.8.2 *The quantum wire*

When two dimensions of a material, say the x and y dimensions, shrink to nanometer range while only the *z* direction remains large, the solid is said to be a quantum wire. The electrons can now move freely along the *z* direction but are confined along the *x* and *y* directions. A quantum wire can be realized in practice by depositing a narrow strip of a low band gap semi-conductor on a high band gap substrate and covering up the strip on the three remaining sides also by the high band semiconductor layer. Structures such as nanowires and nanotubes (*e.g.* carbon nanotubes) are other examples of quantum wires.

The density of states for the case of a quantum wire can be obtained by proceeding as in the case of the quantum well. Here the distribution of states along the k_z direction corresponding to the axis of the wire is quasi-continuous but the electron states are separated by large amounts, π/l_x and π/l_y respectively, along the k_x and k_y directions, l_x and l_y being the dimensions of the wire in the x and y directions. All the possible states can now be visualized in the k space as lines parallel to the k_z axis, the lines being separated by discrete intervals equal to π/l_x and π/l_y along k_x and k_y. Along any of these lines, the density of states is proportional to the length of the line because the states are quasicontinuously distributed along the length. The DOS is obtained by dividing an element of length dk_z by the interval between two states along k_z, *i.e.*

$$n_{1d}(k)dk = \frac{dk_z}{2\pi / l_z} = \frac{l_z dk_z}{2\pi}$$

The number of states per unit length of the wire, considering that each state can be occupied by two electrons, is

$$D_{1d}(k)dk_z = \frac{dk_z}{\pi}$$

As before, $$D_{1d}(E)dE = D_{1d}(k)dk_z = \frac{2dk_z}{\pi}$$

The factor of two appears because k_z can be either positive or negative corresponding to the two directions of the wire.

Therefore $$D_{1d}(E) = \frac{1}{\pi.dE / dk_z} = \frac{1}{\pi}\frac{m_e^*}{\hbar^2 k_z} = \frac{1}{\pi\hbar}\sqrt{\frac{m_e^*}{2(E - E_{k_x,k_y})}} \tag{3.21}$$

Here E is the total energy given by $E = E_{kz} + E_{kx,ky}$ and $E_{kz} = \frac{\hbar^2}{2m}k_z^2$

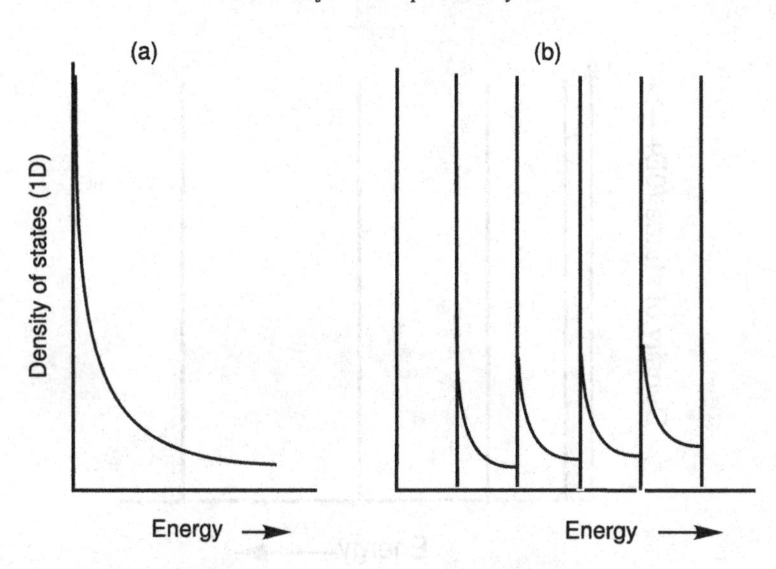

Fig. 3.25. The DOS function for a quantum wire (a) a plot of Eq. 29 gives the density of states *vs.* k for any given values k_x, k_y (b) for each set of k_x, k_y the energy diverges near the band edge.

The density of states, thus, has a $1/\sqrt{E}$ dependence and diverges near the band edges for each set of values of k_x, k_y as shown in Fig. 3.25. The DOS function for quantum wires leads to phenomena such as quantized conductance [54].

3.3.8.3 *The quantum dot*

In a quantum dot all the three dimensions are of the order of the de Broglie wave length of the charge carriers. For semiconductors, this is in the range of 10 to 100 nm. Semiconductor quantum dots can be prepared by wet chemical methods, by lithography as well as by deposition of a semiconductor film on a substrate with a slightly different lattice parameter as discussed in the next chapter.

Assume that the quantum dot is a parallelepiped with dimensions l_x, l_y, and l_z. The electron is trapped inside the box due to high potential at the surface. We had discussed the case of a one dimensional square well

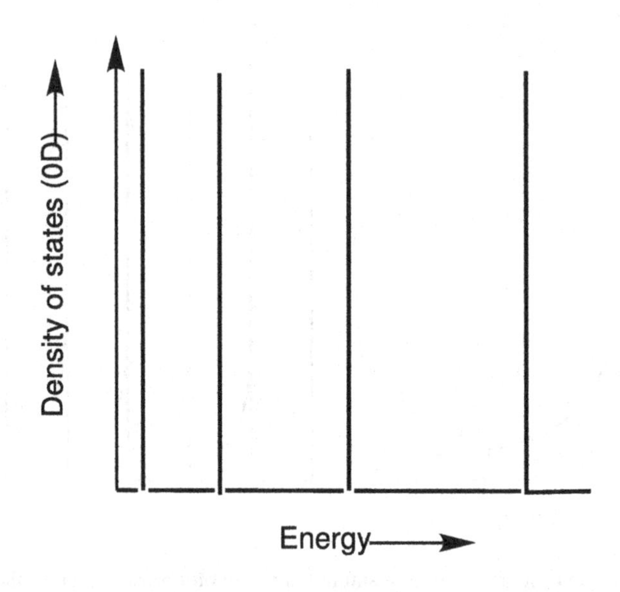

Fig. 3.26. The density of sates *vs.* energy for a quantum dot. In a quantum dot this plot contains only delta peaks which correspond to the individual states and the electrons can occupy only these discrete states.

with infinite potential walls in Section 3.3.1.3. From this we can see that the allowed values of k_x, k_y and k_z are : $k_x = n_1\pi/l_x$; $k_y = n_2\pi/l_y$, $k_z = n_3\pi/l_z$ with n_1, n_2 ,n_3 = ±1, ±2, ±3, ±4, ... The allowed energy values now are

$$E_{0d} = \frac{\hbar^2 k^2}{2m_e^*} = \frac{\hbar^2 \pi^2}{2m_e^*}\left(\frac{n_1^2}{l_x^2} + \frac{n_2^2}{l_y^2} + \frac{n_3^2}{l_z^2}\right) \tag{3.22}$$

For illustration, if we take $l_x = l_y = l_z = l$, then the zero point energy is the minimum energy corresponding to the lowest values of n_x *etc.*, *i.e.*

$$E_{zpe} = \frac{3h^2}{8m_e^* l^2} \tag{3.23}$$

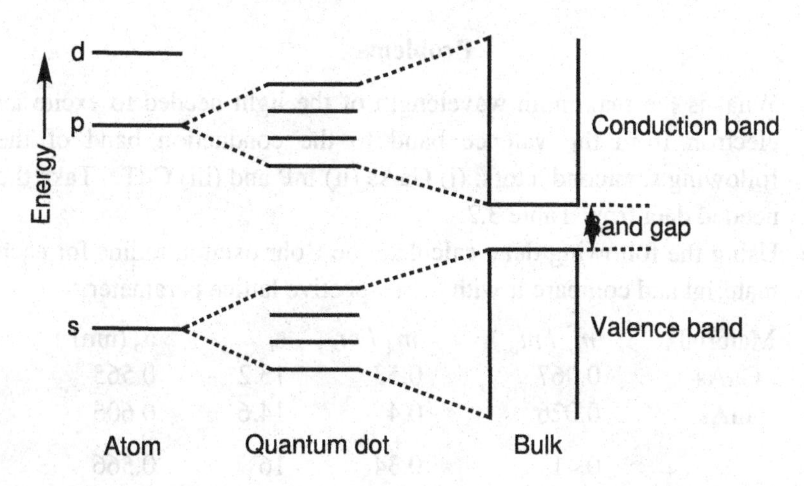

Fig. 3.27. Comparison of the energy levels in an atom, a quantum dot and a bulk semiconductor.

If the quantum dot is taken to be a sphere, then the zero point energy comes out to be slightly higher. In this case, the Schrodinger equation is to be solved using spherical coordinates and, if d is the diameter of the quantum dot, then the zero point energy is

$$E_{zpe} = \frac{h^2}{2m_e^* d^2} \tag{3.24}$$

Thus, in a quantum dot the energy is completely quantized with no free electron propagation, just like the case of atoms (Fig. 3.26). The quantum dots are often called *artificial atoms* because of this similarity. This is brought out in Fig. 3.27 which compares the energy levels in an atom, a quantum dot and a bulk semiconductor.

The quantum dots constitute an important class of nanomaterials because of their several potential applications. They are discussed in detail in the next chapter.

Problems

(1) What is the maximum wavelength of the light needed to excite an electron from the valence band to the conduction band of the following semiconductors: (i) GaAs (ii) InP and (iii) CdTe. Take the needed data from Table 3.2.

(2) Using the following data, calculate the Bohr exciton radius for each material and compare it with the respective lattice parameter.

Material	m_e^*/m_o	m_h^*/m_o	ε_r	a_o (nm)
GaAs	0.067	0.53	13.2	0.565
InAs	0.026	0.4	14.6	0.606
Ge	0.41	0.34	16	0.566

(3) Calculate the first four energy states of an exciton in CdSe and compare them in a diagram with the corresponding energy states of a hydrogen atom. Given that the effective mass of the electron is 0.07 m_o and the effective mass of the hole is 0.35 m_o where m_o is the rest mass of an electron = 9.109×10^{-31} kg, dielectric constant of CdSe = 5.8.

(4) A rectangular space lattice has lattice parameters a = 5 mm, b = 10 mm. Plot the space lattice and its reciprocal lattice to scale. Show the following set of planes in the reciprocal lattice: (1,0); (2,0); (0,1); (0,2); (1,1). What are the interplanar spacings for these set of planes in mm.

(5) In the density of states plot of a quantum well, the density of states jumps at energy values equal to 298.85×10^{-22} J, 1195.4×10^{-22} J, 2689.65×10^{-22} J, *etc.* and the jump in the density of states at each such value is equal to = 0.0365×10^{37} m^{-2}J^{-1}. Find the thickness of the quantum well and the effective mass of the electron in the semiconductor in terms of the rest mass of an electron. Given h = 6.626×10^{-34} Js and the rest mass of electron = = 9.109×10^{-31} kg.

(6) Plot to scale the energy vs. density of states plot for a quantum well made of a semiconductor film 10 nm thick sandwiched between two higher band gap semiconductor films. Be sure to put numbers and the units on both the axes. Take effective mass of the electron to be

0.067(electron rest mass); electron rest mass = 9.109×10^{-31} kg and $h = 6.626 \times 10^{-34}$ Js.

(7) Plot to scale the energy *vs.* density of states plot for a semiconductor quantum wire having a rectangular cross section with sides 10 nm and 12 nm. Be sure to put the units on both the axes. Plot the first four sets of curves. Take effective mass of the electron to be = 0.067 (electron rest mass), electron rest mass = 9.109×10^{-31} kg and h = 6.626×10^{-34} Js.

(8) The figure below shows the optical absorption spectra from a quantum well.

(i) Explain why the spectra has jumps at distinct energy values.
(ii) What is the origin of the sharp absorption peaks at the beginning of each step?
(iii) If it is given that each absorbance unit corresponds to a density of states equal to 10^{17} m^{-2} eV^{-1}, then what is the effective mass of the electron in the semiconductor?

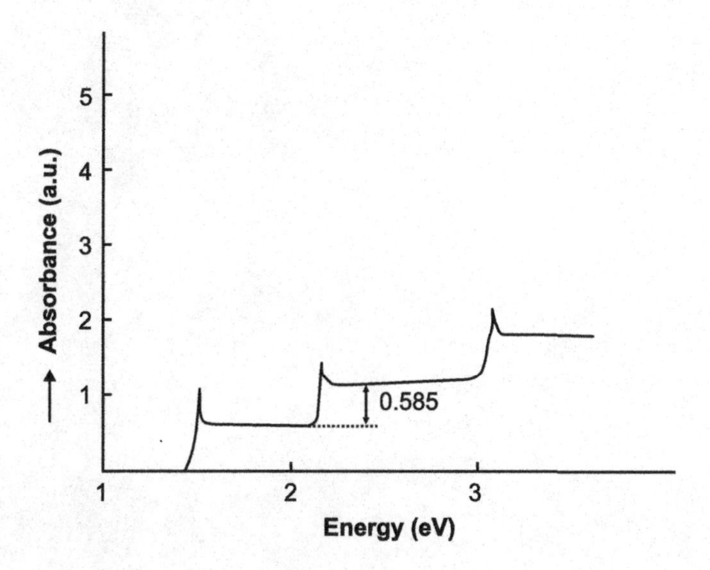

Fig. 3.28. Optical absorption spectra from a quantum well.

Chapter 4

Semiconductor quantum dots

4.1 Introduction

As discussed in Chapter 3, the nanostructures in which the carriers (electrons and holes) are confined within nanometer sized regions in all three dimensions are called quantum dots. Although such a confinement can be effected in semiconductors as well as metals and ceramics, the term quantum dot is more commonly reserved for the semiconductor particles. A nanoparticle of a semiconductor, few nm in size, covered with a layer of a higher band gap semiconductor, is an example of a quantum dot. Such structures can be fabricated by colloidal chemistry, lithography or by epitaxial growth. In the first case the quantum dots are free standing while in the latter two, the quantum dots are fabricated into arrays on a substrate. In this section we discuss the semiconductor quantum dots in some detail because they are an important class of nanostructures.

A quantum dot has to be small enough such that the separation between its energy levels exceeds kT – otherwise the discreteness of the energy levels is smeared out as the thermal energy would be sufficient to excite the electrons to higher energy levels. At room temperature (300 K) kT = 26 meV. As a rule of thumb, the separation between the energy levels should be $3kT$. The spacing between the energy levels, ΔE, varies as $1/L^2$. Calculations show, for example, that for InAs the size of the quantum dot should not be larger than 20 nm.

The lower limit to the size of the quantum dot is fixed by the requirement that at least one electron must be contained in it. This lower critical size is relatively large for semiconductors in comparison to metals and insulators. The optimum size for an InAs quantum dot, for example, is between 4 and 20 nm.

4.2 Energy levels and change in band gap with size in quantum dots

As discussed in Chapter 3, the energy levels in a quantum dot are quantized like in an atom, the difference being that in a quantum dot the separation between the energy levels is smaller than that in the atoms; moreover, this separation depends on the size of the quantum dot, the smaller quantum dots having a larger separation between the energy levels.

Figure 4.1 compares the *E vs. k* dispersion relations for a bulk solid and a quantum dot. In the bulk solid, the points are so closely spaced in the *E vs. k* curve that the curve is quasicontinuous. In the quantum dot, there are discrete and well separated energy levels, with the lowest energy (the zero point energy) being different from zero. As a result, the band gap between the lowest energy levels for the electrons and holes is larger in a quantum dot than in the bulk (see Fig. 4.1).

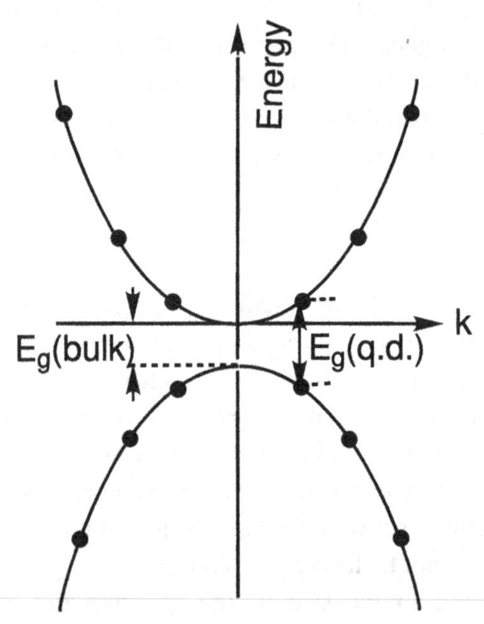

Fig. 4.1. Comparison of the dispersion (E *vs.* k) relations for bulk semiconductor and a quantum dot. In the bulk the states are quasicontinuous and are shown by the continuous parabola. In a quantum dot, the states are discrete as shown by the points on the parabola. The band gap for the quantum dot is larger than that for the bulk.

From Eqs. (3.23) and (3.24) it is clear that a greater confinement (smaller value of l or d) results in a higher zero point energy and a larger separation between the individual energy levels. This also leads to a larger band gap. An approximate expression for the dependence of the band gap on the size of the quantum dot can be derived as follows [55].

The band gap for the quantum dot is equal to the sum of the bulk band gap and the confinement energy for the electron and the hole. The confinement energy is given by

$$E_{conf} = \frac{h^2}{2m^* l^2}$$

where m^* is called the reduced mass of the exciton and is related to the masses of the electron and the hole by the relation

$$\frac{1}{m^*} = \frac{1}{m_e} + \frac{1}{m_h}$$

In addition to the two terms mentioned above, there is a third term which also contributes to the band gap. This is the Coulombic attraction between the hole and the electron and is given by

$$E_{Coul} = -\frac{1.8e^2}{2\pi\varepsilon\varepsilon_o l}$$

The expression for the band gap of a quantum dot can then be written as

$$E_{g(dot)} = E_{g(bulk)} + E_{conf.} + E_{Coulomb} = E_{g(bulk)} + \frac{h^2}{2m^* l^2} - \frac{1.8e^2}{2\pi\varepsilon\varepsilon_o l}$$

Because of the inverse l^2 dependence, the positive term is dominant at low values of l and the band gap increases as the size of the dot decreases.

The band gap of semiconductor quantum dots can be estimated from their absorption spectra. In Chapter 3, the variation of the band gap of CdSe quantum dots with size, estimated from the absorption spectra, is shown in Fig 3.1(c). The band gap calculated using the simple model

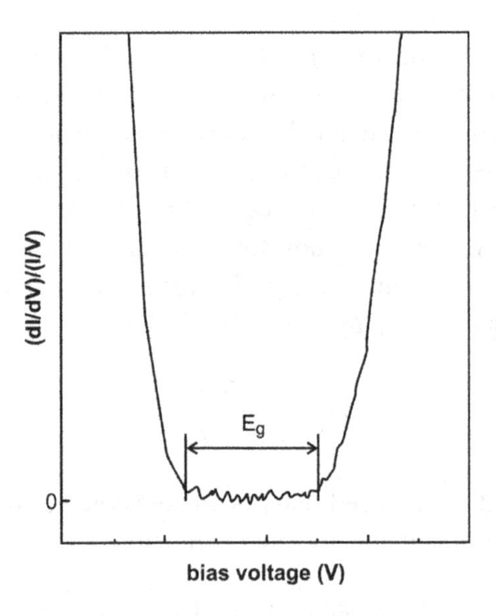

Fig. 4.2. Determination of the band gap of a quantum dot by measuring the quantity *(dI/dV)/(I/V)* as a function of the applied dc bias across the quantum dot using an STM ; the quantity *(dI/dV)/(I/V)*is proportional to the density of states and is zero in the band gap — the extent of the bias over which the density of states is zero gives the band gap, *Eg* ,of the quantum dot [56].

described above agrees with experiments at large sizes but is off for the lower sizes due to the approximations used.

The band gap of a quantum dot can also be measured using an STM. A small AC voltage (20 mV, 6 kHz) is impressed on the tip and the change in current with voltage, *dI/dV* is measured. The quantity *(dI/dV)/(I/V)* is proportional to the density of states. Figure 4.2 shows schematically a plot of this quantity *vs.* the applied dc bias for a quantum dot. As the density of states in the band gap is zero, this plot immediately gives the value of the band gap. Apart from the size effect, the large built in strain in the dot also causes a large change in the effective band gap of the quantum dot. For example the band gap of InAs quantum dot sandwiched between GaAs layers is measured to be 1.25 eV, much larger than the value for the bulk InAs ~ 0.4 eV.

The change in the band gap of the quantum dots with size causes a change in their light absorption spectra as was seen in Fig. 3.1(b). The peaks in the absorption spectra correspond to the band gap of the semiconductor quantum dot. While discussing the excitons in Chapter 3, we had noted that the absorption of a photon with energy $\geq E_g$ (E_g = band gap) produces an electron–hole pair which may recombine if the conditions permit their orbitals to overlap. This recombination leads, under suitable conditions, to release of energy in the form of emission of photons having an energy corresponding to the band gap. The quantum dot will appear to display a color if the energy of the emitted photon is in the visible region. The distribution of the emitted energy is called the *fluorescence spectra*. This spectra would shift to lower wavelengths as the size of the quantum dot decreases as this causes an increase in the band gap of the quantum dot. This shift to lower wavelengths is termed as *blue shift*. Because of the "blue shift" in the absorption spectrum with decrease in size, the color of a suspension of quantum dots shifts from red to blue–green as the size is decreased [57].

Note that to provide effective confinement of electrons, the quantum dots are coated with a higher band gap semiconductor, as in the case of quantum wells. Such quantum dots are called "core–shell" particles where the low band gap semiconductor is the core and the high band gap coating is the shell. A core–shell particle is denoted as "shell @ core"; *e.g.* a ZnS coated quantum dot of CdSe will be denoted as ZnS @ CdSe.

4.3 Coulomb blockade in a quantum dot

Consider an arrangement in which an insulating layer is placed between two conducting electrodes. This arrangement is called a tunnel diode because, if the insulating layer is sufficiently thin, then an electron can tunnel though it from one electrode to the other electrode. If a voltage is applied across the electrodes, a current is observed to flow with the electrons tunneling in and out of the insulating layer. For macroscopic dimensions, the current flow is continuous and the tunneling of individual electrons may appear as a small fluctuation in the current. If, however, the thin film is replaced by a quantum dot, a different phenomenon is observed. We now have an arrangement in which a

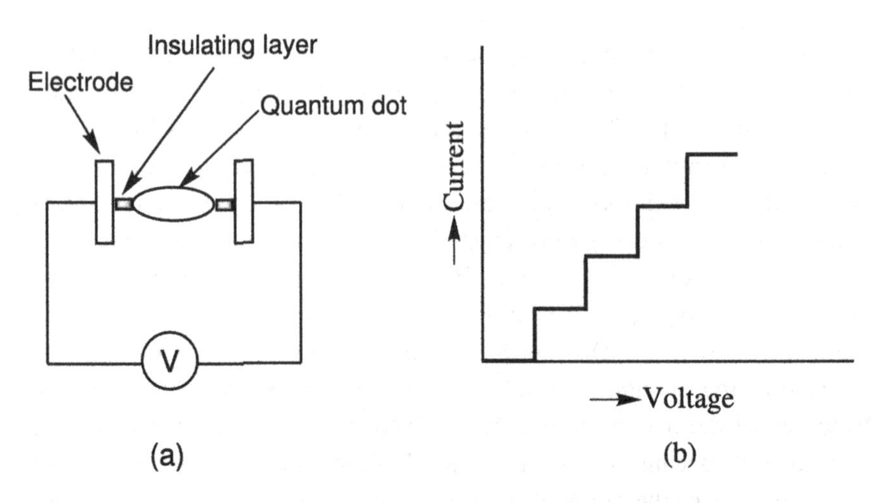

Fig. 4.3. (a) Arrangement to observe the Coulomb blockade (b) Current *vs.* voltage through a quantum dot, illustrating the phenomenon of Coulomb blockade.

quantum dot is connected to two electrodes on either side with each electrode separated from the quantum dot by an insulating layer (Fig. 4.3 a) This arrangement acts as a capacitor also. When an electron tunnels through it, the capacitor is charged with one elementary charge, causing a voltage buildup = e/C where e is the electronic charge = 1.6×10^{-19} Coulomb and C is the capacitance of the tunnel diode. If C is small enough, the voltage build up can be large. At small bias voltage, no current flows through the system, because the probability of the electrons tunneling through the insulating layer is very low. Moreover, the charging energy of $e^2/2C$ of the capacitor also needs to be supplied. Thus below a certain threshold voltage, $V_{th} = e/2C$, no current is observed. When the voltage is increased, the current shows a jump at V_{th}. Thus the current is blocked unless the voltage is incremented by the threshold amount. This phenomenon is called the *Coulomb blockade*.

The capacitance of the junction depends on its size. For the disc and sphere shaped quantum dots, the capacitance is given by $8\varepsilon_0\varepsilon_r r$ and $4\pi\varepsilon_0\varepsilon_r r$ respectively. For the quantum dots, r is small so that the threshold voltage, e/C, becomes measurable.

Consider that an electron enters the quantum dot via the junction on the left (Fig. 4.3 a). This makes it highly energetically favorable for another electron to tunnel out of the quantum dot through the other junction. A current thus flows through the quantum dot. The average number of electrons in the quantum dot at this voltage is raised by one as compared to the zero bias condition. To increase the current, the voltage has to be further raised by V_{th}. A staircase type *I-V* characteristic is therefore observed as shown in Fig. 4.3(b).

Example 4.1. *Calculate the size of a sphere shaped quantum dot of Si that would produce observable single electron effect at room temperature.*
Solution. The energy change on charging of the quantum dot capacitor should be much larger than kT in order to observe the single electron effects.
At 300 K, $kT = 1.38 \times 10^{-23}$ J K^{-1} x 300 K = 414 x 10^{-23} J
Taking 1 eV = 1.602 x 10^{-19} J, kT = 258.43 x10^{-4} eV = 25.84 meV.
The energy change on charging of the quantum dot by a single electron

$$= q^2/2C = e^2/2C$$

The capacitance of the *sphere* shaped capacitor, $C = 4\pi\varepsilon\varepsilon_0 r = 4\pi \times 11.5 \times 8.85 \times 10^{-12} \times r$
(taking the dielectric constant of silicon to be 11.5 and the permittivity of vacuum , $\varepsilon_0 = 8.85 \times 10^{-12}$ F.m^{-1}.)
$C = 1278.294 \times r \times 10^{-12}$ F = $1.278 \times r \times 10^{-18}$ F, when r is taken in nm.
Energy change on charging by a single electron

$$= e^2/2C = (1.6 \times 10^{-19})^2/2 \times 1.278 \times r\ 10^{-18} \text{ J} = = 0.0626/r \text{ eV}$$

This energy should be much larger than kT for the single electron effect to be observable
i.e. $0.0626/r$ eV >> 0.02584 eV at 300K
or $0.0626/r \approx 0.5 \times 0.02584 = 0.1292$ (say) or $r \approx 0.5$ nm
Hence the quantum dot should have a radius of the order of 0.5 nm for this effect to be observable at room temperature. □

Another requirement for observing the single electron effects is that the fluctuations in the number of electrons in the quantum dot should be negligible. The time constant for an R-C circuit is RC. The time taken by an electron to move in or out of a junction should be of this order.

According to the Heisenberg uncertainty principle, the product of the energy change accompanying this transfer and the time taken should be larger than h, the Planck's constant

$$\Delta E.\Delta t > h$$

or

$$(e^2/2C).RC > h$$

or

$$R > 2\,h/e^2 = 51.6\text{ k}\Omega.$$

4.4 Formation of self assembled quantum dots by epitaxial deposition

4.4.1. *Introduction*

The quantum dots can be fabricated by several methods such as colloidal chemistry, lithography and epitaxial growth. The colloidal chemistry methods are suitable for producing a suspension of quantum dots for applications such as biological tagging. The method of epitaxial growth, which results in the formation of an array of quantum dots on a substrate is more compatible with the semiconductor processing practice and is being actively researched with a view to applications in lasers. In the following, after brief remarks about the solution method, we consider in some detail the preparation of self assembled quantum dots by epitaxial deposition.

In the colloidal chemistry method of preparing quantum dots, the reactions between the precursors containing the desired species are carried out in solution. In this method, it is necessary to deposit a layer of suitable molecules on the particle surface as soon as it is formed. Such layers are called capping layers. The primary purpose of a capping layer is to prevent the agglomeration of the particles by steric stabilization or some other mechanism (see chapter 7). In addition, a capping layer may (i) act as a diffusion barrier layer so as to promote a uniform particle size distribution (ii) contain functional groups which make the nanoparticles soluble in different solvents (iii) prevent oxidation of the particles (iv) act as a dielectric barrier at the surface and reduce the density of the surface traps (in case of the semiconductor nanoparticles). The examples of the capping layers are alkane thiols (particularly for the noble metal particles) and long chain alkyl phosphines and alkylphosphine oxides (for semiconductors), phosphates, phosphonates, amides or amines, carboxylic acids and nitrogen containing aromatics, *etc.* The capping layer is bonded to the surface by the relatively weak coordinate bond. This enables the capping molecules to be subsequently easily removed and replaced by other molecules with desired functionality.

The synthesis of CdSe nanocrystals by organometallic route provides an example of the solution synthesis [58]. To achieve a narrow particle size distribution, the nucleation and growth steps are well separated. The precursors dimethyl cadmium, $Cd(CH_3)_2$ and trioctylyphosphine selenide (TOPSe) are injected into a solvent held at 300–320° C. At this temperature, the nucleation rate of CdSe is very high. The temperature is then quickly brought down to 250–300° C where no further nucleation occurs and growth of the already formed nuclei proceeds. The solvents used during the synthesis are trioctylphosphineoxide (TOPO) and trioctylyphosphine (TOP). These are the so called coordinating solvents which also act as the capping agents. The molecules of TOPO and TOP contain an electron rich donating group. The molecule acts as a Lewis base and coordinates to the electron poor metal of the semiconductor such as Cd, Se, In. The TOPO molecule is bonded to the Cd surface site and passivates the dangling orbitals at the surface; the Se surface sites are either unbonded or are passivated by the TOP molecules. The capping layer of these molecules provides a diffusion barrier leading to the diffusion limited growth, which is conducive to a uniform particle size distribution.

4.4.2. *Quantum dots by epitaxial deposition*

The spontaneous formation of an array of quantum dots by heteroepitaxial deposition of a semiconductor thin film on a slightly mismatched substrate has evoked considerable interest because of the ease of integration of this process with the established semiconductor fabrication techniques. Figure 4.4 shows an AFM image of quantum dots grown using this technique.

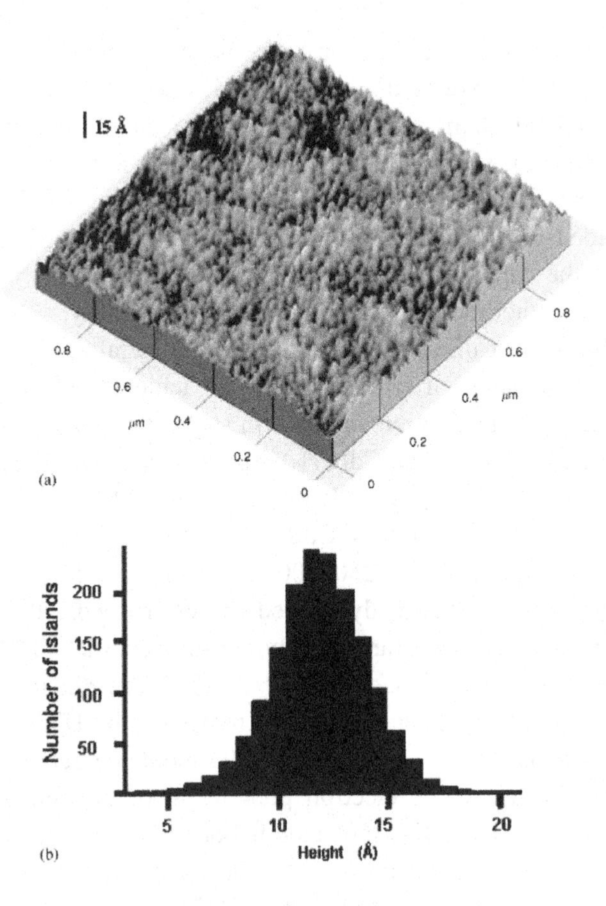

Fig. 4.4. (a) Atomic force microscopy image of self assembled quantum dots of GaAs grown by depositing 2 monolayers GaAs on a Si interlayer. The number density of quantum dots is 1 x 10^{11} cm^{-2} and the average lateral size is 20 nm (b) histogram of the height of the quantumdots — the dispersion in height is centered at about 1.3 nm (reprinted from [59] with permission from Elsevier).

During the epitaxial deposition of a semiconductor film on a slightly mismatched substrate, self assembled islands form after the initial growth of a few (usually one to two) uniform monolayers due to the interplay of the surface and the strain energy. The deposited material is chosen to have a smaller band gap than the substrate. These islands are then covered with another semiconductor layer having a larger band gap than

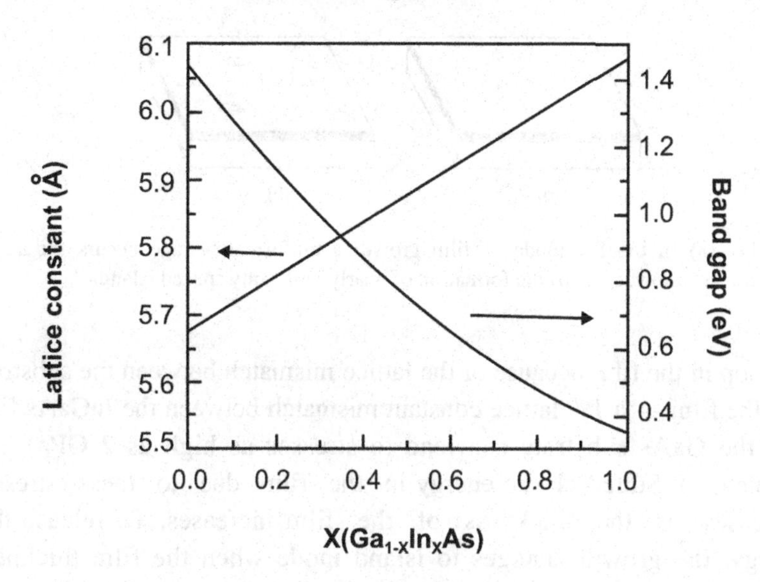

Fig. 4.5. The variation of the lattice parameter and the band gap of the ternary alloy semiconductor $Ga_xIn_{1-x}As$ with composition.

the islands to provide effective quantum confinement. A typical example of this is the deposition of $GaIn_xAs_{1-x}$ on GaAs; here the lattice misfit strain can be continuously varied from 0 to 7% by varying the In concentration, *x*, from 0 to 1 (Fig. 4.5).

In the self organized quantum dots there is a large in-built strain. This large strain causes very significant changes in the band gap as is also shown in Fig. 4.5.

4.4.3. *Energetics of self assembled quantum dot formation — the Stranski–Kranstanow growth*

During the deposition of $In_xGa_{1-x}As$ on GaAs (100), for $x > 0.25$ to 0.3, the growth proceeds by what is known as the Stranski-Krastanow (SK) mode. In this mode of growth, initially for one or two monolayers, the film forms uniformly over the substrate; further growth then occurs in the form of islands which are nearly equally spaced (Fig. 4.6). Initially, as the film grows epitaxially on the substrate (Fig. 4.7 a), large stresses

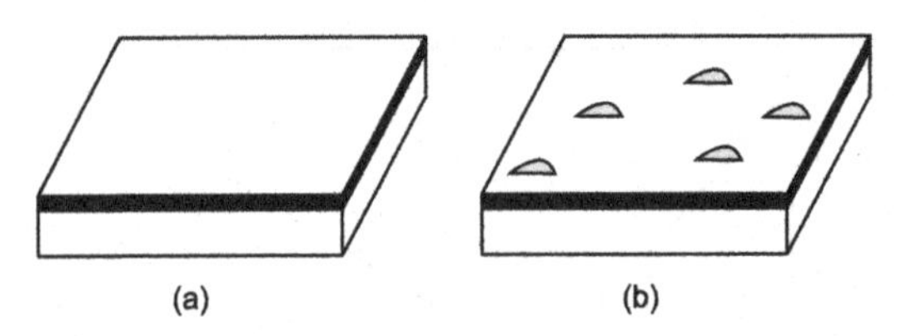

Fig. 4.6. (a) In the SK mode of film growth a uniform coverage occurs for a few monolayers followed by (b) the formation of nearly uniformly spaced islands.

develop in the film because of the lattice mismatch between the substrate and the film — a 1% lattice constant mismatch between the InGaAs film and the GaAs substrate can lead to stresses as high as 2 GPa. (see Problem 7). Stored elastic energy in the film due to these stresses increases as the thickness of the film increases. To release this energy, the growth changes to island mode when the film thickness exceeds a critical value which depends on the value of x. As the edges of an island are not constrained, they can relax and reduce the strain energy (Fig 4.7 b). A net decrease in the free energy would occur if the increase in the surface energy due to the formation of the islands is less than the accompanying reduction in the strain energy. This can be illustrated

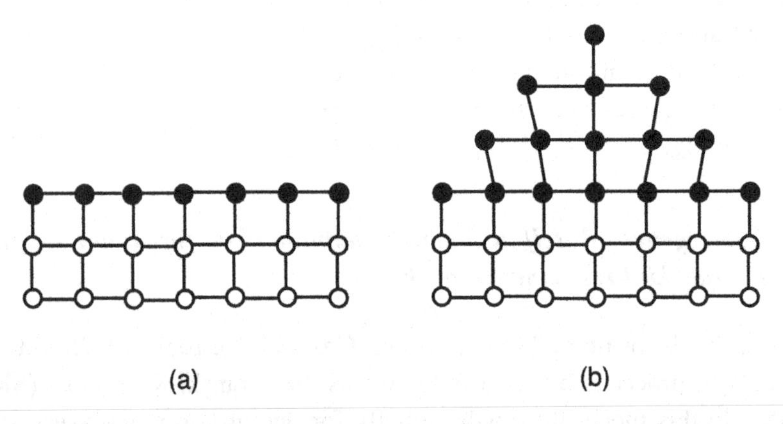

Fig 4.7. (a) In the epitaxial growth of InAs on GaAs, the InAs layer is constrained to match the lattice parameter of the substrate resulting in the generation of large stresses in the InAs layer (b) The island formation is favored because the islands are not laterally constrained and can relax to relieve the strain energy.

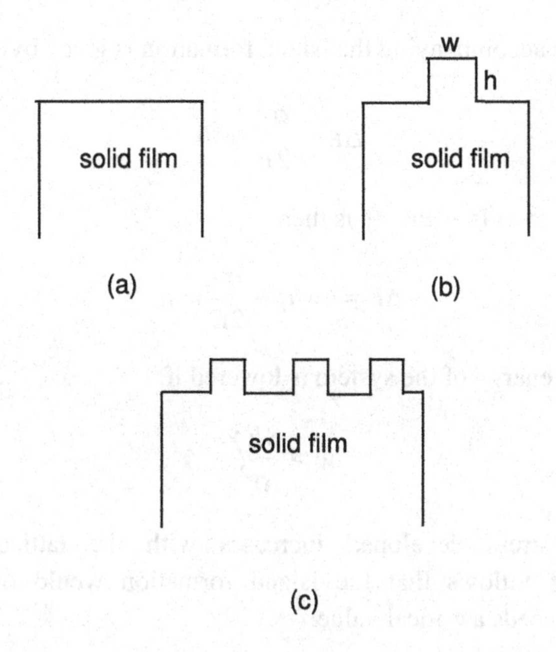

Fig. 4.8. (a) a uniformly stressed film (b) square pyramidal island of width w and height h forms on the film (c) uniformly spaced islands on the film.

using a simple model. Let us assume that the uniform film, before the onset of island mode of growth, is stressed uniformly (Fig. 4.8). Consider now that an island of square cross section forms on the film. If the height and width of the island are h and w respectively, then the formation of the island is accompanied by an increase in the surface energy given by

$$\Delta\Gamma = 4wh\gamma$$

Here, γ is the surface energy per unit area of the surface of the island, which in this case is $In_xGa_{1-x}As$. As the island and the underlying substrate both consist of the same material, there is no interfacial energy term involved.

If initially, the film is assumed to have a uniform stress σ, and it is assumed that the formation of the island results in the complete relaxation of the elastic strain energy, then the decrease in the elastic strain energy accompanying the island formation is given by

$$\Delta\varepsilon = \frac{\sigma^2}{2E} w^2 h$$

The net change in free energy is then

$$\Delta F = 4wh\gamma - \frac{\sigma^2}{2E} w^2 h$$

Thus the net energy of the system is lowered if

$$w > \frac{8E\gamma}{\sigma^2}$$

As the stress developed increases with the lattice parameter mismatch, it follows that the island formation would occur if this mismatch exceeds a critical value.

Example 4.2. *Calculate the in plane stresses developed in a thin film of InAs deposited uniformly on the GaAs substrate. The relevant data are given in Table 4.1 below. Assume that the stress developed in the plane of the film is biaxial and is equal in the x and y directions lying in the plane of the film*

Table 4.1. Lattice parameters, Poisson's ratios and the Young's moduli of GaAs and InAs [60]

Material	Lattice parameter (a_o)(nm)	Poisson's ratio (ν)	Young's Modulus E (GPa)
GaAs	0.565	0.31	8.77
InAs	0.606	0.35	5.22

Solution. The biaxial moduli for the deformation in the plane of the film are given by [61]

$$M = 2\mu(1+\nu)/(1-\nu)$$

where μ is the shear modulus.

Assuming the InAs film to be isotropic, the shear modulus of InAs in terms of the Young's modulus and the Poisson's ratio is

$$\mu = E/2(1+\nu) = 5.22/2(1+0.35) = 1.93 \text{ GPa}$$

Hence, the biaxial modulus is $M = 2 \times 1.935 \times 1.35/0.65 = 8.03$ GPa.

The strain developed in the InAs film deposited on the GaAs substrate is

$$\varepsilon = a_o(\text{GaAs}) - a_o(\text{InAs}) \,/\, a_o\text{InAs} = (0.565\text{-}0.606)\,/0.606 = -0.06766$$

The biaxial stresses *i.e.* σ_{xx} and σ_{yy}, developed in the InAs film therefore are

$$= -0.06766 \times 8.03 = -0.543 \text{ GPa}$$

Hence compressive stresses equal to -0.543 GPa are developed in the film in the x and y directions. □

The model can be extended to the case where an array of islands, rather than a single island, forms [62]. If the islands are assumed to be uniformly spaced, with a uniform spacing λ between the islands (Fig. 4.8 c), then the change in energy in going from a flat film surface to one with the islands, can be shown in a similar manner to be

$$\Delta F = \frac{-\sigma^2}{2E}\frac{\lambda}{2} + 2\gamma$$

The energy of the system will therefore be lowered by island formation if $\lambda > 8\gamma E/\sigma^2$. The array of islands will then form if the stress σ, and so the lattice mismatch, exceeds a critical value. Several systems meet this condition and have been used for growing quantum dots *e.g.* Ge/Si, InGaAs/GaAs, CdSe/ZnSe and GaN/AlN.

4.4.4. *Stacked layers of quantum dot arrays*

The quantum dots are very attractive light sources due to their large radiative efficiency and nearly monochromatic emission. However, the number density of the dots prepared by epitaxial deposition is rather small — *e.g.* only about 10^{10} cm^{-2} for InAs quantum dots on GaAs. This means that only a small fraction of the incident exciting radiation is usefully absorbed; for application in lasers, this also results in a low modal gain. Both these quantities scale as the total number of dots. To take advantage of this scaling, the dots are grown in multiple layers stacked one above the other. As an example, if the Ga(In)As quantum dots are covered by a thin layer of GaAs and another growth cycle for Ga(In)As is carried out, then the dots in the second layer are found to be vertically aligned with the dots in the first layer if the intervening GaAs layer is thin enough and this trend continues. Several layers of quantum dots, with dots in different layers aligned so as to be on top of the previous layer, can be formed. The dots in different layers lose their correlation if the GaAs layer thickness exceeds a critical value. This vertical alignment in the dots and the existence of a critical thickness is believed to result due to the generation of a nonuniform strain distribution in the GaAs layer due to the buried Ga(In)As quantum dot and the elastic interaction between this strain distribution and the strained surface islands [63] as illustrated in Fig. 4.9.

In addition to the applications mentioned in the Sec. 4.1, the self assembled quantum dots are generating great interest due to their potential applications in improved semiconductor diode lasers having low threshold currents in which the wavelength can be changed by changing the dot size, blue lasers, mid-infrared detectors for thermal

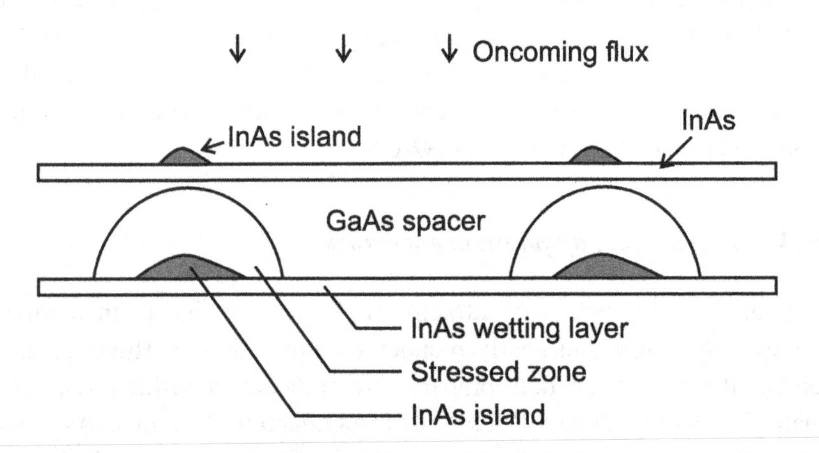

Fig. 4.9. The strain in the quantum dots in each layer produces a stress field which penetrates the GaAs grown immediately after the dot but decays rapidly with distance; if the GaAs layer is not too thick, this stress field has sufficient magnitude to cause preferential nucleation of the dots just above the dots in the previous layer.

imaging, mid infrared lasers, *etc.* However, problems such as size nonuniformity, nonuniform spacing between the dots and the difficulty in accessing a single dot need to be solved before these applications can be realized.

4.5 Applications of quantum dots

The presence of a band gap causes a semiconductor to emit light when excited by a radiation of sufficient energy. In a quantum dot the band gap can be tuned by changing its size which implies that the wavelength (*i.e.* the color) of the emitted light can be changed by changing the size of the quantum dot — the smaller the dot more is the blue shift of the emitted light. Thus a whole range of colors can be obtained from the same material just by tailoring the size of the dots. This makes the free standing quantum dots (*e.g.* the quantum dots prepared by a colloidal process) useful as luminescent tags for biomolecules (*e.g.* DNA strands, drug candidates, agents of biological warfare, *etc.*). Traditionally, dye molecules, which are organic molecules showing photoluminescence ("organic fluorophores"), have been used as tags for molecules of interest. Such organic fluorophores have several drawbacks which limit the advancement of this detection technique. They suffer from photobleaching (*i.e.* lose their luminescent property when excited by light, many doing so within seconds), chemical degradation, toxicity, broad emission spectra, narrow absorption spectra and low quantum yields [64]. In contrast, the quantum dots are much more rugged and so can be used in applications which require larger periods of illumination; they have a wide absorption spectrum extending above the conduction band — the dots emitting in different spectral regions can, therefore, be excited by a single wavelength. Moreover, as mentioned earlier, they have a more nearly monochromatic radiation and a weaker red tail as compared to the dye molecules (Fig. 4.10).

Different kinds of ligands such as antibodies, peptides, proteins, RNA, DNA, viruses and small molecules can be attached (conjugated) to a quantum dot. Such quantum dots can then be used for

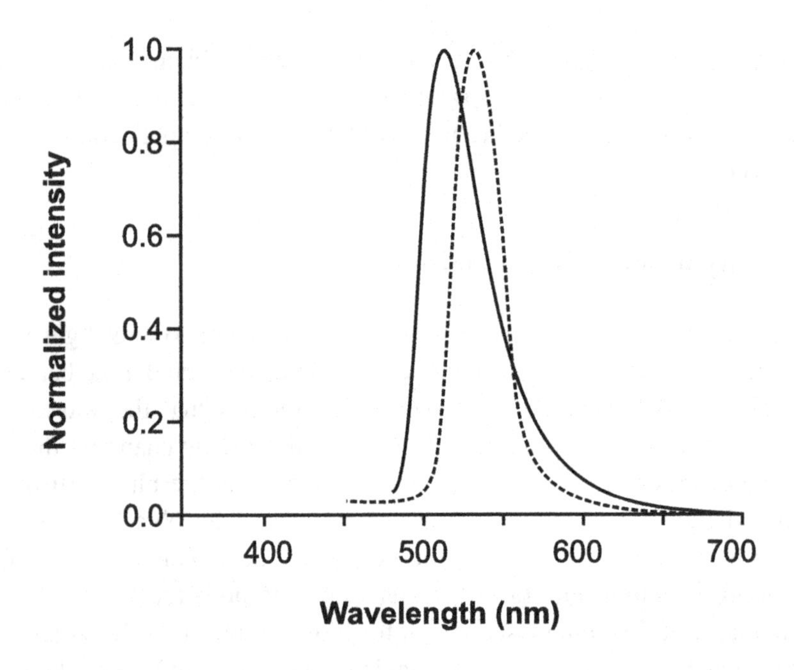

Fig. 4.10. Fluorescence spectra of fluorescein (a dye) and a typical water-soluble nanocrystal (NC) sample (dotted curve). The nanocrystals have a much narrower emission and no red tail [57].

different applications. For example, a quantum dot conjugated with the antibody for a given toxin, say cholera toxin, will attach to the cholera toxin antigen and will reveal its presence. Unlike the dyes, multiple copies of the same ligand or many different ligands can be attached to one quantum dot thus enabling simultaneous assay of a number of toxins. Furthermore, while one ligand may be a targeting ligand which seeks and attaches to the antigen, the other ligand may be a drug or an imaging agent such as a paramagnetic coating for imaging. The quantum dots are therefore being actively explored for applications in targeted drug delivery, for screening drug candidates, detection of toxins and explosives, *etc.*

A good example of the use of quantum dots, which are now replacing the fluorescent dyes, is in the **DNA chip**. A DNA chip is approximately

1 cm x 1 cm substrate on which thousands of different known nucleotide sequences are attached in a grid pattern. If a part of a DNA, having a nucleotide sequence complimentary to that of a sequence on the chip, comes in contact with the latter, hybridization would occur. If the complementary sequence is tagged with a fluorophore, the location of this complementary sequence on the chip can be detected from the light coming out from that spot upon excitation and so the unknown nucleotide sequence can be determined.

A DNA chip can be used, for example, to find out which genes are being expressed by a normal tissue and a cancerous tissue. As discussed later in Chapter 13, DNA is a very long sequence of nucleotides while the gene is a small region of DNA consisting of a definite sequence of nucleotides that encodes a protein or any feature of inheritance. While the DNA in the normal and the cancerous tissue are the same, they use different genes to manufacture proteins. The gene which is used, or expressed, is transcribed into mRNA and so the mRNA's of the two tissues are different. The steps used to detect which genes are expressed by the normal and the cancerous tissues are as follows:

Extract the mRNA from the two tissues.
Convert the mRNA's into complimentary strands of DNA (cDNA) by using the enzyme reverse transcriptase.
Attach different colored fluorescent dyes or quantum dots to the two cDNA's, say red to cDNA from normal tissue and green to cDNA from the cancerous tissue.
Mix and bring in contact with the DNA chip. The cDNA's attach to the complimentary nucleotide sequences on the chip. Wash the excess away.
Scan the chip using suitable excitation. The number of spots on the chip which fluoresce red gives the genes exclusively expressed by the normal tissue, the number of spots which appear green gives the genes exclusively expressed by the cancerous tissue and the number of spots which fluoresce to a mixed color (yellow) gives the genes which are common to both the tissues.

Perhaps the most important potential application of quantum dots is in lasers. The quantum dot lasers have several advantages over the bulk semiconductor lasers. The gain of such a laser can be increased manifold if instead of a single quantum dot, an array of quantum dots is used. The discovery that such an array of quantum dots can form due to surface energy and strain effects during the deposition of a semiconductor film on a suitable substrate, has sparked a great deal of research in this area.

While it is desirable to have all the quantum dots of the same size and shape to get a narrow wavelength laser output, the quantum dots formed by epitaxial deposition usually have a spread in their size. This property is sought to be exploited in using such quantum dot arrays as a source of broad band light. Broad band light is needed in several applications such as optical communication, optical sensors, optical gyroscope, *etc.* [65]. The other applications for which the quantum dots are being explored include LED's to emit white light for use in laptop computers, internal lighting for buildings and cars as well as possible materials for making ultrafast (> 15 tetrabits per second) all-optical switches and logic gates, demultiplexers (for separating various multiplexed signals in an optical fiber), all-optical computing, and encryption.

Problems

(1) The band gap of a semiconductor is 1.5 eV. Calculate the band gap for a quantum dot of diameter 6 nm of this semiconductor. Given that the effective mass of the electron and the hole is 0.1 and 0.5 times respectively of the static mass of an electron. Ignore the Coulombic attraction term. Given 1 eV = 1.602×10^{-19} J and static mass of electron = 9.109×10^{-31} kg.

(2) Calculate and plot the change in the band gap of a quantum dot of CdSe with diameter using the following data: E_g(bulk) = 1.74 eV; effective mass of electrons $m_e = 0.13\ m_o$; effective mass of holes $m_h = 0.4\ m_o$; m_o is the mass of an electron; dielectric constant of CdSe, $\varepsilon_{CdSe} = 4.8$; Compare your results with the value of the bulk band gap and some experimental data on dependence of band gap on size from literature.

(3) The voltage required to inject the first electron in a quantum dot was shown to be $e/2C$, where e is the charge of an electron and C is the capacitance of the quantum dot. What would be the voltages needed to inject the successive electrons in the quantum dot. Give an expression for this in terms of e, C and n, the number of electrons. (Hint: Calculate the energy change when the quantum dot is charged from n to $n + 1$ electrons).

(4) What should the temperature be to observe the Coulomb blockade effect in a quantum dot of Si of radius 15 nm. Take the necessary data from Example 4.1.

(5) A cylindrical rod of diameter 'd' and length 'l' of a material having a Young's modulus 'E' is subjected to a tensile stress σ along its axis. What is the total strain energy stored in the rod? Find the strains developed transverse to the rod axis if the Poisson ratio of this material is ν.

(6) Draw the arrangement of the atoms in the (100) and (110) planes of GaAs.

(7) Answer the following with reference to Example 4.2

a. The stress developed in the plane of the InAs film in Example 4.2 is compressive. What is the magnitude of the stress developed in a direction perpendicular to the film? Is it compressive or tensile?

b. In reality the stresses developed in the film and in the GaAs substrate are not uniform and are more like those shown in the Fig. 4.11 below [66]. The *z* direction is normal to the film and through the apex of the pyramidal quantum dot and the *x* and *y* directions lie in the plane of the film; the strain is plotted along the ordinate.

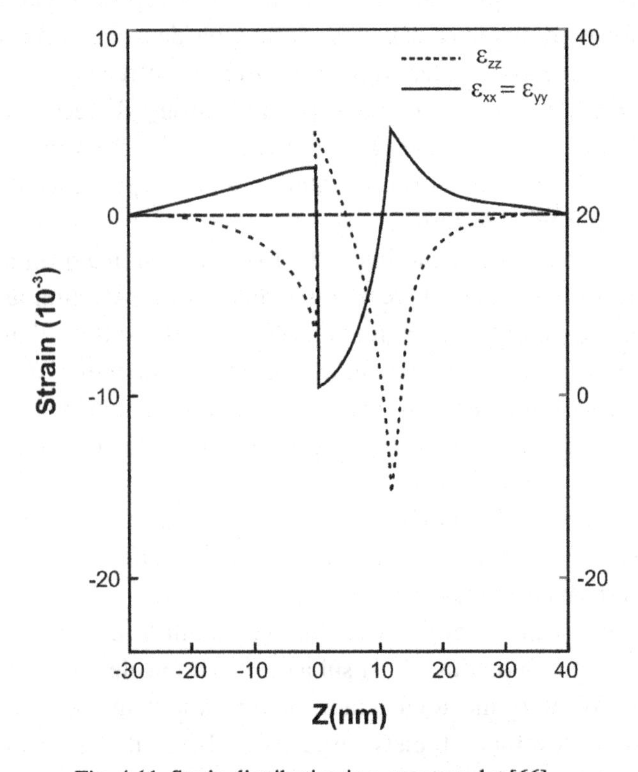

Fig. 4.11. Strain distribution in a quantum dot [66].

Using this figure answer the following questions: Up to what thickness, starting from the substrate, are the stresses in the plane of the film compressive? At what thickness do the in–plane stresses assume maximum tensile value? Where does the maximum compressive stress in the z direction occur? What is the nature of the stress in the x and y directions in the substrate just below the substrate-film interface?

(8) Besides relaxing elastically, what other alternative mechanism can lead to stress relaxation in the quantum dots? What is the effect of this mode of stress relaxation on the quality of the quantum dots?

(9) As the height of the quantum dot increases, the total strain energy stored in it also increases. Beyond a critical thickness, the stress in a quantum dot can relax by the generation of dislocations. However, this degrades the quality of the dots. The formation of dislocations can be avoided if the total strain energy in the dot is not allowed to exceed a critical value. Suggest a way in which this can be done and explain how this limits the strain energy in the quantum dot. (See [67].

Chapter 5

Zero Dimensional Nanostructures II The Metal Nanoparticles

5.1 Introduction

In the last chapter we discussed the effect of the particle size on the properties of the semiconductor quantum dots. One manifestation of this was the change in the band gap with size leading to the appearance of distinct size dependent colors in the suspensions of these particles. The suspensions of the nanoparticles of some metals also show bright colors; however, the source of this is a different phenomenon called the surface plasmon resonance.

Another interesting phenomenon shown by the metal nanoparticles is the so called "metal–insulator" transition at very small sizes when the separation of the energy levels becomes larger than the thermal energy at the temperature of measurement. Several other properties of the metal nanoparticles are also dependent on size. In this chapter, we confine ourselves to the phenomenon of surface plasmon resonance. The phenomenon is responsible for the ruby red color of the gold sols and the colors of the stained glasses containing metal nanoparticles.

5.2 Optical properties of bulk metals

A beam of light incident on a surface can undergo reflection, absorption and transmission to different extents depending on the material and the wavelength of the light. The relative amounts of reflection, absorption and transmission depend on the relative magnitudes of the real and imaginary parts of the complex refractive index function $\tilde{n}$. This function can be written as

$$\tilde{n} = \eta + \iota\kappa$$

The real part of $\tilde{n}$ is η. It is called the refractive index of the material while the imaginary part κ is called the extinction coefficient. If the extinction coefficient is near zero, the light is transmitted through the material while it is reflected if the refractive index is nearly zero. If both are of comparable magnitude, then the light is mostly absorbed [68, 69].

The square of the refractive index function is the dielectric constant

$$\tilde{n}^2 = \varepsilon$$

For metals, the dielectric constant depends on the frequency of the incident radiation as follows

$$\varepsilon(\omega) = \tilde{n}^2 = 1 + \frac{\sigma / \varepsilon_0}{i\omega(1 + i\omega\tau)} \tag{5.1}$$

Here σ is the electrical conductivity, ε_0 is the permittivity of vacuum, ω is the frequency of incident light and τ is the average time between the collisions of the electrons with the scattering centers in the metal such as the lattice vibrations and the impurity atoms *etc.* It is given by

$$\tau = \frac{m\sigma}{Ne^2} \tag{5.2}$$

where m is the mass of an electron, N is the number density of electrons in the metal and e is the charge on an electron.

When the frequency ω is small ($\omega\tau << 1$), Eq. (5.1) can be written as

$$\tilde{n}^2 = -i\frac{\sigma}{\varepsilon_0\omega} \tag{5.3}$$

and using $\sqrt{(-i)} = \sqrt{(e^{-i\pi/2})} = e^{-i\pi/4} = \frac{1}{\sqrt{2}} - \frac{i}{\sqrt{2}}$, we get

$$\tilde{n} = \sqrt{(\sigma/2\varepsilon_0\omega)}.(1 - i)$$

Thus the real and the imaginary parts of ñ have the same magnitude. The large imaginary part leads to rapid attenuation of the wave in the metal. The metal absorbs the radiation in this range. This is the case for $\omega \leq 10^{12}$ Hz for copper, for example.

At very high frequencies ($\omega\tau >> 1$), Eq. (5.1) can be approximated by

$$\tilde{n}^2 = 1 - \frac{\sigma}{\varepsilon_0 \omega^2 \tau}$$

$$= 1 - \frac{Ne^2}{m\varepsilon_0 \omega^2}$$

$$= 1 - (\omega_p/\omega)^2 \quad (5.4)$$

where $\omega_p = (Ne^2/\varepsilon_0 m)$ is known as the plasmon frequency. As will be discussed later, it is the natural frequency of oscillation of a collection of electrons.

From Eq. (5.4) it is seen that for $\omega > \omega_p$, $\tilde{n}^2$ is positive and so the refractive index is real. The light is therefore transmitted through the metal. The plasmon frequency is thus the critical frequency above which the metal becomes transparent to the electromagnetic radiation.

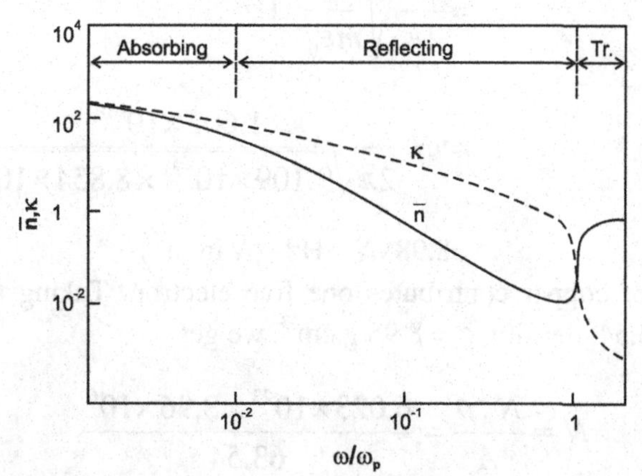

Fig. 5.1 Variation of the index of refraction and the extinction coefficient with frequency for a metal taking $\omega_p\tau = 100$; the absorbing, reflecting and transmitting frequency ranges are also shown [69].

At intermediate frequencies, it can be shown that [69] the refractive index is small in comparison to the extinction coefficient and so almost all the radiation incident on the metal is reflected.

The absorbing, reflecting and transmitting frequency ranges for a metal are shown schematically in Fig. 5.1 together with the variation in the refractive index and the extinction coefficient.

In metals the light is totally reflected in the visible region. Most metals therefore present a shiny appearance. Two exceptions are gold and copper which have distinct colors due to some absorption in the visible range due to inter-band transitions from the filled *d* bands to the *sp* conduction band. Only the complimentary part of the light is visible and so these metals appear colored.

Example 5.1. *Find the plasmon frequency for copper. Take the dielectric constant* $\varepsilon_r = 1$*. Ignore the thermal effects.*

Solution. From Eq. (5.4) the plasmon frequency is given by

$$\omega_p = \sqrt{\frac{Ne^2}{m\varepsilon_r\varepsilon_0}} \quad \text{radians.s}^{-1}$$

$$= \frac{1}{2\pi}\sqrt{\frac{Ne^2}{m\varepsilon_0}} \quad \text{Hz}$$

$$= \sqrt{N}.\frac{1.602\times10^{-19}}{2\pi\sqrt{9.109\times10^{-31}\times8.854\times10^{-12}}}$$

$$= 8.98\sqrt{N} \quad \text{Hz} \quad (N \text{ in m}^{-3})$$

Each atom of copper contributes one free electron. Taking the atomic mass, $A = 63.54$, density, $\rho = 8.96$ g.cm^{-3}, we get

$$N = \frac{N_A\rho}{A} = \frac{6.023\times10^{23}\times8.96\times10^{6}}{63.54}$$

$$= 0.8493\times10^{29} \; m^{-3}$$

where N_A is the Avogadro's number.
Therefore,

$$\omega_p = 8.98 \times \sqrt{0.8493 \times 10^{29}} = 2.6 \times 10^{15}\ Hz \qquad \square$$

5.3 Interaction of light with metal nanoparticles

The smooth metal surfaces totally reflect the visible light and thus appear silvery bright; the metal powders, on the other hand, are dark brown or black. Their large surface area causes the light to be nearly fully absorbed due to repeated reflections. However, if the particle size is reduced to nm range, true colors appear again, notably for gold and silver. Thus a sol of fine gold particles in a liquid has bright red color. The colors of the glass windows of the cathedrals built centuries ago are due to the presence of fine metal particles.

Mie in 1908 provided a solution of the Maxwell's equations for an electromagnetic light wave interacting with small metal spheres. He calculated a quantity called the "extinction cross section" for a dilute solution of spherical metal particles embedded in an isotropic, non absorbing medium of dielectric constant ε_m. The extinction coefficient is a measure of the likelihood of the light being absorbed or scattered by an object. The Mie theory gives the extinction coefficient, C_{ext} to be

$$C_{ext} = \frac{2\pi}{k^2} \sum_{n=1}^{\infty} (2n+1)\,\mathrm{Re}(a_n + b_n) \tag{5.5}$$

Here $k = \dfrac{2\pi\sqrt{\varepsilon_m}}{\lambda}$ and a_n and b_n are scattering coefficients which are functions of particle radius R and wavelength λ in terms of Ricatti–Bessel functions.

When the particle size is small, only the first term in the summation in Eq. (5.5) needs to be considered. In this case the extinction cross section is given by

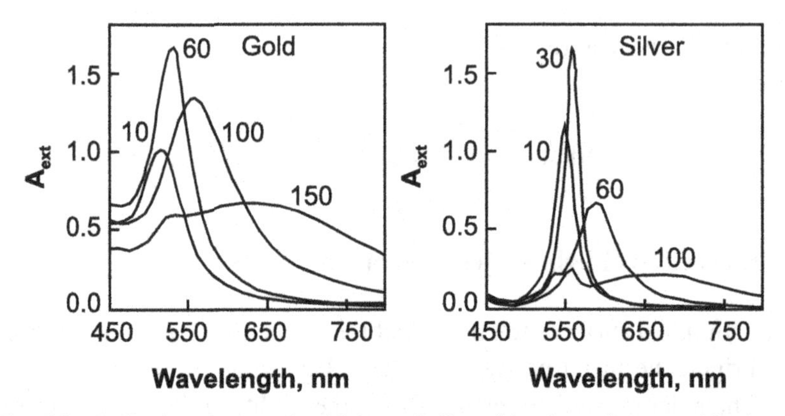

Fig. 5.2. Extinction spectra of gold (a) and silver (b) spheres in water. Numbers near curves correspond to the particle diameter (reprinted from [70] copyright (2010) with permission from Elsevier).

$$C_{ext} = \frac{24\pi^2 R^3 \varepsilon_m^{3/2}}{\lambda} \frac{\varepsilon''}{(\varepsilon'+2\varepsilon_m)^2 + \varepsilon''^2} \tag{5.6}$$

Here R is the particle radius, ε_m is the dielectric constant of the surrounding medium which is related to its refractive index by $\varepsilon_m = \eta_m^2$ and ε is the complex dielectric constant of the metal given by $\varepsilon = \varepsilon' + i\varepsilon''$. A maximum absorption occurs at resonance frequency at which C_{ext} is maximum. This occurs when $\varepsilon' = -2\varepsilon_m$ in the denominator of Eq. (5.6).

To calculate the extinction coefficient using Eq. (5.6), a knowledge of the real and the imaginary parts of the dielectric constant, ε' and ε'' as a function of frequency is needed. Their experimentally determined values can be used if available or, in some simple cases, these can be calculated using models such as the Drude model or the quasistatic approximation [69, 71, 72]. This is illustrated in Fig. 5.2. The figure shows the plots of the extinction coefficients of gold and silver particles in water calculated using the Mie theory [70]. A single narrow peak is observed at small particle sizes. This peak is known as the Mie peak or the "surface plasmon peak", as further discussed below. With an increase in the particle size, the spectra become red shifted and broadened. For the case of the largest sizes (150 nm for gold and 100 nm for silver) multiple peaks can be clearly seen.

5.4 Surface plasmon resonance

5.4.1. *Introduction*

The physical basis of the peak in the extinction coefficient discussed above is the collective oscillation of the free electrons on the surface of a metal particle caused by the electrical field component of the light beam incident on it. The collective oscillations of conduction electrons in metals are called *plasmons*. The oscillation of electrons induced by the light is confined to the surface and in this case the plasmons are called *surface plasmons.* The electric field displaces the electrons to create uncompensated charges at the nanoparticle surface (Fig. 5.3).When the particle size is much smaller than the wavelength of light, $kR \ll 1$, a dipole is created due to the accumulation of the electrons at one end and the depletion of electrons at the other end. This dipole creates a restoring force for the electron cloud. At a certain frequency of the incident light there is a resonance at which there is a maximum absorption of energy from the incident light wave and a peak is observed in the absorption spectra. This peak is called the "*surface plasmon resonance (SPR)*" peak. (The surface plasmons should be distinguished from bulk plasmons (bulk plasma) which have a longitudinal nature and can not be excited by

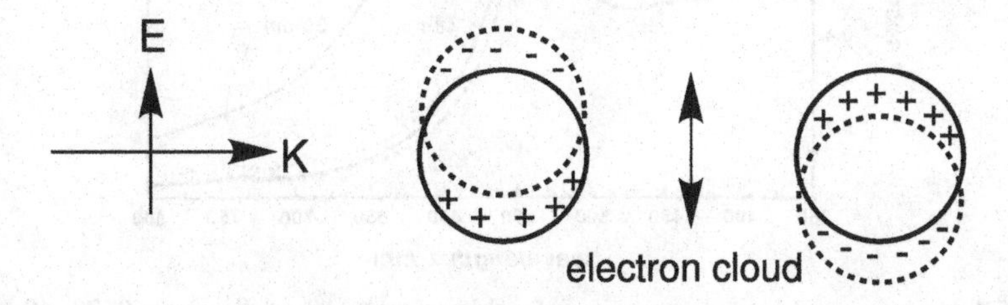

Fig. 5.3. Schematic of the plasma oscillation for a metal sphere showing the displacement of the electron cloud with respect to the nuclei due to the electric field of the incident light.

visible light. One can also have surface propagating plasmons or surface plasmon polaritons in thin films as distinct from the surface localized plasmons which occur at the nanoparticles surface [70]).

For larger particles, the oscillations of electrons become less coherent and higher modes of plasma oscillation (*e.g.* quadrupolar and multipolar) appear. Also when the particle size is much larger than the mean free path of the electrons, electrons can be significantly accelerated within the metal between collisions leading to scattering losses. Due to these factors the surface plasmon peak broadens and additional peaks appear as the particle size is increased.

The SPR induces a strong absorption of the incident light. Gold, silver and copper nanoparticles show strong SPR absorption in the visible region. This SPR absorption is responsible for the appearance of colors in the sols of these metals as only the part of the visible light complimentary to the absorption wavelength reaches the eye. For other metals the absorption band is weak and broad and occurs in the UV region. The SPR absorption peaks for small particles ($5 \leq a \leq 20$ nm) of gold and silver are localized near 530 nm and 415 nm respectively.

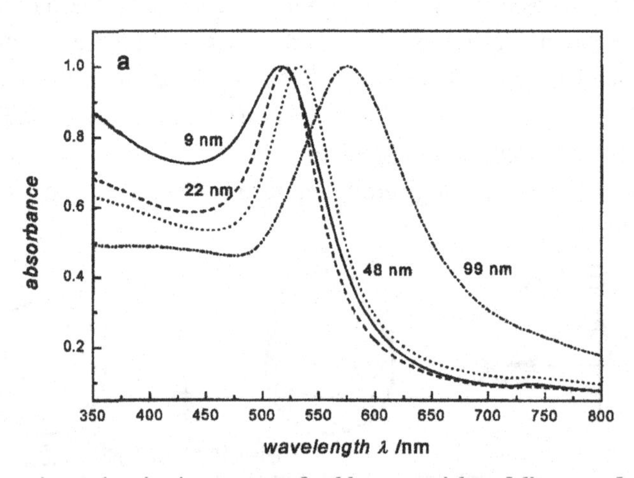

Fig. 5.4. Experimental extinction spectra of gold nanoparticles of diameters 9, 22, 48 and 99 nm. The shift in the peak occurs due to the contributions from quadrupolar higher resonances as well as due to the contribution of the scattering (reprinted with permission from [73] copyright (1999) American Chemical Society).

Figure 5.4 shows the experimental absorption spectra of gold nanoparticles of different sizes [73]. As expected from the theory, a sharp absorption peak appears at low particle sizes and a shift in the position of the peak to the higher wavelengths and a broadening of the peak occurs with increasing particle size due to the contributions of the quadrupolar and higher resonances as well as due to scattering.

The shift in the position of the peak in the extinction coefficient with particle size for the case of gold as determined experimentally by the various researchers has been combined into a single plot by Khlebtsov and Dykman (Fig. 5.5) [70]. The plot can be used as a calibration curve to determine the particle size from the absorption spectra.

Equation (5.6) shows that the extinction cross section varies as R^3. As the number density decreases as R^3 for a given amount of the colloidal material, the absorption coefficient is independent of the particle size in the dipolar approximation (which itself is valid when the particle size is small, $kR \ll 1$). The extinction coefficient is found to be size independent in the size range between approximately 5 nm and 30 nm. Above 30 nm, the quadrupole and the multipole terms in the Mie summation become significant while below 5 nm, the collision of the electrons with the surface of the particle becomes more important as the

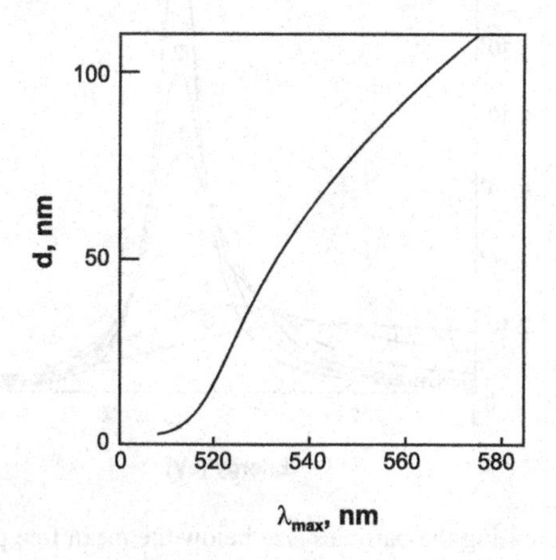

Fig.5.5. Shift in the experimentally determined extinction peak position with particle diameter for gold spheres in water [70].

mean free path becomes smaller than the particle size. The increased rate of electron surface collision in effect changes the dielectric constant of the metal such that the SPR band is largely damped. This usually happens for particle sizes below about 5 nm. The effect is shown schematically in Fig. 5.6 which shows plots of the extinction coefficient *vs.* energy of the incident light for particle sizes smaller than the mean free path.

Nanoparticles of gold and silver show strong plasmon resonance; however, the other metals such as platinum, palladium, iron do not show such resonance. This is because in these metals the resonance is suppressed due to mechanisms such as strong conduction electron relaxation or radiation damping due to scattering.

For the semiconductor nanoparticles, the free electron concentration is order of magnitudes smaller than in metals even in case of heavily doped semiconductors and this absorption occurs in the infrared range. The color of the semiconductor particles is therefore not influenced

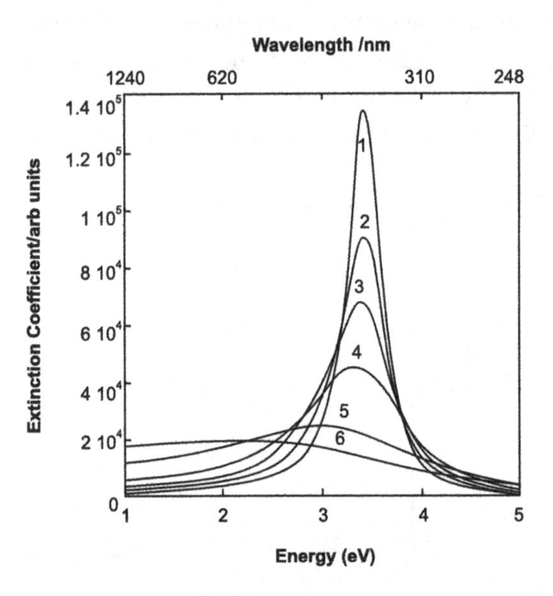

Fig. 5.6. Effect of decreasing the particles size below the mean free path of the electrons in the bulk metal (*i.e.* usually below 5 nm); the particle size is decreasing from 1 to 6 (reproduced from [74] copyright (2001) with permission of John Wiley & Sons, Inc.).

by the surface plasmon resonance. However, when the size of the semiconductor particles decreases further, the quantum effects appear leading to change in the electronic band structure and absorption in the visible region occurs due to electronic transition across the band gap. This has already been discussed in Chapter 4.

5.4.2. *Effect of particle shape*

As noted above, the surface plasmon peak shows a shift with a change in the particle size. This shift is not very large and in effect the color of the nanoparticles is nearly insensitive to their size in the range in which the SPR occurs. This situation is dramatically changed when particles of shape other than spherical are used.

Gans predicted in 1912 that for very small ellipsoids the surface plasmon mode would split into two distinct modes, one corresponding to the electron oscillations along the long axis, called the longitudinal mode and the other a weaker band, called the transverse mode [75, 76]. The transverse mode is a weak mode occurring at a wavelength in the visible region which is nearly same as that for the single mode in the spherical

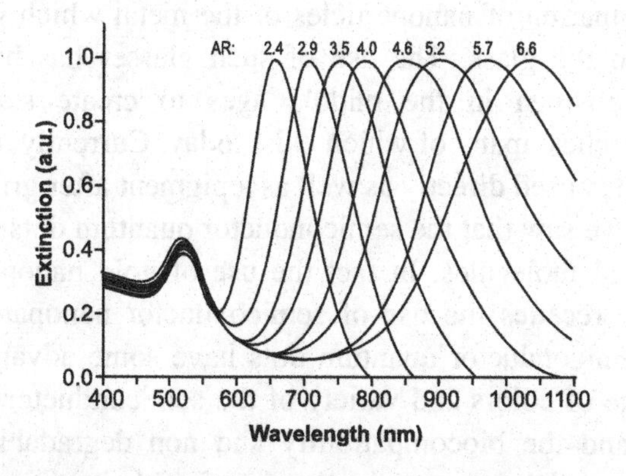

Fig. 5.7. Extinction coefficient of gold nanorods with different aspect ratios; the transverse band at about 520 nm remains nearly unchanged while there is a large shift towards longer wavelengths in the longitudinal mode as the aspect ratio increase (Reprinted with permission from[77] Copyright (2010) Elsevier).

particles while the longitudinal mode is much stronger in intensity and occurs at longer, near infrared (NIR) wavelengths. The splitting of the modes occurs because of the surface curvature which alters the restoring force acting on the electron cloud. The effect becomes stronger as the aspect ratio increases and is found to be very strong in case of nanorods.

Figure 5.7 shows the extinction coefficient of gold nanorods of different aspect ratios. With increasing aspect ratio, there is a large shift to higher wave lengths in the position of the longitudinal mode as the aspect ratio of the rods is increased. The shape therefore provides an additional and effective parameter to control the response of the nanoparticles.

5.4.3. *Applications*

Noble metal nanoparticles have several present and potential applications. Amongst these can be mentioned their use (i) as colors, particularly in glass (ii) as tags in bioassays and (iii) in devices for detecting low concentration of an analyte down to a single molecule.

Their use as pigments for glass has been practiced for thousands of years although it was only in the nineteenth century that Faraday correctly attributed the ruby red color of glasses to the presence of gold nanoparticles. Addition of salt of a noble metal to glass during melting results in the formation of nanoparticles of the metal which give rise to distinct colors in the glass. The use of such glasses has been widely practiced by craftsmen in the middle ages to create stained glass windows for churches, many of which exist today. Currently, the colored glass so prepared is used directly as well as a pigment after grinding it.

In Chapter 4 we saw that the semiconductor quantum dots are used as tags for biological molecules. In fact the use of gold nanoparticles for this application precedes the use of semiconductor nanoparticles [78]. However, the semiconductor quantum dots have some advantages such as a greater range of colors and variety of the semiconductors available. On the other hand the biocompatibility and non degradability of the noble metal nanoparticles may make them preferable in some cases. The principles of application of metal nanoparticles as tags are essentially the same as described in case of the semiconductor quantum dots.

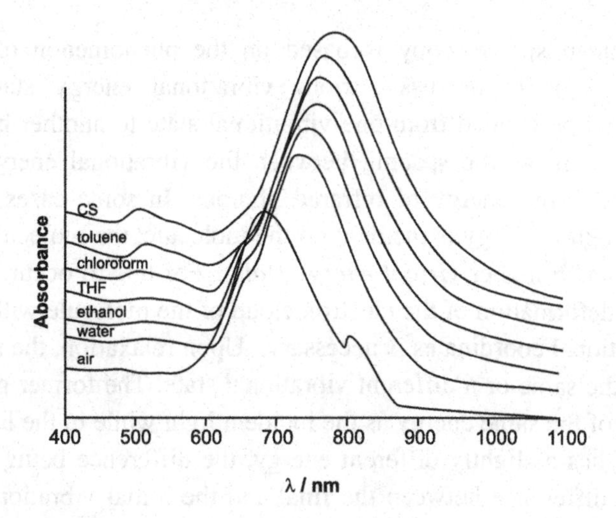

Fig. 5.8. Experimental spectra of gold nanorods in different solvents.(Reprinted from [76] copyright (2005) with permission from Elsevier).

By far the most exciting application of noble metal particles is their potential use in detection of molecules down to a single molecule level. This is directly related to the occurrence of surface plasmon resonance in these particles as discussed below.

From Eq. (5.6) we see that the extinction coefficient is a strong function of the dielectric constant of the surrounding medium, the extinction being maximum when $\varepsilon' = -2\varepsilon_m$ where ε' is the dielectric constant of the metal and ε_m is the dielectric constant of the surrounding medium. A change of the refractive index of the surrounding medium therefore leads to a change in the position of the SPR peak. Figure 5.8 illustrates this effect for gold nanorods.

Efforts have been made to utilize this effect directly by confining a molecule at the tip of a nanoparticle having a double pyramid shape and it has been demonstrated that there is a measurable change in the position of the surface plasmon peak [79]. However, a more powerful technique that has emerged for detection of low concentrations of an analyte down to a single molecule is the *surface enhanced Raman spectroscopy (SERS)* [80].

The Raman spectroscopy is based on the phenomenon of Raman scattering. A molecule has various vibrational energy states. The molecule can be excited from one vibrational state to another by means of infrared light as the spacing between the vibrational energy levels corresponds to the energy of infrared photons. In some cases, when a photon of higher energy is incident on the molecule, the molecule can be excited to much higher *virtual energy states.* For this to occur a certain amount of deformation of the electron cloud of the molecule with respect to its vibrational coordinates is necessary. Upon relaxation, the molecule returns to the same or a different vibrational state. The former gives rise to photons of the same energy as the incident light while in the latter case the photon has a slightly different energy, the difference being equal to the energy difference between the final and the initial vibrational states (Fig. 5.9). If the final state has a higher vibrational energy than the initial state, the shift in the energy of the photon is called the Stokes shift and in the other case, when the final state has a lower energy than the initial state, the shift is called the anti-Stokes shift. In Raman spectroscopy these shifts in energy are detected and used to get information about the

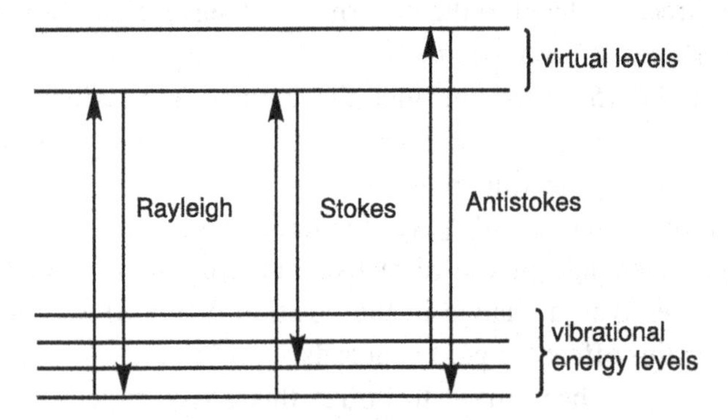

Fig. 5.9. The incident light excites a molecule from one of the vibrational states to an excited state. In Rayleigh scattering, the molecule relaxes back to the same vibrational state and so the energy of the emitted photon is equal to that of the incident photon. If the molecule relaxes to a different vibrational state, the energy of the emitted photon is shifted either down (Stokes shift) or up (antistokes shift) depending on whether the final vibrational state is higher or lower in energy than the initial vibrational state. This is Raman scattering.

molecule. The technique is complimentary to infrared spectroscopy. The Raman signals are weak signals and elaborate instrumentation is needed to obtain a Raman spectrum.

It is observed that very strong Raman signals are obtained when the sample is deposited on a rough film of silver. The enhancement in the signal is found to be several orders of magnitude. It is now generally agreed that this enhancement is due to the surface plasmons in the silver film.

As we saw, the dipole created due to surface plasmons is oriented parallel to the electric field of the incident light. If the light is incident on a plane surface, such dipole can be induced only on a feature which provides a surface perpendicular to the planar surface of the film. The enhanced Raman signal is therefore only seen from surfaces which have been roughened to provide such features at a scale of few nm. This can also be done by depositing nanoparticles of this size on the surface.

The enhancement in the Raman signal occurs due to the enhanced electric field experienced by a molecule deposited next to a feature on the surface. The enhancement in the field occurs due to the dipole created by the surface plasmon. The field outside a sphere, E_{out} due to this is given by [71]

$$E_{out}(r) = E_o + \alpha E_o f(x,y,z)$$

Here E_o is the applied field, $f(x,y,z)$ is a function of the coordinates of the point at which the field is $E_{out.}$ and α is the polarizability of the sphere. The Raman intensity is now determined by $< |E_{out}(\omega)|^2 |E_{out}(\omega'|^2 >$ where ω is the frequency of the incident light and ω' is the Stokes shifted frequency and the brackets denote average over the particle surface. Thus there is an enhancement in the Raman signal by a factor of 10^4 ; however due to the interaction of the field from different points on a rough surface, enhancement by factors in excess of 10^{10} has been reported, Such an enhancement makes it possible to detect very small concentrations of an analyte, down to a single molecule level.

The enhancement in the Raman signal achieved in SERS is illustrated in Fig. 5.10. [81]. In this work, the substrate for SERS was prepared by depositing DNA interlinked Ag nanoparticles on mica. A small drop (20 μl) of a dilute solution (1 x 10^{-5}M) of methylene blue was deposited and Raman spectra were collected from the dried drop. While no spectra could be detected from the sample deposited on bare mica or mica simply immersed in silver nanoparticles suspension and dried, a very strong signal was obtained when the substrate had a controlled roughness due to DNA mediated assembly of the silver nanoparticles.

While in the above example a DNA mediated assembly of Ag nanoparticles on a mica substrate has been used, the substrate with controlled roughness can be prepared by methods as simple as the electrochemical etching of a silver surface. The subject of preparation of the substrate for SERS to maximize the SERS sensitivity is of continuing research interest. SERS has been used to detect a wide variety of analytes present at low concentrations, including pollutants, explosives, chemical warfare agents, and DNA.

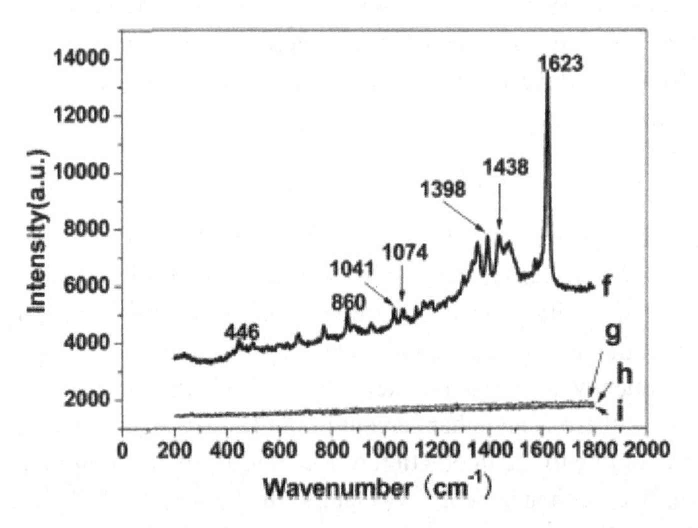

Fig. 5.10. SERS spectra of methylene blue (1 × 10^{-5} M) on (f) DNA network (100 ng/μl) mediated LBL assembled silver film of one layer (g) bare mica, (h) bare mica immersed in silver nanoparticles sol for four hours, and (i) 100 ng/μl DNA network covered bare mica. Laser power, 25 mW, 20%; integration time, 20 s. (reprinted from [81] copyright (2008) with permission from Elsevier).

Problems

(1) It was seen that the metals absorb electromagnetic radiation for frequencies for which $\omega\tau << 1$. Calculate the frequency below which gold metal will absorb the radiation. Assume that there is one free electron per atom. Given the following data for gold: conductivity, $\sigma = 4.52 \times 10^7$ (ohm.m)$^{-1}$ density = 19300 kg.m^{-3}, atomic weight = 196.97, mass of electron = 9.11×10^{-31} kg, Avogadro's number = 6.02×10^{23}.

(2) Calculate the plasmon frequency for gold. Treat the system as consisting of free electrons in vacuum and ignore the thermal effects.

(3) It was mentioned in the text that even in the heavily doped semiconductors, the electron concentration is several orders of magnitude lower than that in the metals and so the surface plasmon absorption occurs in the infrared range. A semiconductor is considered to be heavily doped when there is one dopant atom per ten thousand atoms of the semiconductor. Find the plasmon frequency for Si doped to this level with P. Take the effective mass of electrons in Si to be 1.08 m_0 where m_0 is the rest mass of the electron, and the dielectric constant of Si as 11.9.

(4) Distinguish between the mechanisms responsible for the appearance of colors in the suspensions of metal and semiconductor nanoparticles.

(5) What are the reasons for the broadening of the surface plasmon peak and the appearance of additional peaks as the size of the metal nanoparticles is increased?

Chapter 6

Zero Dimensional Nanostructures III
The Nanoscale Magnetic Structures

6.1 Introduction

Several characteristic length scales in the nanometer range exist in magnetism. These include the exchange length, the domain wall width, maximum equilibrium single domain particle size, the superparamagnetic blocking radius at 300K, *etc.* Some new effects should, therefore, be expected when the object size approaches any of these characteristic lengths. An important length scale for magnetic phenomena at nanoscale is the domain size in ferromagnetic crystals. Typically the domain size is in the range 10–200 nm. One of the important effects that is observed as a result of this is the phenomenon of *superparamagnetism* in nanometer sized ferromagnetic crystals.

Coming to another aspect, the spins of all the electrons in a crystal having only a single domain are aligned in the same direction — either up or down. In an electric field gradient, the electrons move and are able to maintain their spins for a certain distance, called the spin coherence length. If they encounter another ferromagnetic crystal within this distance, then their passage through this second crystal depends on the orientation of the spins in that crystal — if they are in the same direction, the passage is characterized by a low resistance while antiparallel spins lead to an increased scattering at the interface and so to a high resistance. The spin coherence length is of the order of a few nanometers. The phenomenon is utilized by constructing a stack of alternating ferromagnetic and nonmagnetic layers. The direction of magnetization in one of the layers can be changed at will by applying a magnetic field of suitable strength. The difference in resistance between the two states is substantial, justifying the name *"Giant Magneto Resistance (GMR)"* given to the phenomenon. The devices based on GMR have now been used for several years in the read — write heads for the magnetic memory.

The ability to control and manipulate the spins of electrons, has given rise to the whole new field of spin based electronics or *spintronics* where the electron carries not only the charge but also information in the form of its spin.

In the following we concentrate mainly on the first two of the nanoscale magnetic phenomena, namely the superparamgnetism and the GMR and their applications. We first briefly review some basic definitions in magnetism, particularly in ferromagnetism. The idea of magnetic anisotropy is treated in some detail because of its importance to superparamgnetism. An important application of superparamagnetic particles is in designing nanocomposites with high coercivity *together with* a high saturation magnetization. Usually it is hard to have a combination of high coercivity and high saturation magnetization in a single material – hard ferromagnetic materials have high coercivity but low saturation magnetization while opposite is the case in the soft magnetic materials. The chapter ends with a brief discussion of spintronics and the applications of superparamagnetic and other materials.

6.2 Definitions

6.2.1. *Magnetic lines of force*

A magnetic field can be represented by "lines of force" which can be visualized, for example, by sprinkling iron powder near a bar magnet. By convention, the lines are assumed to be emitted outward from the north pole of a magnet. The lines of force are called magnetic flux and are represented by the symbol ϕ. The SI unit of flux is weber and it is equal to 10^8 lines of force. The cgs unit is Maxwell which is equal to 1 line of force. The quantity "lines of force per unit area" is called the magnetic flux density or magnetic induction or simply the "*flux density*" or *magnetic induction, B*.

6.2.2. *Magnetic field*

Every magnet produces a magnetic field around it. In the laboratory, a magnetic field is produced by passing a current through a solenoid. If I is the current, N is the number of turns in the solenoid and L is the length of the solenoid, then the field H produced along the axis of the solenoid is given by

$$H = \frac{NI}{L} \text{ ampere turns m}^{-1} \tag{6.1}$$

The magnetic field strength and the flux density are related as follows

$$B = \mu H$$

Here μ is the permeability of the medium. For vacuum, it is denoted as μ_0 and has a value of 1 in the cgs units and $4\pi \times 10^{-7}$ in the SI units. Thus in the cgs units a unit field strength, 1 Oersted, implies a unit flux or 1 line of force per cm^2 in vacuum (or air). The ratio, $\mu/\mu_0 = \mu_r$, is called the *relative permeability* of the medium.

The cgs units are still used very often in the magnetism literature. Here, we will use primarily the SI units. The two types of units and their conversion factors for some quantities are given in Table 6.1.

6.2.3. *Magnetic moment and magnetization*

It is convenient to define these terms with reference to a bar magnet. A bar magnet suspended in a magnetic field is acted upon by a torque which tends to orient it along the field (Fig. 6.1). If the field is 1 $A.m^{-1}$ and θ, the angle between the field and the axis of the magnet is 90^0 then the moment of this torque, m, is called the magnetic moment of the magnet.

The magnitude of the magnetic moment depends on the volume of the magnet. The magnetic moment per unit volume is called the *magnetization M*.

Table 6.1. CGS and SI units and the conversion factors in magnetism.

Quantity	cgs units	SI units	Conversion factor
Magnetic field, H	oersted	Am^{-1}	1 Am^{-1} = $4\pi \times 10^{-3}$ Oe
Magnetic flux density, B	gauss	tesla (T) Vsm^{-2} Weber m^{-2}	1 tesla = 10^4 gauss
Permeability, $\mu = B/H$	$gauss.oer^{-1}$ (emu)	tesla. $(A.m^{-1})^{-1}$ $henry.m^{-1}$ $VA^{-1}sm^{-1}$ $Wb.A^{-1}.m^{-1}$	$4\pi \times 10^{-7}$ $hen.m^{-1}$ = 1 emu
Relative permeability, (μ/μ_0)	dimensionless	dimensionless	$\mu_{SI} = \mu_{cgs}$
Magnetic moment, m	$erg.oersted^{-1}$ (emu) $maxwell.cm^{-1}$	Am^2 (JT^{-1})	1 Am^2= 10^3 emu
Magnetization, $M=m/V$	$erg.oersted^{-1}.cm^{-3}$ $maxwell.cm^{-3}$	Am^{-1}	1 Am^{-1} = 10^{-3} $emu.cm^{-3}$
Volume susceptibility, χv	$(emu.Oe^{-1}).cm^{-3}$	$Am^2.(Am^{-1})^{-1}m^{-3}$	SI = $10^3/4\pi$ cgs
Mass susceptibility, χm	$(emu.Oe^{-1}).g^{-1}$	$Am^2.(Am^{-1})^{-1}kg^{-1}$	SI = $10^3/4\pi$ cgs
Bohr magneton, μB	9.27×10^{-21} emu	9.27×10^{-24} $A.m^2$	
Permeability of free space (μ_0)	1 $gauss.oersted^{-1}$	$4\pi \times 10^{-7}$ tesla. $(A.m^{-1})^{-1}$	

$$\text{Magnetization } M = \frac{m}{V} \tag{6.2}$$

where V is the total volume of the magnet.

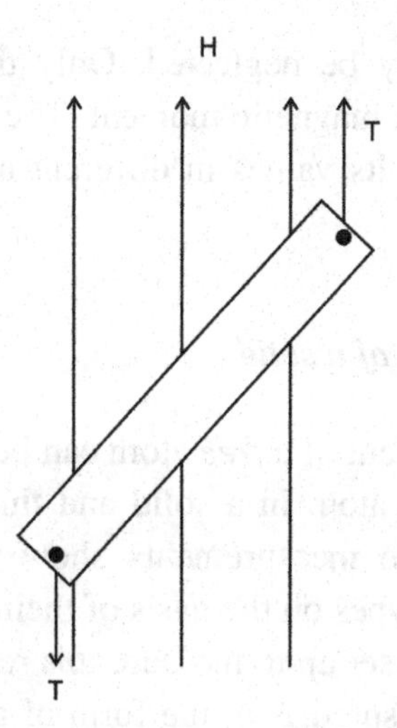

Fig. 6.1. A bar magnet suspended in a uniform magnetic field. A torque acts to align the magnet with the magnetic field. If the field is 1 A.m^{-1} and the angle between the field and the axis of the magnet is 90°, then the moment of this torque, *m*, is called the magnetic moment of the magnet.

6.2.4. *Magnetic moment of an atom*

It is known from atomic physics that the magnetic moment of an atom arises from three sources: the intrinsic spin of the electron, the orbital motion of the electron and the nuclear magnetic moment. The nuclear moments are three orders of magnitude smaller than the electronic

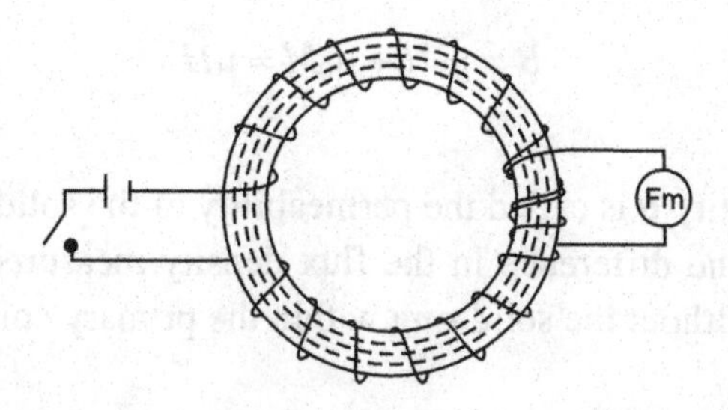

Fig. 6.2. Experimental set up for measuring the magnetic moment of a solid. The solid is taken in the form of a ring for convenience. The flux is measured using the flux meter (Fm) connected to the search coil.

moments and can mostly be neglected. Only the atoms with unfilled electron shells have a net magnetic moment. The unit of atomic moment is a Bohr magneton, μ_B. Its values in different units are given in Table 6.1.

6.2.5. *Magnetic moment of a solid*

While the magnetic moment of a *free* atom can be calculated, it is not yet possible to do so for an atom in a solid and this has to be determined experimentally. Magnetic measurements show that the solids can be classified into different types on the basis of their response to an external magnetic field. A simple set up to measure this response is shown in Fig. 6.2. The solid to be measured is in the form of a solid ring with a coil, called the primary coil, wound around it. Consider first the case with the solid ring taken out of the coil. When a current is passed through the coil, a magnetic field H is generated within it. There now exists a flux density equal to $\mu_0 H$ within the coil. This flux can be detected by a second coil called the search coil, also wound around the solid ring. When the ring of the solid to be measured is now inserted inside the coil, an additional flux is generated in the solid due to its magnetization (this flux can be negative also, see below). The magnitude of this additional flux density is $\mu_0 M$ where M is the magnetization produced in the material. The total magnetic flux density in the material, denoted by B, is given by

$$B = \mu_0 H + \mu_0 M = \mu H \tag{6.3}$$

where the quantity μ is called the permeability of the solid.

Let ΔB be the difference in the flux density measured by the search coil with and without the solid ring within the primary coil so that

$$\Delta B = B_{with\ solid} - B_{air} = \mu H - \mu_0 H = \mu_0 M = \chi\,(\mu_0 H)$$

where χ is called the susceptibility of the solid and is equal to the magnetization produced in the solid by a unit field

$$\chi = \frac{M}{H} \tag{6.4}$$

Having defined the susceptibility, we can now classify the magnetic materials on its basis.

6.3 Classification of magnetic materials

It is found that ΔB and hence the susceptibility, χ, can have a small negative value, a small positive value or a large positive value depending on the material of the ring in Fig. 6.2. Accordingly the various materials are classified as follows.

(1) Diamagnetic if $\Delta B < 0$ (small negative sceptibility)

(2) (i) Paramagnetic or
(ii) Antiferromagnetic if $\Delta B > 0$ (small positive susceptibility)

(3) (i) Ferromagnetic, or
(ii) Ferrimagnetic if $\Delta B >> 0$ (large positive susceptibility)

These different magnetic materials are described briefly below.

6.3.1. *Diamagnetic materials*

The atoms in a diamagnetic material have no net magnetic moment. The negative susceptibility of the diamagnetic materials can be understood in terms of the Lenz's law according to which when the flux in a circuit is changed, an induced current is set up to oppose the change in the flux. An electron orbiting and spinning inside an atom is analogous to a current circuit. All materials therefore have a diamagnetic contribution in their magnetization. However, in materials other than the diamagnetic materials, this is more than compensated by the other contributions to the magnetic moment. Examples of diamagnetic materials are inert gas molecules, alkali halides like NaCl, covalently bonded solids like diamond, Si and Ge and most organic compounds.

6.3.2. *Paramagnetic materials*

Examples of paramagnetic materials are salts of the transition elements, salts and oxides of the rare earths, rare earth elements, and isolated rare earth ions. Oxygen is also paramagnetic. In the paramagnetic materials, the atoms have a small net magnetic moment. In zero field, the various moments point in random directions so that the net magnetic moment is zero (Fig. 6.3 a). An external field tends to align these moments along the field. The thermal energy opposes this and tends to randomize the moments. At any instant only a few of the moments are aligned in the field direction. The net moment and the susceptibility are, therefore, small. Langevin analyzed this problem and gave the following result

$$\frac{M}{M_0} = \coth\frac{\mu H}{kT} - \frac{1}{(\mu H/kT)} = \coth a - \frac{1}{a} = L(a) \qquad (6.5)$$

where $a = (\mu H/kT)$.

Here M is the magnetization produced in the material in an external magnetic field H, M_0 is the maximum possible magnetization which the material can have ($= n\mu$, where μ is the net magnetic moment of an atom, n is the total number of atoms), and T is the absolute temperature.

The function $L(a)$ can be written as a series

$$L(a) = \frac{a}{3} - \frac{a^3}{45} + \frac{2a^5}{945} - \ldots$$

When 'a' is small, ($a < 0.5$), a plot of $L(a)$ *vs.* a is nearly a straight line with a slope of 1/3 whereas at larger a, it tends to 1.

The paramagnetic susceptibility for most materials is in the range 10^{-6}– 10^{-3} at room temperature.

Example 6.1. *The magnetic moment of an oxygen molecule is 2.64 x 10^{-23} Am^2. If the susceptibility is 1.36 x 10^{-6}m^{3}kg^{-1}, what is the ratio M/M_0 in a field of 10^7 Am^{-1}?*

Solution. $M_0 = N\mu = \frac{6.02\times10^{23}\times10^3}{32}(2.64\times10^{-23}) = 497\ \text{Am}^2\text{kg}^{-1}$

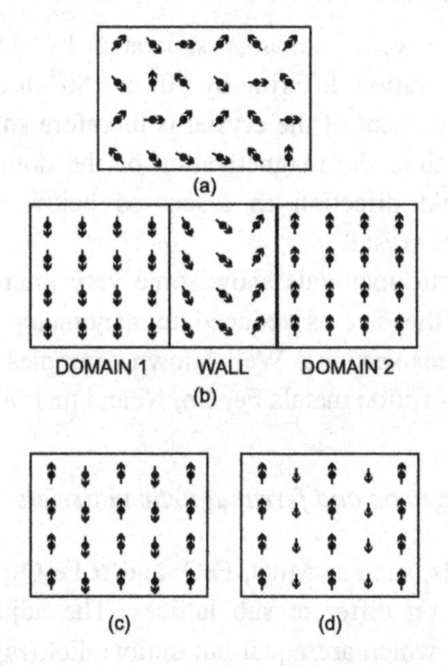

Fig. 6.3. Arrangement of magnetic moments of atoms in different types of materials (a) paramagnetic (b) ferromagnetic (c) antiferromagnetic and (d) ferrimagnetic.

$$M = \chi H = 1.36 \times 10^{-6} \times 10^{7} = 13.6 \ Am^2kg^{-1}$$

Hence $M/M_0 = 0.0274$.

This shows that the thermal effects are so dominant that even strong fields are able to achieve only a small fraction of the maximum possible magnetization. □

6.3.3. *Ferromagnetic materials*

In a ferromagnetic crystal the atoms have a net magnetic moment just like the paramagnetic case; however, unlike a paramagnet in which the moments point in random directions, in the ferromagnet there is an interaction, called the exchange interaction, between the moments on adjacent atoms which causes the moments to line up parallel to one another over large regions called domains as shown in Fig. 6.3 (b). For energetic reasons, a single crystal, or a single grain in a polycrystalline solid, consists of several domains separated by domain walls, the direction of magnetization differing by 90^0 or 180^0 in adjacent domains. The net magnetic moment of the crystal is therefore small or zero. In an external magnetic field the magnetization of the domains tends to get aligned in the field direction as discussed below, producing a net magnetization in the crystal.

The ferromagnetic materials show some very interesting and useful phenomena when the size is reduced to nanometer scale. These are discussed in the later sections. Well known examples of ferromagnetic materials are the transition metals Fe, Co, Ni and their alloys.

6.3.4. *Antiferromagnetic and ferrimagnetic materials*

In some compounds, such as MnO, FeO, and α-Fe_2O_3, the two different atoms are located on different sub lattices. The adjacent atoms have magnetic moments which are equal but antiparallel (Fig. 6.3 c). The net magnetic moment is zero and the material is called *antiferromagnetic*. Some examples of antiferromagnetic materials are, in addition to those mentioned above, CoO, NiO, Cr_2O_3, MnO_2, *etc*.

In another group of materials such as $MnFe_2O_4$, Fe_3O_4 *etc*., the atomic magnetic moments are unequal as well as antiparallel (Fig. 6.3 d). A net magnetic moment is therefore observed. These materials are called *ferrimagnetic*.

6.4 Nanoscale ferromagnetic materials

The ferromagnetic materials were described briefly in Sec. 6.3.3. Before discussing the nanoscale ferromagnetic materials, it is helpful to discuss two other aspects of ferromagnetism – the magnetization curve and the magnetic anisotropy.

6.4.1. *Magnetization curve of a ferromagnetic crystal*

As was mentioned above, the magnetic moments point in different directions in different domains in a ferromagnet. Upon application of a

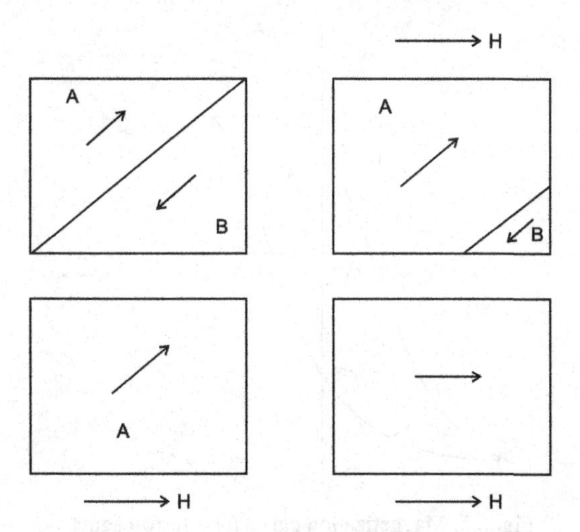

Fig. 6.4. On application of a field H, the domain A grows and the domain B shrinks (top panels); a further increase in the field causes the magnetization vector of the domain to rotate in the direction of the field (bottom panels).

magnetic field, the domains, in which the magnetization is most closely oriented to the direction of the field, grow while the other domains shrink. The net magnetization of the crystal increases. When the domains can no longer grow, a further increase in the field increases the magnetization by rotation of the magnetization vectors of the domains along the field (Fig. 6.4). The magnetization eventually saturates at a value called the saturation magnetization, M_s and no further increase in the magnetization results if the field is increased. On reversal of the field, the domain walls do not fully return to their earlier zero magnetization position due to the presence of defects so that the magnetization stays higher than during the increasing field. A certain magnetization, called the remnant magnetization M_R remains even when the field is decreased to zero. A certain field in the reverse direction is needed to bring the magnetization back to zero. The process is repeated if the field is continued to be increased in the reverse direction (Fig. 6.5).

It is usual to plot the magnetization curve in terms of B rather than M. As the magnetic susceptibility can be very high (~ 10^6) for a ferromagnetic material, $H << M$ and, from Eq. (6.3)

$$B \approx \mu_0 M$$

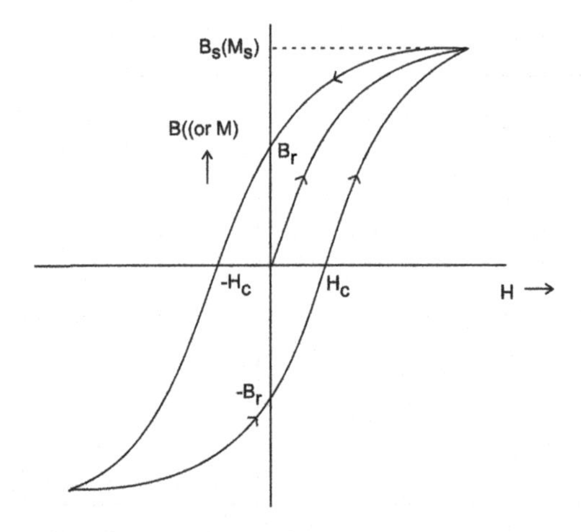

Fig. 6.5. Magnetization curve for a ferromagnet.

The magnetic induction B at saturation is denoted as B_s and the induction retained by the material on reversing the field to zero is called *retentivity* B_r while the field required to bring back B to zero is called *coercivity*, H_C.

The exchange coupling between the moments on adjacent atoms in a ferromagnet can be overcome by the thermal energy. Above a certain temperature called the Curie temperature, the material behaves like a paramagnet. The dependence of the susceptibility on the temperature above the Curie temperature is given by

$$\chi = \frac{C}{T - \theta} \tag{6.6}$$

where C is a constant and θ is the Curie temperature.

6.4.2. *Magnetic anisotropy*

6.4.2.1. *Magnetocrystalline anisotropy*

The ease of magnetization is not same along all the directions in a ferromagnetic crystal. In some directions, the magnetization increases rapidly with the applied field and reaches the saturation M_s at quite low fields while higher fields are required for saturation in the other directions. The primary magnetic anisotropy which is intrinsic to a material is due to its crystal structure. This is termed *magnetocrystalline anisotropy*. Anisotropy can be *induced* by external means such as making the shape nonspherical (*shape anisotropy*) or by plastic deformation, residual stresses, magnetic annealing or irradiation. The magnetocrystalline anisotropy leads to the phenomenon of superparamgnetism discussed below while the shape anisotropy can be used to design novel nanostructures as illustrated later by some examples.

Figure 6.6 shows the magnetization curves for iron and nickel in different directions. The <100> direction in Fe and the <111> direction in Ni are the easiest directions to magnetize and the <111> directions in Fe and the <100> directions in nickel are the hard directions. In cobalt which is hexagonal, the easy direction is [0001] along the C axis while all the directions in the basal plane of the type $[10\bar{1}0]$ are equally hard. Crystals like Co, having only a single easy direction are called *uniaxial* crystals. It is not possible to predict the directions of easy magnetization in a crystal – there is no known correlation between any structural feature such as the atomic density along a direction and the ease of magnetization.

In Fig. 6.4, we considered two domains with spontaneous magnetization in opposite directions. Such a situation can exist in a hexagonal crystal in which the directions parallel to the c axis are the only easy directions. As depicted in Fig. 6.4, on application of a field, the domain oriented favorably to the field direction grows while the other domain shrinks and disappears. A rather large increase in the field is now required to rotate the direction of magnetization in the direction of the field. The energy required to rotate and hold the magnetization vector in the field direction against the force of crystal anisotropy is called the *crystal anisotropy energy*, *E*. For a cubic crystal, if α_1, α_2 and α_3 are the angles made by the magnetization vector M_s with the crystal axes, the magnetic crystal anisotropy energy per unit volume is given by

$$E = K_o + K_1(\cos^2\alpha_1\cos^2\alpha_2 + \cos^2\alpha_2\cos^2\alpha_3 + \cos^2\alpha_3\cos^2\alpha_1) + K_2 \cos^2\alpha_1 \cos^2\alpha_2 \cos^2\alpha_3 + \ldots \quad (6.7)$$

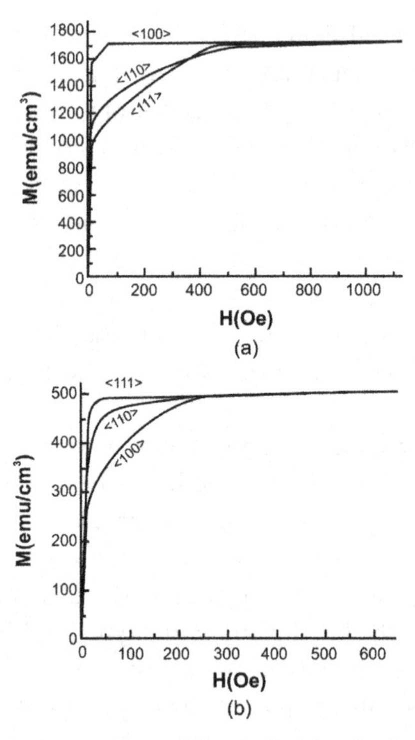

Fig. 6.6. The magnetization curves along different directions in single crystals of (a) iron [82] and (b) nickel [83, 84].

Table 6.2. Anisotropy constants of some important ferromagnetic crystals.

Structure	Substance	K_1 (10^4 J m^{-3})	K_2 (10^4 J m^{-3})
Cubic	Fe	4.8	±0.5
	Ni	- 0.5	-0.2
	$FeO.Fe_2O_3$	-1.1	
	$MnO.Fe_2O_3$	- 0.28	
	$NiO.Fe_2O_3$	- 0.62	
	$MgO.Fe_2O_3$	- 0.25	
	$CoO.Fe_2O_3$	9	
Hexagonal	Co	45	15
	$BaO.6Fe_2O_3$	33	
	YCo_5	550	
	MnBi	89	27

Here K_0, K_1, K_2 are constants for a given material at a given temperature. Usually K_2 and the coefficients of higher terms are small and are neglected. Also, as one is normally interested in the difference in energy, the coefficient K_0 can mostly be ignored. The values of K_1 for some important ferromagnetic materials are given in Table 6.2.

The existence of magnetic anisotropy in the ferromagnetic crystals leads to the phenomenon of *superparamgnetism* in their nanoparticles. This is discussed in a later section.

Example 6.2. *Calculate the difference in the total magnetocrystalline anisotropy energy between the [010] and [110] directions in a 10 nm diameter nanoparticle of (i) $CoFe_2O_4$ (ii) Fe_3O_4. The values of K_1 for the two materials are 90×10^3 and -11×10^3 Jm^{-3}, respectively*

Solution. The expression for the magnetocrystalline anisotropy energy constant K for a cubic crystal is

$K = K_o + K_1(\cos^2\alpha_1\cos^2\alpha_2 + \cos^2\alpha_2\cos^2\alpha_3 + \cos^2\alpha_3\cos^2\alpha_1)$

neglecting the higher order terms. Here α_1, α_2, and α_3 are the angles of the given direction with the cube axis. In the present case these angles are: for [010] $\alpha_1 = 90^0$, $\alpha_2 = 0$ and $\alpha_3 = 90^0$ while for [110] $\alpha_1 = 45^0$, $\alpha_2 = 45^0$ and $\alpha_3 = 90^0$.

The values of K for the two directions are, after substituting the values of the angles,

$$K_{[010]} = K_o$$

$$\text{and } K_{[110]} = K_o + K_1/4$$

Hence the difference in the values of K, $\Delta K = K_{[110]} - K_{[010]} = K_1/4$

The total difference in the anisotropy energy, $\Delta E = \Delta K.v$

where v = volume of the particle.

$$\Delta E = (K_1/4).(4\pi.125/3)10^{-27} = 41.7\,\pi\,K_1 \times 10^{-27}\text{ J}$$

Substituting the values of K_1, ΔE for $CoFe_2O_4 = 1.179 \times 10^{-20}$ J

and for Fe_3O_4, $\Delta E = -0.144 \times 10^{-20}$ J

Hence, the [010] direction is easy to magnetize in case of $CoFe_2O_4$ while the direction [110] is easy to magnetize in case of Fe_3O_4. In general, in the {100} planes, the <100> directions are the easy directions when K_1 is positive while the <110> directions become the easy direction when K_1 is negative (see the problem at the end of the chapter). □

6.4.2.2. *Shape anisotropy*

A polycrystalline specimen, which has no crystallographic anisotropy due to the random orientations of its constituent crystals (grains), can still have magnetic anisotropy due to its shape. This type of anisotropy, called the shape anisotropy, arises due to the demagnetizing field present in a magnetized body.

It is well known that when a body is magnetized, its magnetization causes a field proportional to the magnetization to be set up which opposes the magnetization. This field is called the demagnetizing field

$$H_d = -N_d M \tag{6.8}$$

Here H_d is the demagnetizing field, M is the magnetization and N_d is the proportionality constant, called the demagnetizing coefficient. For a spherical body, the demagnetization coefficient is same in every direction but for a nonspherical object, the magnitude of N_d, and hence the demagnetizing field, is larger along a short axis than along a long axis. As the demagnetizing field opposes the applied field, it is easier to magnetize the body along the long axis.

The energy associated with the demagnetizing field, called the magnetostatic energy, is one half of the dot product of the demagnetizing field and the magnetization vectors

$$E_{MS} = -\tfrac{1}{2}\,\mu_o \mathbf{H_d} \cdot \mathbf{M} \tag{6.9}$$

A model anisotropic shape is an ellipsoid. If the ellipsoid is obtained by rotating an ellipse about its major axis, it is called a prolate ellipsoid while a rotation about the minor axis yields the oblate ellipsoid. Let a prolate ellipsoid (Fig. 6.7) be magnetized making an angle α to the c (major) axis and let N_a and N_c be the demagnetizing factors along the a (minor) and c (major) axes respectively. Its magnetsostatic energy, according to Eq. (6.9) is, in SI units

$$\left| E_{MS} \right| = \tfrac{1}{2}\mu_o[(N_a\mathbf{M_a} + N_c\,\mathbf{M_c}).(\mathbf{M_a} + \mathbf{M_c})] = \tfrac{1}{2}\,\mu_o[N_a M_a^2 + N_c\,M_c^2]$$

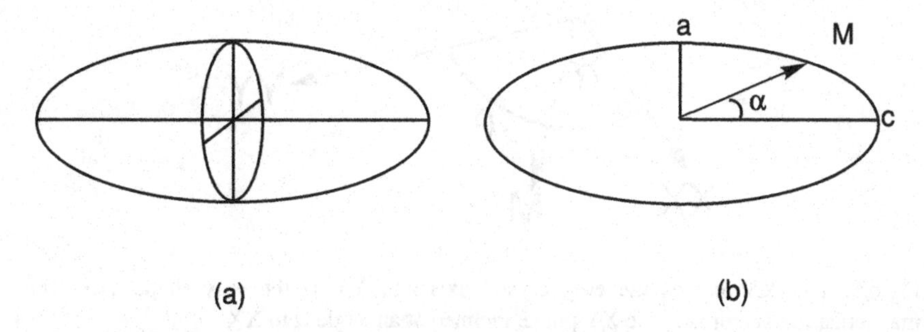

Fig. 6.7. (a) a prolate ellipsoid (b) orientation of the magnetization in the plane of the 'a' (minor) and 'c' (major) axes.

where $\mathbf{M_a}$ and $\mathbf{M_c}$ are the components of $\mathbf{M}$ along the two axes.

Simplifying we get, $E_{MS} = ½\mu_o\,(N_a M^2 \sin^2\alpha + N_c M^2 \cos^2\alpha)$

$$= ½\,\mu_o N_c M^2 + ½\,\mu_o (N_a - N_c)\,M^2 \sin^2\alpha$$

$$= K_o + K_1 \sin^2\alpha \tag{6.10}$$

where the quantities $½\,\mu_o N_c M^2$ and $½\,\mu_o(N_a - N_c)\,M^2$ have been replaced by K_o and K_1 respectively. The magnetostatic energy, therefore, has the same $\sin^2\alpha$ dependence as the magnetocrystalline anisotropy energy

Energy = (*constant*). $\sin^2$ (*angle between* **M** *and the easy axis*)

Note that the value of K_1 depends on the magnetization, M. Depending on the magnitude of $(N_a - N_c)$ and M, the shape anisotropy can be as large as the crystal anisotropy.

A crystal may have both the magnetocrystalline as well as the shape anisotropy with the corresponding easy axes inclined at an angle θ to each other. To illustrate this mixed anisotropy, consider the case where the magnetization vector lies in the plane of the two easy axes and makes an angle α with the magnetocrystalline easy axis (Fig. 6.8).

The energy of the crystal is then

$$E = K_{MC}.\sin^2\alpha + K_{MS}\sin^2(\theta - \alpha) \tag{6.11}$$

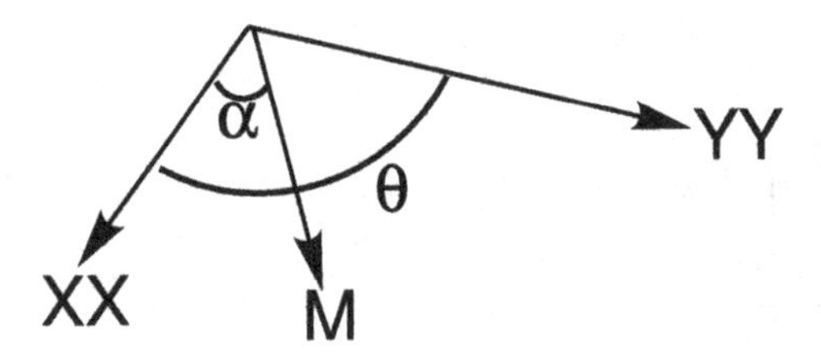

Fig.6.8. The XX axis is the easy crystal axis and YY is the easy shape axis. The magnetization vector is in the XY plane, inclined at an angle α to XX.

To find the resultant direction of easy magnetization, we equate its derivative to zero.

$$\frac{\delta E}{\delta \alpha} = K_{MC}\ 2\sin\alpha\cos\alpha - K_{MS}.2\sin(\theta-\alpha)\cos(\theta-\alpha) = 0$$

or, $$K_{MC}\sin 2\alpha = K_{MS}\sin 2(\theta - \alpha)$$

or $$\alpha = \frac{r}{1+r}\theta \qquad \text{where } r = K_{MS}\,/\,K_{MC}$$

Thus when $r = 1$ *i.e.* two anisotropies are of equal strength, the resultant easy axis bisects the angle between them and has a strength $K = K_{MS} = K_{MC}$ (from Eq. (6.11) above); for $r < 1$, *i.e.* $K_{MC} > K_{MS}$, the resultant easy axis lies closer to XX and has a strength $K > K_{MC}$.

6.4.2.3. *Surface anisotropy*

It was mentioned above that for a nonspherical body, the demagnetizing field is stronger along a shorter axis than along a long axis. An extreme case of this is seen in thin films where the demagnetizing field is very strong along the direction normal to the film plane. As a result, the films get easily magnetized in the plane of the film. For practical applications such as for vertical recording magnetic media and magneto-optic recording, a perpendicular magnetization is needed. One way to achieve this is to exploit the magnetocrystalline anisotropy in thin films of materials such as Co, Cr and $BaFe_{12}O_{19}$ [85, 86]. However, in the thin films, another kind of magnetic anisotropy called the *surface anisotropy* also exists due to the reduced symmetry in the surroundings of the surface atoms. The magnitude of the surface anisotropy constant, K_s, is about ~ 1 mJm^{-2}. The total anisotropy energy is then the sum of those from the volume anisotropy and the surface anisotropy

$$\Delta E = K_v \,.\, \text{volume} + 2K_s \,.\, \text{surface area}$$
$$= K_v At + 2K_s A$$

or,
$$\Delta E/A = Keff \,.\, t$$
$$= K_v t + 2K_s$$

so that
$$K_{eff} = K_v + 2K_s/t$$

Here A and t are the area and the thickness of the film respectively, K_v and K_s are the volume and the surface anisotropy constants and K_{eff} is the effective anisotropy constant. The constant K_v includes contributions from shape, magnetocrystalline and any other volume anisotropy while K_s is from the surface contribution. The factor of 2 arises because of the two surfaces of a free standing film. In a plot of the anisotropy energy per unit area *vs.* the film thickness, the slope is equal to K_v and the intercept at $t = 0$ is $2K_s$.

A surface with high anisotropy energy can be created at the interface of two suitably chosen magnetic and nonmagnetic materials. In a multilayer structure with the films of such materials alternating with each other, a number of such interfaces are created leading to a high total surface anisotropy contribution which can favor the magnetization in a direction perpendicular to the film. This has been demonstrated in a multilayer structure consisting of alternate layers of Co and Pd [87]. An example of this is given in Fig. 6.9. The existence of a small but positive surface anisotropy energy stabilizes the magnetization in a direction perpendicular to the film. The existence of locked–in stresses in thin films is a common occurrence due to the mismatch in the thermal expansion coefficients of the substrate and the film material. The magnetostriction due to these stresses may lead to significant contribution to the magnetic anisotropy energy. The anisotropy energy associated with a stress is given by $K_\sigma = (3/2)\lambda_s\sigma$ where λ_s is the saturation strain due to magnetostriction. Very often the contribution of this term may overwhelm the effect of surface anisotropy.

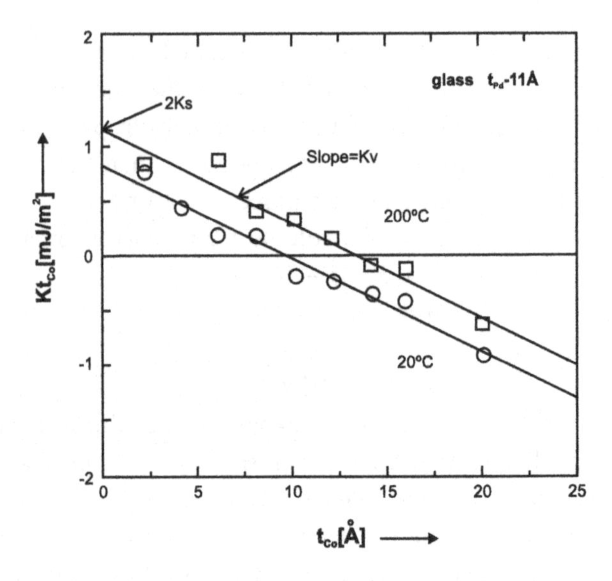

Fig. 6.9. Dependence of Kt_{co} on t_{Co} for polycrystalline Co/Pd multilayers, deposited at T~ = 20 and 200°C (reprinted from [87] copyright (1991) with permission from Elsevier).

6.4.3. *Superparamgnetism*

Superparamgnetism is the name given to the phenomenon observed in ferromagnetic nanoparticles in which the particles behave in many respects like the particles of a paramagnetic material but with high magnetic moments. Figure 6.10 shows schematically the change in coercivity of a single ferromagnetic particle with the change in size. At large sizes, the particle consists of several domains and the remnance and the coercivity are independent of the particle size. Here the magnetization changes by the movement of the domain walls followed by the rotation of the magnetization at high fields. As the particle size is reduced, a stage is reached when the particle consists of a single domain. In a single domain particle, the rotation of the magnetization is the only mechanism available for change in the magnetization. Such rotation has to act against the magnetic anisotropy. This leads to a very high coercivity in the single domain particles. At this size, the coercivity

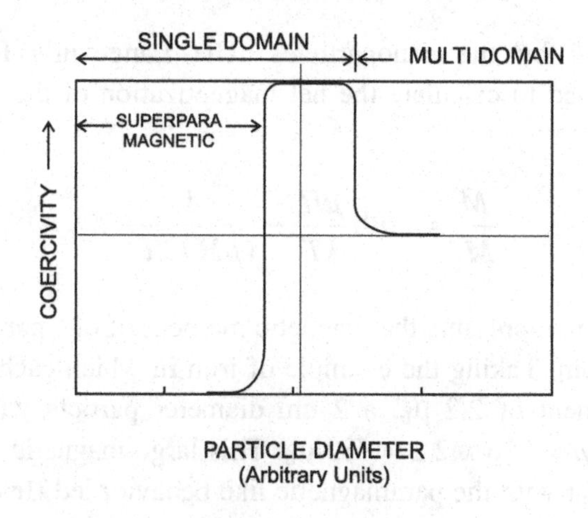

Fig. 6.10. Change in the coercivity of a ferromagnetic particle with size.

increases sharply and stays at a maximum as the particle size is decreased. Below a critical size, it drops sharply to zero as explained below.

The unit of the anisotropy constant K is Jm^{-3}. The total energy due to anisotropy is therefore Kv where v is the volume of the particle. If the size of single domain particle is reduced, the quantity Kv may become small enough to be comparable to the thermal energy kT at some size depending on the temperature. The thermal energy can then cause the magnetization vector in the nanoparticle to flip from one easy direction to another *i.e.* no external field is required to change the magnetization direction and the coercivity becomes zero. The temperature at which this happens is called the *blocking temperature*, T_B.

Now consider an assembly of single domain particles at a temperature $T > T_B$. Each particle has its magnetic moment oriented along one of the easy directions at any given instant. As this direction can be changed by the thermal energy, the magnetization points in different directions in different particles with the assembly as a whole having no net magnetic moment. An external magnetic field, if now applied, will tend to align the magnetization in all the particles along the field direction while the thermal energy will tend to redistribute them amongst the different easy directions. This is similar to the case for a paramagnetic material with atoms replaced by the nanoparticles. The Langevin's formula can therefore be used to calculate the net magnetization of the assembly of nanoparticles

$$\frac{M}{M_0} = \coth\frac{\mu H}{kT} - \frac{1}{(\mu H / kT)} \tag{6.5}$$

Note that μ now represents the magnetic moment of one particle and not that of one atom. Taking the example of iron in which each atom has a magnetic moment of 2.2 μ_B, a 2 nm diameter particle will have 356 atoms and so $\mu = 356 \times 2.2 = 783\ \mu_B$. This large magnetic moment per particle together with the paramagnetic like behavior led Bean *et al.* [88] to coin the term "*superparamagnetism*" to describe this phenomenon.

The blocking temperature, above which the thermal energy can cause flipping of the magnetic moment vector, depends not only on the particle size but also on the time taken for a measurement, as the following discussion shows.

Consider that the rotation or flipping from one easy direction to another of the magnetization vector in a single domain particle is a thermally activated process with an energy barrier equal to Kv. Its frequency can then be written as

$$f = f_0 \exp(-Kv / kT)$$

Here f_0 is a proportionality constant. It is usually assigned a value of $\sim 10^9$ $\sec^{-1}$. If τ is the average time it takes for the magnetization to change from one easy direction to another, then

$$f = \frac{1}{\tau} = f_0 \exp(-Kv / kT)$$

The time τ is called the relaxation time. If the time taken for measurement is smaller than τ, then the superparamagnetic behavior will not be observed because the magnetic moments will not be able to flip within the measurement time. In this case, the particle is ferromagnetic.

For a given measurement time, the minimum temperature for the particle to be paramagnetic can be calculated for a given particle size (*i.e.*

given v). This is the blocking temperature, T_B, for that particle. Thus, for a measurement time of 60 s, and taking f to be 10^9, one gets

$$\frac{1}{60} = 10^9 \exp(-\frac{Kv}{kT_B}) \qquad \text{or} \quad Kv \approx 25\, kT_B$$

and so the superparamagnetism will be observed only above a temperature $T_B = Kv/25$ k.

Figure 6.11 shows schematically the magnetization curves for an assembly of particles at different temperatures across the superparamagnetic region. Above the blocking temperature, at temperatures T_1 and T_2, the curves are characterized by (i) no hysteresis, (ii) no linear region in agreement with the Langevin relation (Eq. 6.5) and (iii) an increasing saturation magnetization with decreasing temperature, again as predicted by Eq. (6.5). The curve at T_3, which is below T_B, is characteristic of ferromagnetic materials.

According to Eq. (6.5), M should be independent of temperature when plotted against H/T *i.e.* all the curves in Fig. 6.11 should collapse into a single curve at temperatures above T_B. This is indeed found to be the case.

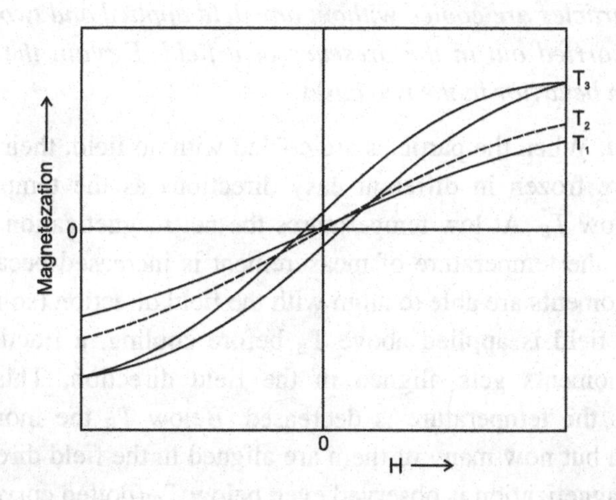

Fig. 6.11. Magnetization curves for an assembly of ferromagnetic particles at temperatures spanning the blocking temperature T_B. Temperatures T_1 and T_2 are above T_B while T_3 is below T_B.

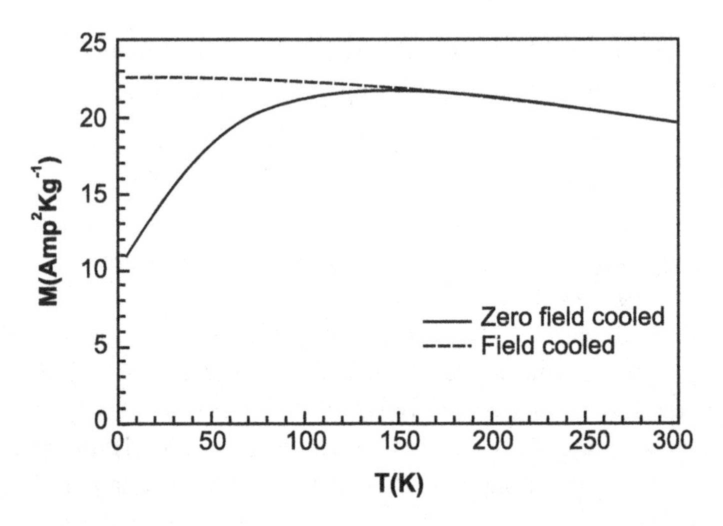

Fig. 6.12. Magnetization at 0.05 *T* for an assembly of γ–F_2O_3 particles; the solid curve is for the sample cooled to a low temperature in zero field and measurements performed while increasing the temperature while in dotted curve a field was on during cooling [89].

Example 6.3. *The figure above (Fig. 6.12) shows the magnetization at a fixed field for an assembly of γ-Fe_2O_3 particles [89] for two cases, one when the particles are cooled without any field applied and two when the cooling is carried out in the presence of a field. Explain the observed difference in behavior in the two cases.*

Explanation. When the particles are cooled with no field, their magnetic moments are frozen in different easy directions as the temperature is lowered below T_B. At low temperatures the net magnetization is low. It increases as the temperature of measurement is increased because more and more moments are able to align with the field direction (solid curve).

When a field is applied above T_B before cooling, a fraction of the magnetic moments gets aligned in the field direction. This fraction increases as the temperature is decreased. Below T_B the moments are frozen again but now many of them are aligned in the field direction and so a high magnetization is observed even below T_B (dotted curve). □

The susceptibility of a superparamagnetic material at low fields can be obtained using the Langevin equation. For low values of $mH/kT = a$, Eq. (6.5) becomes

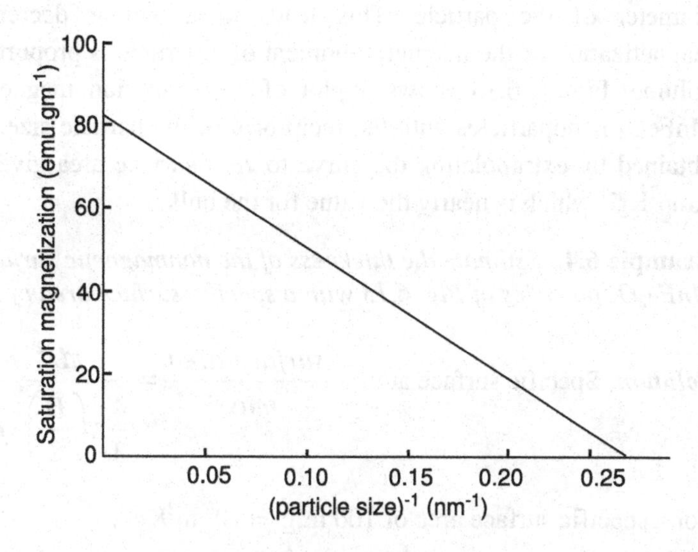

Fig. 6.13. Saturation magnetization vs. reciprocal of particle size for MnFeO4 nanoparticles [90].

$$M = nm[\frac{a}{3} - \frac{a^3}{45} + \frac{2a^5}{945} - ....]$$

$$\approx \frac{nma}{3} = \frac{nm^2H}{3kT} \quad \text{(for small } H\text{)} \qquad (6.12)$$

Hence, susceptibility $\chi = \frac{\delta M}{\delta H} = \frac{nm^2}{3kT}$ (6.13)

Equation (6.13) shows that (i) the susceptibility increases as the square of the magnetic moment m of a particle (ii) it increases as square of the particle volume because $m \sim v$.

As the size of a particle is reduced, its surface becomes increasingly important. So is the case with the ferromagnetic nanoparticles. In the surface layer of a ferromagnetic particle, the spins are not as well aligned as in the interior. This nonferromagnetic layer causes a reduction in the magnetization which becomes more and more significant as the particle size is reduced. If the thickness of this layer is t, then the effective magnetic volume is reduced by a factor = $(D - 2t)^3/D^3$ where D is the diameter of the particle. This leads to a similar decrease in the magnetization as the magnetic moment of a particle is proportional to its volume. Figure 6.13 shows a plot of the saturation magnetization of $MnFeO_4$ nanoparticles with the reciprocal of the particle size. The value obtained by extrapolating the curve to zero surface area gives $M_S = 81$ Amp^2Kg^{-1} which is nearly the value for the bulk.

Example 6.4. *Estimate the thickness of the nonmagnetic surface layer of $MnFe_2O_4$ particles of Fig. 6.13 with a specific surface area of $100\ m^2g^{-1}$.*

Solution. Specific surface area = $\dfrac{surface\ area}{mass} = \dfrac{\pi D^2}{\frac{4}{3}\pi\left(\frac{D}{2}\right)^3.\rho} = \dfrac{6}{\rho D}$

For a specific surface area of $100\ m^2g^{-1} = 10^5\ m^2Kg^{-1}$,

$$D = \frac{6}{\rho s} = \frac{6}{10^5 \times 5000} = 12\ nm$$

(taking density of $MnFe_2O_4$ to be $5000\ Kg.m^{-3}$).
From Fig. 6.13, the saturation magnetization, M_s, at 12 nm ($D^{-1} = 0.0833$) is 55 emu g^{-1}.
If t is the thickness of the nonmagnetic surface layer, and taking M_s of the bulk to be 81 emu.g^{-1} from Fig. 6.13, we get

$$\left(\frac{D-2t}{D}\right)^3 \times 81 = 55\ \text{emu.g}^{-1}$$

or $t = 0.73$ nm □

Example 6.5. *A magnetic hard disk contains a total of 10^{11} single domain bits of size 30 x 30 x30 nm^3. The crystalline anisotropy energy is $K = 5 \times 10^6$ ergs.cm^{-3}. What is the reversal frequency of the bits for the whole disk (i.e. the number of bits reversing per second and losing the* *information) at 300 K? How would it change if the bit volume is reduced by a factor of 50 and the total number of bits increased to 10^{12}? Will the error rate be significant in any of the two cases? Take the frequency factor f_0 to be $10^{10}s^{-1}$.*

Solution. Frequency of reversal for a single bit $f = f_o \exp(-Kv/kT)$
Taking f_0 as $10^{10}s^{-1}$, we get for the first case, $f = 10^{(-1406)}$
Frequency of reversal for the whole disk = $10^{(-1406)}$ x 10^{11} = 10^{-1395}.
Hence it takes 10^{1395} s for a single bit in the disk to reverse and lose information. This is for ever, considering that the age of the universe is 5 x 10^{17} s.
For the second case f = 4.8 x 10^{-19} s^{-1}.
For the whole disk, the reversal frequency =10^{12}x4.8x10^{-19}=4.8x$10^{-7}$$sec^{-1}$.
The time for one bit to reverse and lose the information = 2x10^{6} s = 23.2 days.
This may be a significant error rate as the information usually needs to be stored for much longer periods. □

6.4.4. *Superparamagnetic materials: Ferrofluids*

A well established application of ferromagnetic nanoparticles is in ferrofluids. A ferrofluid is a stable dispersion of ferromagnetic particles (*e.g.* γ Fe_2O_3, Fe_3O_4) in a liquid. The liquid is usually oil or water. A typical ferrofluid usually contains 3 to 8 vol. % nanoparticles and about 10 % surfactant. The particles in the ferrofluid remain suspended due to the Brownian motion and are prevented from coagulation by the surfactant coating on their surfaces. In the absence of a magnetic field, the net magnetic moment of the assembly is zero because their moments point in random directions. In a magnetic field gradient, the whole fluid moves to the regions of the highest flux. A magnetic field can therefore be used to position and hold the ferrofluid at a desired location. An application of this effect is the use of ferrofluids as bearings for a rotating shaft (Fig. 6.14).

Another important property of the ferrofluids is a significant increase in viscosity in the presence of a magnetic field. As the liquid is sheared, the particles tend to keep their moments aligned in the field direction and offer resistance to the shear force. This property is used for

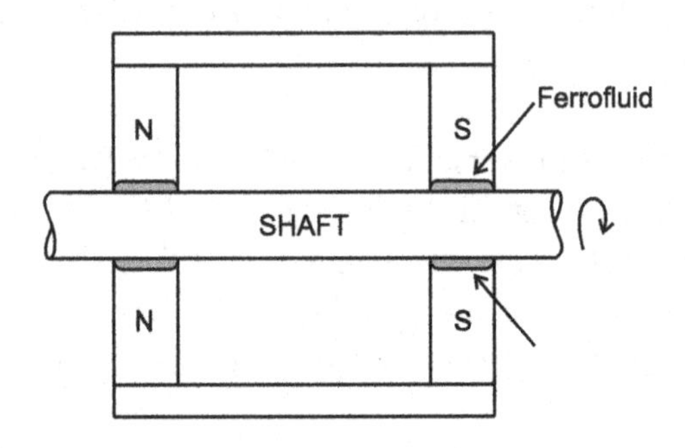

Fig. 6.14. Ferrofluid as bearing for a rotating shaft.

damping applications in the shock absorbers of automobiles and in CD and DVD player systems. A commercially successful application of ferrofluids is in loudspeakers. By placing the voice coil of the loudspeaker inside the ferrofluid surrounded by the field of a permanent magnet, three effects are achieved simultaneously: (a) positioning of the coil in the center position (b) damping of the membrane and (c) conducting away of the heat generated by the coil (Fig. 6.15).

The ferrofluids have found several applications in addition to those indicated above *e.g.* for forming seals between regions of different pressures, magnetic ink printing, electromagnetic drug delivery, *etc.*

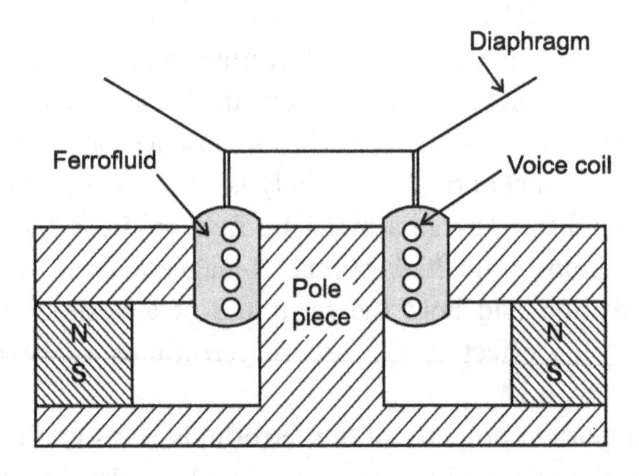

Fig. 6.15. Use of a ferrofluid in a loudspeaker.

6.4.5. *Nanostructures exploiting the shape anisotropy*

The shape anisotropy provides a powerful means to control the magnetic properties of nanostructures. An example is shown in Fig. 6.16 which shows the results for nanolithographically produced rods of cobalt and nickel. These rods were about 1 μm long and 35 nm thick. Despite being polycrystalline, the shape anisotropy forces the magnetization in all the grains to be along the long axis of the rod even in the absence of an external magnetic field *i.e.* effectively the bar is a single domain. There are only two stable states of magnetization, both of them along the axis of the bar but in opposite directions. The field required to switch the magnetization from one easy direction to another depends strongly on the aspect ratio of the bar as it changes the shape anisotropy (Fig. 6.16).

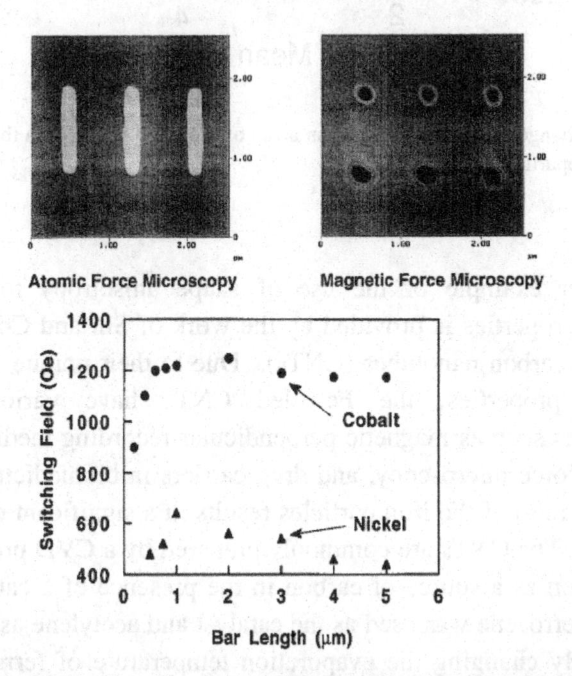

Fig. 6.16. (Top) Atomic force image and the magnetic force image of three single domain nickel bars that are 100 nm wide, 1 μm long and 35 nm thick; the magnetic force microscopy shows that there are two poles at the opposite ends of the bars and no poles in between. (Bottom) Dependence of the switching field of the bars of Ni and Co on the bar length (reprinted with per permission from [91] copyright (1996) American Institute of Physics).

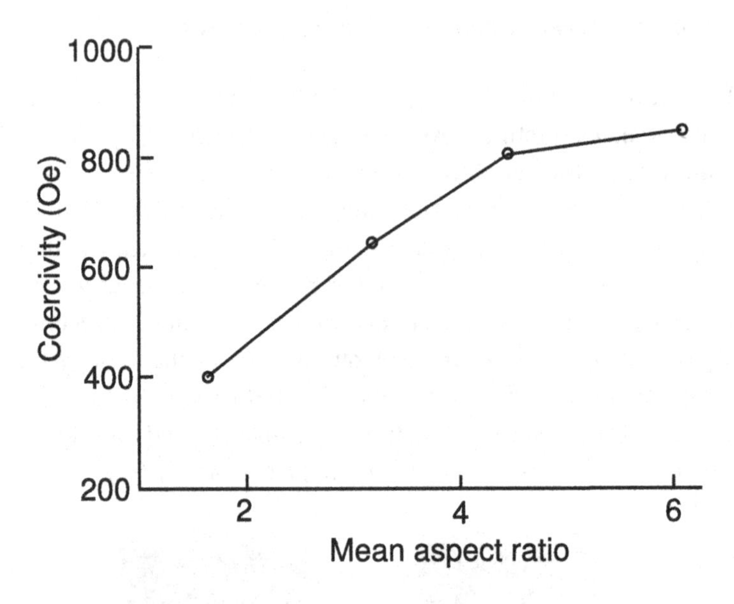

Fig. 6.17. Change in the coercivity of an array of Fe filled CNTs with the aspect ratio of the iron nanoparticles [92].

Another example of the use of shape anisotropy to control the magnetic properties is provided by the work of Shi and Cong [92] with iron filled carbon nanotubes (CNTs). Due to their unique electrical and magnetic properties, the Fe-filled CNTs have various potential applications such as magnetic perpendicular-recording media, probes for magnetic force microscopy, and drug carriers in biomedicine. Changing the aspect ratio of the iron particles results in a significant change in the coercivity. The CNTs are commonly prepared by a CVD process using a hydrocarbon as a source of carbon in the presence of a catalyst. In this case, the ferrocene was used as the catalyst and acetylene as a source of carbon. By changing the evaporation temperature of ferrocene, CNTs filled with iron particles of different aspect ratios were obtained. Figure 6.17 shows the effect of the aspect ratio on the coercivity of the filled CNTs.

6.4.6. *Spin dependent scattering: The Giant Magnetoresistance (GMR)*

The term Giant Magneto Resistance (GMR) refers to a large increase in the resistance of certain ferromagnetic thin film structures on changing the applied magnetic field. Magnetic thin film structures consist of multilayers, called *heterostructures*, in which the different layers can all be ferromagnetic or a combination of ferromagnetic and nonmagnetic or antiferromagnetic. When the layers are all epitaxial, the structure is called a *superlattice*.

The phenomenon of GMR is quite different from the ordinary magneto- resistance which is the property of all metals and refers to the increase in the resistance of metals in the presence of an external magnetic field. In this case, the magnetic field causes the electrons to move in a helical trajectory with the magnetic field direction as the helix axis. This increases the path length and, so, the number of scattering events. However, the field must be strong enough to curve the helix to have a pitch smaller than the mean free path of the electrons. All metals show magnetoresistance but, except for the ferromagnetic metals, the effect is quite small. In the ferromagnetic metals the change in resistance can be up to 2% when saturation fields are used. The early read heads for the magnetically stored data were based on the magnetoresistance of the Ni-Fe alloy thin films.

The GMR effect, on the other hand, is based on the spin dependent scattering of the electrons from the interface of a normal and a ferromagnetic metal. One of the first reports of this phenomenon is from Baibitch *et al.* in 1988. Their structure consisted of a stack of 30 to 60 bilayers of Fe and Cr, *e.g.* a configuration with 30 bilayers with each bilayer consisting of a 30 A^o layer of Fe next to a 9 A^o layer of Cr is represented by (Fe 30 A^o / Cr 9 A^o)$_{30}$. The change in the resistance of the stack depended on the thickness of the Cr layer and changed by as much as a factor of 2 on applying a magnetic field. Their results are shown in Fig. 6.18. The effect has come to be known as giant magneto resistance (GMR).

The main mechanism to explain the GMR in the transition metal based layers is considered to be the spin-dependent scattering of the electrons at the interfaces. Consider a configuration consisting of two ferromagnetic layers separated by a thin nonmagnetic metal layer. When an electron in the nonmagnetic layer approaches the upper magnetic

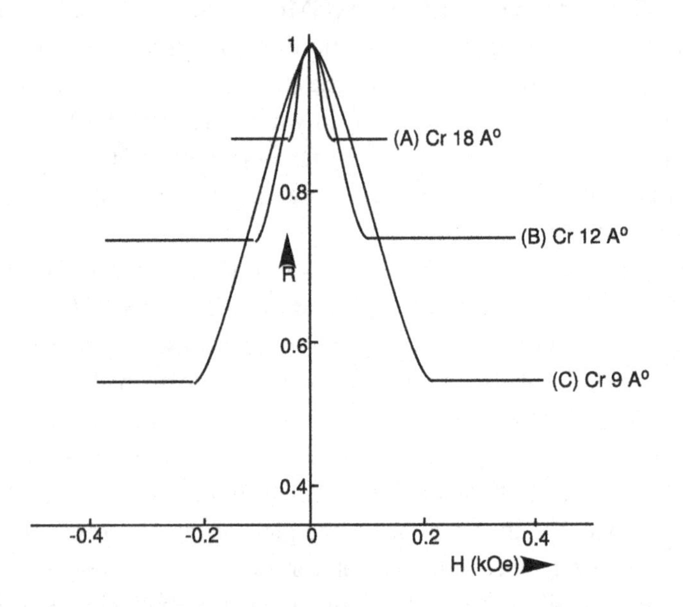

Fig. 6.18. Relative resistance R (resistance/resistance at zero field) *vs.* magnetic field for three Fe/Cr superlattices at 4.2 K. A, B and C denote superlattices with the arrangements $(Fe\ 30\ A^o/Cr\ 18\ A^o)_{30}$, $(Fe\ 30\ A^o/Cr\ 12\ A^o)_{35}$ and $(Fe\ 30\ A^o/Cr\ 9\ A^o)_{60}$ respectively. The current and the applied field are along the same [110] axis in the plane of the layers. Note that the saturation field H_s and the drop in the resistance at H_s increase as the Cr layer thickness is decreased from 18 A° to 9 A° [93].

layer, it is scattered at the interface if its spin is antiparallel to the magnetic layer. In this case the resistance will be high. On the other hand, if the spin of the electron is parallel to the direction of magnetization, its chances of getting scattered are low and the resistance is reduced. The resistance therefore can be controlled by controlling the spin of the electrons incident on the interface and the direction of magnetization of the top layer. A method is therefore needed to control both.

In a ferromagnetic metal, the number of electrons with their spin in the direction of magnetization (say the "up ↑" direction) significantly exceeds the number of electrons with spin in the "down ↓" direction.

This phenomenon is referred to as "spin polarization". The strength of the polarization, P, is given by $P = (n\uparrow - n\downarrow)/(n\uparrow + n\downarrow)$ where $n\uparrow$ and $n\downarrow$ denote the number of electrons with spin in the up and down directions respectively. In metals such as Fe, Co, Ni and their alloys, P ranges from 0.4 to 0.5.

In the arrangement shown in Fig. 6.19, the electrons emanating from the bottom ferromagnetic layer, will have their spins predominantly parallel to the direction of magnetization of that layer. Provided that the spin can be maintained during their transport, the resistance of the device will be controlled by the direction of the magnetization of the top layer as explained above.

The relative orientation of the two magnetic layers can be controlled in several ways. When the two ferromagnetic layers are separated by a thin nonmagnetic layer, as in Fig. 6.19, a phenomenon known as interlayer exchange coupling occurs. By controlling the thickness of the nonmagnetic spacer layer, this coupling can be chosen to be ferromagnetic, antiferromagnetic or zero *i.e.* the moments in the two magnetic layers can be parallel, antiparallel or uncorrelated [94]. In the work of Baibich *et al*, shown in Fig. 6.18, there is a strong antiferromagnetic coupling when the Cr layer thickness is 9 A°. This necessitates the application of a large field to make the magnetizations in the two layers parallel and thereby reduce the resistance.

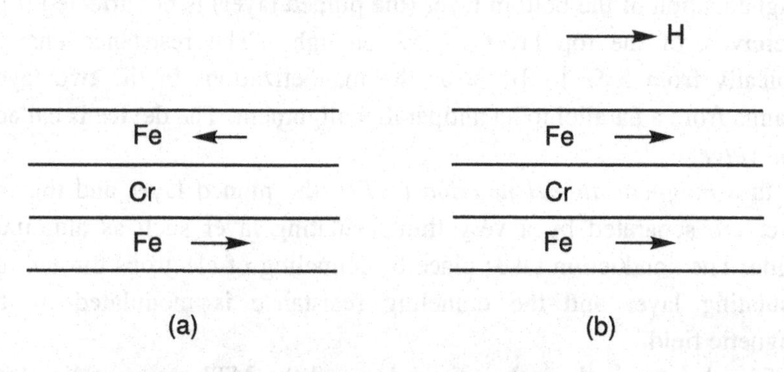

Fig. 6.19. Two layers of a ferromagnetic metal separated by a layer of a nonmagnetic metal (Cr). In (a) the spins in the two Fe layers are antiparallel while in (b) they are made parallel by an applied magnetic field H.

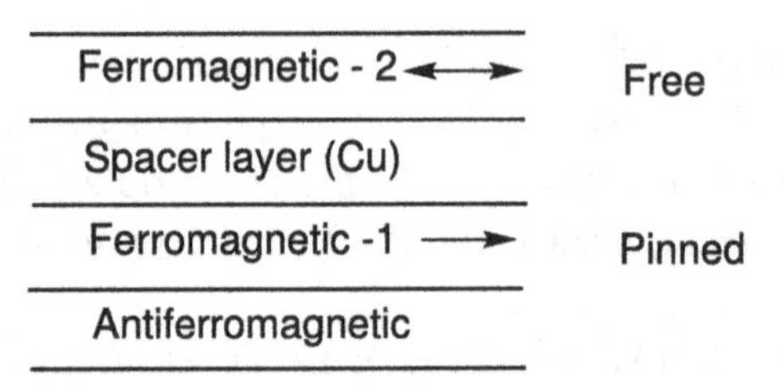

Fig. 6.20. The antiferromagnetic layer pins the magnetization in the ferromagnetic-1 layer next to it. The magnetization direction in the top ferromagnetic layer can be changed by an external magnetic field if its coercivity is lower than that of the lower layer.

Another scheme which can be used to control the relative orientation of the magnetizations in the two ferromagnetic layers is shown in Fig. 6.20. In this case, an antiferromagnetic layer is first deposited on the substrate before depositing the first ferromagnetic layer followed by the normal layer and the second ferromagnetic layer as before. The antiferromagnetic layer interacts with the ferromagnetic layer next to it and forces its magnetization in a certain direction *i.e.* the magnetization direction in the lower ferromagnetic layer is effectively pinned due to the antiferromagnetic layer next to it. The magnetization direction of the top layer, called the free layer, can now be controlled by means of an external magnetic field. Because of the pinning effect, the direction of magnetization of the bottom layer (the pinned layer) is not affected if the coercivity of the top layer is low enough. The resistance changes typically from 5 % to 10 % as the magnetizations in the two layers change from a parallel to an antiparallel alignment. The device is called a *spin valve*.

In a *magnetic tunnel junction (MTJ)*, the pinned layer and the free layer are separated by a very thin insulating layer such as aluminum oxide. The conduction takes place by tunneling of electrons through the insulating layer and the tunneling resistance is modulated by the magnetic field.

The devices such as the spin valve and the MTJ are examples from the new field of *spintronics* in which the spin of the electrons is used to develop new devices. In the traditional electronics the spin of the electron is completely ignored. But if one can control the spin of the electrons, inject these *spin polarized* electrons into a circuit, control the lifetime for which the spin is maintained and otherwise manipulate the spin using stimulants like a magnetic field or light then the spin becomes available as an additional variable. The field of spintronics seeks to use the spin of electrons to develop new nonvolatile devices in which logic operations, storage and communication are combined.

Problems

(1) What is the flux density produced in vacuum by a field of 1 Am^{-1} according to Eq. (6.2)? Using the conversion factors in Table 6.1, show that it is equivalent to 1 line cm^{-2} from a field of 1 oersted.

(2) In a thin layer on the surface of a ferromagnetic particle the spins are not well aligned. This is equivalent to the presence of a thin nonmagnetic layer at the surface. Because of this the magnetization of the nanoparticles is effectively reduced in proportion to the volume of the nonmagnetic surface material. Calculate the bulk saturation magnetization for $MnFe_2O_4$ if its powder sample with a surface area of 100 m^2g^{-1} has a saturation magnetization of 45 amp^2kg^{-1}. Assume that the particles have a 1 nm thick nonmagnetic surface layer and the density of $MnFe_2O_4$ is 5000 kg m^{-3}.

(3) The magnetcrystalline anisotropy energy for a hexagonal crystal is given by

$$E = K_0 + K_1\sin^2\theta + K_2\sin^4\theta + ..$$

Show that the c axis is the only axis of easy magnetization when K_1 is positive and $K_2 > - K_1$. In which materials is this condition satisfied?

(4) For the case when the crystalline easy axis and the shape easy axis are at right angles to each other, show that the resultant easy axis is stronger of the two. Also show that there is no anisotropy if the two axes have equal strengths.

(5) Plot schematically the variation in the anisotropy constant K with direction in the (001) plane for a cubic crystal for the following cases: (i) K_1 positive (ii) K_2 negative. Which are the easy directions in the two cases?

(6) Compare the frequency of flipping of spins in nanoparticles of 10 and 20 nm diameter at 300 K. Given, crystalline anisotropy energy constant $K = 0.25 \times 10^6$ Jm^{-3} and the frequency factor $f_0 = 10^9$ s^{-1}.

Chapter 7

Zero Dimensional Nanostructures IV Colloids and Colloidal Crystals

7.1 Introduction

Colloids can be defined as systems which are microscopically heterogeneous and in which one component has at least one of its dimensions in the range between 1 nm and 1 μm. Examples of colloids are milk, mists, smoke *etc.* Many industrial products such as paints, pastes, foams contain colloids. For the purpose of this chapter, a model colloidal system is a suspension of particles having diameters ranging from 1 nm to 1 μm in a liquid. Such suspensions are of interest in nanotechnology in the preparation of nanoparticles, their modification and their assembly to yield nanomaterials and nanodevices.

One kind of force which is always present between any two particles in a suspension is the van der Waal's force. This force is attractive in nature and varies as d^{-n} where d is the distance between the two particles and n is equal to ~ 2 for two equal spheres. Thus the van der Waal's attraction between particles can be significant at large distances (of the order of their own dimensions). As the submicron particles suspended in a liquid constantly undergo Brownian motion, they will occasionally come very close to each other if there is no repulsive barrier present. The particles would agglomerate due to the van der Waal's attraction and eventually settle down *i.e.* the stability of the suspension would be lost.

In order for the particles to remain suspended (*i.e.* for the dispersion to be stable) for any useful length of time, the presence of a repulsive force between them is essential. Such repulsive interaction is produced either by the presence of electrical charges or by the adsorption of some polymers on the particle surface. The stabilization of dispersion by these techniques is called *electrostatic stabilization* and *steric stabilization* respectively.

In this chapter the interactions responsible for the stability of a suspension of nanoparticles are first discussed. Secondly, the example of colloidal crystals is used to illustrate the use of nanoparticles in the fabrication of nanostructures.

7.2 The van der Waal's forces

7.2.1 *Introduction*

When two molecules approach each other, an attractive force, called the van der Waal's force, begins to act between them when the separation between the molecules reaches about 1 nm. The van der Waals forces are weak forces as compared to the forces involved in the covalent or ionic bonds (Table 7.1). However, the latter come into play only at much closer distances of the order of 1A°. In addition to the stability of a suspension of particles, the van der Waals interactions play an important role during the fabrication and manipulation of nanostructures. The hydrophobic interactions (which are crucial in protein folding, in the holding together of the cell membranes and in other biological structures) are dependent on the van der Waals interactions. Such biological structures serve as models for the fabrication of useful nanostructures by supramolecular chemistry. As another example, the lateral van der Waals interactions between nanowires (or nanotubes) cause them to grow in a self organized parallel aligned fashion *e.g.* growth of parallel aligned wurtzite ZnO nanowires and carbon nanotubes. Formation of the

Table 7.1. Comparison of the various interactions with the van der Waals interaction

Type of bond	Magnitude of interaction (eV)
Covalent and ionic bonds	2 – 10
Hydrogen bond	0.05 – 0.3
van der Waals	0.004 – 0.04

colloidal crystals from their suspensions occurs because of a balance between the attractive van der Waals interactions and the repulsive interactions due to the electrical double layer or the steric forces; such a balance is also believed to play a crucial role in the growth of curved forms of mesoporous silica from aqueous solutions and in the biomimetic growth. To understand these forces we first consider the forces between two molecules.

The long range (~1 nm) van der Waals interactions between two molecules actually comprise of three different interactions: (i) London dispersion interaction (ii) dipole – dipole or Keesom interaction and (iii) dipole – induced dipole or Debye interaction. The last two arise when one or both the molecules have a permanent dipole moment. The dispersion force arises due to the interaction between the fluctuating dipoles on the two molecules and is the most important of the three — the other two are, in most cases, minor by comparison. All the three forces together are referred to as van der Waals force.

The explanation for the long range attractive force between molecules was first given by London. Due to the constant motion of the electrons in a molecule, the electron distribution in it at any instant is asymmetric and is constantly changing. Due to this the molecule behaves as a fluctuating dipole. This fluctuating dipole affects the fluctuating dipole moment of a nearby molecule through electromagnetic interaction. The net result is an attractive force between the two molecules (Fig. 7.1). This attractive force is called the dispersion force because London made use of the dispersion relation between the wavelength and frequency. For two molecules, it is given by

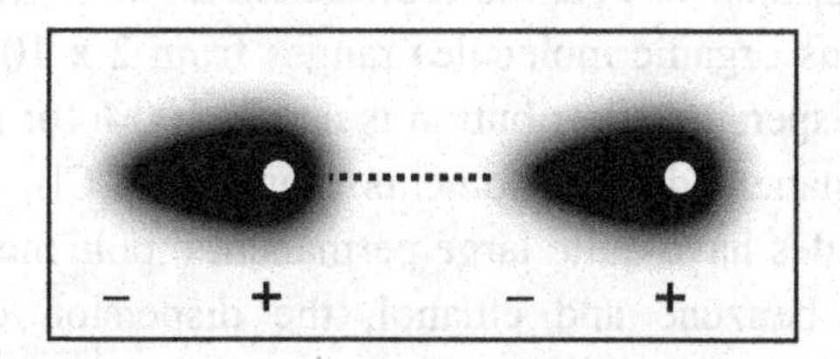

Fig. 7.1. Fluctuation of the electron clouds around the nuclei of two atoms induces temporary local positive and negative charges which cause an attraction between the molecules.

$$\phi = -\frac{3}{2}\left(\frac{h\nu_1 h\nu_2}{h\nu_1 + h\nu_2}\right)\frac{\alpha_1\alpha_2}{r^6} = -\frac{\lambda_{12}}{r^6} \tag{7.1}$$

Here the subscripts 1 and 2 refer to the two molecules at a separation r, $h\upsilon$ is the characteristic energy of a molecule (nearly equal to the ionization energy) α is the polarizability and λ_{12} is the London constant. Note that the unit of ϕ is Joule and that of λ is J m^6.

If the relevant quantities ($h\upsilon$ and α) for two different molecules are known (usually available in tables), then λ_{12} can be approximated as

$$\lambda_{12} = (\lambda_{11}\,\lambda_{22})^{1/2} \tag{7.2}$$

A well known empirical equation for the potential energy Φ of two molecules separated by a distance r is the Lennard Jones potential

$$\phi(r) = 4\phi_0\left[\left(\frac{x_0}{r}\right)^{12} - \left(\frac{x_0}{r}\right)^{6}\right] \tag{7.3}$$

Here the negative term corresponds to the attractive dispersion force while the positive term with inverse 12th power dependence becomes dominant when the molecules approach so close that their electron clouds begin to interpenetrate. A plot of this function is shown in Fig. 7.2. The parameter Φ_0 is the minimum value of Φ which occurs at $r = r_0 = (2)^{1/6}x_0$.

The dispersion force between two similar molecules is given by $\Phi = -\lambda_{11}/r^6$ (Eq. 7.1). If the other two contributions, Keesom and Debye are also included, then this can be represented as $\Phi = -\beta_{11}/r^6$. The value of β_{11} for various organic molecules ranges from 2 x 10^{-77} Jm^6 to 15 x 10^{-77} Jm^6. The dispersion contribution is nearly 100% for molecules with little or no permanent dipole moments (benzene, CCl_4, toluene). Even when the molecules have quite large permanent dipole moments, such as t-butanol, chlorobenzene and ethanol, the dispersion contribution is nearly 50% or more. Only in case of watér, with a very high permanent dipole moment, does the dispersion contribution go down to a low value of 10%.

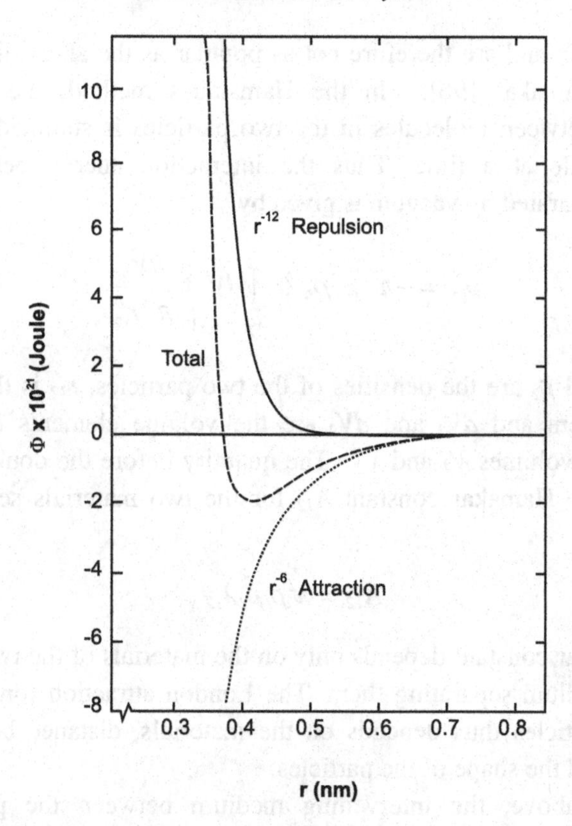

Fig. 7.2. Potential energy between two methane molecules according to the Lennard Jones potential.

7.2.2 *The van der Waal's forces between particles*

The van der Waal's force between the particles is only due to the London dispersion interaction between fluctuating electron cloud on each particle. Even in particles consisting of dipolar molecules, the dipole–dipole forces are negligible as the molecules line up to minimize the energy and the net dipole moment on the particle is zero.

The fluctuation in the electron cloud of a particle is reflected in the frequency dependence of its dielectric constant. Thus the dispersion force between two particles can be calculated if this dependence is known. This has been done by Lifshitz [95]. However, these calculations are very complex and are therefore not as popular as the alternative method used by Hamakar [96]. In the Hamakar's method, the dispersion attraction between molecules in the two particles is summed up taking one molecule at a time. Thus the interaction energy between two particles separated by vacuum is given by

$$\phi_{12} = -\pi^2 \rho_1 \rho_2 \lambda_{12} \oint_{V_1} dV_1 \oint_{V_2} \frac{dV_2}{\pi^2 r_{12}^6} \tag{7.4}$$

Here, ρ_1 and ρ_2 are the densities of the two particles, r_{12} is the distance between them and dV_1 and dV_2 are the volume elements of the two particles of volumes V_1 and V_2. The quantity before the double integral is called the Hamakar constant A_{12} for the two materials separated by vacuum

$$A_{12} = \pi^2 \rho_1 \rho_2 \lambda_{12} \tag{7.5}$$

The Hamakar constant depends only on the materials of the two particles and the medium separating them. The London attraction force between the two particles thus depends on the materials, distance between the particles and the shape of the particles.

In the above, the intervening medium between the particles is vacuum. For any other medium (*e.g.* water) the Hamakar constant can be obtained as follows [97]. Consider a pseudochemical reaction in which solid particle 1 and solid particle 3 with their associated (satellite) solvent particle 2 react to form a doublet of 1 and 3 and a doublet of the solvent particles (Fig. 7.3). The change in energy for this reaction is given by

$$\Delta\Phi = \Phi_{13} + \Phi_{22} - \Phi_{12} - \Phi_{32}$$

All the terms in the above equation depend in an identical manner on the size and distance parameters; they differ only because of the material properties which are contained in the Hamakar constants. So one can write

$$A_{123} = A_{13} + A_{22} - A_{12} - A_{32}$$

where A_{123} is the Hamakar constant for the particles of materials 1

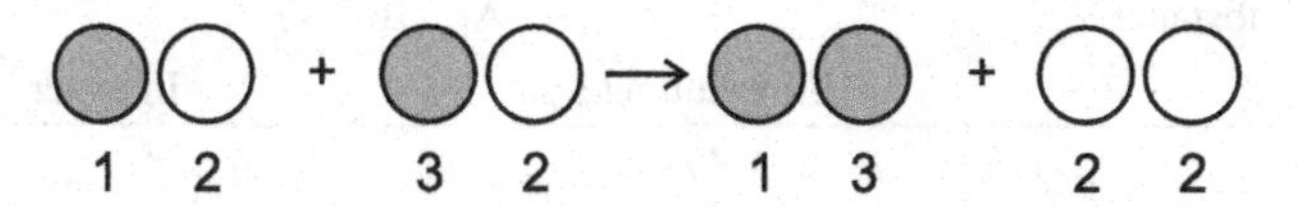

Figure 7.3. A pseudochemical reaction in which the solid particles 1 and 3 (shaded circles) with their associated solvent particles react to form a doublet of 1 and 3 and a doublet of the solvent particles [97].

and 3 separated by a liquid 2. If the two particles are of the same material, then

$$A_{12} = A_{11} + A_{22} - 2A_{12}$$

If the approximation (7.2) is used, then

$$A_{12} \sim (A_{11}A_{22})^{1/2}$$

so that $A_{123} = (A_{11}A_{33})^{1/2} + A_{22} - (A_{11}A_{22})^{1/2} - (A_{22}A_{33})^{1/2}$

$$= (A_{11}^{1/2} - A_{22}^{1/2})\,(A_{33}^{1/2} - A_{22}^{1/2})$$

If A_{22} lies between A_{11} and A_{33}, then A_{123} is negative so that the particle of 1 and 3 will repel each other in medium 2. This will result for example, in the detachment of particles from a surface when immersed in medium 2.

The Hamakar constant A_{121} can also be similarly approximated as

$$A_{121} \sim (A_{11}^{1/2} - A_{22}^{1/2})^2$$

Thus A_{121} is always positive and so two similar particles will always have an attractive dispersion force.

The values of Hamakar constant for different materials and intervening media are readily available in literature [21, 98]. Some values are listed in Table 7.2. It can be seen that conductors with free electrons and high polarizabilities have the highest values while the insulators have the lowest values.

Table 7.2. Hamakar constants for some materials in air and water

Substance	A_{11} (10^{-30} J)	
	In vacuum or air	In water
Graphite	47.0	
Gold	45.5	33.5
Copper	28.4	17.5
Rutile (TiO_2)	43	26
Zirconia (ZrO_2)	27	13
Iron oxide (Fe_3O_4)	21	
Magnesia	10.5	1.6
Alumina (Al_2O_3)	15.4	4
Mica	1.0	2.3
Quartz	7.93	1.08
Silicon	25.5	11.9
Polyethylene	10.0	0.4
Polyvinyl alcohol	8.84	0.54
Natural rubber	8.58	
Polyethylene oxide	7.5	
Toluene	5.4	

7.2.3 *Effect of the particle geometry*

The dispersion force between the particles or other bodies depends on the material, shapes of the bodies and the distance between them (Eq. 7.4). The integrations in Eq. (7.4) have been carried out for several geometries. The results for three important cases are given below.

7.2.3.1 *Plane parallel plates*

For two thick plates separated by a distance R, the free energy per unit area is

$$\Delta G = -A/12R^2 \tag{7.6}$$

7.2.3.2 *Two equal spheres*

For two equal spheres of radius a with their centers separated by a distance R, the total free energy is

$$\Delta G = -\frac{A}{6}\left[\frac{2a^2}{R^2-4a^2}+\frac{2a^2}{R^2}+\ell n\left(1-\frac{4a^2}{R^2}\right)\right] \tag{7.7}$$

When the spheres are very close, *i.e.* $R \sim 2a$, the above expression reduces to

$$\Delta G = -\frac{Aa}{12(R-2a)} \tag{7.7A}$$

7.2.3.3 *Sphere and a plate*

For a sphere and a thick plate, the total free energy change is

$$\Delta G = -\frac{A}{6}\left[\frac{a}{R-a}+\frac{a}{R+a}+\ell n\frac{R-a}{R+a}\right]$$

Here R is the distance from the center of a particle of radius a to the plate. When the particle is very close to the plate surface, then $R = a + \Delta x$ and the above equation reduces to

$$\Delta G = -\frac{Aa}{6(R-a)} = -\frac{Aa}{6\Delta x} \tag{7.8}$$

The above equations show that for macroscopic bodies such as particles and plates the van der Waal's attraction decays much more slowly (*1/distance* or *1/distance*^{2}) as compared to that for the molecules (*1/distance*^{6}). Thus the range of attractive force between particles is quite large which causes rapid flocculation of a suspension in the absence of adequate repulsive interaction. Furthermore, it can also be inferred from the above equations that small particles are attracted more to larger particles than to each other. This is because a large particle can be viewed as behaving as a plate with respect to a small particle and, as Eqs. (7.7A) and (7.8) show, the attractive energy between a plate and a particle is twice as large as that between two particles.

How does a gecko (the household lizard) stick to a vertical surface? The answer is the van der Waal's attraction. Each foot of a gecko has about 500,000 hairs or *setae*. Each seta ends in 1000 fine tips, 200 nm across. Calculations and experiments [99] show that each seta exerts enough force to lift an ant (20 mg). Thus the maximum potential force that can be supported by the 2000,000 setae on the four feet of gecko is 40 kg while the weight of a gecko is only 50 – 150 g!

7.2.4 *Charge on the particle surfaces*

The stabilization of a suspension primarily due to the presence of charges on the particle surface is called *electrostatic stabilization*. Particles of oxides such as silica (SiO_2) almost invariably have charges on their surfaces when suspended in water. This is easily demonstrated by viewing such a suspension through an optical microscope while a DC voltage is applied across the suspension (Fig. 7.4). The particles are found to move towards one of the electrodes and the direction of motion changes on reversing the field indicating that all the particles have similar charge. This phenomenon is called electrophoresis.

As the suspension has to be electrically neutral as a whole, it must also contain counterions having a charge opposite to that on the particles so that the total net charge in the suspension is zero. Why do the particles and the counterions not attract each other and get neutralized? The answer to this question is found by answering another question: why do the ions of opposite charge in an aqueous solution not get similarly neutralized? This is because the ions are hydrated and have a layer of water molecules bound to them. If the thickness of this water layer is such that the attractive energy between the ions does not exceed their thermal kinetic energy at the point of closest approach, then the ions can overcome the attraction and stay separated. In a like manner, the surfaces of particles and the counterions in a suspension are hydrated by layers of water molecules and so the two do not neutralize each other.

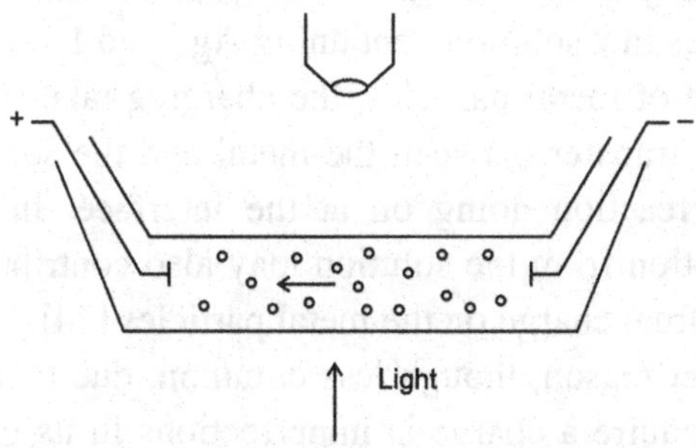

Fig. 7.4. Particles of an oxide suspended in water are observed to move towards one of the electrodes when an electric field is applied; here the particles are negatively charged and move towards the positive electrode.

Nevertheless, the attraction between the charged particles and the oppositely charged counterions causes the latter to form a layer around the particles. Due to the thermal energy, the layer of the counterions is not a compact layer but is a diffuse layer. The net charge in the diffuse layer decays to zero at some distance away from the particle. It is the repulsion between the diffuse layers of the counterions surrounding each particle which is responsible for the particles staying dispersed in a suspension. The repulsion extends to larger distances as the thickness of the diffuse layer increases.

There are several mechanisms by which the surface of a particle can acquire charges.

(i) In an aqueous suspension, the particles of oxides (or metal particles having an oxide layer) can preferentially absorb OH^- or H^+ ions

$$MOH + OH^- \rightarrow MO^- + H_2O$$

$$MOH + H^+ \rightarrow MOH_2^+$$

Depending on the pH, the particles can acquire either a net negative or a net positive charge.

(ii) When a solution contains ions which make up the solid, these ions can adsorb on the particle surface in unequal amounts, imparting a net charge to the particle as in the case of AgI particles in a solution containing Ag^+ and I^- ions.

(iii) In case of metal particles, the charging takes place due to the charge transfer between the metal and the solution due to the redox reaction going on at the interface. In addition, ionic adsorption from the solution may also contribute significantly to the final charge on the metal particles [74].

(iv) Another reason, though less common, due to which a particle may acquire a charge is imperfections in its crystal structure. This mechanism is responsible for the presence of a net charge on the faces of layers of clay particles in water. During the formation of clay minerals such as kaolinite, a part of Si^{4+} is sometimes substituted by Al^{3+} while a part of Al^{3+} may be substituted by Mg^{2+}. This leaves a net negative charge on the layer surface which is compensated by the adsorption of larger cations which can not be accommodated in the lattice of the mineral. When the clay is suspended in water, these compensating cations go in the liquid, leaving a fixed charge on the clay surface. On the other hand, the charges on the edges of the clay particle are governed by the mechanism described above for oxides. Thus it is possible for a clay particle to have opposite charges on the edges and the surfaces. This leads to the so called "card house structure" in clay suspensions in which the edges of the particles are attracted to the surfaces forming a very open structure.

(v) Charging of particles in nonpolar media: The mechanisms discussed above are all applicable to aqueous suspensions. Unlike water, free electric charges are not present in nonpolar media such as most organic liquids. However, it is still possible for the particles to get charged. A surfactant is usually added to such liquids to keep the particles dispersed. The excess surfactant molecules associate to form inverse micelles (Chapter 14) with their polar heads turned inwards and nonpolar tails extending in the liquid. Any water and ions present in the liquid are held within the polar interiors of the micelles. The micelles can exchange charge with one another so that at any time there will be some micelles which are charged due to unequal number of positive and negative ions. These charged micelles can in turn exchange charge with the suspended particles and make them charged [100].

7.2.5 *Zeta potential and the isoelectric point*

Consider the case in which the particles are charged by adsorption of ions from the liquid. In this case, the sign of the net charge on the particle surface can be changed by changing the concentration of the ions in the solution. For example, the oxide particles change from a positively to a negatively charged surface as the pH is increased starting from a low acidic value. The pH at which the net charge on the particle surface is zero is called the point of zero charge (PZC).

The ions which adsorb on the solid surface (*e.g.* H^+ and OH^-) and change its charge are called the potential determining ions; an electrolyte such as NaCl which does not affect the charge on the particle surface is called an indifferent electrolyte. However, as we shall see later, these electrolytes have dramatic effect on the stability of the suspension by compressing the thickness of the electrical double layer.

As will be discussed later, the potential drops exponentially as one moves away from the particle surface into the surrounding liquid. While the surface potential, ψ_o, is difficult to measure, the potential at a point very close to the surface can be obtained by measuring the velocity of the particles towards the counter electrode under an electrical field (Fig. 7.4). This potential is called the *zeta potential* and is given by

$$\zeta = \frac{f_h \eta v}{\varepsilon_r \varepsilon_0 E} \text{ volts} \tag{7.9}$$

Here η is the viscosity of the liquid, v is the measured particle velocity under a field E, ε_r is the relative permittivity of the liquid (78.54 for water at 298 K), ε_o is the permittivity of vacuum (8.85×10^{-12} coulomb meter) and f_H is a parameter whose value depends on the particle diameter — for large particles ($a\kappa > 100$, where κ is the reciprocal of the thickness of the double layer and a is the particle radius) $f_H = 1$ and for small particles ($a\kappa < 1$), $f_H = 1.5$.

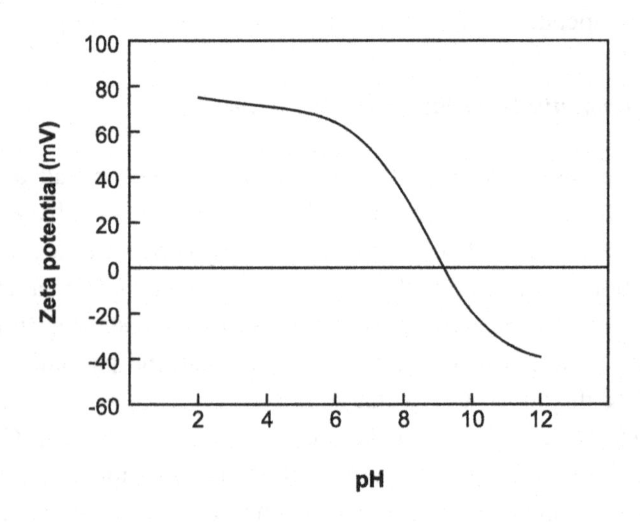

Figure 7.5. Variation of zeta potential with pH for Al_2O_3 particles

During the electrophoretic measurement of the zeta potential, a small layer of the liquid adheres to the particle as it moves *i.e.* the shear of the liquid does not take at the particle surface but at a small distance removed from it. The zeta potential is the potential at this shear plane.

Figure 7.5 shows the variation of the zeta potential with pH for Al_2O_3 particles. The particles are positively charged below pH = 9 and negatively charged above it. The pH at which the zeta potential is zero is called the isoelectric point, IEP. It is close, but not equal, to the PZC. The isoelectric points for some oxides and other materials are given in Table 7.3.

7.2.6 *Electrostatic stabilization : The electrical double layer*

7.2.6.1 *Introduction*

As was discussed above, the counterions form a diffuse layer around the charged particles. Some of the counterions may be strongly adsorbed on the surface. This layer of adsorbed counterions and strongly polarized water molecules is called the Helmholtz or Stern layer. The layer of electrical charge on the particle surface together with the diffuse layer of counterions in the liquid around it is termed the *electrical double layer* (Fig. 7.6).

Table 7.3. Isoelectic points of some oxides and other materials.

Material	Isoelectric point (pH)	
	Range	Average
Quartz, SiO_2		2.0
Zirconia, ZrO_2		7.6
Titana, TiO_2	4 – 6	5.7 (rutile), 6.2 (anatase)
Kaoline (edges)	5 -7	
Alumina, Al_2O_3	8 - 9.5	9.1
Magnesia, MgO		12.4
Fe_2O_3		8.2
Fe_3O_4		8.6
Fluorapatite, $Ca_5(PO4)_3(F,OH)$		6
Hydroxyapatite, $Ca_5(PO4)_3(OH)$		7
Silver iodide, AgI		pAg = 5.6

When two particles, with their respective double layers, approach each other, they experience a repulsive force. The distance at which this repulsion becomes significant depends on the "thickness" of the double layer. We now calculate this thickness for some simple cases.

7.2.6.2 *Thickness of the double layer*

The double layer can be modeled as consisting of a uniform charge density σ on the solid surface and a distribution of point charges in the liquid near it. Let ψ_o denote the electrical potential at the surface. Then $\psi(x)$, the potential at a distance x from the surface is related to the charge density $\rho(x)$ by the Poisson's equation

$$\frac{d^2\psi}{dx^2} = -\frac{\rho}{\varepsilon} \tag{7.10}$$

The concentration of ions at any position x in the solution can be determined from the Boltzmann equation. Thus if n_i is the concentration (number vol^{-1}) of the ith ion in the bulk of the solution ($x = \infty$), then its concentration at x is given by

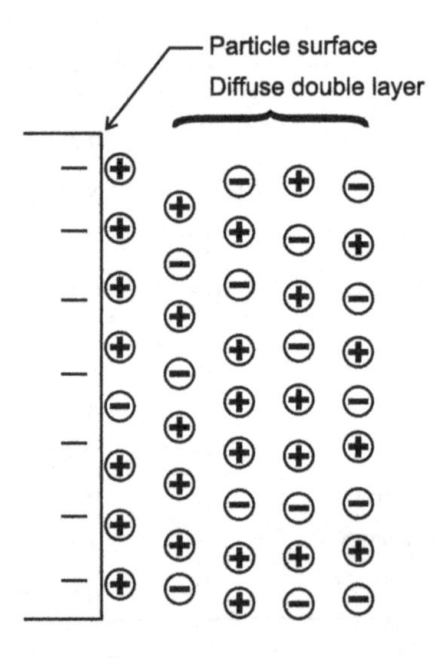

Fig. 7.6. A schematic representation of the electrical double layer

$$n_i = n_{i\infty} \exp(-\frac{Z_i e\psi}{kT}) \tag{7.11}$$

where Z_i is the degree of ionization of the ith ion, e is the charge on an electron and T is the temperature. The charge density ρ at any point is then given by

$$\rho = \Sigma n_i Z_i e = \sum_i Z_i e n_{i\infty} \exp(-\frac{Z_i e \psi}{kT}) \tag{7.12}$$

Combining Eq. (7.10) and Eq. (7.12) we get the Poisson-Boltzmann equation

$$\frac{d_2 \psi}{dx^2} = -\frac{e}{\varepsilon} \sum_i Z_i n_{i\infty} \exp(-\frac{Z_i e \psi}{kT}) \tag{7.13}$$

This equation can be easily solved if the potential ψ at every point is assumed to be small so that $ze\psi/kT << 1$. The using $e^{-x} \sim 1-x$ for $x<<1$, we get

$$\frac{d^2 \psi}{dx^2} = -\frac{e}{\varepsilon} \sum_i Z_i n_{i\infty} (1 - \frac{Z_i e \psi}{kT})$$

The electroneutrality condition requires that $\sum_i Z_i n_{i\infty} = 0$, so that

$$\frac{d^2 \psi}{dx^2} = [\frac{e^2}{\varepsilon kT} \Sigma Z^2_i n_{i\infty}] \psi = \kappa^2 \psi \tag{7.14}$$

where

$$\kappa^2 = \frac{e^2}{\varepsilon kT} \sum_i Z_i^2 n_{i\infty} \tag{7.15}$$

Equation (7.14) can be solved using the boundary condition $\psi = \psi_o$ at $x = 0$ and $\psi = 0$ at $x = \infty$. This gives

$$\psi = \psi_0 \exp(-\kappa x) \tag{7.16}$$

This shows that the potential drops exponentially and, at a distance κ^{-1} from the particle surface, it reduces to $1/e$ of its value ψ_o at the surface of the particle. This distance is called the *thickness of the double layer* δ (Fig. 7.7).

For high potentials (ψ_o >> 25 mV/Z at 298 K), Eq. (7.14) can still be solved to yield a relation between functions of ψ and ψ_0 given below [101]

$$\tanh\left(\frac{Ze\psi(x)}{4kT}\right) = \tanh\left(\frac{Ze\psi_0}{kT}\right)\exp(-\kappa\psi) \qquad (7.17)$$

This shows that the thickness of the double layer is still equal to κ^{-1}.

In the above treatment we have assumed a flat double layer. This is only valid when the particle size is large, $a \gg \kappa^{-1}$ or $\kappa a \gg 1$. For the nanoparticles this may not be true. The double layer then must be considered spherical and spherical coordinates must be used. For small particles, this gives

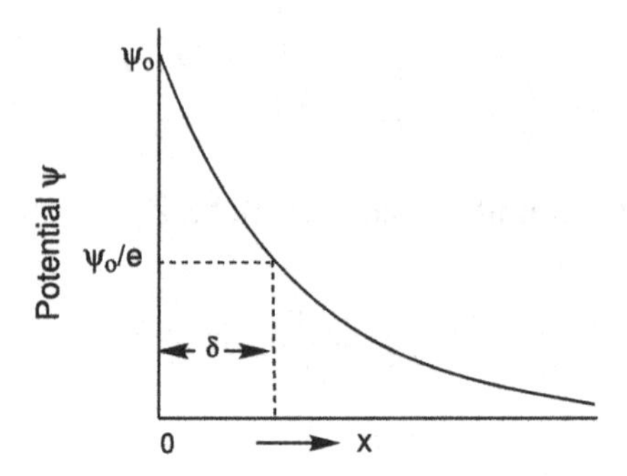

Fig. 7.7. Variation of potential with distance from the particle surface; the thickness of the double layer is defined as the distance ($\delta = \kappa^{-1}$) at which the potential falls to 1/e of its value ψ_0 at the surface.

$$\psi(r) = \psi_0 \frac{a}{r}\exp[-\kappa(r-a)]$$

The above derivation has been done assuming low potential at the particle surface. However, low potentials (ψ_o < 25 mV at 298 K) are not sufficient to prevent particle agglomeration. For high potentials numerical methods must be used to arrive at a solution [102].

Example 7.1. *Calculate the thickness of the double layer in a 0.01M solution of a 1:2 electrolyte (e.g. $CaCl_2$) in water at 298 K assuming that the electrolyte is fully ionized.*

Solution. We have, from Eq. (7.15)

$$\kappa^2 = \frac{e^2}{\varepsilon kT}\sum_i Z_i^2 n_{i\infty} = \frac{e^2}{\varepsilon kT}[Z_1^2 n_{1\infty} + Z_2^2 n_{2\infty}]$$

Now 1 M = 1 mole $l^{-1} = 10^3$ N ions m^{-3} where N = Avogadro's number.

Thus

$$\kappa^2 = \frac{e^2}{\varepsilon kT}.1000N[Z_1^2 M_1 + Z_2^2 M_2]$$

The ionic strength I is defined by the relation

$$\Sigma Z_i^2 M_i = 2I$$

Thus

$$\kappa^2 = 1000\frac{e^2}{\varepsilon_r \varepsilon_0 kT}.N.2I$$

Now a 0.01 M solution of a 1:2 electrolyte (*e.g.* $CaCl_2$) would dissociate to give 0.01 M ions with $Z = Z_1 = 2$ (*e.g.* Ca^{2+}) and 0.01 x 2 M ions with $Z = Z_2 = 1$ (*e.g.* Cl^-).

Hence $2I = \Sigma Z_i^2 M_i$ = (2x2x0.01x1 + 1x1x0.01x2) = 0.06

so that

κ^2 = [1000 x $(1.6\times10^{-19})^2$ (6.023 x 10^{23})0.06] / [78.54 x 8.85×10^{-12} x 1.38 x 10^{-23} x 298]

which gives $\kappa^{-1} = \delta$ = thickness of the double layer = 1.76 nm. □

As mentioned earlier, a repulsive interaction is needed to overcome the van der Waal's attractive interaction to make the suspension stable. If the particles are charged, a repulsive interaction is produced when the double layers around two particles overlap with each other. It can be shown that the repulsive energy decays exponentially with distance r from the particle surface. The expression for two spheres with radius a and zeta potential ξ for low values of potential is given by [97]

$$V_r = 4\pi\varepsilon_r\varepsilon_0\xi^2\frac{a^2}{r}\exp[-\kappa a(\frac{r}{a}-2)], \quad r >> a \tag{7.18}$$

Furthermore, the van der Waal's interaction energy for two spheres of radius a with a separation r is given by

$$V_{atr} = -\frac{A}{6}\left[\frac{2a^2}{r^2-4a^2}+\frac{2a^2}{r^2}+\ln(1-\frac{2a^2}{r^2})\right] \tag{7.19}$$

where A is the Hamakar constant for the given solid-liquid-solid system.

A schematic plot of the two interaction energies is given in Fig. 7.8.

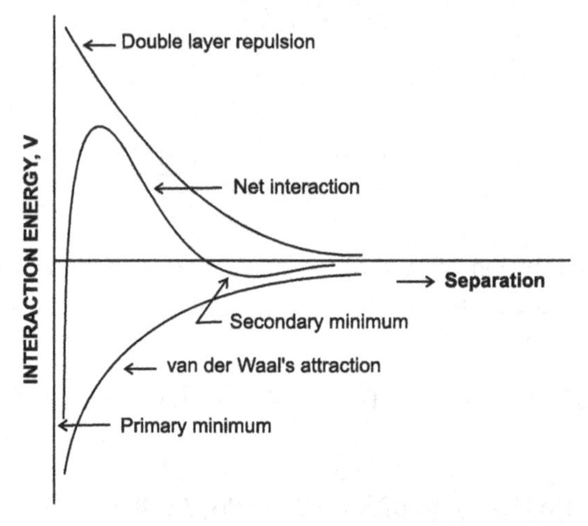

Fig. 7.8. The interaction energy between two spheres vs distance between the particle surfaces.

The net interaction energy obtained by the summation of the two curves is also shown. In the most general case, the net curve is characterized by a deep minimum near the particle surface, called the *primary minimum*, a maximum called the *repulsive barrier* and a shallow minimum called the *secondary minimum.* As the particles approach each other, they will have to overcome the repulsive barrier in order to get bound in the primary minimum. If the height of the repulsive barrier is more than ~ 20 kT, then the thermal energy (kT) is not sufficient to overcome the barrier and the dispersion is stable.

7.2.6.3 *Controlling the stability of the electrostatically stabilized suspensions*

The thickness of the electrical double layer and hence the stability of the suspension is controlled primarily by changing the pH and the electrolyte concentration. However, several factors play a role in the electrostatic stabilization.

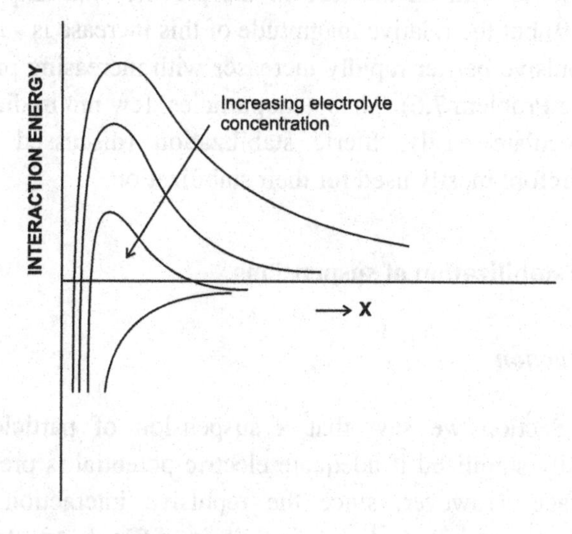

Fig. 7.9. Effect of the electrolyte concentration on the net interaction energy between the particles.

(a) *Zeta potential*: A high zeta potential produces a high repulsive interaction (Eq. 7.18). The potential can be adjusted by adjusting the concentration of the potential determining ions *e.g.* pH in case of aqueous suspensions of oxide particles.

(b) *Thickness of the double layer*: A large thickness produces an increased barrier (Eq. 7.18). The value of $\delta = 1/\kappa$ decreases as the ionic strength increases (Eq. 7.15). Hence a low electrolyte concentration favors a stable dispersion while the addition of an electrolyte to a stable dispersion leads to rapid flocculation. This phenomenon is used sometimes in water purification where the addition of a small quantity of alum leads to rapid flocculation and settling of the particulate matter in the water. Figure 7.9 shows the effect of electrolyte concentration on the repulsive interaction energy. Similarly, an increasing valence of the either ion in the electrolyte leads to a decrease in the repulsive interaction.

(c) *Particle size*: Both the attractive and the repulsive interactions increase with an increase in the particle size (Eqs. 7.18 and 7.19) but the relative magnitude of this increase is such that the repulsive barrier rapidly increases with increasing particle size (see Problem 7.6). Thus nanoparticles, few nm in diameter can flocculate easily. Steric stabilization (discussed below) is therefore mostly used for their stabilization.

7.3 Steric stabilization of suspensions

7.3.1 *Introduction*

In the last Section we saw that a suspension of particles can be electrostatically stabilized if adequate electric potential is present at the particle surface. However, since the repulsive interaction decreases rapidly with decreasing particle size, it is difficult to stabilize the *nano*particles electrostatically. Furthermore, the conditions required for electrostatic stabilization (pH *etc.*) are often so harsh as to rule it out. There is another method, known as *steric stabilization* in which polymeric or surfactant molecules (dispersants) are used to stabilize a suspension. As an example, in the preparation of noble metal nanoparticles by wet chemistry methods, the particles are sterically stabilized by alkanethiols soon after their generation (Chapter 14). The method is in fact quite common in other fields also and a large number of dispersants are commercially available for different applications.

7.3.2 *Mechanism*

The ideal situation for steric stabilization is one in which one end of a long chain molecule is anchored to the particle while the rest of the chain sticks out in the surrounding liquid (Fig. 7.10). As two such particles with "hairs" approach each other, the two chains begin to interpenetrate. The number of different configurations which each chain can adopt decreases because of the constraint of the other chain; this causes a decrease in the configurational entropy. There is also an increase in the enthalpy because the chains come in close proximity to each other which is a higher energy state compared to each chain being surrounded by the solvent only. The net free energy change is given by

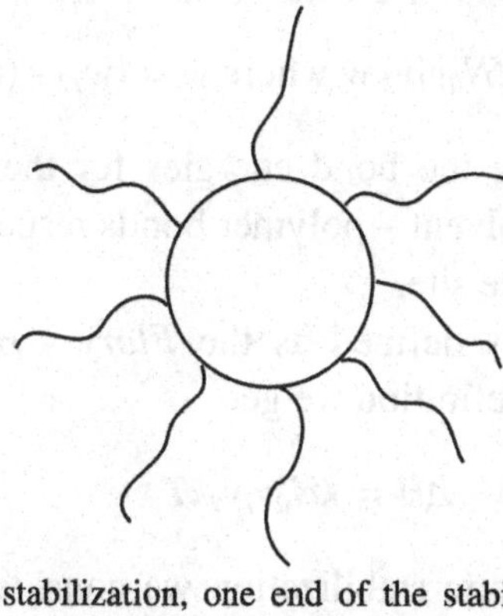

Figure 7.10. For steric stabilization, one end of the stabilizing chain should be firmly anchored to the particle and the remaining part should stick out in the solution.

$$\Delta G = \Delta H - \Delta S$$

As the enthalpy term is positive and the entropy term is negative, the coming together of the particles is accompanied by an increase in the free energy and is not favored. The suspension is thereby stabilized. (If the solvent is not a very good solvent for the polymer, then the enthalpy term can be negative and can lead to agglomeration).

The expressions for the quantities ΔH and ΔS have been developed using a simple model of polymer solution [97] in which the solvent molecules and the polymer chain segments are imagined to randomly occupy points on a three dimensional lattice, each lattice point having an equal volume. The ΔS term, denoted by entropy of mixing ΔS_m is given by

$$\Delta S_m = -k[(N_1 \ln v_1 + N_2 \ln v_2)] \tag{7.20}$$

Here N_1 and N_2 are the number of lattice sites occupied by the solvent and the polymer chain segments respectively and N_0 is the total number of sites ($N_0 = N_1 + N_2$) and v_1 and v_2 are the volume fractions of the solvent and the polymer.
The enthalpy of mixing, ΔH_m, is given by

$$\Delta H = ZN_0 v_1 v_2 (w_{12} - (w_{11} + w_{22})/2)$$

$$= ZN_0 v_1 v_2\, w \text{ where } w = (w_{12} - (w_{11} + w_{22})/2)$$

Here w_{11}, w_{22} and w_{12} are the bond energies for the solvent – solvent, polymer – polymer and solvent – polymer bonds respectively and Z is the coordination of each lattice site.

The quantity Zw/kT is defined as the *Flory - Huggins interaction parameter* κ. Using this definition we get

$$\Delta H = \kappa N_0 v_1 v_2 kT \tag{7.21}$$

For the case of the steric stabilization we consider a single polymer chain attached to each of the two particles. The polymer chain occupies a volume V_d which can be estimated to be $4/3\pi R_g^3$ where R_g is the radius of gyration of the polymer chain (see Chapter 13). In this domain, $N_2 = 1$ and $N_1 >> N_2$. It can be shown [97] that the change in free energy due to the interpenetration of the two chains attached to the two particles is

$$\Delta G_{overlap} = 2kT\underline{v_2}^2 (1/2 - \kappa)/V_d\, \underline{v_1} \tag{7.22}$$

where $\underline{v}_1$ and $\underline{v}_2$ are the partial volumes of the solvent molecule and the polymer chain segments respectively.

A positive value of $\Delta G_{overlap}$ implies repulsion between the particles and leads to steric stabilizationx. For this κ should be < ½. Note that, since

$$\kappa = Zw/kT \text{ with } w = [w_{12} - (w_{11} + w_{22})/2]$$

a negative value of w implies that the polymer and solvent molecules have greater liking to each other than to the molecules of their own kind. In this case the solvent is a good solvent for the polymer. Even if κ has a small positive value (< 0.5), the $\Delta G_{overlap}$ is still positive and the steric stabilization is still possible. The situation for which κ = ½ corresponds to the so called θ point and the solvent is called a *theta* solvent. If κ > ½, there is a tendency for the solution to phase separate and this may lead to coagulation of particles even if the Hamakar constant for them is very small.

The effects of the relative magnitudes of the entropy and enthalpy change and the temperature are shown in Fig. 7.11. It shows that a stable

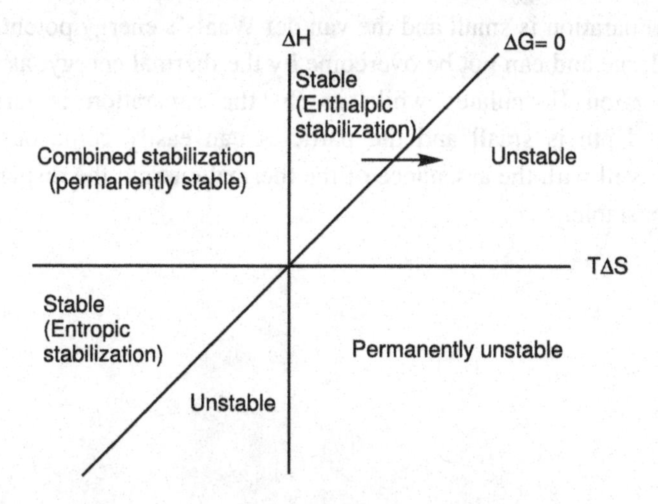

Fig. 7.11. Effects of the relative magnitudes of the entropy and enthalpy changes and temperature on the stability of a sterically stabilized suspension

suspension may become unstable and vice versa upon a change of temperature [100]. This is directly related to the change in the solubility of the polymer chain in the liquid. If the solubility decreases, the polymer chains will no longer extend to sufficient distance in the solution and destabilization would occur. Very often a polymer is soluble between two temperatures called the upper critical temperature and the lower critical temperature. The sterically stabilized suspensions thus display two flocculation temperatures, the upper and the lower flocculation temperatures (UCFT and LCFT). These are nearly the same as the critical temperatures for the solution. A suspension can be flocculated or deflocculated by changing the temperature or by controlling the ratios of the various solvents in the solvent mixture.

7.3.3 *Requirement on the chain length*

For effective steric stabilization, the protruding length of the polymer chain should be sufficiently long such that at the distance of closest approach, the van der Waal's attraction is small enough to be overcome by the thermal energy. This is illustrated in Fig. 7.12. In Fig. 7.12 (a) the particle separation is small and the van der Waals's energy potential well depth is large and can not be overcome by the thermal energy; as a result the dispersion flocculates while in (b) the separation is large, the potential depth is small and the particles can easily come out of the potential well with the assistance of the thermal energy; the dispersion is, therefore, stable.

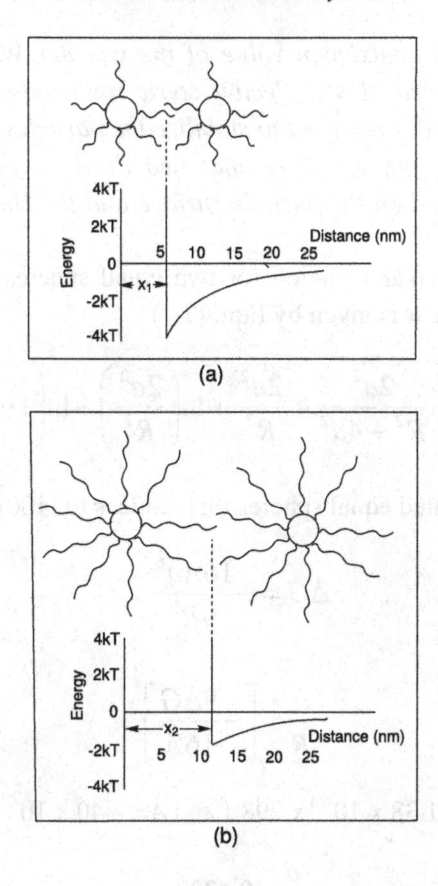

Fig. 7.12 (a) the steric barrier is represented by the vertical line at distance x_1; at x_1, the van der Waal's interaction energy curve has a large magnitude which can not be overcome by kT so that the dispersion flocculates (b) the steric barrier is at x_2 where the attractive interaction is small and can be easily overcome by thermal fluctuation; the suspension can be stabilized in this case (a potential well of –3kT is sufficient to flocculate the particles).

Example 7.2. *If the maximum value of the van der Waal's attraction energy can not exceed kT for effective steric stabilization, calculate the length of the molecules required to stabilize the particles of size (a) 2 nm and (b) 10 nm at 298 K. Given that two third of the length of the molecule is adsorbed on the particle surface and the Hamakar constant is* $A_{121} = 40 \times 10^{-20}$ *J.*

Solution. The Hamakar equation for two equal spheres of radii a with centers at a distance R is given by Eqn. (7.7)

$$\Delta G = -\frac{A}{6}\left[\frac{2a^2}{R^2-4a^2}+\frac{2a^2}{R^2}+\ln\left(\frac{2a^2}{R^2}\right)+\ln\left(1-\frac{4a^2}{R^2}\right)\right]$$

For two well separated equal spheres this reduces to [100]

$$\Delta G = -\frac{16Aa^6}{9R^6}$$

or

$$\frac{a}{R} = \left[-\frac{9\Delta G}{16A}\right]^{\frac{1}{6}}$$

Taking $\Delta G = kT = 1.38 \times 10^{-23} \times 298$ J and $A = -40 \times 10^{-20}$ J, we get

$$\frac{a}{R} = 0.4237$$

The values of R for $2a$ = 2, 10 nm are 2.36, and 11.8 nm respectively. These are the maximum values of R for the van der Waal's attraction energy not to exceed kT.

The protruding length of the molecule should be at least one half of the distance between the surfaces of the particles *i.e.* one half of $(R\text{-}2a)$. The minimum total length of the molecule for effective steric stabilization should therefore be $(3/2)(R\text{-}2a)$ *i.e.* 0.54 and 2.7 nm for particles of 2 and 10 nm diameter respectively. □

Thus the length of the molecule necessary of steric stabilization depends on the strength of the van der Waal's interaction. As it is lower for smaller particles, small molecules such as some surfactant molecules may be adequate to stabilize the nanoparticles. It should be remembered, however, that a large part of the molecule is usually adsorbed on the surface of the particle and only a fraction, protruding into the liquid, is effective for steric stabilization.

An ideal situation for steric stabilization is the one in which one end of the molecule is firmly anchored to the particle surface and the rest of it protrudes out into the solvent. Homopolymers, random copolymers or block copolymers can all be adsorbed on the particle surface to impart steric stabilization; however, the block copolymers are most effective as the block copolymers can be custom designed to contain the anchoring and the stabilizing groups. Examples of anchoring and stabilizing groups in block copolymers are given in Table 7.4 [100].

One end of the molecule can be firmly anchored by "grafting" polymers on the particle surface [103]. For this, some functional group is to be attached to the particle surface as well as to the chain end. Some examples of these groups are hydroxides, silanes, silanols, siloxanes, alcohols, acids and amines. An example is SiH group on the particle surface and vinylic double bonds on the chain end. The SiH and C=C are then coupled by hydrolysation on platinum catalyst. In this kind of grafting, termed "grafting onto", the number of chains per particle is low due to the steric hindrance (Fig. 7.13 a).

Another way to graft a polymer is to impart polymerization initiator functionality on the particle surface. The monomer is then added and the polymerization carried out. Polymer chain initiates on the particle surface. In this case the number density of the chains is high but there can be significant distribution in the chain lengths (Fig. 7.13 b).

Grafting is also quite useful for the preparation of polymer – nanoparticle composites — a nanoparticle can be made compatible with the desired polymer matrix by grafting a suitable polymer on the particle surface.

Table 7.4. Some examples of the anchoring groups and stabilizing groups in block copolymers

Anchoring groups	Stabilizing groups
	I. Aqueous suspensions
Polystyrene	Polyethylene oxide
Polyvinyl acetate	PVA
PMMA	Polyacrylic acid
	II. Nonpolar suspensions
Polyacrylonitrile	Polystyrene
Polyethylene oxide	Polylauryl methacrylate
Polyethylene	Poly (1,2-hydroxystearic acid)

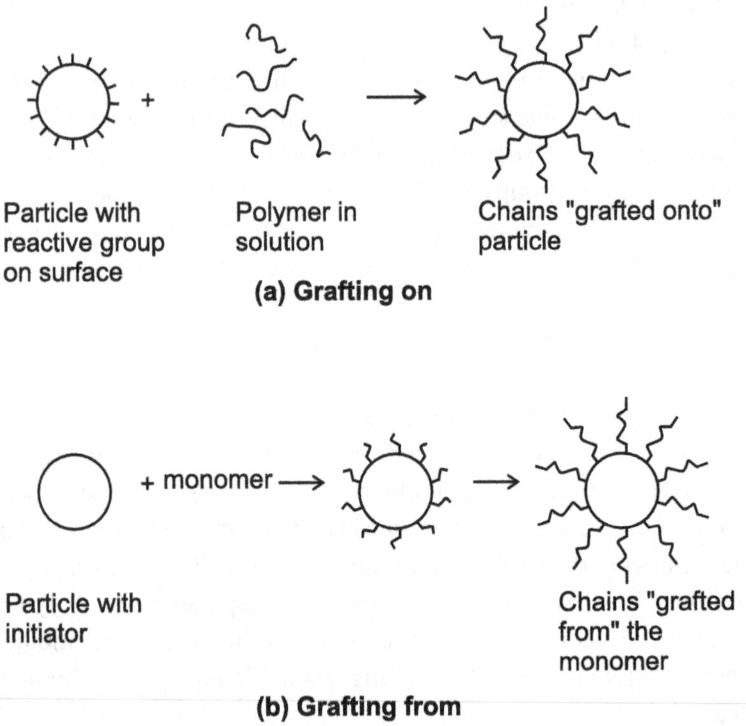

Fig. 7.13 Schematic showing the grafting of polymer chains on a particle surface.

Configurations of the adsorbed polymers — homopolymer, random copolymer, block copolymer and grafted polymer are shown schematically in Fig. 7.14. In case of the homopolymers, in the beginning only a part of the polymer chain is adsorbed on the surface, with the rest extending into the solution and providing the stability; with time the adsorbed length increases and the stability may be lost. High molecular weight homopolymers are therefore needed to prolong the life of the suspension.

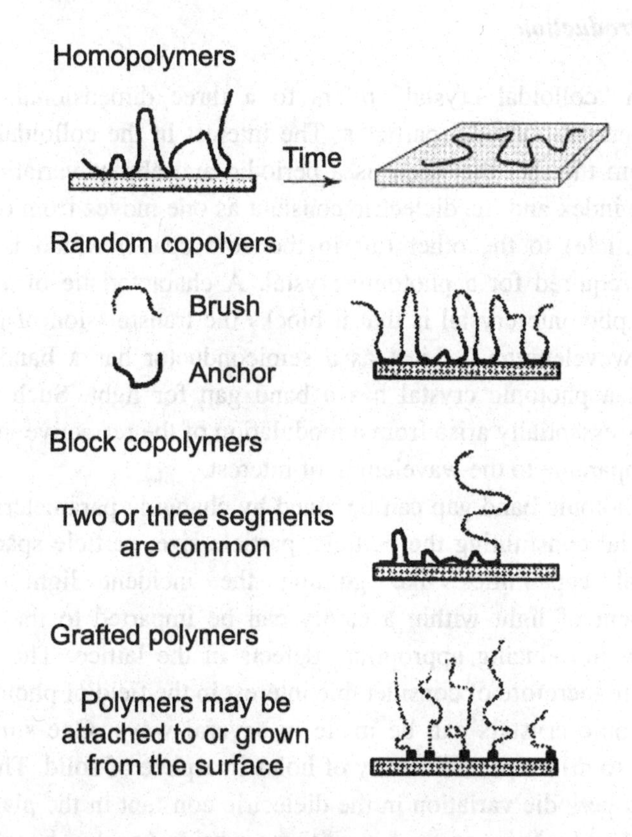

Fig. 7.14. Schematic representation of the configuration in the adsorbed state of the various kinds of polymers used for steric stabilization (reproduced from [100] copyright (2002) with permission from John Wiley, Inc.).

An important example of steric stabilization is the use of alkane thiols to stabilize the suspension of noble metal particles such as gold during their preparation and subsequent handling. The size of the particles can also be controlled between 1 to 5 nm by controlling the amount of thiol. The alkane thiols form a self assembled monolayer on the gold surface. This is discussed in detail in Chapter 14.

7.4 Colloidal crystals for photonic applications

7.4.1 *Introduction*

The term "colloidal crystal" refers to a three dimensional, periodic arrangement of colloidal particles. The interest in the colloidal crystals stems from the fact that there is a periodic and abrupt variation in the refractive index and the dielectric constant as one moves from one phase (solid particle) to the other (air in the void space) which is just the property required for a photonic crystal. A characteristic of a suitably designed photonic crystal is that it blocks the transmission of light in a band of wavelengths *i.e.* just as a semiconductor has a band gap for electrons, a photonic crystal has a band gap for light. Such photonic properties essentially arise from a modulation of the refractive index on a scale comparable to the wavelength of interest.

The photonic band gap can be tuned by changing parameters such as the material constituting the particle, particle size, particle spacing, *etc.* Additional capabilities like guiding the incident light and the confinement of light within a cavity can be imparted to the photonic crystal by introducing appropriate defects in the lattice. The photonic crystals are therefore of considerable interest in the field of photonics.

The photonic crystals can be made in several ways. One simple way would be to drill a periodic array of holes in a plate of solid. This would produce a periodic variation in the dielectric constant in the plane of the plate. Other techniques such as lithography have also been used to produce the photonic crystals. All these techniques involve considerable effort and resources. The formation of colloidal crystals by self assembly

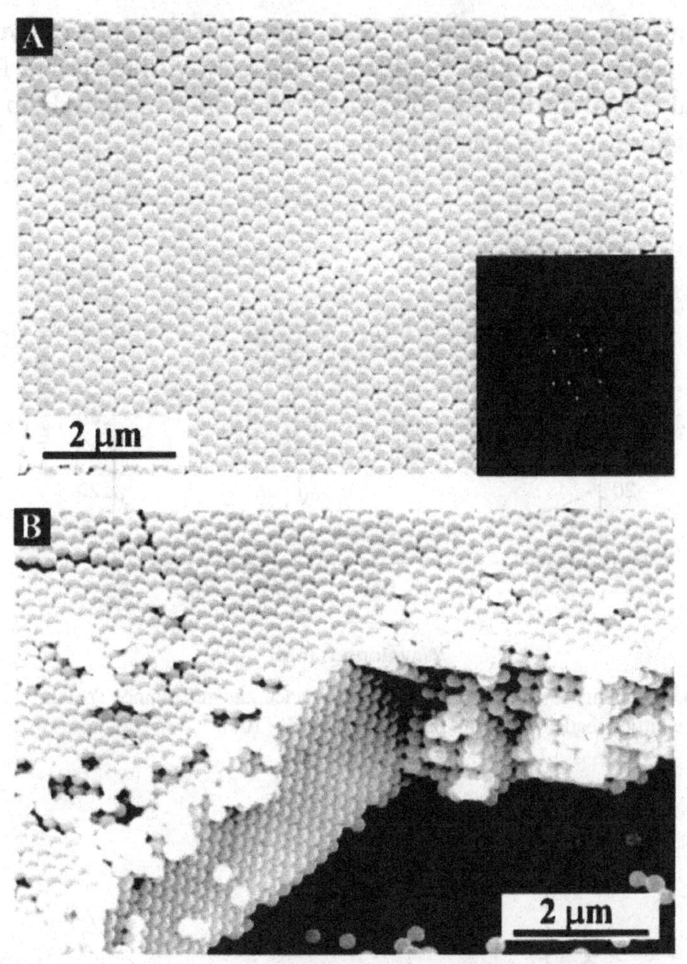

Fig. 7.15. SEM image of a colloidal crystal made from 298.6 nm silica spheres (A) top view; the inset shows a Fourier transform of a 40 x 40 nm region. (B) SEM side view of the same sample (Reprinted with permission from [104] copyright (1999) American Chemical Society).

of the particles from a suspension, on the other hand, provides an easier route. Spherical particles of silica and some polymers with a very uniform size are now routinely available. Ordered three dimensional and two dimensional arrangements of such spherical particles can be rather easily obtained starting from their suspension and allowing the particles to settle by sedimentation or by evaporation on a substrate (although translating this into practical devices is still a challenge). An example is shown in Fig. 7.15 while Fig. 7.16 shows a uv–visible spectra from a colloidal crystal made from polystyrene beads. A pronounced dip in the transmittance at ~ 565 nm can be seen.

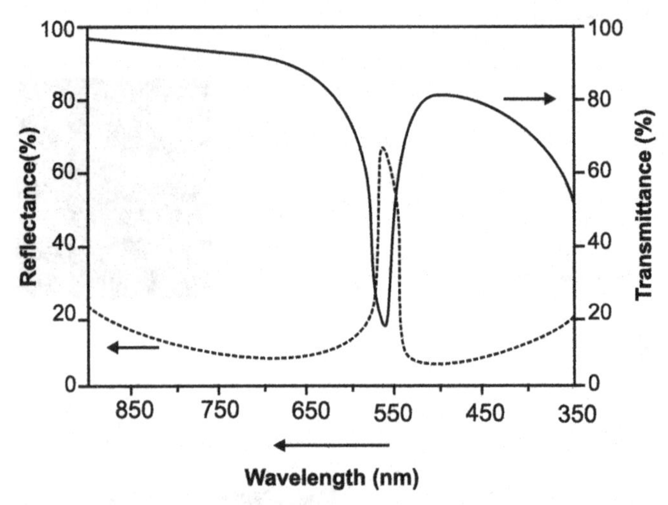

Fig. 7.16. A schematic of the UV-Vis transmittance and reflectance spectra from a 3D crystalline lattice made of 208 nm polystyrene beads [105].

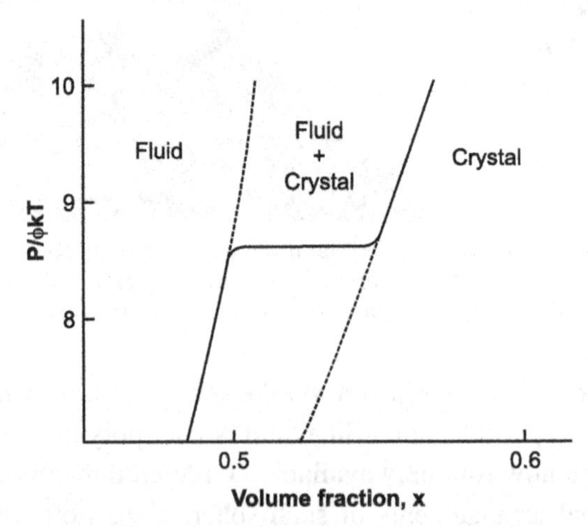

Fig. 7.17. The phase transition from a fluid to colloidal crystal phase for the hard sphere interaction; the phase diagram is presented as pressure (normalized by kT and ϕ, the FCC close packing density) and the volume fraction x of the particles [106].

The colloidal crystals assume different crystal structures (*e.g.* fcc, bcc) depending on the experimental conditions. As the interaction between the particles can be easily tuned from "hard" to "soft" (discussed below), the colloidal crystals, in addition to their technological potential, provide a model system to study the phase transformations under different conditions.

7.4.2 *Soft and hard interactions*

As we have seen, the interactions between the particles in a suspension can be tuned easily by changing the pH, the electrolyte concentration *etc.* Let us consider two situations. In one, there is no repulsive interaction between the particles at all and the van der Waal's interaction is also made negligible by a suitable choice of the suspension fluid so that the van der Waal's attraction between the particles is balanced by the particle–solvent interaction. The particles experience no interaction at all and can approach each other until they touch after which they can not come any closer *i.e.* the particles behave as hard spheres. As there is no interaction between the particles, the only contribution to the free energy comes from the change in entropy. Intuitively it would seem that a random arrangement of hard spheres would have higher entropy; however, it is in fact the case that the particles gain entropy by arranging themselves equidistantly from one another as this maximizes the space in their vicinity. This compels the particles to order. The transition from a state in which the particles are dispersed and undergo a random Brownian motion to an ordered state with particles held on a crystal lattice, can be looked upon as a solidification phase transition. The computer simulations predict this first order transition to start at a volume fraction of solid, $x = 0.494$ and be complete at $x = 0.545$ with the two phases coexisting in between, as shown in Fig. 7.17. This transition is now well known as the *Kirkwood-Adler transition* [106].

The second case corresponds to a situation in which the long range repulsion ("soft" repulsion) between the particles is introduced either as a double layer electrostatic repulsion or as a steric repulsion due to the presence of the capping layers on the particles. This keeps the particles well separated so that they do not experience the van der Waal's attraction. In this case the Kirkwood-Adler transition also occurs but at a lower volume fraction which further decreases as the repulsive interaction is increased. The ordered "soft" particles also assume cubic structures like the hard particles. An fcc structure is favored when the volume fraction is high or the repulsion is short range. Long range repulsion or a lower volume fraction favors the less dense bcc structure.

In Sec. 7.2.6.2 we saw that the repulsive interaction energy due to the overlap of the double layers around two particles decays exponentially with distance r from the particle surface. The expression for two spheres with radius a and zeta potential ξ is given by

$$V_r = 4\pi\varepsilon_r\varepsilon_0\xi^2 \frac{a^2}{r}\exp[-\kappa a(\frac{r}{a}-2)], \qquad r \gg a \qquad (7.18)$$

$$\approx U_0 \exp[-\kappa r],$$

The total energy for a colloidal suspension containing N particles is obtained by summing the interaction energy between each pair of particles. This results in the following expression

$$V_t = \frac{1}{2N} U_0 \sum_{\substack{i,j=1 \\ i\neq j}}^{N} \exp(\kappa r_{ij})$$

Here r_{ij} is the distance between the ith and j^{th} particle. First consider the case when $T = 0$. As the entropy term in the free energy, $T\Delta S$ becomes zero, the stable configuration would be the one for which the total energy, V_t is minimum. When the summation is performed for different crystal structures [107] , it is found that the bcc phase is stable (*i.e.* the energy is minimum) when $\kappa a < 1.72$ or for large value of the double layer thickness and the fcc phase is stable for $\kappa a > 1.72$ (a is the particle radius).

For temperatures other than zero, it is necessary to calculate the free energy of the system taking the entropy into account. Such calculations are done using techniques such as molecular dynamics or lattice

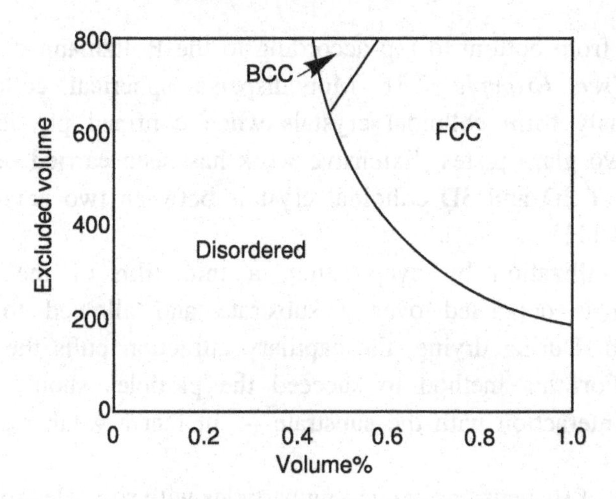

Fig. 7.18. Phase diagram for a suspension of particles interacting through a soft potential [106].

dynamics [108]. These calculations reveal that, as in the case of hard spheres, the particles assume cubic arrangements. A generalized phase diagram based on such calculations is shown in Fig. 7.18. An fcc phase is stable at high particle concentrations or at a low value of the double layer thickness while the less densely packed bcc structure prevails at low particle concentrations or high electrostatic repulsion.

7.4.3 *Preparation of colloidal crystals*

Both the hard sphere repulsion and the soft sphere repulsion conditions can be utilized for the formation of the colloidal crystals. Techniques for the formation of colloidal crystals in which the hard sphere interaction prevails are (i) sedimentation under gravity (ii) crystallization under physical confinement and (iii) crystallization by evaporation. Sedimentation under gravitation results in the formation of a dense sediment at the bottom and a clear liquid above. For the colloidal crystals to form by sedimentation, the particles should be large (> 100 nm), spherical and uniform in size. For smaller particles, sedimentation results in a well dispersed equilibrium state with the particle number density increasing from bottom to top according to the Boltzmann distribution function (*see Example 7.3*). Monodisperse spherical colloids can spontaneously form colloidal crystals when confined physically *e.g.* between two glass plates. Extensive work has been carried out on the formation of 2D and 3D colloidal crystals between two parallel glass plates [109-111].

In crystallization by evaporation, a thin film of the colloidal suspension is deposited over a substrate and allowed to dry by evaporation. During drying, the capillary attraction pulls the particles together. For this method to succeed the particles should have no attractive interaction with the substrate — in fact a weak repulsion is favored.

The use of suspensions containing particles with soft (electrostatic or steric) interactions can be a powerful route for the formation of large colloidal crystals.

Example 7.3. *Show that in a well dispersed suspension of small particles, the particle number density follows the Boltzmann distribution.*

Solution Consider a suspension of particles contained in a cylinder (Fig. 7.19). Across any plane, there is a downward flux of particles due to gravity. The resulting gradient in concentration of the particles is opposed by a diffusive flux upwards. At equilibrium the total flux is zero so that

$$c\frac{dx}{dt} - D\frac{dc}{dx} = 0 \tag{7.23}$$

where c is the concentration at any depth x and D is the diffusion coefficient due to the Brownian motion. It is given, in terms of the particle radius and the fluid viscosity η by the following expression:

$$D = \frac{kT}{6\pi\eta\, r} \tag{7.24}$$

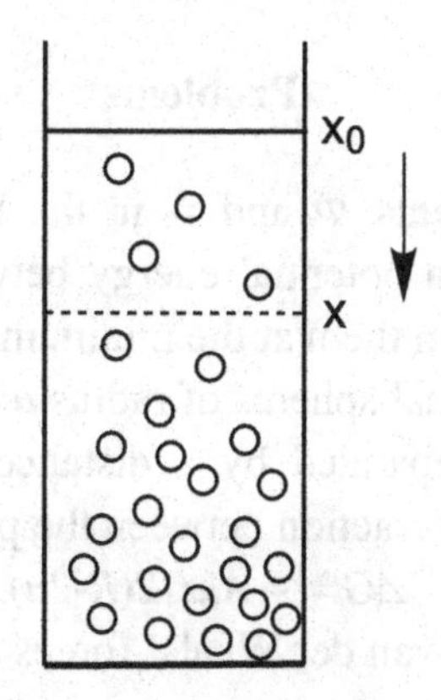

Fig. 7.19. Sedimentation of particles undergoing Brownian motion.

In a suspension, the particles settle with a terminal velocity given by

$$\frac{dx}{dt} = \frac{2}{9} r^2 (\rho - \rho_0) \frac{g}{\eta} \tag{7.25}$$

Combining Eqs. (7.23) – (7.25) one gets

$$\frac{dc}{c} = \frac{m}{kT}(1 - \frac{\rho_0}{\rho}) g dx$$

or

$$\ln \frac{c}{c_0} = \frac{m}{kT}(1 - \frac{\rho_0}{\rho}) g (x - x_0)$$

or

$$c = c_0 \exp(-\omega / kT) \tag{7.26}$$

where ρ and ρ_0 are the densities of the particle and the liquid respectively and ω is the potential energy difference between the particle at height x_0 and x. The particles therefore assume a Boltzmann like distribution at equilibrium.

Problems

(1) Show that the constants Φ_0 and x_0 in the Lennard Jones Potential refer to the minimum potential energy between the molecules and the separation between them at the minimum respectively.

(2) Show that for two equal spheres of radius *a* very close to each other, with their centers separated by a distance *R*, the free energy of London dispersion interaction between the particles becomes

$$\Delta G = -Aa/12(R-2a).$$

(3) Considering only the van der Waals' forces and no other interaction, explain the following

(a) In a suspension containing small and large particles, the small particles are attracted more towards the large particles than to one another.

(b) Under certain conditions, particles adhering to a surface may detach when the surface is brought into contact with a different liquid.

(4) In an electrophoretic cell, the particles are found to travel a distance of 200 μm in 15 sec when a voltage of 30 V is applied across two electrodes 5 cm apart. The particles are in aqueous suspension. Calculate the zeta potential. Assume $f_H = 1$.

(5) The points of zero charge of certain oxides are as follows

Al_2O_3 8.5; ZrO_2 5.0; SiO_2 2.0; MgO 10.0

(a) What is the nature of the charge (+ or –) on the surface of these oxides when suspended in water at pH 7.0?

(b) Show schematically the variation of potential as a function of distance between two particles in each of the suspensions of pH 7.0.

(c) Explain the curves in (b).

(d) Arrange the suspensions in the order of the thickness of the double layer around the particles.

(6) Answer the following questions

(a) Show that the thickness of the double layer at 20 ^{0}C in an electrolyte with valence ratio Z:Z and molar concentration M is $(3 \times 10^{-8})/Z\sqrt{M}$ cm. Take the dielectric constant of water at 20 ^{0}C as 78.

(b) What concentration (mol l^{-1}) of $CaCl_2$ is required to have the same thickness of the double layer as that with 0.01 M $AlCl_3$? Assume complete dissociation of the electrolyte in both cases.

(c) Why is the electrostatic stabilization not very effective for nanoparticles?

(7) Calculate numerically and plot the net interaction energy normalized by kT vs. separation between two spherical gold particles in a 0.001M aqueous NaCl solution for particles of radii 1, 3, 10 and 30 nm. Take the temperature as 300 K. Comment on the changes in the height of the repulsive barrier and the depth of the secondary minimum as a function of the particle size. Given Hamaker constant $= 25 \times 10^{-20}$ J, $\xi = 0.1$ V.

(8) Answer the following

(a) A surfactant is needed to stabilize nanoparticles in an *organic solvent*. From amongst the groups below, chose the groups which can form the end of the surfactant chain that is to be attached to the particle. Justify your answer.
Polystyrene, polyacrylic acid, polyvinyl acetate, PMMA, polyethylene oxide, PVA.

(b) Spherical particles are to be sterically stabilized by using a polyethylene chain. The C-C bond length is 0.154 nm and the bond angle is 109.5°. The characteristic ratio is 6.9 and the chain expansion factor α is 3.2. One half of the chain length is adsorbed on the surface of the particles while the other half protrudes in the solvent. The van der Waals's interaction energy between the particles is given by $V = (2.4 \times 10^{-63} / R^6)$ J, where R is the distance between the *surfaces* of the particles. If the maximum value of the van der Waal's energy can not exceed 5 kT for effective steric stabilization, find the minimum molecular weight of the polymer which is needed to sterically stabilize the particles at 27^{0} C.

(9) Answer the following

(a) Derive an expression for the terminal velocity of a spherical particle of density ρ settling in a liquid of density ρ_0. State the assumptions made.

(b) What are the diameters corresponding to settling times of (i) 30 minutes (ii) 2 hours in water at 20^0C through a distance of 20 cm for alumina spheres . Given: density of alumina 3.98 x 10^3 kg m^{-3}, density of water 1 x 1000 kg m^{-3}, viscosity of water at 20 C is 1 mPa s.

(10) Using the expression derived in the text, plot and compare the relative concentration of particles in a suspension vs. height for particles of (i) molar mass $M = 10^9$ and particle radius r = 60 nm (ii) $M = 10^7$ and r = 12.5 nm and (iii) $M = 10^5$ and r = 2.7 nm (molar mass = mass of a particle x Avogadro's number).

Chapter 8

Carbon Nanostructures

8.1 Introduction

Carbon is perhaps the most important element on earth because without it the life as we know it would not be possible. Carbon forms the backbone of the organic and biological molecules. Many biological molecules have dimensions in the nanometer range so that an important area of nanotechnology deals with biological molecules and their manipulation to yield newer applications and products.

Carbon in the inorganic form, though less ubiquitous, is perhaps no less important. Diamonds, graphite fibers, amorphous carbon filters, plastics, paints and lubricants, all contain inorganic carbon. The smallest dimension for the inorganic carbon structures was limited to about a micron until the discovery of fullerene in 1985 by H. W. Kroto and R. E. Smalley and their coworkers [112]. The fullerenes are nanometer sized hollow cages whose walls are formed by a fixed number of carbon atoms arranged in hexagons and pentagons. After their discovery in the laboratory, it has been realized that they in fact form naturally during high temperature and pressure reactions such as lightening strikes and combustion processes.

The next carbon nanostructured material to be discovered in laboratory was the carbon nanotubes (CNT) which are single walled or multiwalled hollow cylinders of carbon. The multiwalled CNT were first reported by Sujimo Iijima of NEC laboratories in Japan in 1991[113]. In 1993 Bethune *et al.* [114] reported the preparation of single walled CNT.

The carbon nanostructures such as fullerenes and nanotubes can be thought of as derived from the carbon sheets which make up the structure of graphite. The graphite structure is made of sheets of carbon atoms stacked one above the other, atoms in each sheet being arranged in a hexagonal arrangement. Each such sheet is called a *graphene* sheet. For many years, graphene was merely a concept used to describe the structure of the other carbon forms and as a model for solid state theoretical studies. Although, efforts to produce single or a stack of at most a few layers of graphene can be traced back to the 1960's, it was only after the discovery of fullerene and the carbon nanotubes that efforts to prepare single sheet graphene intensified. In 2004, Giem's group succeeded in extracting and studying the electrical properties of a single graphene layer by repeatedly peeling off layers of graphene from a graphite crystal by an adhesive tape [115]. For this A. K. Geim and K. S. Novoselov were awarded the noble prize in Physics in 2010. Because of its extraordinary properties such as high electrical and thermal conductivity together with optical transparency and high mechanical strength, graphene has generated a great deal of excitement amongst the theorists as well as the experimentalists.

Graphenes, fullerenes and carbon nanotubes have remarkable properties making them useful for a large number of potential applications. In the following we describe the structure, preparation, properties and some potential applications of these nanostructures.

8.2 Structure

8.2.1 *Structure of graphite*

As mentioned above, all the nanostructures of carbon can be thought of as derived from the sheets of carbon atoms which make up the structure of graphite. In graphite, these sheets, called the graphene sheets, are stacked one above the other. Both the fullerene and the carbon nanotubes can be imagined to form by folding a graphene sheet in different ways. It is useful therefore to consider the structure of graphite in some detail.

Carbon is the sixth element in the periodic table. Its electronic structure is $1s^2 2s^2 2p^2$. The two core electrons in the 1s orbital are strongly bound to the nucleus while the four electrons in the $2s$ and $2p$ orbitals are weakly bound. The energy difference in the upper $2p$ energy levels and lower $2s$ level is small so that the four electronic orbitals

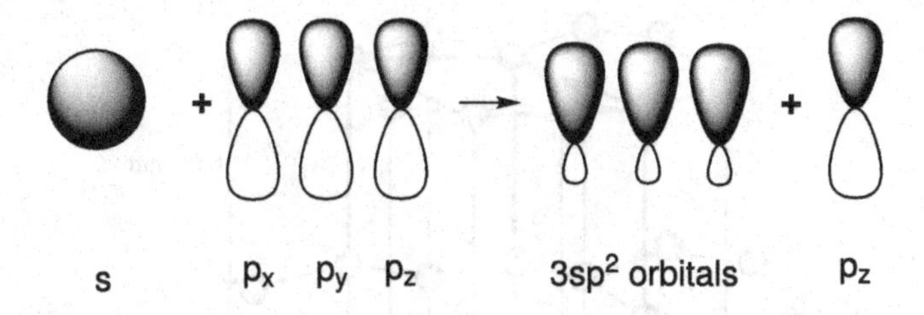

Fig. 8.1 The $2s$ orbital and two of the three $2p$ orbitals (p_x and p_y) combine to produce three sp^2 orbitals which overlap to produce three in plane bonds per carbon atom. The remaining p_z orbital, oriented perpendicularly to the plane of the sheet, overlap in graphite to produce a conducting path for electrons which is responsible for the high conductivity of graphite.

($2s$, $2p_x$, $2p_y$, $2p_z$) can easily hybridize to enhance the binding energy between the carbon atoms. Three types of hybridizations are possible in the carbon structures. These are the sp^1, sp^2 and sp^3 hybridizations corresponding to the mixture of a single $2s$ orbital with one, two or three $2p$ orbitals respectively. Depending on the type of hybridization, each carbon forms strong bonds with two (sp^1), three (sp^2) or four (sp^3) other carbon atoms giving rise to chain, sheet (graphene and graphite) or tetrahedral (diamond) structures.

In graphene and in graphite there is sp^2 hybridization. The one s orbital and the $2p$ orbitals (p_x and p_y) combine to produce three orbitals which overlap to produce three in plane bonds per carbon atom (Fig. 8.1). The remaining p_z orbitals, oriented perpendicularly to the plane of the sheet, overlap in graphite to produce a conducting path for electrons which is responsible for the high conductivity of graphite.

In graphite the alternate sheets of graphene in the stack are perfectly aligned with one another while the adjacent sheets are displaced with respect to each other by a fixed amount as shown in Fig. 8.2. In the ideal graphite (called Bernal graphite), the stacking is ABABA.. and the inter – sheet distance is 0.334 nm. In less perfect graphite (i) there are stacking faults and (ii) the interplanar spacing is larger. When the interplanar spacing increases to 0.344 nm, the correlation between the layers is essentially lost. Such graphite is called *turbostratic* graphite. As

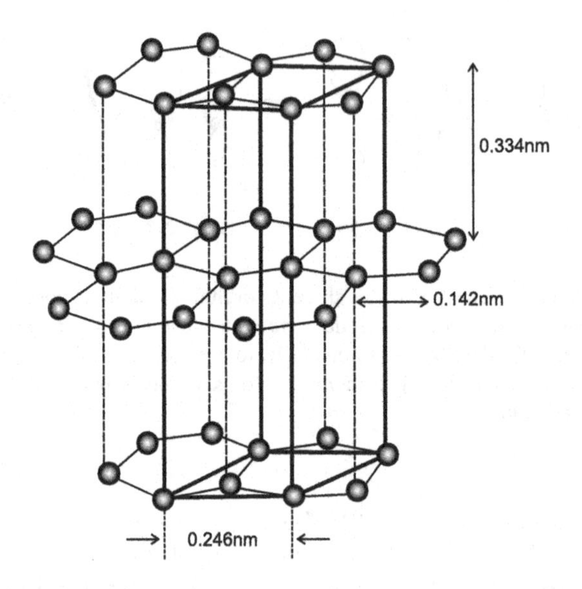

Fig. 8.2. A schematic showing the relative positions of the alternate layers in the graphite structure; the unit cell of graphite has been outlined.

we shall see later, in the multiwall carbon nanotubes, because of the different diameters of the adjacent tubes, the graphene sheets have stacking sequence similar to turbostratic graphite with little bonding between the adjacent sheets.

The structure of (Bernal) graphite can be described using a hexagonal unit cell with $a = b \neq c$, $\alpha = \beta = 90$, $\gamma = 120$ (Fig. 8.2). The number of atoms per unit cell is 4 (1 in the centre + (two at the faces)/2 + (12 at the sheet corners)/6) = 4 atoms/ cell).

8.2.2 *Structure of graphene*

As described above, each graphene sheet is a single atom thick sheet of carbon atoms. The carbon atoms are placed at the corners of hexagons which share the sides (Fig. 8.3). This can also be viewed as two interpenetrating triangular sub–lattices with the atoms at the corners of one sub–lattice being at the centers of the other sub–lattices. The inter–atom carbon to carbon distance is 0.142 nm.

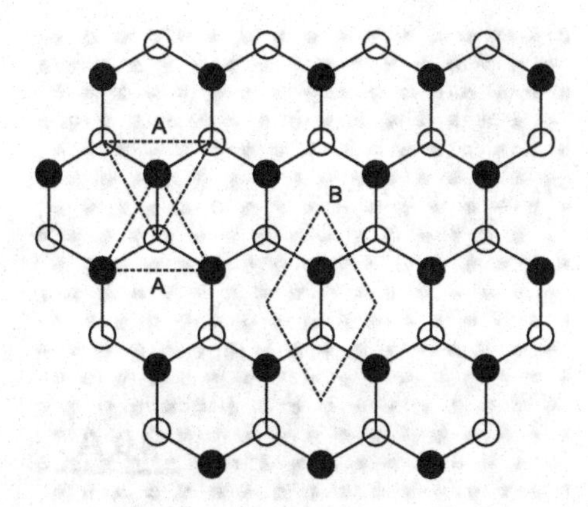

Fig. 8.3. The structure of graphene consists of carbon atoms located at the corners of hexagons, each carbon atom being shared by three hexagons. The unit cell is a hexagon with 6/3 = 2 atoms per unit cell. The structure can also be looked at as consisting of two interpenetrating triangular lattices (A) and the unit cell can be taken as the parallelogram shown at B.

Even though only one atom thick, it is possible to see a piece of graphene in the optical microscope if it is placed on a suitable substrate. High resolution transmission electron microscope is able to resolve the hexagonal arrangement of atoms in the graphene (Fig. 8.4) while the atomic force microscope can reveal the contrast between the regions of different thicknesses.

An important technique to characterize the various carbon structures is Raman spectroscopy (Chapter 5). The technique can distinguish between the various forms of carbon and can also reveal the number of layers in graphene. Figure 8.5 shows the Raman spectra from a number of carbon structures. The main features in the spectra are the *G* and *2D* peaks at 1560 and ~2670 cm^{-1}. The *G* band originates from the in plane vibrations of the sp^2 atoms. The 2D band is also ascribed to the sp^2 atoms and originates from a so-called "2 phonon double resonance process" [116]. These two bands are dominant in the carbon structures derived from graphene. The peak due to the sp^3 hybridization, as in diamond, occurs at 1332 cm^{-1}. However, even in the amorphous carbon with a high sp^3 content, the spectra is dominated by the sp^2 vibrations.

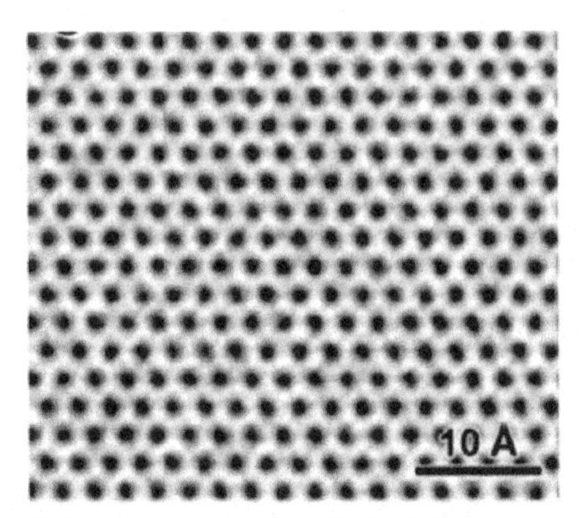

Fig. 8.4. A high resolution electron micrograph of graphene; atoms are the white dots (reprinted with permission from [117] copyright (2008)American Chemical Society).

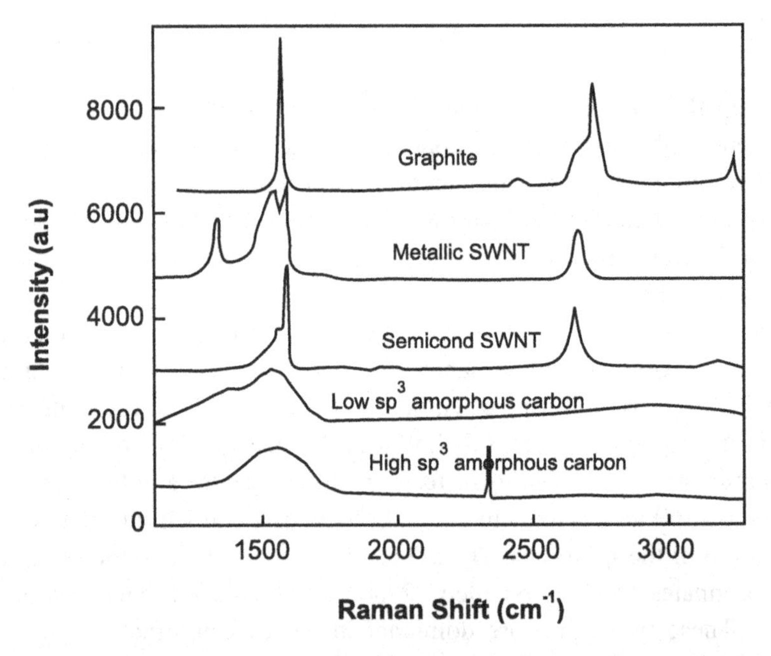

Fig. 8.5. Raman spectra of graphite, metallic and semiconducting carbon nanotubes, low and high sp^3 amorphous carbons. (Reprinted from [118] copyright (2007) with permission from Elsevier).

The *G* and *D* bands are present in all poly-aromatic hydrocarbons where the origin of these two bands is well established — the *G* peak is ascribed to the bond stretching of all pairs of sp^2 atoms while the *D* peak originates from the breathing modes of sp^2 atoms in the rings. However, in the solid carbon structures, the association of these bands with different modes is not so straightforward and some aspects still remain to be clarified [118]. Nevertheless, these characteristic bands have proved to be very useful for identifying the various forms of carbon.

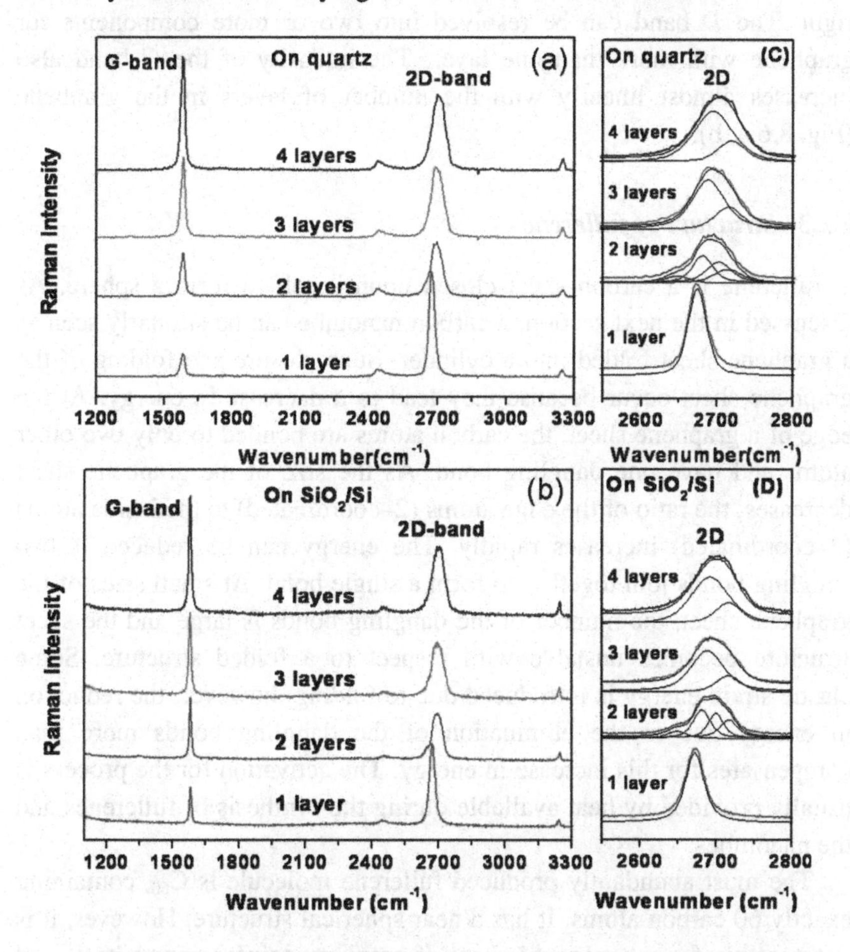

8.6. (a, b) Raman spectra of graphene with 1, 2, 3, and 4 layers. (c, d). The enlarged 2D band regions with curve fitting. Reproduced with permission from [119] Copyright 2008 American Chemical Society).

Figure 8.6 shows the Raman spectra from single layer and multilayer graphene on quartz as well as the standard SiO_2 (300 nm)/Si substrate. The *2D* band for single layer graphene is sharp and symmetric (in contrast with that for graphite which is broad and can be fitted with two peaks, Fig. 8.5). Figure 8.6 also demonstrates the power of Raman spectroscopy in differentiating between graphenes with different number of layers. The shape and the position of the *D* band change very significantly with the number of layers as shown in the figures on the right. The *D* band can be resolved into two or more components for graphene with more than one layer. The intensity of the *G* band also increases almost linearly with the number of layers in the graphene (Fig. 8.6 a, b).

8.2.3 *Structure of fullerene*

A fullerene is a carbon sheet closed upon it self to form a sphere. As discussed in the next section, a carbon nanotube can be similarly seen as a graphene sheet folded into a cylinder. Such closure and folding of the graphene sheet occur because they lead to a decrease in energy. At the edge of a graphene sheet, the carbon atoms are bonded to only two other atoms and have one dangling bond. As the size of the graphene sheet decreases, the ratio of the edge atoms (2–coordinated) to the inside atoms (3–coordinated) increases rapidly. The energy can be reduced if two dangling bonds join together to form a single bond. At small sizes of the graphene sheet, the number of the dangling bonds is large and the sheet structure becomes unstable with respect to a folded structure. Some elastic strain energy is introduced due to folding; however, the reduction in energy due to the elimination of the dangling bonds more than compensates for this increase in energy. The activation for the process is usually provided by heat available during the synthesis of fullerenes and the nanotubes.

The most abundantly produced fullerene molecule is C_{60}, containing exactly 60 carbon atoms. It has a near spherical structure. However, it is not possible for a hexagonal sheet of graphene to close upon itself and form a spherical structure consisting entirely of hexagons as this goes against the well known Euler's theorem according to which the numbers of vertices (v), edges (e) and faces (f) in a polyhedron are related by the following equation

$$v - e + f = 2$$

The relation can be satisfied only by a structure consisting of precisely 12 pentagons and 20 hexagons. Furthermore, the pentagons must be isolated from each other to minimize the strain. The structure can be thought of as derived by truncation of 12 points of an icosahedron (Fig. 8.7). The C_{60} is popularly called buckminsterfullerene or buckyball after the famous architect known for the design of the geodesic dome. Fullerenes are important for nanotube structure, because the ends of the nanotubes are capped by half of a fullerene.

Besides C_{60}, a number of other fullerenes with higher number of carbon atoms are known. C_{70} is much less dominant than C_{60}. Its structure is similar to that of C_{60} with an extra band of 10 carbon atoms around the waist. The other species, with order of magnitude less abundance, are C_{76}, C_{78}, C_{84}, C_{90}, C_{92}, C_{96}, *etc.*

The C_{60} molecule can be dissolved in nonpolar solvents like toluene and can then be deposited into a film. The C_{60} molecules arrange themselves in an FCC structure in the film with a spacing of 0.29 nm.

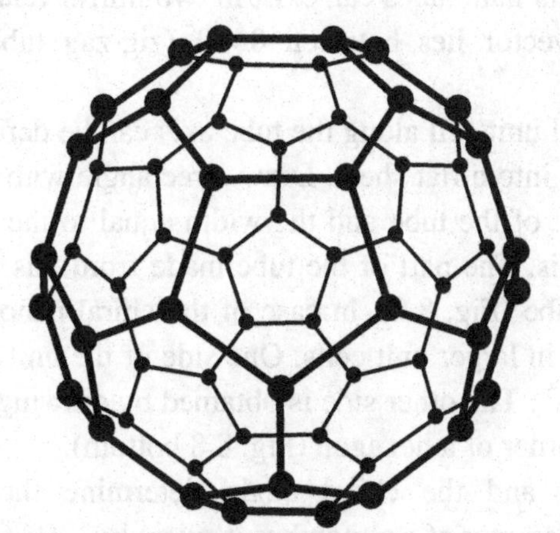

Fig. 8.7. Structure of C_{60}; the interatomic distance is 0.14 nm for two carbon atoms shared by adjacent hexagons and 0.145 nm for atoms shared by a hexagon and a pentagon.

This material is called fullerite. The bonding between the fullerenes is van der Waal's bonding so that the film is mechanically weak.

8.2.4 *Structure of carbon nanotubes*

A single wall carbon nanotube can be imagined to be made by taking a rectangular strip of graphene sheet, rolling it into a cylinder and welding the two edges. Figure 8.8 shows the arrangement of hexagons in a graphene sheet with two unit vectors $\bar{a}_1$ and $\bar{a}_2$. Any lattice point in the graphene sheet can then be obtained by a translation by integral numbers of $\bar{a}_1$ and $\bar{a}_2$. A carbon nanotube is produced by choosing a vector $\bar{c} = n\bar{a}_1 + m\bar{a}_2$ and rolling the sheet so that the origin coincides with the end point of this vector, called the chiral vector. The angle between the chiral vector and $\bar{a}_1$ is called the chiral angle. If $\bar{c} = n\bar{a}_1$ then the hexagons in the resulting tube are arranged in a zig-zag fashion along the axis of the tube. The chiral angle in this case is zero. Such a tube is called a "zig-zag" tube (Fig. 8.8 middle). If $\bar{c} = n\bar{a}_1 + n\bar{a}_2$ (*i.e.* $n = m$), then the tube obtained is called an armchair tube as the hexagons are arranged in an armchair fashion along the axis of the tube (Fig. 8.8 top). In this case the chiral angle is 30°. In all the other cases, the tube is called a chiral nanotube and the hexagons are arranged along a helix on the surface of the tube. The chiral nanotubes can exist in two mirror related forms. Any arbitrary chiral vector lies between $\theta = 0°$ (zig-zag tube) and $\theta = 30°$ (armchair tube).

A translational unit cell along the tube axis can be defined as follows. Open up the tube into a flat sheet. Draw a rectangle with length equal to the circumference of the tube and the width equal to the repeat distance along the tube axis. The part of the tube made from this rectangle is the unit cell of the tube (Fig. 8.8). In case of the chiral nanotube, the lower symmetry results in larger unit cells. One side of the unit cell is made by the chiral vector $\bar{c}$. The other side is obtained by drawing a normal to $\bar{c}$ until it meets a corner of a hexagon (Fig. 8.8 bottom).

The diameter and the chiral angle determine the properties of nanotubes. The diameter of a nanotubes is given by

$$d_t = (\sqrt{3}/\pi)a_{c\text{-}c}(m^2 + mn + n^2)^{1/2}$$

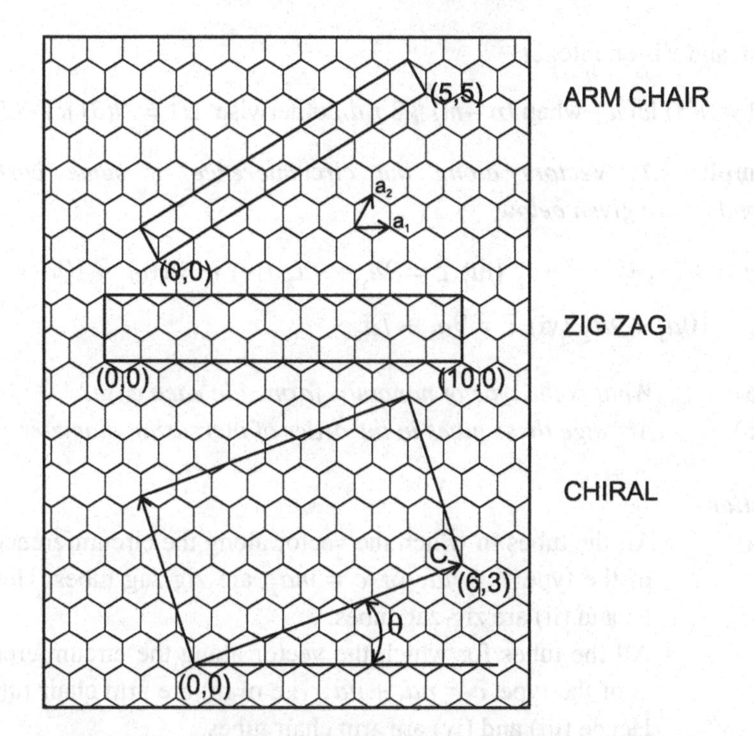

Fig. 8.8. Arrangement of hexagons in a graphene sheet; a_1 and a_2 are the lattice vectors with $|a_1| = |a_2| = a = 0.246$ nm; each side of the hexagon is 0.142 nm; if the sheet is rolled so that the origin coincides with the point *(n,n)*, *e.g.* (5,5), then the hexagons are arranged in an armchair fashion along the axis of the tube and the tube is called an armchair nanotube (top); if the sheet is rolled so that the origin coincides with the point *(n,0)* ,*e.g.* (10,0), then the hexagons in the tube will be arranged in a zig-zag manner along the axis (vertical direction) of the tube and the tube is called a zig-zag tube (middle); rolling along any other direction $na_1 + ma_2$ with $n \neq m \neq 0$ gives a chiral nanotubes in which the hexagons are arranged along a helix on the surface of the tube (bottom).

where $a_{c\text{-}c}$ is the distance between neighboring carbon atoms in the graphene sheet.

The chiral angle, ϕ, is given by;

$$\phi \;=\; \tan^{-1}(\sqrt{3}n/(2m + n))$$

The magnitude $|T|$ of the side of the unit cell, normal to c, is given by the following relation in which d_H is the highest common divisor of n

and m and r is an integer

$|T| = \sqrt{(3)}\ |c|\ d_H$ when $(n - m) \# 3\ r\ d_H$; otherwise $|T| = \sqrt{(3)}\ |c|\ /3d_H$

Example 2.1. *Vectors along the circumference of some carbon nanotubes are given below.*

(i) $\bar{c} = 9\bar{a}_1$ (ii) $\bar{c} = 9\bar{a}_2$ (iii) $\bar{c} = 9\bar{a}_1 + 9\bar{a}_2$ (iv) $\bar{c} = 10\bar{a}_1 + 10\bar{a}_2$

(v) $\bar{c} = 10\bar{a}_1 + 9\bar{a}_2$ (vi) $\bar{c} = 9\bar{a}_1 + 7\bar{a}_2$

(a) *What is the type of nanotube formed in each case?*
(b) *Arrange these tubes in the order of increasing diameter.*

Solution.

(a) All the tubes in which the vector along the circumference is of the type $\bar{c} = n\bar{a}_1$ or $\bar{c} = m\bar{a}_2$ are zig-zag tubes. Hence (i) and (ii) are zig-zag tubes.
All the tubes for which the vector along the circumference is of the type $\bar{c} = n\bar{a}_1 + n\bar{a}_2$ (i.e n=m) are arm chair tubes. Hence (iii) and (iv) are arm chair tubes.
All the other tubes, *i.e.* for the tubes for which $n \neq m \neq 0$ are chiral nanotubes. Hence (v) and (vi) are chiral nanotubes.

(b) The diameter of a nanotube is given by
$d_t = (\sqrt{3}/\pi)a_{c\text{-}c}(m^2 + mn + n^2)^{1/2}$
Therefore, the tube with the larger value of the quantity $(m^2 + mn + n^2)$ has a larger diameter.
The tubes therefore have increasing diameters in the following order: (i), (ii), (vi), (iii), (v), (iv) □

The earliest nanotubes produced were multiwalled nanotubes (MWNT). In these tubes, more than one graphene cylinders are arranged concentrically. Subsequently, single walled nanotubes (SWNT), consisting of a single tube, were also produced. The single wall nanotubes require a catalyst for their growth while the multiwall tubes can be grown without a catalyst. The conventional MWNT have inner diameters ranging from 1.5 to 15 nm and outer diameters from 2.5 nm to

30 nm. The SWNT usually have diameters ranging from 0.7 to 1.6 nm. The dimensions of the SWNT depend sensitively on the catalyst used and the preparation conditions. Single wall nanotubes with much larger diameter, up to 6 nm, can be produced when, in addition to the catalyst (*e.g.* Co), some sulfur as elemental S or as metal sulfide is introduced in the anode.

Example 2.2. *Assume that a multiwall carbon nanotube is made of two concentric cylinders with separation between the adjacent cylinders approximately equal to that between the sheets of hexagonal graphite i.e. 0.334nm. Show that this separation is possible for the armchair nanotubes but not for the zig-zag nanotubes. The length of the side of a hexagon in the graphene sheet is 0.142 nm.*

Solution. Figure 8.9 below shows the arrangement of hexagons along the circumference in the arm chair and the zig-zag tubes.

Repeat distance along the circumference of an armchair tube

$$X_1 = a + a + a = 3a = 3 \times 0.142 = 0.416 \text{ nm}$$

Difference in the circumference of the two cylinders

$$2\pi\,(r_2 - r_1) = 2\pi \,.\, 0.334 = 2.1 \text{ nm}$$

where r_1 and r_2 are the radii of the inner and the outer tubes respectively. As 2.1 nm is an integral multiple of 0.416 (since 0.416 x 5 ≈ 2.1 nm), 5 repeat units can exactly accommodate the difference in the circumference and so this separation is possible for the armchair tubes. For the zig-zag tubes, the repeat distance along the circumference is

$$x_2 = 2a\cos 30 = a\sqrt{3} = 0.142\sqrt{3} = 0.246 \text{ nm}$$

Since 2.1 is not an integral multiple of 0.246, the desired separation is not possible in the zig-zag tubes. □

The cross sectional shape of the MWNT is frequently observed to be noncircular. The spacing between the cylinders in the MWNT is usually uniform but sometimes uneven spacing is also observed. The magnitude of this interlayer spacing is ~ 0.344 nm which is similar to that found in

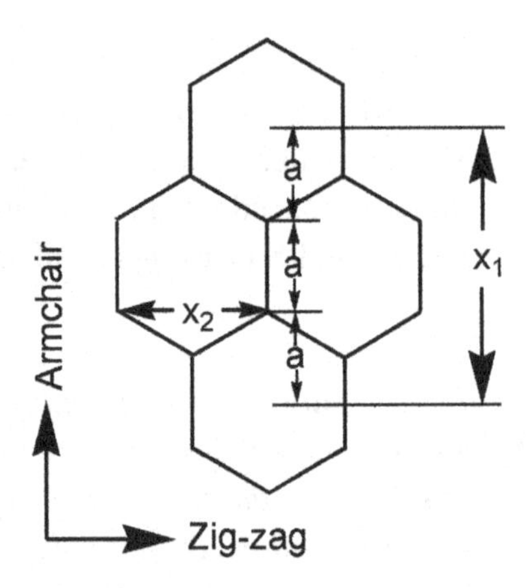

Fig. 8.9. The arrangement of hexagons along the circumference of a zig-zag and an armchair tube.

turbostratic graphite; however, values ranging from 0.342 to 0.375 have also been noted. An interlayer spacing of 0.334 nm, equal to that in graphite is possible only for the arm chair tubes and not for the zig-zag tubes. In the arm chair tube the ABAB.. stacking of graphite can be maintained. For the zig-zag tube, extra rows of atoms have to be introduced so that the ABAB.. stacking is present only in a part of the circumference.

The ends of all the nanotubes, the SWNT as well as the MWNT, are almost always closed. As a hexagonal lattice of any size or shape can only be closed by inclusion of precisely 12 pentagons (Euler's law), each such cap must contain six pentagons. To minimize the strain, these pentagons must be isolated from each other. It has been shown by Fujita *et al.* (1992), that the smallest nanotubes which can be capped with isolated pentagons are the two archetypal tubes (5, 5) and (9,0). For these tubes there is only one possible cap — one half of a C_{60} molecule divided along one of the three fold axes for the zig-zag (9,0) tube and along the five fold axis for the armchair (5,5) tube. For the larger tubes, the number of the possible caps increases rapidly with increasing diameter.

In MWNT the caps can be symmetric or asymmetric. A symmetric cap is nothing but half of a fullerene. The fullerenes become less

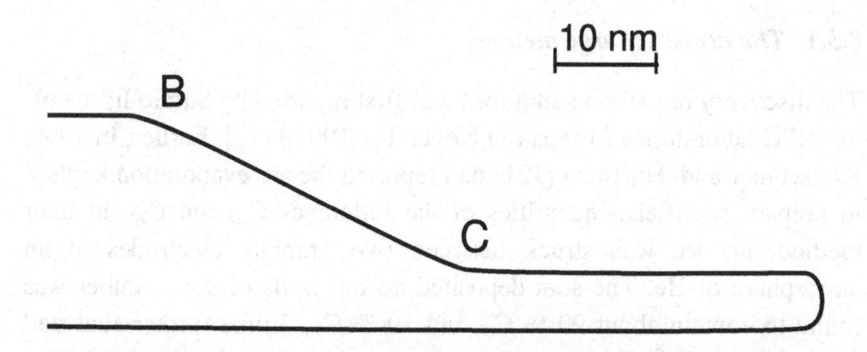

Fig. 8.10. Multiwall CNT with an asymmetrical cone cap; a pentagon at B and a heptagon at C give the tube positive and negative curvatures respectively (adapted from [120] copyright (1993) with permission from Elsevier).

spherical in shape as their size increases. Amongst the asymmetrical caps, more common is the asymmetric cone shape (Fig. 8.10). Note that the presence of a pentagon is required to impart a positive (convex) curvature (at B in Fig. 8.10) to the tube while a heptagon is needed to make the tube to bend in a concave manner (at C in Fig. 8.10). Another remarkable structural feature of the MWNT is that one or more layers may close midway or form compartments.

8.3 Preparation of Carbon nanotubes

In all the methods for the preparation of carbon nanotubes, carbon is first brought into gaseous phase and then condensed. The source of carbon is either solid graphite or a gaseous compound such as a hydrocarbon. A catalyst is usually needed to obtain single walled tubes while the multiwall tubes can be grown without a catalyst. Some of the methods reported for the preparation of single and multiwalled tubes are described below.

8.3.1 *The arc–discharge method*

The discovery of carbon nanotubes was first reported by Sumio Iijima of the NEC laboratories in Japan in November 1991 [113]. Earlier, in 1990, Kratschmer and Huffman [121] had reported the arc evaporation method to prepare significant quantities of the fullerenes C_{60} and C_{70}. In their method, an arc was struck between two graphite electrodes in an atmosphere of He. The soot deposited on the walls of the chamber was found to contain about 90 % C_{60} and 10 % C_{70}. Iijima further analyzed the soot as well as the deposit produced on the cathode. While nothing remarkable was found in the soot, he found that the fibrous core of the deposit contained aligned microfibrils made of carbon nanotubes.

In the arc evaporation method, an arc is struck between two graphite electrodes. The electrodes are water cooled. The chamber is evacuated to about 10^{-6} torr (diffusion pump vacuum). A continuous flow of helium at ~500 torr is maintained. The anode is consumed and is continuously advanced. A cylindrical deposit is produced on the cathode. Efficient cooling of the electrodes produces more cylindrical and homogeneous deposit. The quantity of the nanotubes in the cylindrical deposit increases with He pressure, the optimum pressure being ~500 torr. Ebbesen and Ajayan (1992) subsequently optimized the arc discharge method to obtain high quality MWNT at the gram level [122].

The nanotubes prepared as above are multiwalled. One of the first groups of researchers to report the preparation of single walled carbon nanotubes were Bethune *et al.* at IBM [114]. With a view to obtain carbon coated clusters of magnetic metals like Fe, Co, Ni, they used an arc evaporation method in which the graphite electrodes were impregnated with the desired metal. Instead of carbon coated metal clusters, they found single walled carbon nanotubes (together with amorphous graphite, particles of metal and metal carbides) in the soot which hung from the chamber after the arc evaporation experiment. The metals impregnated into the electrode act as catalysts, promoting the growth of SWNT. Journet *et al.* [123] optimized the growth of SWNT by the arc discharge method using a carbon anode containing 4.2 at % Ni and 1.0 at % Y as catalysts.

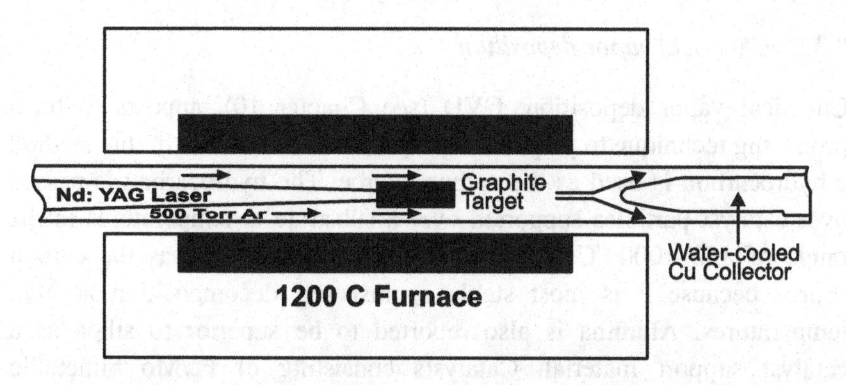

Fig. 8.11. Schematic of the laser evaporation method for the preparation of multiwalled nanotubes (adapted with permission from [124] copyright (1995) American Chemical Society).

8.3.2 *Laser vaporization*

In 1995, Smalley's group reported the preparation of MWNT by using a laser beam to vaporize a graphite target [124]. Later in the same year, this group also reported the preparation of single walled nanotubes by laser ablation of a composite metal–graphite target containing 0.5 at % of Ni and Co. The target was placed in a tube furnace heated to 1200°C. An inert gas was passed through the furnace and the grown nanotubes were collected on a cold finger kept downstream (Fig. 8.11). A high yield (70 to 90 % of the graphite target) of uniform sized single wall tubes was obtained when the catalytic mixture comprised equal amounts of Co and Ni. The individual tubes have a tendency to form bundles which were called nanotubes ropes. The ropes consist of tens of nanotubes held together in a hexagonal packing by van der Waal's interactions. The ropes are between 10 and 20 nm across and up to 100 μm long. Each rope is found to consist of a bundle of single-wall carbon nanotubes aligned along a single direction. The diameters of the single-wall nanotubes have a narrow distribution with a strong peak. The diameter of the tubes can be controlled by the type of catalyst used and the furnace temperature.

8.3.3 *Chemical vapor deposition*

Chemical vapor deposition, CVD (see Chapter 10), appears to be a promising technique to prepare large quantities of SWNT. In this method a hydrocarbon is used as the carbon source. The hydrocarbon is passed over catalyst particles supported over a substrate at temperatures in the range of 850–1000 °C. Dai *et al.* [125] used methane as the carbon source because it is most stable against self decomposition at high temperatures. Alumina is also reported to be superior to silica as a catalyst support material. Catalysts consisting of Fe/Mo bimetallic species have been reported to be more efficient for the production of the nanotubes. The yield of the nanotubes can be increased by uniform dispersion of unagglomerated particles of the catalyst. Use of an aeogel support for this purpose results in the production of 2 grams of nanotubes for every gram of the catalyst. The amount of catalyst required per gram of nanotubes produced is an important issue in the bulk production of SWNT.

Both the MWNT as well as the SWNT can form in the CVD reactor depending on the temperature and the size of the catalyst particles. Higher temperature (900–1200 °C) is needed for SWNT as compared to 600–900 °C for the formation of MWNT. By using plasma enhanced CVD (PECVD) these temperatures are lowered to 350 °C and 120 °C respectively which is attractive for electronic applications.

The nanotubes prepared using CVD show poorer mechanical properties as compared to the nanotubes prepared by arc discharge indicating that the former have much larger number of defects.

The ends of the SWNT are invariably capped by hemispherical caps. Fujita, Dresselhaus *et al.* [126] have shown that the smallest nanotubes that can be capped are the archetypal (5,5) armchair and (9,0) zig-zag tube. From Euler's law stated earlier, each cap must contain exactly 6 pentagons. For a zig-zag tube, the cap is produced by cutting a C_{60} fullerene parallel to one of the three fold axes; for the arm chair tube the cut is along one of the five fold axes. The number of the possible caps increases rapidly with increasing tube diameter.

8.3.4 *Patterned growth of carbon nanotubes*

For the utilization of nanotubes in devices and other structures, it is important to be able to grow the nanotubes at specific desired locations on a substrate. The fact that the metal catalysts facilitate the growth of nanotubes provides a way to achieve this goal. Typically, a patterned layer of the suitable metal catalyst is first deposited on the substrate. Growth of CNT is then carried out by CVD or other suitable technique. The carbon nanotubes grow out preferentially over the metal catalyst pattern [127]. The catalyst chosen should have a strong interaction with the substrate so that the catalytic particles stay anchored and do not sinter at high temperatures used during the growth. Porous silicon has been found to be an ideal substrate for this purpose. Fig. 8.12 shows an array of self-oriented MWNT grown in this way. The MWNT are held together by the van der Waals's forces which help in the growth direction remaining perpendicular to the substrate. Patterns of SWNT have also been obtained by first patterning the catalyst by electron beam lithography and using methane as the precursor in CVD.

Fig. 8.12. Carbon nanotube arrays grown on n+type porous silicon ; the substrate was patterned with an array of Fe thin film squares (reprinted from [127] copyright (2000) with permission from Elsevier).

8.3.5 *Theories of nanotube growth*

Several models have been proposed to explain how the nanotubes form and grow. The growth of nanotubes by CVD using a catalyst is believed to proceed in the following manner. The hydrocarbon decomposes at the surface of the catalyst; the carbon released is then dissolved in the catalyst particle. When the metal particle gets saturated with carbon, the precipitation of carbon from the metal particle occurs, leading to the formation of tubular nanotubes (Fig. 8.13). This model is termed as Vapor–liquid–solid (VLS) model [128, 129]. Significant solubility of carbon in metals like Fe, Co, Ni, Mo supports this mechanism. The model is discussed in more detail in the next chapter in connection with the growth of one dimensional nanostructures of inorganic materials. The tube formation is favored over other forms such as a sheet because the tube contains no dangling bonds. The mechanisms involved in the growth of carbon nanotubes are at present poorly understood.

8.4 Preparation of graphene

The technique of extracting graphene from graphite using the adhesive tape as pioneered by Novosolev and Geim [115] is called "micromechanical exfoliation". The interlayer bonding energy in graphite is about 2 eV nm^{-2} which translates into a force of 300 nN μm^{-2} which is easily achieved during peeling of an adhesive tape. The graphite

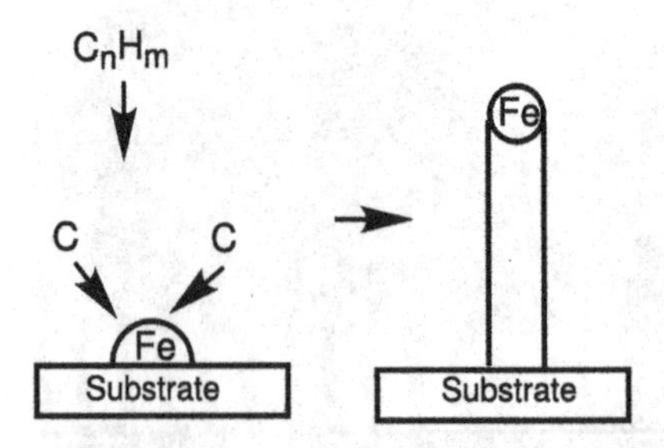

Fig. 8.13. Carbon from the decomposition of the hydrocarbon dissolves in the catalyst particle; this is followed by the precipitation of carbon at the catalyst surface leading to the formation of the nanotube.

and graphene pieces sticking to the tape can be easily transferred to a substrate by pressing the tape against it. The quality of graphene is superior to that produced using other techniques. The technique is good for preparation of small samples for study.

For applications such as in composites and transparent conducting thin films, graphene samples in large quantities are needed. For electronic applications, graphene needs to be deposited and patterned on suitable substrates, preferably on silicon or silicon carbide to make it compatible with the well established silicon technology. Several approaches are being pursued to meet these objectives [130].

The exfoliation of graphite to yield graphene can also be effected by chemical means. A product called *expandable graphite* has been known since 1958 and is commercially available. It is made by intercalating an acid such as sulfuric acid, nitric acid or acetic acid in between the layers of graphite in the presence of an oxidant like H_2O_2, HNO_3 or $KMnO_4$. The intercalation causes an increase in the interlayer spacing between the sheets of graphite from 0.34 nm to 0.65 to 0.75 nm. A rapid annealing to 1050° C generates high pressure CO_2 between the layers, causing them to separate resulting a in a product having a few layers or even a single layer. However, the oxidation and exfoliation step converts a large fraction of sp^2 to sp^3. The sp^2 character can only be partially recovered by a reducing treatment. Despite this partial sp^2 character, a coating of the reduced graphite oxide made by spin coating from a suspension is transparent and has a conductivity which is higher by four to five orders of magnitude as compared to graphene oxide. There is thus much interest in using it as a transparent conductor to replace indium tin oxide. Graphene can be deposited on metallic substrates (Pt, Ni, Cu, Ru, Ir, Co) and on carbides (TiC, TaC) by chemical vapor deposition using a hydrocarbon as a source of carbon. The formation and growth of graphene on metals is believed to proceed by the dissolution of carbon in the metal followed by its precipitation as graphene, as in the VLS mechanism for the carbon nanotubes. Some of the problems encountered in the production of graphene film by this method are island formation instead of a continuous film, small crystallite size of the film and its strong chemisorption to the metal substrate which makes it difficult to transfer the film to another substrate such as silicon. However, Reina *et al.* [131] have demonstrated the growth of centimeter sized continuous graphene films with domains having one to few layer thick graphene on nickel polycrystalline film substrates by CVD growth. Limited adhesion of the film to nickel makes it possible to dissolve out nickel by an acid and to transfer the film to another substrate. Mobility as high as 4000 $cm^2 V^{-1}s^{-1}$ has been measured for these films.

Another promising technique for the growth of graphene supported on SiC is to heat SiC in vacuum or in an inert gas atmosphere at temperatures between 1200–1600 °C. As the sublimation rate of silicon is higher than that of carbon, excess carbon is left behind on the surface which rearranges to form graphene. The technique produces graphene varying in thickness from one to few layers. Shivraman *et al.* [132] deposited graphene on SiC using this technique and, patterned it into doubly clamped resonators up to 20 μm in length. The high quality of graphene prepared using this method, as evidenced by its high electron mobility (2000 $cm^2 V^{-1}s^{-1}$ at 27 °K), is suitable for fabricating field effect devices.

Procedures for producing graphene by cutting the carbon nanotubes along their lengths are also being pursued. The resulting material in most cases consists of graphitic nanoribbons and partially opened tubes in addition to the graphene flakes.

8.5 Properties

8.5.1 *Mechanical properties*

The carbon–carbon bond is one of the strongest bonds known. It is not surprising therefore that the carbon fibers constitute a class of strongest materials available and are used extensively in structural components of aircraft and space vehicles where light weight together with high strength and modulus are needed. In the carbon fibers the basal planes are aligned along the fiber axis. The full potential of the C–C bond is, however, not realized in the carbon fibers due to the presence of defects. The carbon nanotubes are found to have much higher strength; this means that the number and severity of defects in the nanotubes are much smaller as compared to the carbon fibers. There is much interest therefore in using carbon nanotubes for producing composites with high strength and modulus.

For the following discussion of the mechanical properties of the carbon nanotubes and for the material covered in Chapters 10–12, some knowledge of the mechanical properties of materials is needed. For the readers not very familiar with this topic, an introduction to the mechanical properties is given in Appendix 2.

8.5.1.1 *Theoretical modeling of the modulus, strength and fracture*

The modulus of graphene and the carbon nanotubes has been successfully modeled; however, the presence of defects makes the prediction of strength difficult. Using molecular structural mechanics and other approaches [133, 134] the modulus and strength of graphene sheet are estimated to be ~1 TPa (10^{12} Nm^{-2}) and 140 to 177 GPa respectively. Most calculations show the modulus of the SWNT to be close to that of the graphene sheet *i.e.* close to 1 TPa (Table 8.1) Thus the molecular dynamics (MD) simulations by Liew *et al* [135] have shown the modulus of one to four walled CNT to be between 1.04 to 0.93 TPa. Most of the other simulation results also arrive at a number close to 1 TPa for the modulus of SWNT and MWNT, although values as low as 0.76 TPa and as high as 3.62 TPa have also been reported (Table 8.1).

Theoretical estimates of the strength of CNT have also been made. These calculations assume a defect free structure and calculate the strength and the strain to fracture. The generation of defects in the perfect tubes during loading and their growth leading to fracture is also of interest in such models. Some results on the calculated strength of SWNT are also given in Table 8.1. The values are of the order of 10 GPa.

The maximum elongation of CNT before fracture has been calculated to be ~20 % [136]. The fracture is believed to be preceded by growth of a type of defects called the Stone–Wales defect. The defect consists of pairs of pentagons and hexagons located next to each other amongst the hexagons constituting the tube. The formation of this defect can be

Table 8.1 Some theoretical estimates of the modulus and the strength of CNT.

Sample	Technique	Assumptions	E (TPa)	σ (GPa)	Reference
Graphene sheet	Theory	Thickness =0.34 nm	1.1		[137]
SWNT	First principles calculation		0.76	6.25	[138]
SWNT	Molecular dynamics		3.62	9.6	[139]
SWNT	Molecular dynamics		1.24 –1.35		[140]
MWNT	Molecular dynamics		1.05		[141]
SWNT	Tight binding	Dia. 0.3 –13.5 nm	0.97		[142]

visualized to occur by rotation of a carbon – carbon bond as depicted in Fig. 8.14. These defects can be produced during synthesis or may nucleate on applying a stress to the nanotube. The nucleation and growth of Stone-Wales defect in a carbon nanotube leading to its fracture is shown schematically in Fig. 8.15.

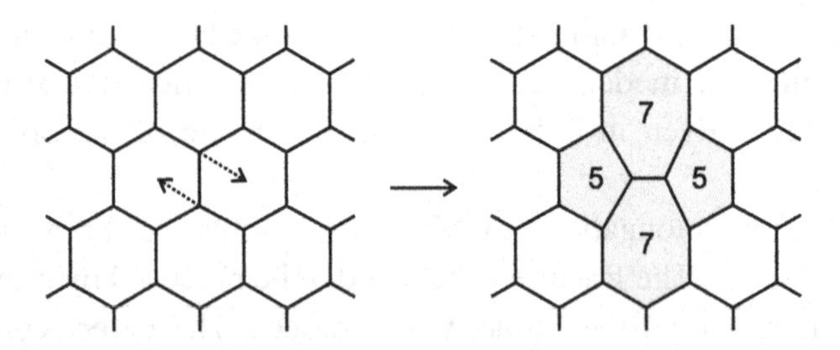

Fig. 8.14. Formation of the Stone–Wales defect in a carbon nanotube; the process can be visualized to occur by the rotation of the vertical bond on the left hand side resulting in the formation of two heptagons and two pentagons.

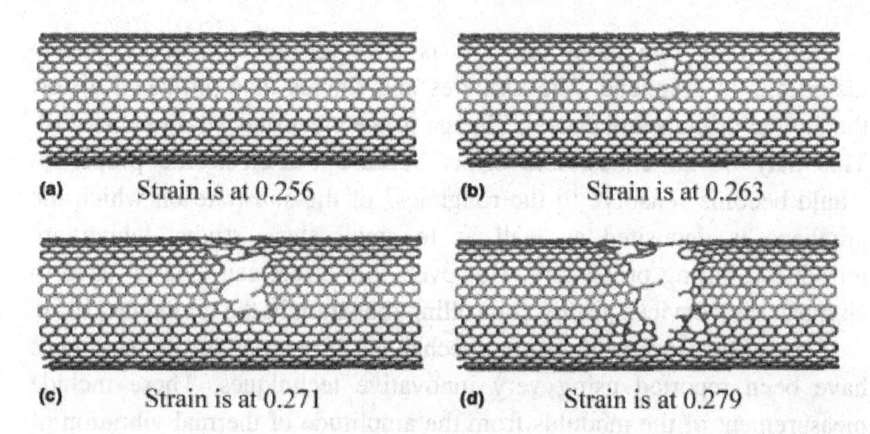

Fig. 8.15. The brittle fracture process of the outermost layer of four-walled CNT. (a) Formation of initial fracture. (b) Fracture propagation in circumferential direction. (c) Fracture propagation in the axial and circumferential direction. (d) The CNT is almost completely broken. (Reprinted from [135] copyright (2004) with permission from Elsevier.)

8.5.1.2 *Experimental measurements of modulus and strength*

The experimental measurements of the mechanical properties of the carbon nanostructures are difficult due to their small size. Graphene was discovered after the carbon nanotubes; its properties have, therefore, been much less investigated. Hone and coworkers [143] measured the Young's modulus of single graphene sheet. They prepared an array of wells, 1 and 1.5 μm in diameter and 500 nm deep, in Si with a top SiO_2 layer by nanoimprint lithography and reactive ion etching. The graphene flakes were mechanically deposited over the substrate. The graphene adhered strongly around the holes. An atomic force microscope was used to indent the graphene lying over a hole and record the force vs. displacement. From these a value of 1 ± 0.1 Tpa for the Young's modulus and 130 ± 10 GPa for the strength was obtained. The latter is close to the theoretical strength of a defect free sheet.

The graphene oxide sheets are much easier to prepare in large quantity as compared to graphene. Despite their defects, their modulus has been measured to be 0.25 TPa [144]. This makes graphene oxide an attractive material as a reinforcement in composites.

Being only one atom thick, it is easy to induce "out of plane" distortions in graphene. The wrinkles and ripples thus produced distort the atomic orbitals and cause a change in the electronic structure as well. This may be an undesirable effect because the electronic properties would become sensitive to the roughness of the substrate on which the graphene is deposited as well as to small shear strains which are introduced during processing. However, this effect can also be used to engineer new devices just by controlling the strain in the graphene [145].

Several measurements of the mechanical properties of the nanotubes have been reported using very innovative techniques. These include measurement of the modulus from the amplitude of thermal vibration of the free end of a tube in the TEM and by bending of a tube lying across a nanopore in an Al membrane by an AFM tip. The modulus values tend to be very close to the value for the graphene sheet in most of the cases. Some of these are summarized in Table 8.2 and a plot of some of the experimental and theoretical values of the modulus is shown in Fig. 8.16. The current estimates of the experimental modulus are 1 TPa for both the SWNT and the MWNT as shown in Table 8.2.

Table 8.2. Some experimental values of the modulus of carbon nanotubes.

Sample	Technique	E(TPa)	Comments	Reference
MWNT	Amplitude of thermal vibration in TEM	1.8 average (0.4–4.15)		[146]
MWNT	Bending of tip of anchored CNT by AFM tip.	1.28 ± 0.5	No dependence on tube dia.	[147]
MWNT	Bend CNT lying across a hole in an Al membrane	0.81 ± 0.41 0.001– 0.005	Arc discharge grown. Catalytic decomposition of acetylene (higher density of defects).	[148]
MWNT	Direct tensile testing.	0.91 ± 0.8 Range 0.1–1.8	Arc discharge grown.	[149]

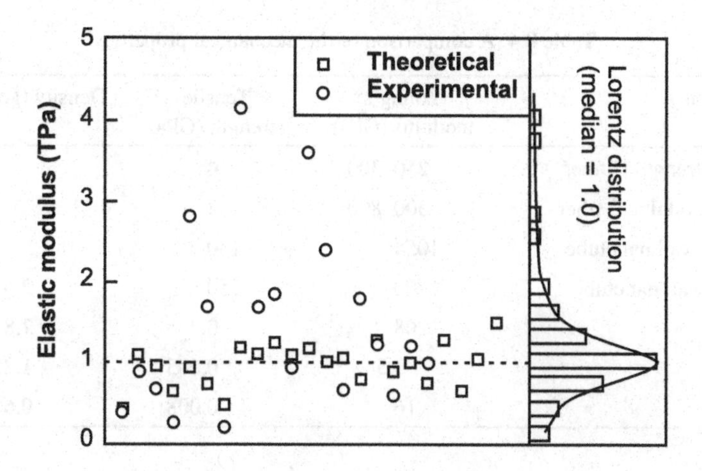

Fig. 8.16. A plot of the available experimental and theoretical values of the modulus of both the SWNT and MWNT (reproduced from [150] with permission from Elsevier).

Theoretical estimates of the strength of CNT as presented earlier have been made assuming no preexisting defects. However, the strength of brittle solids, unlike the modulus, is determined by the nature and density of defects. Even though CNT can show large elongations before failure, the essential failure mechanism is the breakage of the C–C bond, a process which is accompanied by no plastic deformation. Hence the strength of CNT is also determined by the nature and the intensity of

Table 8.3. Weibull parameters from the strength easurements.

Material	Weibull scale parameter (GPa)	Weibull shape parameter
CVD grown MWCNT	109.3	1.7
Arc discharge MWCNT	31.5	2.4
SWNT bundles	33.9	2.7
Pitch based carbon fibers Low modulus	1.6	4.0
Medium modulus	1.6	3.2
High modulus	1.6	2.1

Table 8.4. A comparison of the mechanical properties

Material	Young's modulus (GPa)	Tensile strength (GPa)	Density (gcm^{-3})
High strength C fiber	250–300	4	
High modulus C fiber	300–800	2	
Single wall nanotube	1054	150	
Multi wall nanotube	1200	150	2.6
Steel	208	0.4	7.8
Epoxy	3.5	0.005	1.25
Wood	16	0.008	0.6

defects. A knowledge of these defects is required for a reliable estimation of the strength. Presence of defects also introduces considerable variation in the measured strength. A few measurements of the strength of CNT have been made in an SEM and the data has been fitted to the Weibull distribution. The parameters of the distribution obtained for CNT are listed in Table 8.3; the data for the carbon fibers is also included for comparison [151].

The arc discharge grown CNT are more regular and have fewer defects as compared to the CVD grown tubes. This is reflected in their higher shape parameter — a higher value of this parameter indicates smaller scatter in the strength data. The Weibull scale parameter is a measure of the average value of the strength. The higher value for the CVD tubes is ascribed to a greater degree of stress transfer from the outer tubes to the inner tubes due to irregularities in the tube.

The early measurements of the strengths of the CNT were carried out on a few samples and were not fitted to the Weibull distribution. On the basis of these measurements the strength of the SWNT as well as the MWNT was placed at 150 GPa. However, the more thorough measurements reported above show that the strength is probably closer to 100 GPa for the CVD grown tubes and much lower for the arc discharge grown tubes.

An interesting feature of the nanotubes is that under compression they buckle to a significant degree but then unfold to recover the original shape on removal of the load [152].

Table 8.4 compares the modulus and strength of the CNT with carbon fibers and some other conventional materials.

8.5.2 *Electrical properties*

8.5.2.1 *Electrical properties of graphene*

In Chapter 3 we saw that in metals the valence and the conduction bands overlap while in the semiconductors and the insulators there is a finite gap between the valence and the conduction bands. The band gap has a small value in case of the semiconductors and is much larger in insulators. The band structure of solids is presented as E vs k plots. It is customary to show such plots with reference to the important high symmetry points of the reciprocal lattice (see Chapter 3).

The graphene has been a much studied model two dimensional solid structure for solid state theoreticians for several decades prior to its extraction in the laboratory in 2004. Figure 8.17 shows the unit cell, a section of the reciprocal lattice and the first Brillouin zone for graphene. The calculated band structure of graphene is shown in Fig. 8.18 with reference to the high symmetry points of the Brillouin zone. As can be seen from this figure, in graphene the valence band (the π band) and the conduction band (the π^* band) cross at points K and K′. As there is no

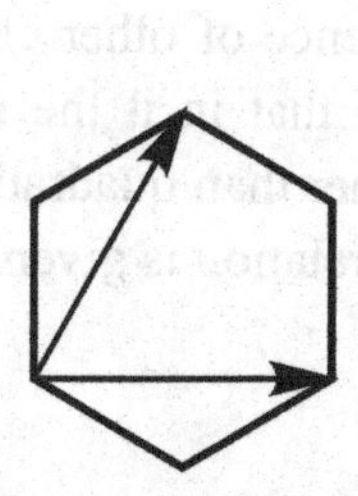

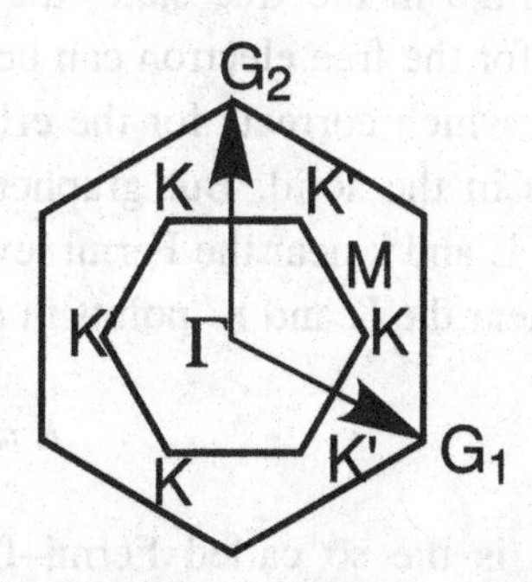

Fig. 8.17. The figure shows the unit cell of graphene with the primitive lattice vectors and the corresponding reciprocal lattice with primitive vectors G_1 and G_2. The first Brillouin zone is obtained following the procedure described in Chapter 3 i.e. by joining the normal bisectors of the reciprocal lattice vectors The points Γ, K and K′ are the high symmetry points of the Brillouin zone.

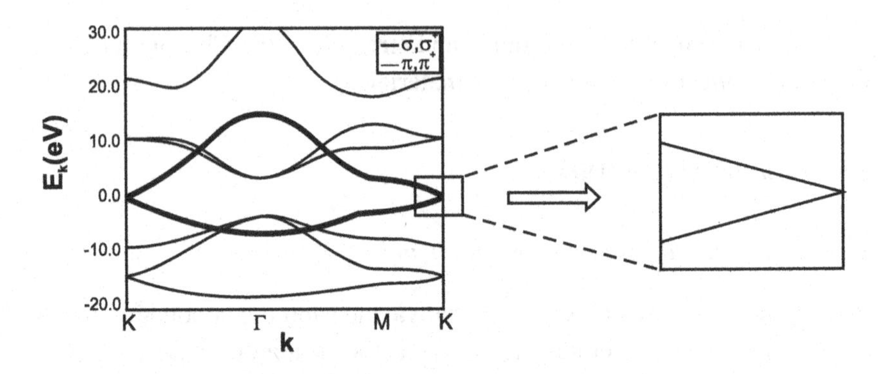

Fig. 8.18. (a) The band structure of graphene at 0 °K. The valence band (π) and conduction band (π^*) are shown by bold lines. The valence and the conduction bands cross at points K and K′. (b) In the vicinity of K and K′ the energy varies linearly with k.

gap between the valence and the conduction bands at these points, the graphene behaves as a metal.

Graphene has several unique properties. One of them is that the electrons in graphene near the Fermi energy are not described by the Schrodinger equation but instead follow another equation, the so called relativistic Dirac equation. As we saw in Chapter 3, in the free state the energy of an electron varies as square of k (*i.e.* $E = \hbar^2k^2/2m$). Inside a solid, unlike in the free state, the potential V is not zero; even so the relation for the free electron can be used with m replaced by an effective mass m^* which corrects for the effect of the presence of other electrons and ions in the solid. But graphene is unique in that in it the relation between E and k near the Fermi level is linear rather than quadratic. This is seen near the K and K′ points in Fig. 8.18. This relation is given by

$$E = \pm v|p|$$

where v is the so called Fermi–Dirac velocity which depends on the material properties. In graphene v is equal to about c/300 where c is the velocity of light. This implies that the quantum mechanical description of graphene is identical to that of relativistic particles with nearly zero mass which are described by the so called relativistic Dirac equation. For this reason the electrons in graphene are called Dirac Fermions. This leads to several interesting phenomena such as the ability of the electrons to tunnel through barrier of any height (the Klein paradox)[130, 153, 154].

The electrons in graphene have extremely high mobilities. A value of 2000 $cm^2V^{-1}s^{-1}$ has been measured at room temperature for micromechanically deposited graphene. With improved sample preparation and treatment, this can be increased to 200,000 $cm^2V^{-1}s^{-1}$ in annealed and suspended graphene [130]. In annealed samples the mean free path of the electrons reaches 1 μm which is larger than the size of many devices. This implies that the ballistic transport of electrons and holes can be achieved in the devices made out of graphene.

The band structure of graphene changes with the number of graphene layers as well as the stacking of the layers. Usually the layers in multilayer graphene are stacked as ABAB…i.e. the carbon atoms in a B layer are located in one set of hollows in the A layer. Hence the band structures have mostly been calculated for this kind of stacking. Other possible stackings are ABCABC.. and a stacking with no particular order as is the case with the turbostratic graphite.

Calculations show that the band structure of bilayer graphene has parabolic bands; so, unlike the single layer graphene, the electrons in this case are not Dirac fermions. The bands touch at the Fermi level but a gap can be introduced by applying an electric field. The presence of a band gap is necessary for graphene to be used in devices. As the number of layers increases, the band structure becomes more complicated and the conduction and the valence bands begin to overlap, the band structure gradually changing to that of graphite with an overlap of ~ 40 meV between the valence and conduction bands.

For device purposes, a band gap and a non linear relation between *E* and k is needed for applications such as signal amplification. As the band structure is sensitive to lattice symmetry, a change in symmetry can be used to introduce a band gap in graphene. Such change of symmetry can be produced by an electric field which can be applied externally, by doping, by introducing defects or even by the effect of the substrate on which graphene has been deposited [155]. Thus the partial hydrogenation of graphene produces sp^3 hydrocarbon defects and leads to the appearance of a small band gap. Graphene deposited on Ni and Ru is strongly chemisorbed resulting in a band gap of ~ 1–2 eV while epitaxial graphene grown on the silicon rich face of SiC has a band gap of 260 meV.

A much more direct effect leading to band gap is the lateral confinement of electrons which is present in the ribbons of graphene. The ribbons are perfectly metallic when their edges have the zig-zag arrangement of hexagons but are semiconducting for the armchair arrangement at the edges. The gap of these ribbons varies with width, length and topology [156].

8.5.2.2 *Electrical properties of carbon nanotubes*

The single wall carbon nanotubes have diameters of the order of 1nm. At this size the quantum effects become important. It was shown theoretically in 1991 that SWNT can have either metallic conductivity or be semiconducting depending on their structure The E vs k relations for three different nanotubes are shown in the Fig. 8.19. For the case of the armchair tube (Fig. 8.18a) the valence and the conduction bands touch at $-(2/3)(\pi/a)$. All armchair tubes have similar dispersion relations so that all the armchair tubes are metallic.

In case of the zig-zag tubes, the valence and the conduction bands touch at $k = 0$ for the tube for which n is divisible by 3. Hence such tubes are metallic. This is not the case for a tube for which n is not divisible by 3, *e.g.* for a (10, 0) tube. For such a tube there is an energy gap between the conduction and the valence band at k=0 and so the tube would be expected to be a semiconductor.

It has been shown that the chiral nanotubes are metallic when $n–m = 3q$ where q is an integer.

Thus all the armchair nanotubes are metallic while those of the zig-zag tubes $(n,0)$ for which the index n is divisible by 3 are also metallic; so are the chiral tubes for which $n - m = 3q$. All the other tubes are semiconducting.

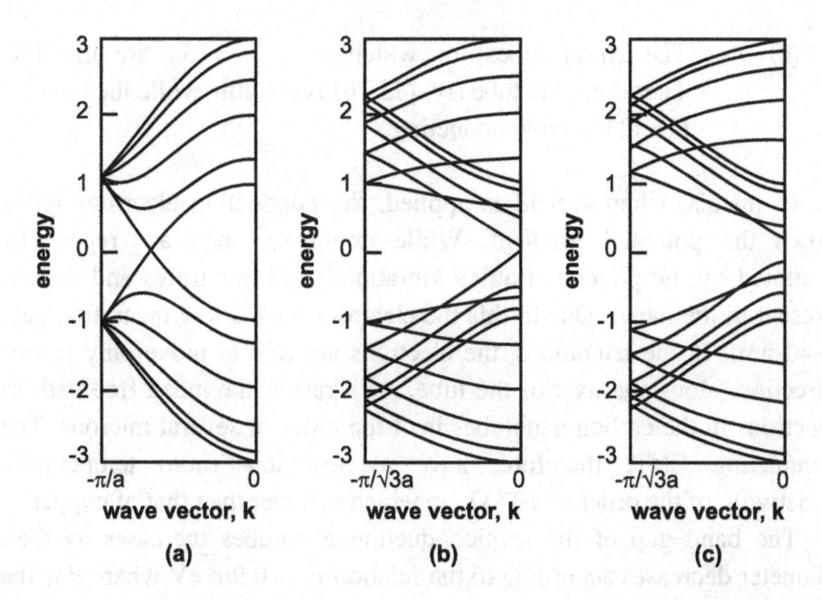

Fig. 8.19. *E vs. K* relations for (a) armchair (b) zig-zag (with *n* divisible by 3) and (c) semiconductor nanotubes (see text).

(Note: The electron–electron interaction in nanotubes has been neglected in the above discussion. It has been reported that this interaction can overwhelm the kinetic energy of the electron gas and introduce a small band gap (10 to 100 meV), making the tubes semiconducting [157]. This effect is called the Mott metal–insulator transition. Thus in reality, no carbon nanotube is metallic. However, for practical purposes, due to the very small magnitude of this effect, the above description remains valid).

Example 8.3. *Which of the following carbon nanotubes are expected to have metallic conduction behavior?*

(i) (6,6) (ii) (9,0) (iii) (14,0) (iv) (13,10) (v) (16,11)

Solution.

(a) All the armchair tubes are metallic. Therefore, the tube (i) (6, 6) is metallic.

(b) All the zig-zag tubes with index divisible by 3 are metallic. Therefore, (ii) (9, 0) is metallic while (iii) (14, 0) is semiconducting.

(c) The chiral tubes for which $n - m = 3q$ are metallic. Therefore, the tube (iv) (13,10) is metallic while the tube (v) (16,11) is semiconducting. □

In metals, when a field is applied, the conduction electrons move down the potential gradient. While doing so, they are repeatedly scattered by the phonons (lattice vibrations) and impurities and defects present in the metal. Due to this the electrons have a low mean free path (~40 nm). In the nanotubes, the electrons are free to move only in one direction, along the axis of the tube. As a result, the mean free path of electrons in the carbon nanotubes is of the order of several microns. The conducting CNT, therefore, have a very low room temperature resistivity, of the order of 10^{-6} Ω.cm which is lower than that of copper.

The band gap of the semiconducting nanotubes increases as their diameter decreases according to the relation $E_g = 0.9/d$ eV where d is the diameter of the tube in nm.

The picture regarding the electrical behavior of MWCNT is not yet clear. Some of the calculations have shown that the interaction between the different tubes does not affect the behavior of individual tubes — thus a metallic tube would remain metallic and a semiconducting tube would remain semiconducting. Other calculations have showed two metallic tubes can become semiconducting depending on their separation due to interlayer interactions.

8.6 Applications of carbon nanotubes

8.6.1 *Composites*

Because of the excellent mechanical properties of the carbon nanotubes, their application in load bearing composites is of much interest. This topic is discussed in detail later in chapter 12.

8.6.2 *Field emission*

In a metal, the electrons occupying energy levels at the top of the conduction band can be ejected from the metal if an additional energy equal to or greater than the work function of the metal is imparted to the electron. The work function is the energy difference between the top of the conduction band and the vacuum level. In metals, the energy corresponding to the top of the conduction band is the Fermi energy (Fig. 8.20). The work function of most metals, the difference between the Fermi level and the vacuum level, is in the range 2.5–5.5 eV.

The energy required for the emission of the electrons from the metal can be provided by different means. These include light, heat and electric field. The corresponding emissions of electrons are called photoemission, thermionic emission and field emission respectively. In case of the field emission, an applied field lowers the energy barrier to the vacuum level enabling the electrons to tunnel through (Fig. 8.20).

The field emission of electrons is used in several devices where an intense source of electrons is needed. An example is the electron microscope. Several compounds are known to have large field emission currents at low fields and are used as field emission sources. In the electron microscopes, a commonly used source is lanthanum hexaboride.

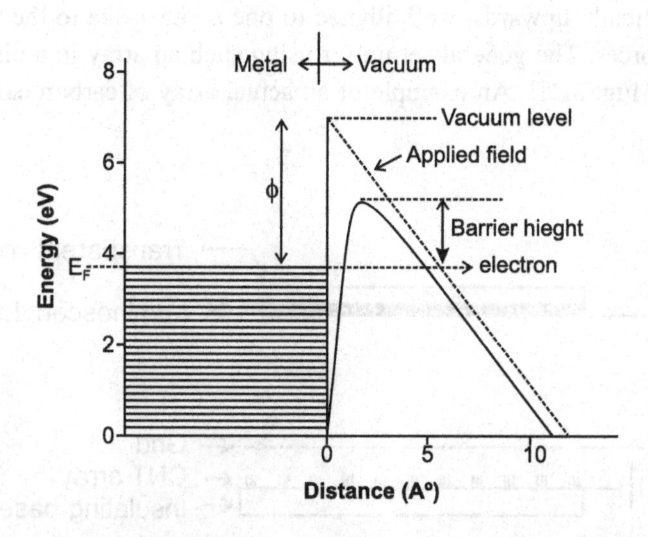

Fig. 8.20. The top of the conduction band in a metal is the Fermi level. The energy difference between the Fermi level and the vacuum level is the work function Φ of the metal. The applied field lowers the barrier between the Fermi level and the vacuum level, enabling the electrons to tunnel through.

The first requirement for a material to be useful as a field emission source is that it should produce a large electron current density at low applied fields. To generate high field, the source is shaped as a wire with a fine tip. Because of the very small diameter of the carbon nanotubes, high fields are generated at their ends when a voltage is applied. The carbon nanotubes have several desirable characteristics as a field emission source. These are: a low turn on electric field and low threshold electric field, a high field enhancement factor, high current density and high current stability with a low degradation rate. The metallic nanotubes show extremely high field emission electron current density at low threshold fields (> 4000 mA cm^{-2} at 1.5 $volts.mm^{-1}$) as compared to other materials (*e.g.* 10 mA cm^{-2} at 50 to 100 V mm^{-1} for Mo or Si). Because of these features, the carbon nanotubes, besides replacing the traditional materials, are being intensively explored for use in displays. For this purpose, the carbon nanotubes have to be deposited in vertically aligned columns arranged in an array on an insulating substrate. This is accomplished by first depositing a metal catalyst such as Fe as an array of spots on which the nanotubes are then grown by CVD. The nanotubes grow vertically upwards, well aligned to one another due to the van der Waal's forces. The general set up for using such an array in a display is shown in Fig. 8.21. An example of an actual array of carbon nanotubes

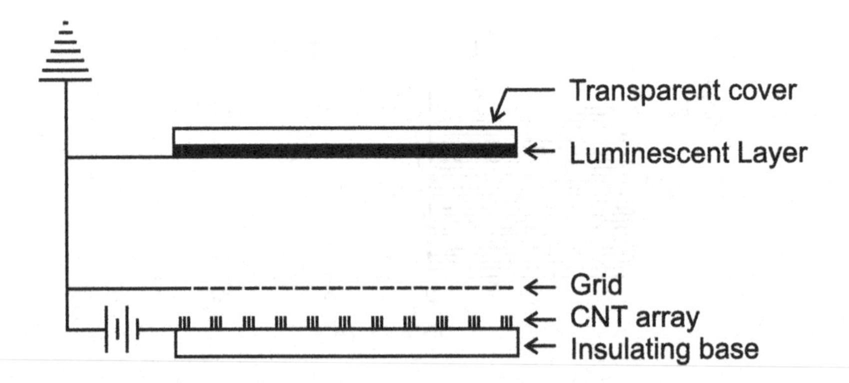

Fig. 8.21. Schematic figure showing the set up for using the carbon nanotube arrays as field emission source in a display.

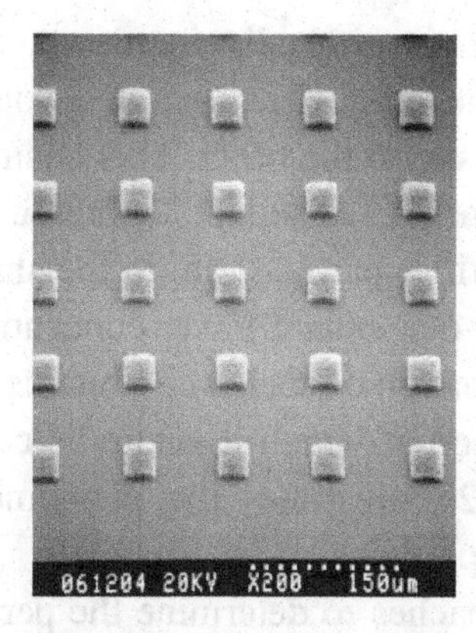

Fig. 8.22. Scanning electron micrographs showing , on the left, the array of carbon nanotubes deposited on a porous silicon substrate patterned with iron islands of size 30 μm x 30 μm and, on right, a magnified view of one such array of nanotubes (reprinted from [158] copyright (2001) with permission from Elsevier.

from the work of Sohn *et al.* [158] is shown in Fig. 8.22. The array was grown on a porous silicon substrate which had earlier been patterned with 30 mm x 30 mm Fe islands arranged in a square array at a pitch distance of 125 mm. The scanning electron micrograph at the right shows that the tubes are self oriented and well aligned. The measured current density was 1 $mA.cm^{-2}$ at a field of 2 V μm^{-1} and 80 $mAcm^{-2}$ at 3 $V\mu m^{-1}$.

The application of displays for television and computer monitors using carbon nanotubes as field emission sources is actively being pursued. In addition there are several other applications such as microwave amplifiers, small x-ray tubes *etc.* where CNT can be used as field emission source.

8.6.3 *Conductive composites*

Another application of carbon nanotubes, which takes advantage of their twin properties of metallic conductivity and high aspect ratio, is in conductive polymer composites. Conducting polymers have numerous applications as in electromagnetic shielding, electrostatic dissipation coating, antistatic coating *etc.* If conducting filler is introduced in a polymer matrix and its concentration is gradually increased, at some concentration a continuous chain of the conducting particles would form extending from one end of the composite to the other. This chain acts as an easy conducting path and there is a sudden increase in the conductivity of the composite. If the filler particles are in the shape of a rod, then the jump in the conductivity occurs at a lower concentration of the particles. The phenomenon of formation of such a continuous chain is called percolation and the concentration at which it occurs is called the percolation threshold, Φ_c. Figure 8.22 illustrates the phenomenon of percolation for the case of cylindrical particles.

There are several theoretical approaches to determine the percolation threshold. For the nonspherical particles, the concept of "excluded volume" is commonly used. The excluded volume is defined as the volume around an object in which the center of another similarly shaped object is not allowed to penetrate. For randomly oriented cylinders, the excluded volume is given by

$$V_{ex} \approx \pi L^2 D/2$$

where L is the length and D is the diameter of the cylinder. The relation between the percolation threshold and the aspect ratio $\eta = L/D$ has been derived to be [159, 160]

$$\Phi_c = 0.7/\eta \tag{8.1}$$

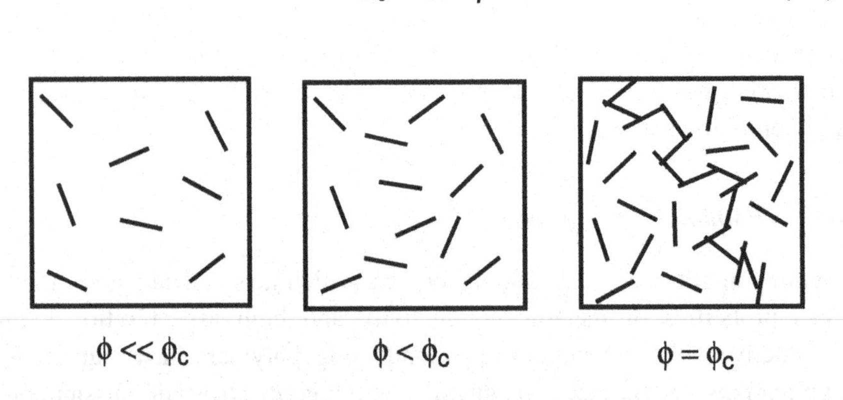

Fig. 8.23. Formation of a continuous chain of particles at the percolation threshold Φ_c.

For the carbon nanotubes, the aspect ratio is very high (~ 1000) so that a low percolation threshold is expected. This is borne out by the experimental data on conductive polymer–CNT composites reviewed by Bauhofer and Kovacs [159]. Although there is some variation in the values of Φ_c determined by different researchers, it appears that for nearly all the polymer–CNT combinations, a value of ~ 0.1 wt % can be achieved with good dispersion of the nanotubes in the matrix. This also agrees with the theoretical prediction of the percolation threshold given above if an aspect ratio of ~ 1000 is used.

The expression (8.1) for the percolation threshold has been derived for a statistical distribution of the filler particles. In some cases, a much lower percolation threshold is seen due to "kinetic percolation". This refers to the ability of the particles to move during the processing of the composite due to diffusion, convection, shearing or some external field and form a conducting chain at lower concentrations. An example of the

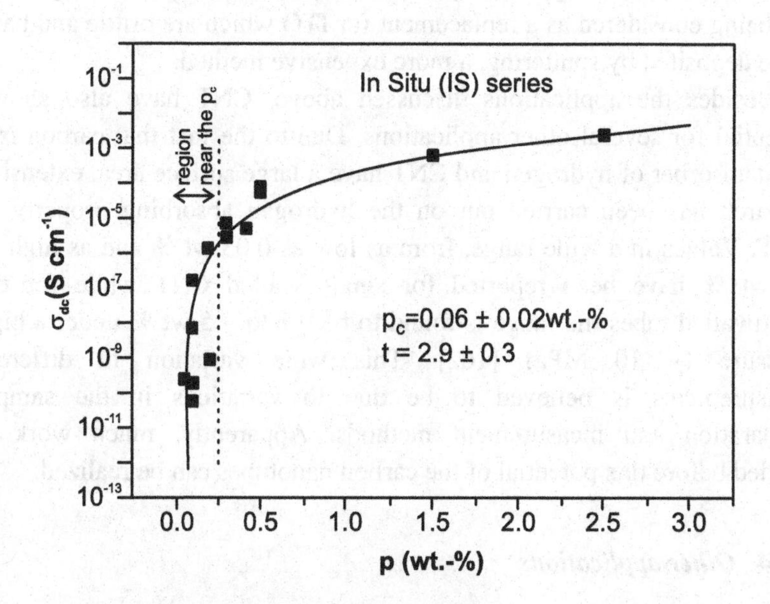

Fig. 8.24. Change in conductivity with wt % of multiwall carbon nanotubes in a PET–MWCNT composite; the percolation threshold is seen to be at 0.06 wt % (reprinted from [161] copyright (2010) with permission from Elsevier).

phenomenon of percolation is shown in Fig. 8.24 for a Polyethylene terephthalate – MWCNT composite made by in situ polymerization [161]. The conductivity increases sharply at about 0.06 wt % which is identified as the percolation threshold. This apparently is a case of kinetic percolation, because before the polymerization occurs, the nanotubes have ample opportunity to move. For the same composites prepared by melt mixing a percolation threshold between 0.1–0.2 wt percent was obtained which is the value expected for a non kinetic percolation.

Because of the large aspect ratio of the carbon nanotubes, the weight fraction of the nanotubes required for achieving reasonable conductivity is very small. At these small contents of the filler, transparent polymer matrices retain their transparency making it possible to produce transparent conductive coatings to replace coatings such as indium tin oxide (ITO). As the polymer–CNT composite coatings are much more flexible and can be deposited by the simple process of spin coating , they are being considered as a replacement for ITO which are brittle and have to be deposited by sputtering, a more expensive method.

Besides the applications discussed above, CNT have also shown potential for several other applications. Due to the fact that carbon is a good absorber of hydrogen and CNT have a large surface area, extensive research has been carried out on the hydrogen absorbing property of CNT. Values in a wide range, from as low as 0.03 wt % and as high as 20 wt % have been reported for single walled CNT while for the multiwalled tubes the range is found to be 0.1 to ~ 5 wt % under a high pressure (~ 10 MPa) [162]. This wide variation in different measurements is believed to be due to variations in the sample preparation and measurement methods. Apparently, much work is needed before this potential of the carbon nanotubes can be realized.

8.6.4 *Other applications*

In addition, a myriad of applications are being envisaged for the carbon nanotubes. Their very high thermal conductivity is already being exploited in heat sinks in electronics. Their high electrical conductivity, high surface area and linear geometry make them useful as electrode materials for batteries. Electrical conductivity and nano dimensions are proposed to be exploited in their use as connectors in molecular electronics while the semiconducting nanotubes are being looked at for use as active components. Other potential applications include catalyst supports, air and water filters, as well as biomedical applications.

Problems

(1) (a) Draw a graphene sheet structure. Choose an origin. Show the locations of the points (*n, m*) for values of n and m from 0 to 5 *i.e.* (0,0), (0,1), ...(1,0), (1,1) *etc.* in terms of unit vectors $\bar{a}_1$ and $\bar{a}_2$.

(b) Draw the unit cell of a (6, 6) tube and a (9, 0) tube. What kind of tubes are these?

(2) Given that the size of a hexagon in a graphene sheet is 0.142 nm, prove the following relations

(a) $|\bar{a}_1| = |\bar{a}_2| = 0.246$ nm

(b) If $\bar{c}$ is any vector given by $\bar{c} = n\,\bar{a}_1 + m\,\bar{a}_2$, then $\bar{c} = 0.246\,(\sqrt{n^2 + mn + m^2})$

(c) $Sin\,\theta = (\sqrt{3}\,m) / 2\sqrt{(n^2 + mn + m^2)}$ where θ is the chiral angle.

(3) Vectors along the *circumference* of some nanotubes are

$$6a_1 + 6a_2;\ 5a_1;\ 5a_1 + 5a_2;\ 6a_1 + 3a_2;\ 11a_1 + 7a_2$$

(a) Write the type of nanotube in each case.

(b) Which of these tubes is metallic and which is semiconducting? Why?

(4) Answer the following questions

(a) In the multiwall carbon nanotubes are the individual layers arranged in a scroll like fashion or as concentric cylinders?

(b) How does the spacing between the individual tubes in multiwall tubes compare with the spacing between the graphene sheets in graphite?

(c) Does the arc discharge method produce multiwall tubes or single wall tubes.

(d) What is the usually observed range of the diameters of (a) single wall and (b) multiwall carbon nanotubes?

(e) What is the effect of the presence of pentagons and heptagons on the shape of the cap of a carbon nanotube?

(5) You are required to grow single walled carbon nanotubes on the surface of commercially available carbon fibers. Describe how you will accomplish this.

(6) Describe the steps involved in depositing CNT columns on Si pillars having square top surfaces, 2 μm x 2 μm, starting from the fabrication of Si pillars.

(7) (a) A multiwall nanotube is pinned at one end and is pushed by the AFM tip at the other end. The free length of the tube is 1 μm. It can be approximated to a cylinder of wall thickness 5 nm and inner diameter 5nm. If the modulus of the tube is 1100 GPa, calculate the force required to deflect the tip by 10 nm.

(b) Describe a procedure to anchor the nanotubes for this kind of experiment.

(c) In a measurement of the strength of CNT, the Weibull modulus for the arc discharge produced CNT is 5 while for the CVD produced CNT it is 3.5. Explain what does Weibull modulus signify and why it is higher in the former case.

(8) Explain, using the wave vector vs. energy (*k vs. E*) diagrams, the electrical conducting properties of carbon nanotubes.

Chapter 9

Other One Dimensional Nanostructures

9.1 Introduction

In Chapter 8 we saw that the layer structure of graphite is conducive for the growth of fullerenes and one dimensional carbon nanotubes. By folding into tubes or spheres the dangling bonds at the edges of a graphene sheet are eliminated making the structure more stable. This holds true for other layered structures also. The chalcogenides of metals such as tungsten or molybdenum (having the general formula MX_2, with M = W, Mo, Ta, Nb, Hf, Ti, Zr, Re and X = S, Se or Te) are well known inorganic layered structures. Indeed the first noncarbon nanotubes to be observed consisted of MoS_2 and WS_2. In these compounds each layer consists of three atomic sub layers denoted as X-M-X. The overall structure can be represented by

(X-M-X) (X-M-X) (X-M-X) (X-M-X) …..

While there is a covalent bonding between X and M layers within a sublayer, the bonding between X and X from the two adjoining layers is weak van der Waal's bonding.

Besides the metal dichalcogenides many other inorganic compounds have a layer structure and are natural candidates for the formation of nanotubes. These include metal dihalides (*e.g.* $NiCl_2$), metal oxides (V_2O_5, $H_2Ti_3O_7$) and many ternary and quaternary compounds.

After the discovery of carbon nanotubes, efforts were intensified to prepare nanotubes of other materials. Although, the layered inorganic compounds were the first ones to be explored because of the similarity with graphite, at present synthesis strategies are available to produce one dimensional nanostructures from almost any material — metals, semiconductors, oxides, carbides, nitrides, polymers, *etc.*

In addition to the tubular morphology, nanostructures having several other one dimensional morphologies have been observed depending on the growth technique and the growth conditions. These include shapes such as nanorods (or nanowires), nanobelts (or nanoribbons), nanosprings, dendritic shapes, *etc.* Figure 9.1(a) shows schematically these nanostructures while Fig. 9.1(b) gives examples of some of the various nanostructures of one of the important oxides, ZnO. In the discussion below we confine ourselves to nanowires and nanorods which can be defined as solid cylinders with diameter less than 100 nm and with aspect ratios of > 20 and < 20 respectively.

9.2 Growth of one dimensional nanostructures

Most of the synthesis strategies exploit one of the following features of the material or the growth process [163-165].

(i) Layer structure of the material
(ii) Liquid metal assisted epitaxial growth
(iii) Template assisted growth
(iv) Exploitation of inherent or induced surface energy anisotropy

In the following we briefly describe these strategies. The vapor-liquid-solid growth technique is described in some detail because of its versatility and because it is the most well understood mechanism of growth.

9.2.1 *The vapor-liquid-solid (VLS) nanowire growth process*

As mentioned above, a single sheet of a layered structure such as graphite or MoS_2 tends to fold upon itself to reduce the energy by eliminating the dangling bonds at its edges. The folded structure, such as a nanotube or a fullerene has some strain in it. The reduction in energy on folding should more than compensate for this elastic strain energy. This condition is satisfied when the size of the sheet is sufficiently small so that the ratio of the dangling bonds to the total bonds is high. An activation energy has to be supplied to overcome the strain energy

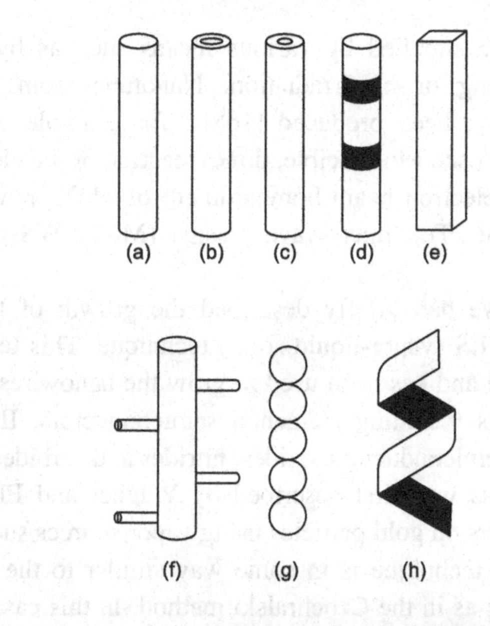

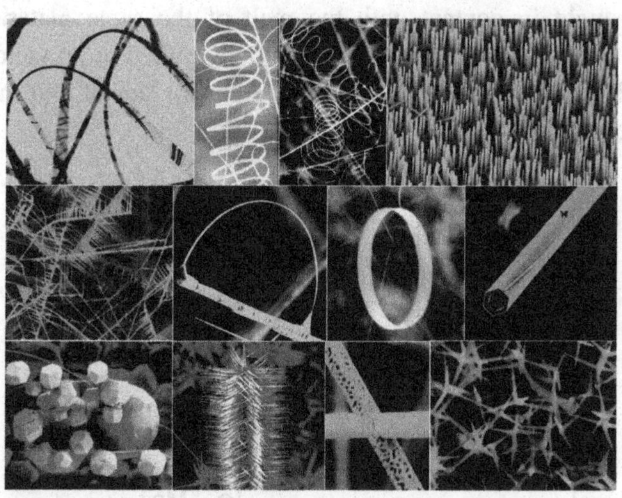

Fig. 9.1. (top) Schematics of the various one dimensional nanostructures (a) nanowires and nanorods (b) core–shell wires or rods (c) nanotubes and hollow nanorods (d) axial heterostructures (e) nanobelts/nanoribbons (f) dendrites (g) nanosphere assembly (J) nanosprings [166] (bottom) Examples of various nanostructures of ZnO synthesized by thermal evaporation under different conditions (Reproduced from [167] copyright (2004) with permission of Elsevier).

barrier. This can be supplied by various means such as by chemical reactions, by heating or by irradiation. Nanotubes from the layered inorganic solids have been produced [168] , for example, by heating MoS_2 powder in a closed Mo crucible, direct reaction of the elements Nb and Se at 800° C, electron beam bombardment of MoX_2 powder (M = Mo, Nb; X = S, Se, Te), microwave plasma (MoS_2, WS_2) and laser ablation of SnS_2.

In Chapter 8, we had briefly described the growth of the carbon nanotubes by the VLS (vapor-liquid-solid) technique. This technique is in fact quite general and has been used to grow the nanowires of a large number of materials including elemental semiconductors, II–VI semiconductors, III–V semiconductors, oxides, nitrides and carbides [163].

The VLS process was first described by Wagner and Ellis for the growth of Si whiskers on gold particles using vapor sources such as $SiCl_4$ or SiH_4 [169]. The technique is in some way similar to the growth of crystals from a melt as in the Czochralski method. In this case, a crystal seed is brought in contact with the melt of the same material. After equilibrating with the melt, the seed is slowly pulled up to yield a large, cylindrical single crystal while the temperature gradient and the rate of pulling are properly controlled (Fig. 9.2).

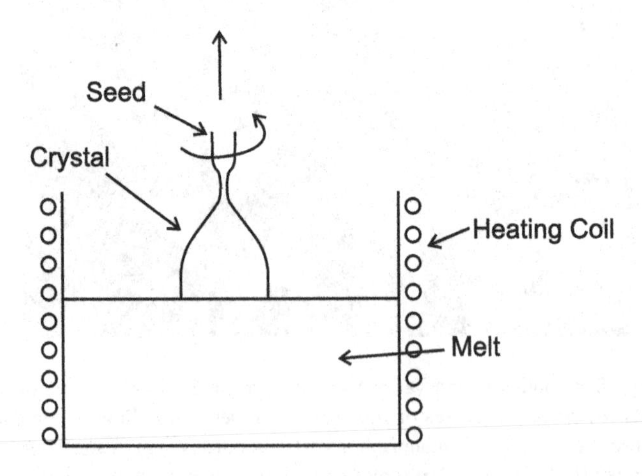

Fig 9.2. The Czochralski method of crystal growth. A seed is brought in contact with the melt of the same material and after equilibrating, the seed is pulled up at a controlled rate to yield a long single crystal.

The VLS process can be illustrated with reference to the growth of Si nanowire on gold particles. The prospect of using silicon nanowires as components for future electronic and optoelectronic devices has drawn considerable attention to them. In the most studied VLS process for Si, gold nanoparticles are first deposited on a substrate, usually silicon. This can be accomplished in several ways. One way is to deposit a thin layer of gold on the substrate followed by annealing at a temperature of about 700°C. During annealing, the gold film breaks up to form nanometer sized gold islands such that their contact angle with the substrate is equal to the equilibrium contact angle. An example of this is shown in Fig. 9.3. Another method, which is more amenable to precise control, is to use lithography which consists of the following steps: (i) deposit a layer of a photoresist on the substrate (ii) expose the photoresist and develop to form the desired pattern of holes (iii) evaporate gold to form gold dots at the bottom of the exposed wells and (iv) lift off the photoresist.

Growth of the silicon nanowires on a silicon substrate with gold islands on it is carried out at a temperature of about 400°C. At this

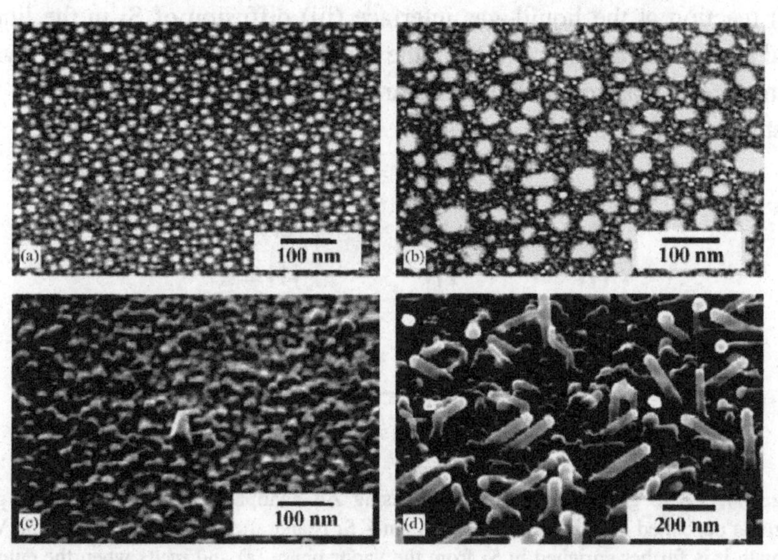

Fig.9.3: SEM images of Au–Si islands grown on Si (001) substrates for (a) 1.2 nm and (b) 3.0 nm of Au deposited at room temperature followed by annealing at 700° C (reprinted from [170] copyright (2007) with permission of Elsevier).

temperature silicon from the substrate diffuses into gold to form a eutectic alloy with a low melting point of 363.8 °C. The temperature is maintained such that the gold–silicon alloy particle is in liquid state. Subsequently precursor gases ($SiCl_4$ and H_2 or SiH_4) are introduced in the chamber. The precursor gas is reduced to atomic Si over the liquid drop. The atomic Si readily diffuses into the liquid drop. The liquid eventually becomes supersaturated with Si and the excess Si precipitates as a solid at the interface between the molten drop and the substrate. As the solid grows, the liquid droplet rises from the substrate surface. The diameter of the growing nanowire is determined by the diameter of the droplet (Fig. 9.4).

The catalysis is a physical catalysis because the metal (Au or other) is always expected to be present in atomic amounts on the sides of the nanowire also; a much lower sideways growth rate of the nanowire implies that it is the liquid surface which is responsible for the high axial growth rate.

The overall growth process can be seen to consist of four steps (i) diffusion of Si species in the gas phase to the liquid surface (ii) reaction at the liquid–gas interface (iii) diffusion of Si in the liquid and (iv) incorporation of Si from the liquid into the lattice of the growing nanowire. It was concluded by Givargizov [171] that amongst these, the last step is the rate controlling step.

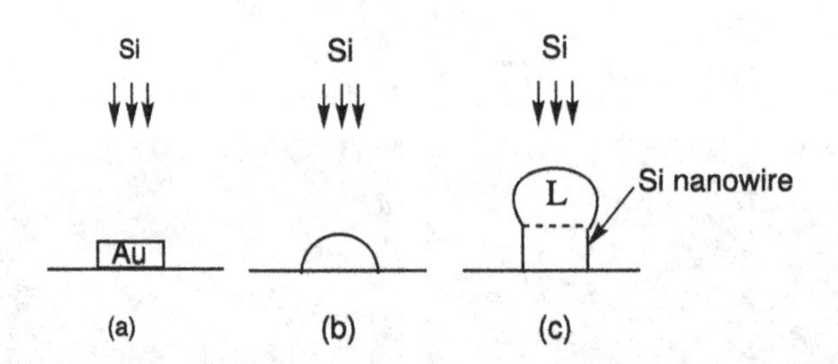

Fig. 9.4. V-L-S growth of Si nanowire using Au nanoparticles as "catalyst"; a gold particle deposited on a substrate forms a Au – Si island after annealing; during the VLS growth it is further enriched in Si from the vapor phase (a) and melts when the eutectic composition is reached (b); continued diffusion of Si into liquid supersaturates it causing the precipitation of solid Si at the substrate-liquid interface (c) leading to the growth of a nanowire whose diameter is determined by the diameter of the liquid droplet.

The VLS process has been used to produce nanowires from a variety of materials using appropriate metals as catalysts. The metal should preferably form a low melting eutectic with the material of interest. We have already described the growth of carbon nanotubes using transition metals Fe, Ni, Co as catalysts. In some cases, the catalyst is supplied in the vapor phase and subsequently forms metal particles to promote the growth of nanowirs. Thus for GaN nanowires, Fe is used as a catalyst and is supplied as $Fe(C_5H_5)_2$ vapor.

It was mentioned above that the diameter of the nanowire is determined by the diameter of the liquid drop — a larger liquid drop produces a larger diameter wire. It is also found that the growth rate of a nanowire depends on its diameter, the growth being faster for the larger diameters. Furthermore, the VLS process may not always result in the formation of nanowires; instead, only hillocks may form. These two aspects are discussed below.

9.2.1.1 *Dependence of growth rate on the diameter*

The dependence of the growth rate on the diameter of the nanowire is similar to that found for the growth of whiskers (micron thick single crystals), by Givargizov [171] and others in the 1960's. Although the experimental conditions in the nanowire growth are somewhat different than those in case of the whisker growth, the explanation in the two cases is essentially the same and is based on the Gibbs-Thomson effect i.*e.* the effect of the curvature of a surface on the chemical potential and the vapor pressure. The driving force for the nanowire growth is the difference in the chemical potential between the initial state (Si in the vapor phase) and the final state (Si in the nanowire). As we saw in Chapter 2, the difference in the chemical potential of a species on a flat surface (μ_o) and on a curved surface (μ) is given by

$$\mu - \mu_o = \gamma\Omega(\frac{1}{R_1} + \frac{1}{R_2})$$

where R_1 and R_2 are the radii of curvature of the curved surface, γ is the surface energy of the solid and Ω its atomic volume. For the cylindrical

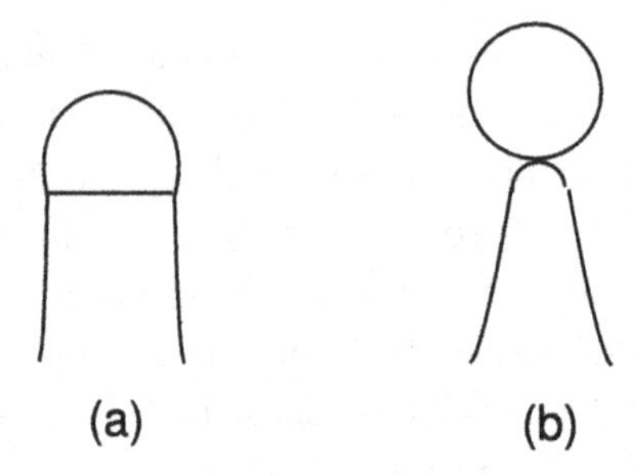

Fig. 9.5. Effect of the line tension on nanowire growth (a) nanowire growth occurs when the line tension is very small or negative (b) only a hillock is produced when the line tension is large.

surface of the nanowire, $R_1 = R$ = radius of the cylinder and $R_2 = \infty$ so that

$$\mu - \mu_0 = \frac{\gamma\Omega}{R} = \frac{2\gamma\Omega}{d}$$

In the above equation d is the diameter of the nanowire. If μ_{vapor} is the chemical potential of the same species (Si) in the vapor phase, then we can write

$$\frac{2\gamma\Omega}{d} = \mu - \mu_o = \mu - \mu_{vapor} - \mu_o + \mu_{vapor}$$

$$-\frac{2\gamma\Omega}{d} = (\mu_{vapor} - \mu) - (\mu_{vapor} - \mu_o) = \Delta\mu_{wire} - \Delta\mu_o$$

Or

$$\Delta\mu_{wire} = \Delta\mu_0 - \frac{2\gamma\Omega}{d} \tag{9.1}$$

where $\Delta\mu_{wire}$ is the difference in the chemical potentials of Si in the wire and the vapor phase and $\Delta\mu_o$ is the same for the vapor phase and a flat surface. The former is the magnitude of the driving force for the wire growth and according to Eq. (9.1), is larger for wires of larger diameters, leading to a faster growth as observed.

9.2.1.2 *Conditions for nanowire growth*

The growth of the nanowires is very sensitive to the surface energies of the various interfaces present during the growth process. These include the liquid–substrate, liquid–wire, liquid–vapor, wire–vapor and the substrate–vapor interfaces. In addition to the surface energies, another term — the line tension of the substrate–liquid–vapor contact line, which is the periphery of the liquid drop on the substrate, also becomes important when the drop size is small, as is the case for nanowires. This line tension controls the contact area of the drop on the substrate. This area shrinks if the line tension increases. In the limit, the drop assumes the shape of a sphere with only a point contact with the substrate. In this state, there is no wire growth and only hillocks are produced (Fig. 9.5). As is discussed below, the line tension should have a small positive value or be negative for the nanowire growth to occur.

Figure 9.6 shows schematically the Au/Si liquid droplet on the substrate (a) and the initial stage of the nanowire growth (b). In Fig. 9.6(b) α is the angle of inclination of the nanowire flank. It is zero initially and increases as the wire grows. It is clear from Fig. 9.6(b) that an increase in α is accompanied by increase in β and a reduction in the liquid–solid contact radius *r*. The final radius of the nanowire is therefore smaller than the initial radius.

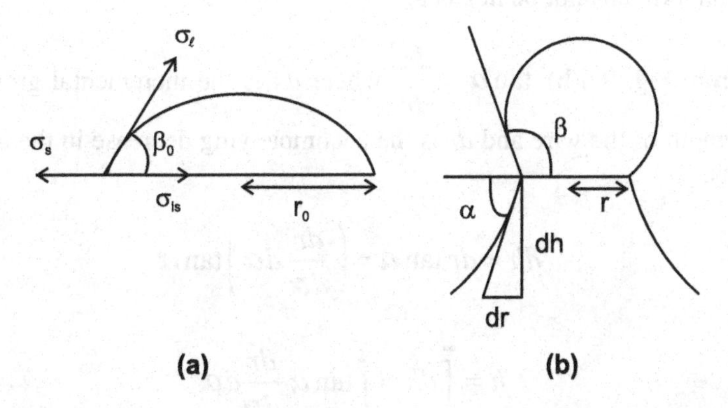

Fig 9.6. (a) A Au/Si droplet on the substrate before the start of the nanowire growth (b) initial stage of the nanowire growth [172].

The growth of the nanowire under these conditions can be calculated as follows [172]. From Fig. 9.6(a), the Young's equation (see Chapter 2), taking into account the line tension, can be written as

$$\sigma_l \cos\beta_o = \sigma_s - \sigma_{ls} - \frac{\tau}{r} \tag{9.2}$$

Volume of the droplet, assuming it to be a spherical cap, is given by

$$V = \frac{\pi}{3}\left(\frac{r_0}{\sin\beta_0}\right)^3 (1-\cos\beta_0)^2(2+\cos\beta_0) \tag{9.3}$$

From Fig. 9.6(b), the angle β can be related to the surface tension, the line tension and the angle α by the following relation

$$\sigma_l \cos\beta = \sigma_s \cos\alpha - \sigma_{ls} - \frac{\tau}{r}$$

Here σ_l, σ_s and σ_{ls} are the surface tensions for the liquid–vapor, solid substrate–vapor and the liquid–substrate interfaces respectively, τ is the line tension of the liquid–vapor–substrate contact circle and r is the radius of this circle. Normally, the line tension term, being very small, is neglected. However, for small droplets, r is also small and so the line tension term can not be neglected.

From Fig. 9.6(b) $\tan\alpha = \frac{dh}{dr}$ where dh is the incremental growth in the length of the wire and dr is the accompanying decrease in the contact radius.

or $$dh = dr\tan\alpha = \left(\frac{dr}{d\alpha}d\alpha\right)\tan\alpha$$

$$h = \int_0^\infty dh = \int_0^\infty \tan\alpha \frac{dr}{d\alpha} d\alpha \tag{9.4}$$

Using Eqs. (9.2) and (9.3), the integral (9.4) can be evaluated numerically if the surface energies and the line tension τ are known. While the values of the surface energies are available or can be estimated, the line tension is not generally known. Schmidt *et al.* [172] have calculated the height of the nanowire as a function of radius r using the following values of the surface energies and angle β: $\sigma_l = 0.85$ Jm^{-2}, specific surface energy of the (111) Si surface, $\sigma_s = 1.24$ Jm^{-2}, $\beta_o = 43^o$ and $\sigma_{ls} = 0.62$ Jm^{-2} and taking different values of the line tension τ. The results are shown in Fig. 9.7. Here τ is expressed as a dimensionless parameter defined as

$$\tau' = \frac{\tau}{\sigma_l r_0}$$

For $\tau' = 0.08$ and 0.12, as r decreases from r_o to about 0.05 r_o, the curves flatten out implying no nanowire growth. Only small positive

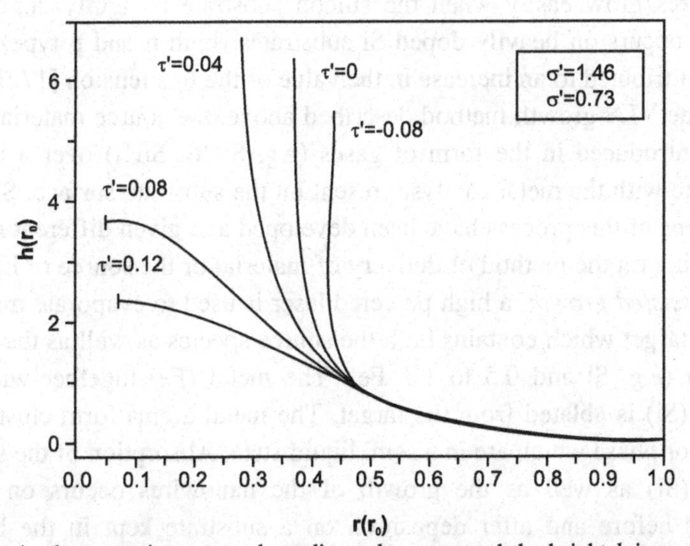

Fig. 9.7. As the nanowire grows, the radius r decreases and the height h increases. For line tension values given by $\tau' = \tau/\sigma_l r_0 = 0.08$ and 0.12, there is no nanowire growth while for $\tau' = 0.04$, 0 and – 0.08, the nanowire growth does occur with h increasing rapidly for $r < 0.5$; the quantities σ_s' and σ_{ls}' are defined by $\sigma_s' = \sigma_s/\sigma_l$ and $\sigma_{ls}' = \sigma_{ls}/\sigma_l$ (reproduced from [172] copyright (2005) with permission from Springer).

($\leq$0.04) or negative values of τ' lead to nanowire growth, with h increasing rapidly when r has reduced to ~ < 0.5 (Fig. 9.7). It should be noted that the line tension is not a commonly used or measured quantity. Just as the surface energy is a measure of the excess energy associated with the interface, the line tension is the excess free energy concentrated in the contact line between the three phases – liquid, vapor and the substrate. Unlike the surface energy, which is always positive, the line tension can be either negative or positive. It has been shown [173] that the line tension can be expressed as follows in terms of the solid – vapor surface energy and an elementary thickness l_o which is equal in magnitude to the atomic layer distance in the wire growth direction

$$\tau = l_o \, \sigma_{vs}$$

From the discussion above, it can be seen that the diameter of the nanowire decreases as it grows up from the substrate until it reaches a constant value provided the line tension is not large. This line tension is very sensitive to small changes in composition. Thus while the nanowires grow easily when the silicon substrate is lightly doped, no growth occurs on heavily doped Si substrates (both n and p type). This can be attributed to an increase in the value of the line tension [172].

In the VLS growth method described above, the source material (*e.g.* Si) is introduced in the form of gases (*e.g.* $SiCl_4$, SiH_4) over a heated substrate with the metal catalyst present on the substrate surface. Several variations of this process have been developed and given different names depending on the method of delivery of material or the source of heat. In *laser assisted growth,* a high powered laser is used to evaporate material from a target which contains both the source species as well as the metal catalyst (*e.g.* Si and 0.5 to 1% Fe). The metal (Fe) together with the source (Si) is ablated from the target. The metal atoms form clusters in the vapor phase which are in a semi liquid state. Absorption of the source atoms (Si) as well as the growth of the nanowires occurs on these clusters before and after deposition on a substrate kept in the heated chamber [174]. One unique advantage of this approach is that nanowires with complex chemical composition can be synthesized because it is not necessary to prepare the target in the form of a uniform alloy – a simple mixture of the elements in the target is enough.

The VLS growth can be carried out using *molecular beam epitaxy* (MBE) to deliver the source species to the substrate. High quality semiconductor nanowires and heterostructures in which the location of the junctions is well controlled can be prepared by this method.

A similar technique is *chemical beam epitaxy* (CBE) which also works at high vacuum like MBE but instead of generating vapors from the effusion cells, the gaseous source materials are introduced in the reaction chamber in the form of a beam. High quality heterostructures have been prepared using this technique [175] .

Growth from the vapor phase can also occur in some cases where apparently there is no catalyst (metal) present. Thus the nanowires of ZnO, SnO_2, In_2O_3 and some semiconductors can be made by simple *thermal evaporation* and condensation of the vapors over a substrate kept in the cooler section of the chamber. The nanowires are produced directly from the vapor phase without the assistance of any catalyst, the process therefore being called *vapor–solid growth*. The mechanism of this process is not well understood.

The metal catalyst assisted growth of nanowires has also been carried out in a liquid medium. To maintain high temperatures necessary for the growth, the liquid has to be subjected to a high pressure. An example of this is the growth of Si nanowires from thiol capped gold nanocrystals dispersed in a solution of diphenylsilane in supercritical hexane at 500 °C and 200 -270 bar pressure. At this temperature, diphenylsilane decomposes to yield element Si which dissolves in gold. When the supersaturation is reached, Si precipitates to form nanowires as in the case of the VLS growth. Silicon nanowires of 4 to 5 nm diameter, with very narrow diameter distribution and an aspect ratio of greater than 1000 were prepared using this technique [176].

9.2.2 *Template assisted growth of nanowires*

The template assisted growth of nanowires consists in filling the cylindrical pores of a template by electrodeposition, sol-gel or from a solution and removing the template to recover the nanowires. The common templates used are anodic alumina membrane (AAM), track etched polycarbonate or mesoporous materials with nanoscale channels.

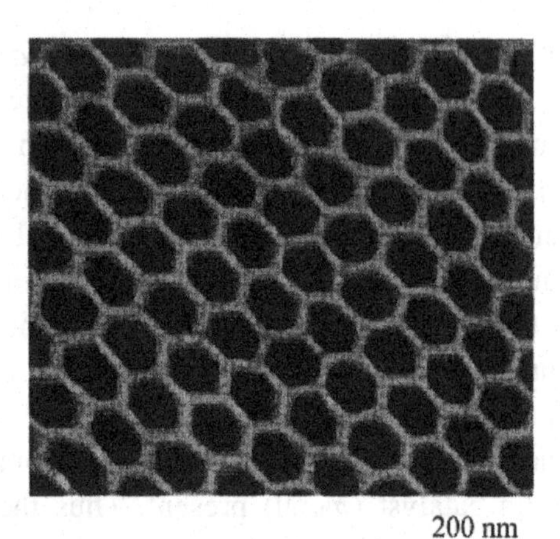

Fig. 9.8. AFM image of a porous alumina membrane (reprinted from [177] copyright (2001) with permission of Elsevier).

Anodized alumina membranes are prepared by anodization of aluminum foil. At present they are commercially available. Figure 9.8 shows an AFM image of an anodized alumina membrane. It possesses a hexagonal porous structure. The pores size (4 to 200 nm) and separation (50 to 420 nm) can be varied by controlling the preparation conditions [178].

Electrodeposition was used by Zheng *et al.* to fill the pores of an AAM with indium. Subsequent oxidation in air at 973 K–1073 K and removal of the membrane yielded In_2O_3 nanowires [177]. Nanowires, nanorods and tubules of metals, semiconductors, carbon and other materials have been successfully fabricated using anodized alumina membranes [178].

9.2.3 *Exploitation of inherent or induced surface energy anisotropy*

During the growth, a crystal tries to minimize its total surface energy by growing in such a way that the highest energy crystal planes are not exposed. However, the difference in surface energies of different crystal planes is usually not large enough to suppress the growth in the lateral directions to the extent necessary for the formation of nanowires. Nevertheless, in some cases the anisotropy in surface energy does lead to one dimensional structures. During the growth of the TiO_2 nanocrystals from aqueous solutions, it was found [179] that the nanocrystals fuse together along the [001] direction to produce nanowires. This occurs because the {001} planes have the highest surface energy and by fusing together along the [001] direction, these planes are not exposed. Similar growth of nanowires has been observed in CdTe, PbSe, ZnS and ZnO [163].

In many cases, the preferential absorption of some molecules on some surfaces of a crystal can be exploited for nanowire growth. Such absorption is called *capping* of the surface by these molecules. The capping drastically reduces the growth rate of the capped surfaces leading to one directional growth. Thus nanocrystals of silver can be produced by reducing $AgNO_3$ with ethylene glycol at 160°C. When poly(vinyl pyrrolidone) (PVP) is added to the reaction mixture, anisotropic growth of nanocrystals occurs so that silver nanowires with high aspect ratio are produced (diameter 30 to 60 nm, length up to 50 μm). This is believed to occur by selective binding of PVP on the {100} faces of silver so that the growth takes place along <111>. ZnO nanowires have also been grown hydrothermally from zinc salts by using amines or hexamethylenetetramine [163].

9.3 Selected properties and applications

One dimensional nanostructures are being investigated for applications in sensors, photoelectric energy generation, field effect transistors and other devices. They have a large surface area and, in comparison to the nanoparticles, are easier to handle and to assemble. As sensors, they offer special advantages for detecting biomolecules in liquid environment. For this purpose electrochemical cells are used and the sensing element is one of the electrodes in the cell. A few nanowires realized on a chip can act as a working electrode. Due to their small size, the nanowires can respond to fast electron kinetics. Furthermore, due to the low surface area, their capacitance, and so the time constant *RC*, is small leading to a capability to detect faster reactions. Figure 9.9 shows an example of a device to detect DNA hydbridization.

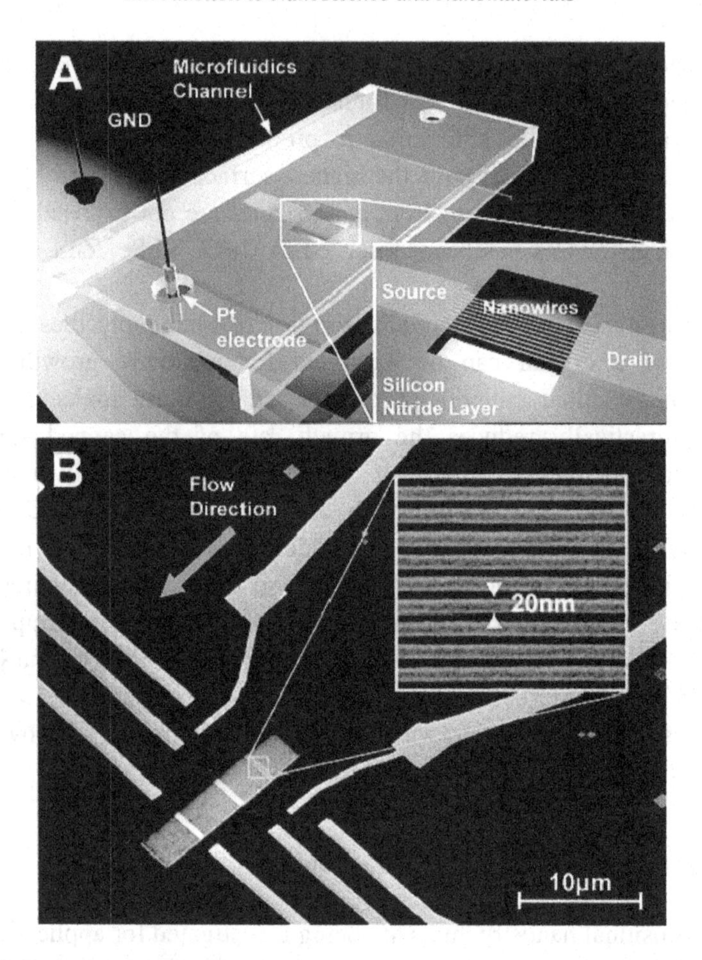

Fig. 9.9. Example of a nanowire sensor to detect biomolecules : (A) the device and (B) SEM picture showing the details; a single section of the device has three groups of 10 Si nanowires in a microfluidic channel . Inset in (B) shows a high resolution image of 20 nm Si nanowires (reproduced from [180] copyright (2006) with permission of American Chemical Society).

The semiconductor nanowires have generated a great interest because of the feasibility of incorporating them in electronic circuits. This is because the semiconductor nanowires have certain features which if successfully exploited will lead to their integration with the silicon technology to give feature sizes below 22 nm [181].

Two of the unique advantages which semiconductor nanowires offer are the following

(1) Heterostructures consisting of different semiconductors along the axis of the nanowire can be grown (Fig. 9.10). Usually, the lattice mismatch between different semiconductors which can be incorporated in a heterostructure is small to limit the strain and avoid defects like dislocations which form when excessive strain is relieved. However, in a nanowire much larger mismatch between adjacent structures is tolerated because the strain is relieved due to its small dimension and one dimensional geometry. This offers more flexibility in the selection of the constituents of the heterostructure and makes it possible to engineer the bandgap of the device to lead to new or better functionality.

(2) The linear geometry of a nanowire lends itself easily to the formation of coaxial layers of different materials *i.e.* a shell of a

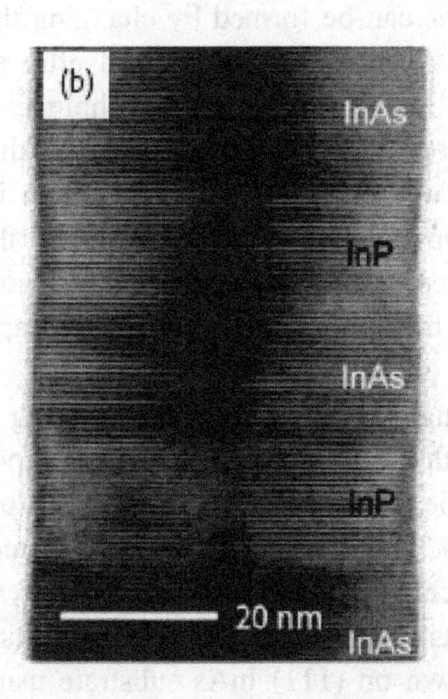

Fig. 9.10. High resolution transmission electron micrograph of a nanowire heterostructure consisting of alternate layers of two different semiconductors (InAs and InP) along its axis (reproduced from [182] copyright (2006) with permission of Elsevier).

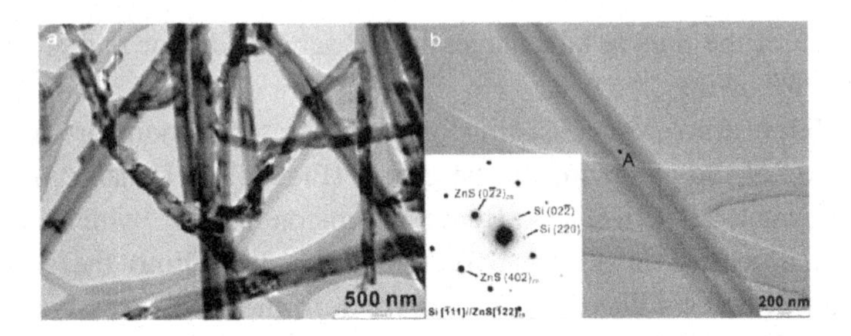

Fig. 9.11. TEM image of a ZnS(core)/Si(shell) nanowire (a) at low magnification and (b) at high magnification; the inset in part (b) is the small area electron diffraction (SAED) pattern of the nanowire (reproduced from [183] copyright (2008) with permission from Elsevier).

second material can be formed around the nanowire. It is also possible to modulate the concentration of dopants in the radial direction, rather than forming a layer of a second material over the nanowire. Such core shell nanowires can be formed by changing the precursor gases during the growth or by forming the shell after the growth of the nanowires by using any other deposition technique. For example, ZnS nanowires with a core of Si have been grown by thermal evaporation in which the source ZnS powder was first kept in a crucible in a thermal growth apparatus. After the growth of the nanowires, the crucible was replaced with another crucible containing Si powder. Fig. 9.11 shows a transmission electron micrograph of such a core-shell nanowire.

This feature is sought to be exploited in making better field effect transistors in which the gate wraps around the nanowire providing a better control over the performance of the transistor. Nanowire field effect transistors with wrap around gate are expected to yield several advantages. These include a low "off" leakage current, increase in the "on" current, high density lateral and vertical integration, *etc.* Nanowires of InP have been grown on (111) InAs substrate using Au as catalyst. Deposition of a dielectric layer of SiN_x followed by Ti metal as gate was carried out subsequently by inductively coupled plasma technique. Such wrap around gate leads to low leakage currents in the OFF position of the FET and also to better ON/OFF current ratio [184].

Several issues regarding the integration of semiconductor nanowires in the Si based technology remain to be solved. However, devices based on one dimensional nanostructures such as semiconductor nanowires and carbon nanotubes are forecast to be realistic additions by the *International Technology Roadmap for Semiconductors* [185]. Other applications, envisaged for these structures are in biosensors, light emitting diodes, lasers and photodetectors.

Nanowires of semiconductors and metals are also being looked at as building blocks for negative – index materials i.e. materials with a negative index of refraction which might lead to the realization of invisible cloaks [186] !

Problems

(1) Name the various techniques for the preparation of naorods and classify them into the top – down and the bottom – up techniques.

(2) In addition to the alumina membranes, polycarbonate membranes are also used for the template based synthesis of nanorods and nanowires. Find out how these polycarbonate membranes are made and compare the pore size distributions of these membranes with the alumina membranes.

(3) Name the various steps that occur in the VLS growth starting from the diffusion of the Si to the surface of the gold-silicon alloy droplet. Which of these is the slowest and so the rate controlling step?

(4) In the VLS growth, what is the effect of the contact angle of the liquid with the substrate on the diameter of the growing wire?

(5) Why does the growth rate of a nanowire increase with the diameter of the wire in the VLS technique? Explain.

(6) Explain the effect of the line tension of the liquid – substrate contact line on the growth of the nanowires.

Chapter 10

Two Dimensional Nanostructures

10.1 Introduction

The two dimensional (2D) nanostructures are defined as structures in which two dimensions (say along the x and y axes) are large while the third dimension along the z axis is less than 100 nm. To really justify as a "nanostructure" it is desirable that some nano-effects manifest themselves in such structures. For this the thickness of the structure should be of the order of some length scale such as the Bohr exciton radius, the de Broglie wavelength of the carriers, the critical size for a defect to exist, *etc*. We have already seen two important examples of such nanostructures — quantum wells and graphene in Chapters 3 and 8 respectively.

Preparation of the 2D nanostructures is carried out by the techniques of thin film deposition. Thin film deposition is a key process in nanotechnology. In addition to the direct formation of the 2D nanostructures, thin film deposition is used for the preparation of core–shell particles, hollow shells, functionalization of particles and surfaces, and other miscellaneous structures. Physical vapor deposition (PVD) techniques such as evaporation and sputtering have been routinely used in industry for the deposition of metals and their compounds *e.g.* in the fabrication of integrated circuits. Amongst the chemical deposition techniques, the chemical vapor deposition (CVD) technique is also well established. The other chemical deposition techniques such as the atomic layer deposition (ALD) and layer-layer-deposition (LBL) have also been in existence for many decades and are now being actively investigated for the preparation of novel nanostructures consisting of both inorganic as well as organic constituents. In this chapter the various deposition techniques and their applications in 2D nanostructures are briefly described and one important example, multilayers for mechanical engineering applications, is discussed in some detail.

10.2 Thin film deposition

10.2.1 *Some considerations in thin film deposition*

Prior to the deposition of a thin film, the substrate preparation is necessary. Various techniques are used to clean the substrates which include the use of acids, bases and oxidizing agents as well as the physical methods such as ion beam etching and reactive plasma cleaning.

To promote the adhesion of the film with the substrate, it may be necessary to introduce an intermediate adhesion layer between the substrate and the film *e.g.* a layer of Ti or Cr is deposited on oxide substrates before deposition of a noble metal such as gold. The gold layer forms strong metallic bonds with the Ti or Cr layer if the deposition is sequentially carried out in vacuum without allowing the Ti or Cr layer to oxidize. Because of the tendency to oxidize easily, the Ti and Cr layers form strong bonds with the oxide substrate also.

In the dry deposition methods, the depositing particles arrive at the substrate having a random trajectory and random points of contacts with the substrate. The film structure depends to a very large extent on the mobility of the particles on the substrate. The mobility is directly proportional to the ratio of the substrate temperature during deposition to the melting point of the depositing species. When the mobility is low, arriving particles tend to bond to the point of first contact. In this case a porous and amorphous film is obtained. If the mobility is high, the particles can adjust their positions to assume a crystalline structure. A polycrystalline film results in this case. Films having porosity ranging from micro to nano range can be prepared by suitably adjusting the deposition conditions.

Depending on the deposition conditions, the film growth can occur in broadly three different ways. These are: (i) Island or Volmer- Weber growth (ii) Layer or Frank-van der Merwe growth and (iii) Island–layer or Stranski–Kranstonov growth. The island growth occurs when the

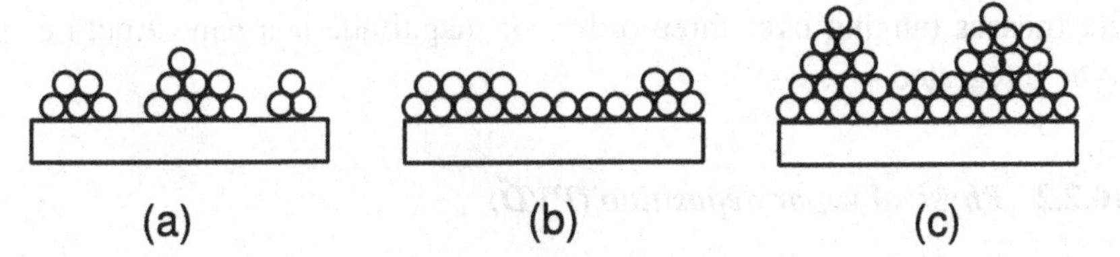

Fig. 10.1. The three modes of thin film growth (a) Island or Volmer–Weber growth (b) Layer or Frank-van der Merwe growth and (c) Island-layer or Stranski–Kranstonov growth.

bonding energy between the particles themselves is higher than that between the particles and the substrate. In this case, the particles bind to each other and form islands. These islands then grow to form a continuous film. This is usually the case for the deposition of metals on insulating substrates such as alkali halides, mica, graphite, *etc.* In the layer-by-layer growth, the particles bond more strongly to the substrate than to themselves. A layer covering the whole substrate is deposited first. This is followed by the deposition of the next layer. This is the case in the epitaxial growth. The island-layer growth is an intermediate case. The growth of self assembled quantum dots, discussed in Chapter 4, is an example of this mode of growth. The three modes of growth are illustrated in Fig. 10.1.

The epitaxial growth is required in many applications such as in the preparation of self assembled quantum dots in which a stack of semiconductor layers needs to be deposited (see Chapter 4). For epitaxial deposition, the substrate is chosen to be a single crystal and the depositing species should have the same crystal structure and nearly the same lattice spacing as the substrate. In addition, the depositing particle should have enough mobility on the substrate so that they preferentially bond to the growing lattice steps and not to the lattice plane. A single crystal film having the same structure as the substrate is obtained. The film may contain built in stresses due to the lattice mismatch.

In the deposition of a stack of layers, several additional considerations come into play. These are inter-diffusion between the layers, stresses developed due to processing or thermal expansion mismatch, differences in the etching rates, extension of grain boundaries or complex morphological parameters at the interface, need to integrate thicknesses ranging over three orders of magnitude if a nanostructure is to be integrated.

10.2.2 *Physical vapor deposition (PVD)*

In the PVD methods, atoms from a source (called target) are removed and deposited on the substrate. The removal of the atoms is effected either by heating the target so as to vaporize it or by impinging energetic ions on it to knock out its atoms [187, 188].

Three of the main methods employing the heating of a target for thin film deposition are thermal evaporation, electron beam evaporation and pulsed laser deposition. Figure 10.2(a) shows schematically the arrangement to deposit thin films by *thermal evaporation*. In this case the target is heated by resistive heating of a boat containing the target material. The deposition is carried out at pressures ranging from 10^{-3} to 10^{-10} torr such that the mean free path of the residual gaseous species in the deposition chamber is much larger than the target to substrate distance and the atoms from the target arrive at the substrate without suffering any collisions. The deposition is along the line of sight from the source to the substrate and so the conformal coverage is difficult.

Instead of resistive heating, electron beam can also be used for heating. In the electron beam evaporation (Fig. 10.2 b), an electron beam is focused on the target. Temperatures as high as 3500 °C can be achieved making it possible to deposit films of refractory metals such as Ta, W, *etc.* and many alloys and compounds. The electron beam's path and spread can be manipulated by using magnetic fields so that the rate of evaporation can be varied over a very wide range — from one nanometer per minute to a few microns per minute. By cooling the crucible, the contamination from the crucible can be eliminated, making it possible to produce coatings of extremely high purity.

In the pulsed laser deposition technique, short (~ 10 ns) and high power (> 100 $w.cm^{-2}$) laser pulses are used to ablate material from the surface of a solid target. Extremely high heating rates of the target surface, of the order of hundred million degrees per second, are achieved. The material is ablated from the target in the form of a plume of atoms, ions, electrons, atomic clusters and molten droplets. The process can be carried out in the presence of a variety of gases, making it useful for deposition of oxides, nitrides and other compounds. Due to the extremely high heating rates, species having widely different melting or sublimation temperatures are ablated simultaneously, replicating the stoichiometry of the target. The technique is therefore very useful for depositing films of complex stoichiometry such as oxide superconductors (*e.g.* $YBa_2Cu_3O_7$), ferroelectrics, certain magnetic materials, *etc.*

Process has to be controlled to minimize the formation of particulates to yield smooth films. Also the rotation of both the target and the substrate may be necessary to minimize the problems due to the narrow cone in which the ablated material is confined and due to the formation of pits on the target surface by the laser beam.

Molecular beam epitaxy (MBE) is a very precise technique by which it is possible to grow thin films in which the structure is controlled atomic monolayer by atomic monolayer. Such control is required to grow semiconductor quantum wells, super-lattices, spin valve structures consisting of metallic or magnetic multilayers, quantum wires, etc.

Molecular beam epitaxy is essentially an evaporation technique. Evaporation is carried out by resistive heating of cells in which the source material (usually elements such as Ga, As, Al, Sn *etc.*) is placed (although gaseous sources can also be used) (Fig. 10.2c). The cell is called an effusion cell or a Knudsen cell. The deposition is carried out in an ultrahigh vacuum chamber (vacuum better than 10^{-10} torr). At this vacuum, the mean free path of gas molecules is of the order of 100 m while the source to substrate distance is of the order of 30 cm. The atoms emitted from the source cells therefore reach the substrate without suffering any collision (hence the name molecular beam). The substrate is usually a thin wafer (~ 0.5 mm) heated to a suitable temperature (about 550 °C for GaAs). A number of cells are positioned in the chamber, all pointing towards the substrate. Shutters of the cells containing the desired species are opened for deposition, while the other shutters remain closed. Thus for the deposition of GaAs monolayers, the shutters of the cells containing Ga and As are opened. The atoms reach the substrate, interact and form an epitaxial semiconductor monolayer. The growth rate

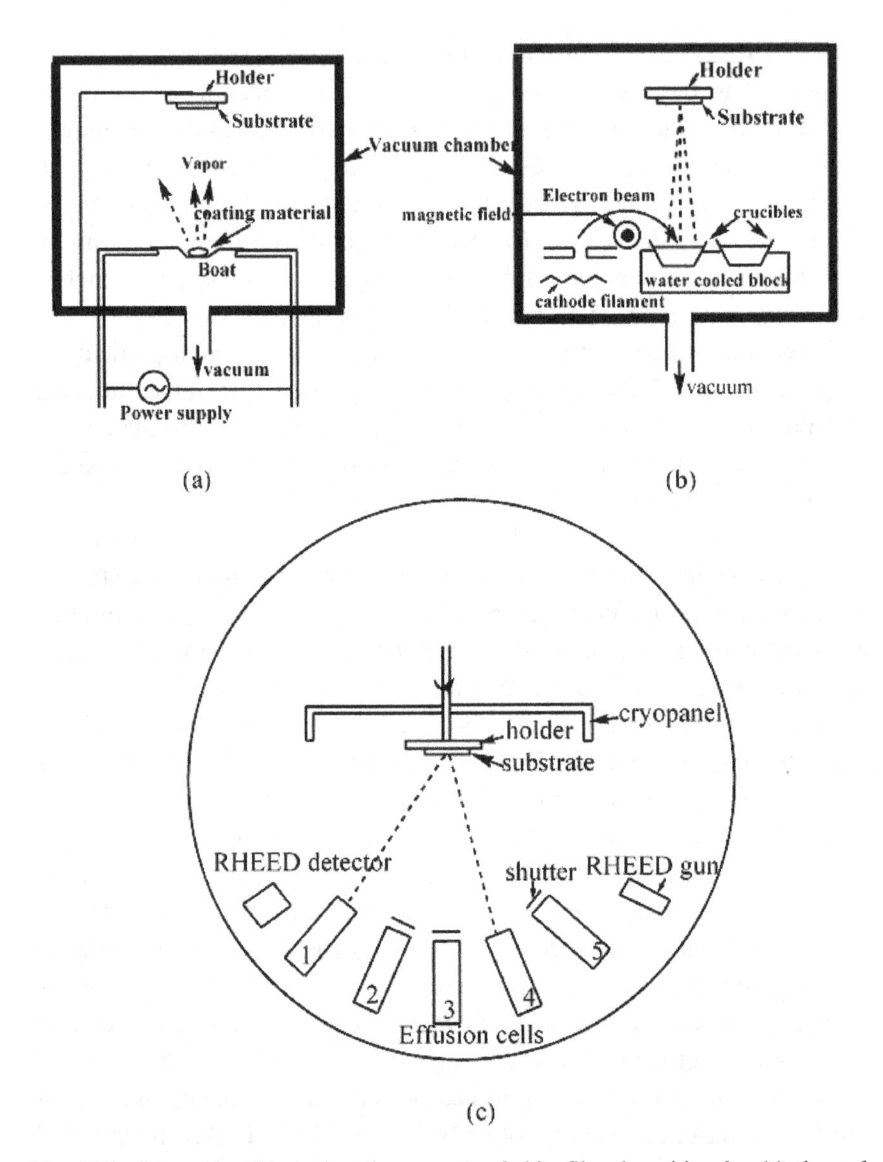

Fig. 10.2. Schematics illustrating the process of thin film deposition by (a) thermal evaporation (b) electron beam evaporation and (c) molecular beam epitaxy.

is very slow, of the order of 1 μm h^{-1} so that the atoms have sufficient time to diffuse on the substrate and form a defect free monolayer. If it is desired to next grow the monolayers of AlAs, the shutter of the Ga cell is closed and that of the Al cell is opened. As the closing and opening of the shutters can be carried out in much less time than it takes to deposit a monolayer, very sharp interfaces between the monolayers can be achieved.

One or more characterizing instruments to monitor the process and the film growth in real time are incorporated in the deposition chamber. These can be RHEED (reflection high energy electron diffraction) system, Auger electron spectroscope, x-ray photoelectron spectroscope (XPS), mass spectrometer, etc.

In the sputtering technique, the atoms from the target are knocked out by ions of an inert gas such as argon and deposited on the substrate. In the simplest case, the target and the substrate are positioned opposite to each other; an inert gas, usually argon is introduced in the chamber at a pressure of a few millitorr and a field of several kilovolts per centimeter

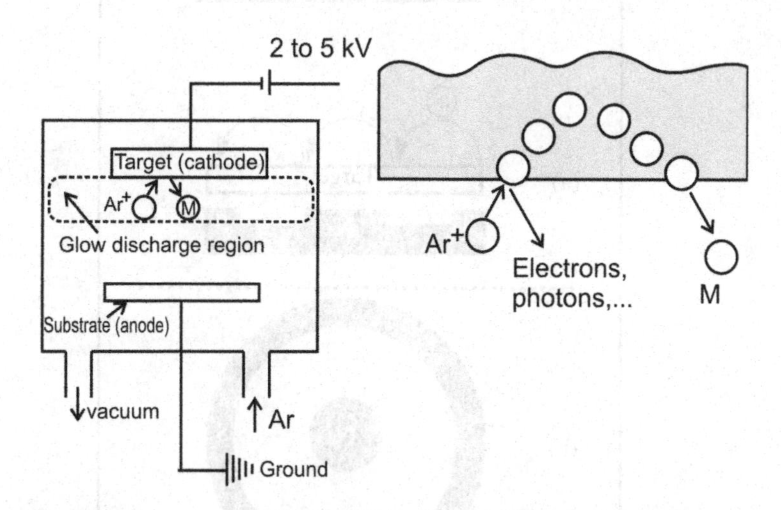

Fig. 10.3. Schematics illustrating the process of thin film deposition by sputtering; M denotes the atoms knocked out of the target by the energetic argon ions.

is applied across the substrate with target as the cathode (Fig. 10.3). A glow discharge is initiated due to the electrons in the chamber getting accelerated under the field and ionizing the gas molecules, thereby producing gas ions and more electrons. Under suitable conditions of gas pressure and the applied field, a stable glow discharge is sustained. The gas ions accelerate towards the target, and following a series of events, target atoms are knocked out and get deposited on the substrate. The process of sputtering imparts a high energy (1 to 10 eV) to the target atoms arriving at the substrate which is 10 to 100 times larger than the energy of the atoms in thermal evaporation (~0.1 eV). This produces films with better adhesion and a higher density and may also lead to some other changes like a different texture, phase changes, *etc.*

Besides the knocking out of the target atoms, the impact of the gas ions with the target also produces secondary electrons, photons and x-rays. The electrons are needed to sustain the plasma. However, not all the

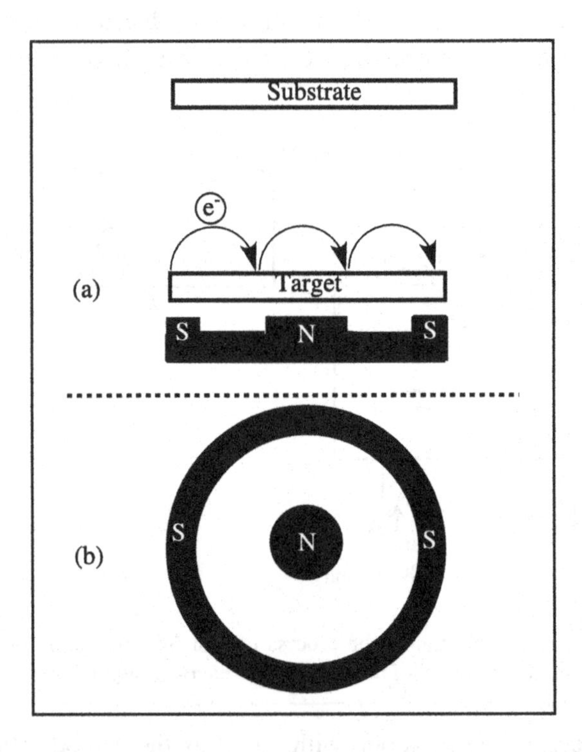

Fig. 10.4. (a) In magnetron sputtering, magnets placed below the target trap the electrons generated due to the impact of ions with the target to a region near the target and produce a more concentrated plasma (b) plan view of the magnets.

electrons escaping the target contribute to the ionization of the gas atoms; most of them are accelerated away from the target. To improve the process efficiency, an array of magnets is placed behind the target with their north and south poles arranged to produce a magnetic field perpendicular to the electric field between the target and the substrate, Fig. 10.4. This confines the electrons to the space near the target (Fig. 10.4 a) and leads to the generation of more ions and a higher deposition rate at a lower gas pressure. The process is termed as *magnetron sputtering*.

A modification of the magnetron sputtering is the unbalanced *magnetron sputtering*. In the magnetron sputtering as described above, the magnetic field is balanced *i.e.* the north and the south magnets behind the target are of equal strength. This leads to the creation of dense plasma which is confined in the vicinity of the target. By using stronger magnets on the outside, the magnetic field is unbalanced. As a result, the plasma extends further and may reach right up to the substrate (Fig. 10.5 b). The advantage is that as the substrate is surrounded by the plasma, there is a greater bombardment of the substrate with ions, improving the film quality. For the sputter deposition of insulating films, a radio frequency (RF) field, in the range of 5 to 30 MHz (usually 13.56 MHz, as assigned by the US Federal Communications Commission for this purpose) is used instead of a dc field. The target is coupled to the RF generator by a

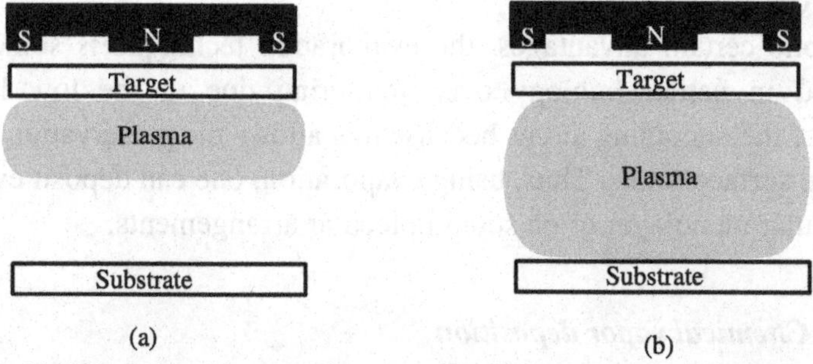

Fig. 10.5. (a) In the balanced magnetron sputtering the N and S magnets are balanced (*i.e.* have equal strengths) and the plasma extends only part of the distance to substrate (b) in the unbalanced magnetron sputtering the strengths of N and S poles are unequal and adjusted so that the plasma extends nearly all the way to the substrate.

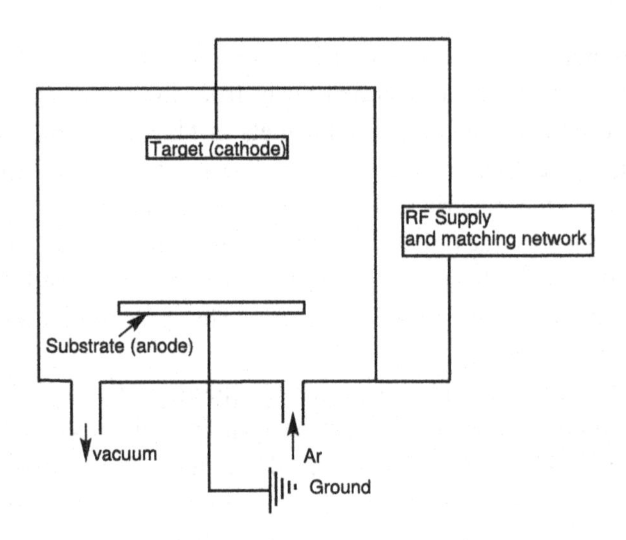

Fig. 10.6. A schematic of RF sputtering.

low RF impedance capacitor (Fig. 10.6). As only the electrons due to their lighter mass, and not the ions, are able to follow the RF field, a self negative bias is generated on the target leading to sputtering of the target atoms. The process is called RF sputtering.

Besides the thin film deposition, sputtering can also be used for a variety of purposes by varying the ion energy (1 eV to 10 MeV). These include surface cleaning and etching, sequential removal of layers for through-the-depth analysis, ion implantation and generation of ions in ion sources.

Despite certain advantages, the evaporation technique is sometimes preferred in nanotechnology over sputtering due to the low kinetic energy of the oncoming atoms because this allows the preservation of the sensitive surface states. Thus, using evaporation, one can deposit even on a molecular monolayer or on supramolecular arrangements.

10.2.3 *Chemical vapor deposition*

The chemical vapor deposition (CVD) technique for film deposition [189] is also used, like the PVD techniques, for the deposition of inorganic materials. In the CVD process, reactants in the vapor or the gaseous form are brought near a heated substrate and react to form a solid film on the substrate (Fig. 10.7). Thus a film of silicon can be deposited from $SiCl_4$ and H_2 using the following reaction

$$SiCl_4 + 2H_2 \rightarrow Si_{(s)} + 4HCl^{\uparrow}$$

An essential condition for film deposition is that the reaction should occur heterogeneously on the substrate and not homogeneously in the gaseous phase — the latter will produce powder rather than a film. The extent of homogeneous nucleation increases with temperature and partial pressure of the reactants.

Different kinds of reactions including pyrolysis (thermal decomposition), reduction, oxidation, compound formation, *etc.* can be used as shown by the following examples [190, 191].

Pyrolysis

(i) $SiH_4 + O_2 \rightarrow Si_{(s)} + 2H_2O$ $\quad T > 777^\circ C$

(ii) $Ni(CO)_{4\,(g)}$ (Nickel carbonyl) $\rightarrow Ni_{(s)} + 4CO_{(g)}$

Reduction

(iii) $SiCl_{4(g)} + 2H_2 \rightarrow Si_{(s)} + 4HCl_{(g)}^{\uparrow}$ $\quad 900\text{-}1200^\circ C$

(iv) $WF_{6(g)} + 3H_2 \rightarrow W_{(s)} + 6HF$ $\quad 500^\circ C$

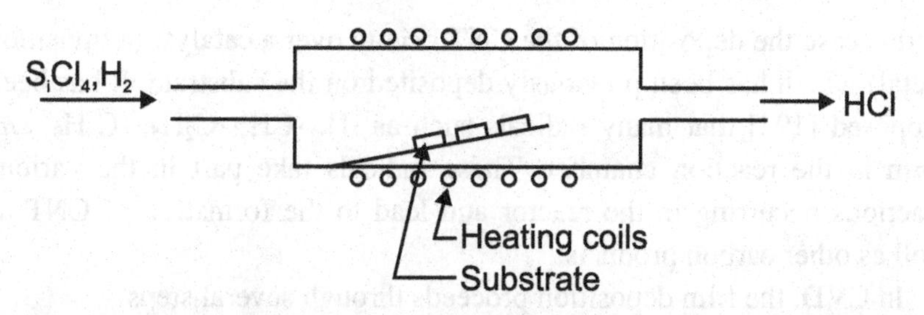

Fig. 10.7. Deposition of a Si film by CVD using $SiCl_4$ and H_2 as reactants; the substrates are heated to ~ 1200° C using external heating coils.

Oxidation

(v) $SiH_4 + 2O_2 \rightarrow SiO_{2(s)} + 2H_2O$ $\quad 350^\circ C$

(vi) $Zn(C_2H_5)_2 + 4O_2 \rightarrow ZnO_{(s)} + 5H_2O + 2CO$ $\quad 250^\circ C\text{-}550^\circ C$

Compound formation

(vii) $2AlCl_3 + 3CO_2 + 3H_2 \rightarrow Al_2O_3 + 6HCl + 3CO$ 800°C-1150°C
(viii) $TiCl_4 + \frac{1}{2} N_2 + 2H_2 \rightarrow TiN + 4HCl$ 1200°C
(ix) $3SiH_4 + 2N_2H_4 \rightarrow Si_3N_4 + 10H_2$ 800°C

Disproportionation

(x) $3GaCl \rightarrow 2Ga + GaCl_3$
(xi) $SiCl_4 + CH_{4(g)} \rightarrow SiC_{(s)} + 4HCl_{(g)}$ 1400°C

It should be noted that although the CVD process is conveniently expressed by the reactions given above, the actual reactions occurring in a CVD reactor are quite complex. As an example, for the deposition of Si by the reduction of chlorosilane

$$SiCl_{4(g)} + 2H_2 \rightarrow Si_{(s)} + HCl_{(g)}\uparrow$$

at least eight gaseous species can exist in the reaction chamber depending on the deposition conditions: $SiCl_4$, $SiCl_3H$, $SiCl_2H_2$, $SiClH_3$, SiH_4, $SiCl_2$, HCl and H_2. As another example of the complexity of the CVD reactions, consider the deposition of carbon nanotubes (CNT) from the decomposition of ethylene:

$$C_2H_4 \rightarrow 2C + H_2$$

In this case the deposition of the CNT occurs over a catalyst (a transition metal) which has been previously deposited on the substrate. It has been proposed [192] that many radicals such as $\cdot H$, $\cdot CH$, $\cdot C_2H_5$, $\cdot C_4H_9$, *etc.* form in the reaction chamber. These radicals take part in the various reactions occurring in the reactor and lead to the formation of CNT as well as other carbon products.

In CVD, the film deposition proceeds through several steps

(i) Diffusion of the reactants to the surface.
(ii) Adsorption.
(iii) Surface events such as chemical reaction, surface motion, lattice incorporation.
(iv) Desorption of products/ byproducts.

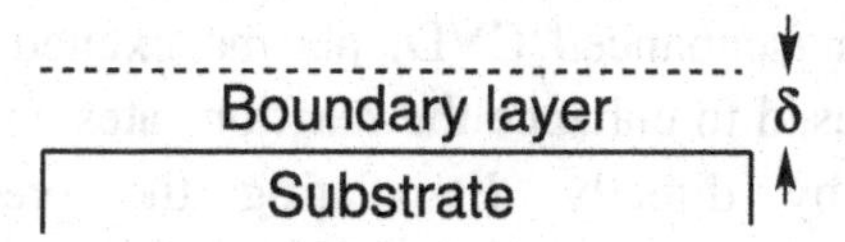

Fig. 10.8: Due to the laminar flow conditions that exist within a CVD reactor, a stagnant boundary layer of thickness δ develops above the substrate.

(v) Diffusion of products away from the surface.

Of the above steps, the slowest step controls the film deposition rate. If the substrate temperature is high enough so that the reactions at the substrate are not rate limiting, the first step *i.e.* the diffusion of reactants to surface may control the rate of deposition. Low gas pressures, usually 0.01 atm., are used in CVD to enhance the rate of diffusion. At these pressures, the gas velocities are low and the mean free path for the precursor molecules is larger than the reactor length and the flow is laminar. A stagnant boundary layer of thickness δ develops over the substrate (Fig. 10.8). The flux of the growth species through the boundary layer is then proportional to the pressure gradient across the boundary layer

$$J \sim D(P_i - P_o)/\delta$$

Here P_i and P_o are the pressures of the growth species in the bulk of the gas phase and next to the growing film surface respectively. The diffusivity depends on the pressure as follows

$$D = D_o (P_o/P)(T/T_o)^n$$

where T_o and P_o are the standard temperature (273 K) and pressure (1 atm.) and D_o is the corresponding diffusivity; the exponent n is found to be approximately 1.8. The diffusivity, and so the flux, decreases as the pressure increases. If the diffusion is the rate limiting step, then the growth rate can be enhanced by using low pressures.

Several variants of the CVD process have been developed giving rise to as many acronyms:

- MOCVD: Metalorganic CVD; metalorganic compounds are used as precursors.

- PECVD: Plasma enhanced CVD; plasma excited by an rf field or microwave is used to enhance the reaction rates either by heating the substrate or by directly dissociating the precursor molecules photolytically.
- AACVD: Aerosol assisted CVD; to use a nonvolatile liquid as a precursor, it is fed as an aerosol to the CVD reactor.
- LPCVD: Low pressure CVD as discussed above.

The ability of the gaseous species to penetrate small pores imparts a unique capability to the CVD process. This ability is used to deposit a solid phase on highly porous substrates or inside porous media. The technique is then called chemical vapor infiltration (CVI). Its major conventional application is in depositing carbon in the voids in porous graphite and fibrous mats to make carbon-carbon composites.

An example of the use of CVI in nanotechnology is in the preparation of crack free silica photonic crystal film. A silica photonic crystal film consists of a close packed assembly of silica microspheres. Typically this ordered assembly is obtained utilizing the repulsive double layer interactions (Chapter 7) between the silica spheres in an aqueous suspension. On sedimentation, or using some other technique, an ordered and close packed film of the silica spheres is obtained. The silica spheres have a layer of water on them. Evaporation of this layer generates capillary stresses leading to the formation of cracks in the

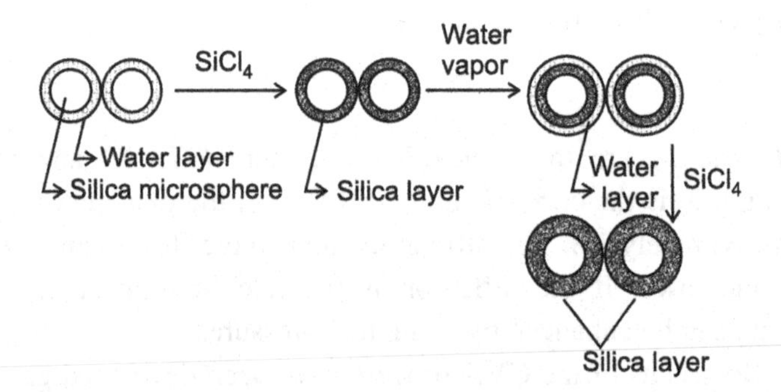

Fig. 10.9: The silica spheres in the as deposited photonic crystal silica film have a layer of water. To use the CVI to replace this layer by a coating of solid silica, chlorosilane and water vapor are alternately passed over the film.

assembly. Using CVI, the hydration layer around the silica spheres can be replaced by solid silica by alternately passing $SiCl_4$ and water vapor over the film (Fig. 10.9).

A striking application of CVD is in the growth of patterned arrays of carbon nanotubes (CNT) as already discussed in Chapter 8 and shown in Fig. 8.12.This was accomplished by depositing a film of iron, which is a catalyst for CNT growth, on a porous silicon substrate through a shadow mask to form an array of 38 μm squares. Ethylene was used as a precursor for the deposition of CNT. The carbon nanotubes grew on the catalyst particles in each square and were held together in square blocks due to the van der Walls forces between them. Straight, square bundles of tubes having an aspect ratio as high as 5 could be obtained.

10.2.4 *The sol – gel process*

10.2.4.1 *Introduction*

The sol-gel process can be used for depositing thin films as well as for the preparation of particles, fibers and monoliths (*i.e.* solid pieces). Because of its versatility it finds many uses in nanotechnology.

A sol is defined as a suspension of colloidal (1–1000 nm) particles in a liquid. The particles can be dense solid particles or they can be oligomers (a polymer molecule containing only a few monomers) formed from some precursors due to polymerization reactions. The particles can be prepared in situ or they can be prepared separately and then dispersed in a liquid. In this size range the gravitational forces are negligible; the particles undergo Brownian motion. To keep the particles suspended for a reasonably long period, the van der Waal's attractive interaction between the particles is to be counterbalanced by a repulsive interaction which can be due to the surface charges or by steric forces. The two most important starting points for the preparation of sols are (i) solution of a metal salt in water and (ii) a metal alkoxide dissolved in a suitable solvent. The former is mostly used for the preparation of particles while the latter is more versatile and more amenable to control so that not only particles, but films, coatings, fibers and even monoliths can be obtained.

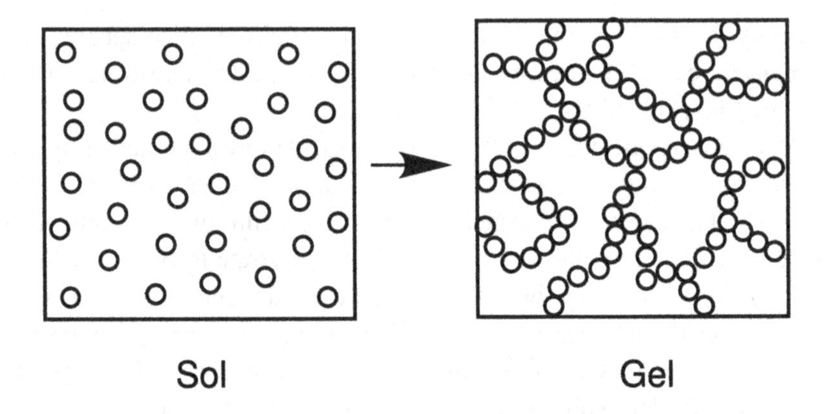

Fig. 10.10. A sol is a suspension of colloidal particles in a liquid; gel is a rigid mass in which the particles are connected together in a single skeleton with the pores filled with the liquid.

Under suitable conditions, a sol can transform to a gel. In defining the gel, we can distinguish between two situations. In one, the gel is defined as a rigid mass in which the particles of colloidal dimensions are connected together in a single skeleton; the pores in the skeleton are also of colloidal dimension and are filled by the liquid (Fig. 10.10). The sol to gel transformation is accompanied by a rapid rise in the viscosity. The point at which the gel is said to have formed can be identified with the formation of a single spanning cluster which reaches across the vessel. Subsequently more clusters can link to it and thereby the rigidity of the gel can continue to increase. While a sol can be poured out of a beaker, a gel behaves like a rigid solid. If the sol is applied to a substrate as a thin coating, the evaporation of the solvent brings the particles into close proximity and rapid gelation occurs. Subsequent drying and heat treatment of the gel film produces a dense oxide or glass film.

The name gel is also given to the fluffy precipitate which can form from a sol due to precipitation, destabilization or super saturation. In this case the amount of the liquid phase is very high so that the viscosity does not reach a value equal to that of solids.

10.2.4.2 *Sols from metal salts*

Sols of oxide nanoparticles can be prepared from solutions of their salts *e.g.* a sol of hydrated ZrO_2 particles can be prepared from a solution of $ZrOCl_2.8H_2O$.

When a salt is dissolved in water, each cation is surrounded by water molecules with their oxygens turned towards the metal. The resulting water cation complex has usually four to six water molecules in it, with six being the more common as in $[Al(H_2O)_6]^{3+}$. Depending on the charge z on the cation and the pH of the solution, the cation-water complex can lose a proton; this is called *hydrolysis* and occurs at a high pH

$$[M(OH_2)_N]^{z+} + H_2O \rightarrow [M(OH)(OH_2)_{N-1}]^{(z-1)+} + H_3O^+$$

Repeated hydrolysis leads to the formation of complexes represented by

$$[MO_x(OH)_y\,(OH_2)_{N-x-y}]^{(z-y-2x)+} \qquad \text{(A)}$$

The values of x and y depend on the charge on the cation and the pH. Thus Li^+ with low z forms $[Li(OH_2)_4]$ while S with high z forms SO_3.

Several of the species of type (A) next join together to form the building block for the formation of the particle. This joining together is called *condensation*. A simple type of condensation reaction is as follows

$$\text{M-OH} + \text{OH-M} \rightarrow \text{M-O-M} + \text{OH}$$

Here the two cations are joined together by an oxygen. This is called an '*oxo*' bond and the condensation reaction is called oxolation.

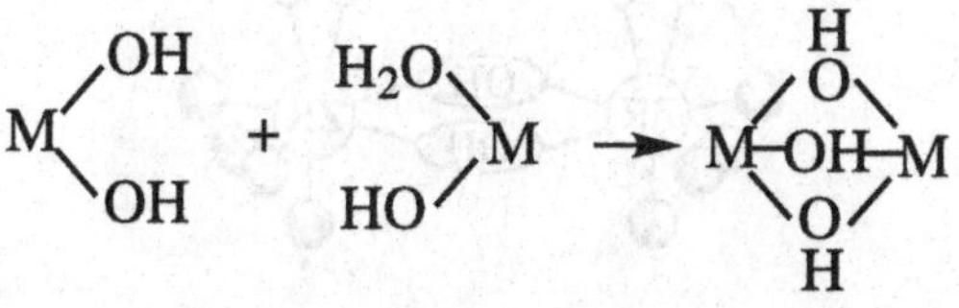

$$\text{M}(\text{OH})_2 + (\text{H}_2\text{O})(\text{HO})\text{M} \rightarrow \text{M}\langle(\text{OH})(\text{OH})(\text{OH})\rangle\text{M}$$

Fig. 10.11. The olation reaction

Another type of condensation reaction is called *olation* in which the bonding is through an OH group (Fig. 10.11).

A well known example of the polymeric complex formed by olation is $[Zr_4(OH)_8.16H_2O]^{8+}$, formed [193, 194] in an aqueous solution of $ZrOCl_2.8H_2O$ (Fig. 10.12).

The complexes formed in the solution can undergo further condensation. The suspension containing the complexes and their condensation products is now a sol. The rate and extent of condensation is high near the isoelectric point (IEP). As a result either precipitation or gel formation occurs. Formation of precipitates is more common due to high condensation rates. As an example, when a solution of ammonium hydroxide is added to a solution of zirconium oxychloride in water, a high pH is produced which causes hydrolysis to occur rapidly. The condensation reactions are also rapid so that immediately upon mixing, a white fluffy suspension is produced. At this point the dimensions of the species produced are quite large so that the suspension appears white due to the scattering of light. The size can be brought back to the colloidal dimensions by adding a small amount of acid, which leads to a transparent sol.

A clear sol of cerium oxide particles can be prepared by adding ammonium hydroxide to a Ce(III) salt solution. This produces a Ce(IV) hydrate precipitate. Addition of a small amount of nitric acid results in the formation an aqueous sol of CeO_2.

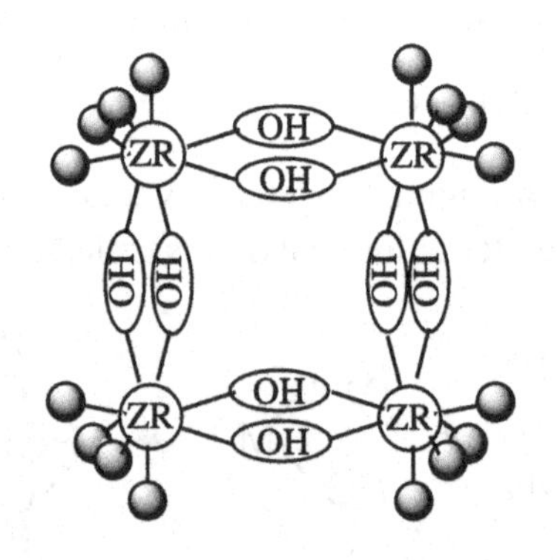

Fig.10.12. The complex formed as a result of hydrolysis and condensation in a solution of the salt $ZrOCl_2.8H_2O$.

10.2.4.3 *Sols from alkoxides*

A very useful route to the formation of sols is to start from metal alkoxides. The metal alkoxides have a general structural formula given by $M\text{-}(OR)_n$ where R is an alkyl group like CH_3, C_2H_5 *etc.* and OR is called an alkoxy group. Some common alkoxy groups and examples of corresponding alkoxides are given in Table. 10.1.

By addition of a small amount of water and adjusting the pH, the alkoxides can be made to undergo the hydrolysis and condensation reactions leading to an M-O-M network.

The process can be illustrated using tetraethylorthosilicate (TEOS), the most studied alkoxide, as an example. First the alkoxide is mixed with a solvent, which is a solvent for water also. The solvent is usually an alcohol. In the case of TEOS, ethyl alcohol is mostly used. The water and a small amount of an acid (*e.g.* nitric acid) or a base (*e.g.* ammonium hydroxide) are then mixed into it. The hydrolysis of the alkoxide molecules occurs. For the case of TEOS, the overall reaction can be written as shown in Fig. 10.13.

The mechanism of hydrolysis depends on the pH of the solution. Under basic conditions, reaction is by nucleophilic substitution. The negatively charged hydroxide attacks the positively charged silicon forming a five coordinated intermediate. The attacking and the leaving groups are on the opposite sides of silicon. This provides the maximum separation of charge (Fig. 10.14).

The attack by the hydroxyl will be greater if

(i) The silicon is more positively charged i.e. if highly basic (electron rich) alkoxy groups (OR) are replaced by OH.

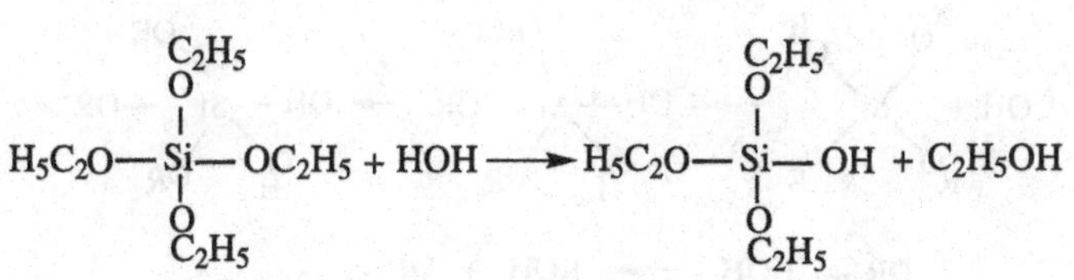

$$H_5C_2O\text{—}Si(OC_2H_5)_2\text{—}OC_2H_5 + HOH \longrightarrow H_5C_2O\text{—}Si(OC_2H_5)_2\text{—}OH + C_2H_5OH$$

Fig. 10.13. Hydrolysis reaction for TEOS.

Table 10.1. Some common alkoxide groups and examples of corresponding alkoxides.

Alkoxy ligand		Example of alkoxide
Methoxy	$\bullet OCH_3$	Sodium methoxide, Na OCH_3
Ethoxy	$\bullet OCH_2CH_3$	Tetraethylorthosilicate (TEOS), $Si(OCH_2CH_3)_4$
n-propoxy	$\bullet O(CH_2)_2CH_3$	Titanium propoxide ,$Ti(O(CH_2)_2CH_3)_4$
iso-propoxy	$H_3C(\bullet O)CHCH_3$	Aluminium isopropoxide, $Al(OCH_3CHCH_3)_3$
n-butoxy	$\bullet O(CH_2)_3CH_3$	Titanium butoxide, n-butoxy, $Ti(O(CH_2)_3CH_3)_4$
sec-butoxy	$H_3C(\bullet O)CHCH_2CH_3$	Aluminum sec butoxide, $Al(OCH_3CHCH_2CH_3)_3$
iso-butoxy	$\bullet OCH_2CH(CH_3)_2$	Potassium isobutoxide, K $OCH_2CH(CH_3)_2$
tert-butoxy	$\bullet OC(CH_3)_3$	Titanium tert butoxide, $Ti(OC(CH_3)_3)_4$

(ii) Bulkiness of the alkoxy group is not conducive to attack i.e. the attack is favored if the OR is replaced by OH.

Because of the above factors, under basic conditions, the hydrolysis of a single TEOS molecule is most difficult and becomes easier as it gets more and more hydrolyzed. The condensation of the hydroxyl groups results in the formation of highly cross linked structures and even particles when the reaction is carried out under basic conditions. Despite this, some organic matter is retained in the base catalyzed product. This is because if the condensation occurs between monomers which were incompletely hydrolyzed, the hydrolysis of the remaining alkoxy groups is retarded *i.e.* polymers which are incompletely hydrolyzed result.

Under the acidic conditions, the mechanism of hydrolysis is by electrophilic substitution as illustrated in Fig. 10.15. A protonated water molecule is attracted to the oxygen atoms in the basic, alkoxide groups.

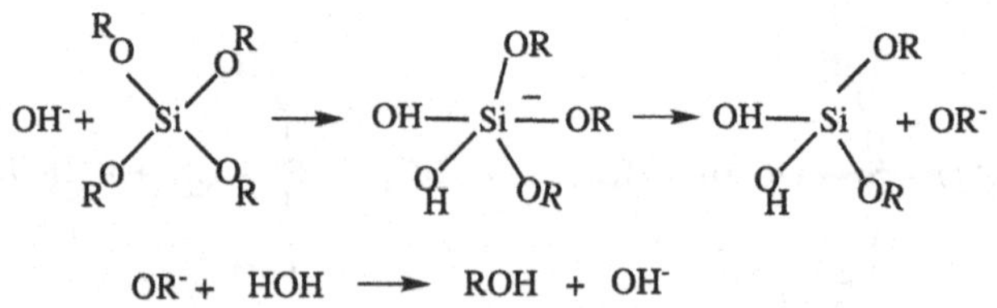

$$OR^- + HOH \longrightarrow ROH + OH^-$$

Fig. 10.14. Hydrolysis of TEOS under basic conditions.

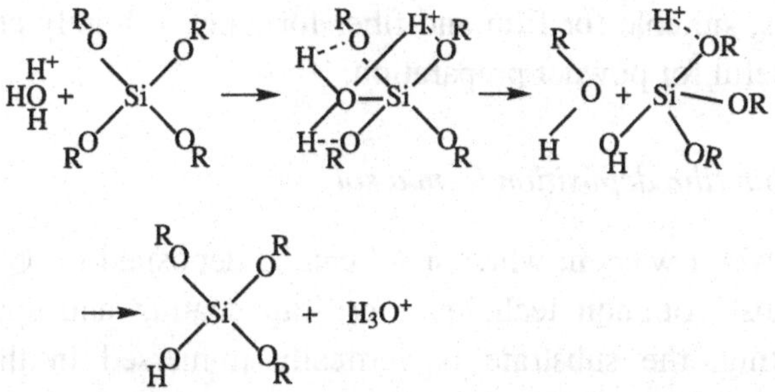

$$H^+ HOH + Si(OR)_4 \rightarrow [H\text{-}O(H^+)\cdots Si(OR)_4]^\ddagger \rightarrow ROH \cdot H^+ + Si(OH)(OR)_3$$

$$\rightarrow Si(OR)_3(OH) + H_3O^+$$

Fig.10.15. Hydrolysis of TEOS under acidic conditions; the dashed lines represent the partial bonds.

Dashed lines represent the partial bonds which form. One set of partial bonds strengthens faster, forming the alcohol molecule and the silanol group. The positive charge on Si is highest in the monomer and decreases as the hydrolysis proceeds. The reactivity of the hydrolysis reaction is therefore highest for the monomer and decreases with hydrolysis. Also, if there is no repulsion between the attacking group and the oligomer, the attack will be more when the number of the OR groups is more *i.e.* the reactivity goes as

Monomer > end groups on chains > side groups on chains

Because of the above considerations, the acid catalysis produces less cross linked, nearly linear chains.

In the acid catalysis, the leaving group departs from the same side as the protonized water attacks — no inversion of the molecule occurs (Fig. 10.15). This type of attack is not affected much by the degree of polymerization since the hydrolysis can go on even when the condensation has occurred *i.e.* the hydrolysis can go to completion in the acid catalyzed gel. However, the reesterification *i.e.* the reverse of hydrolysis is also easier and in fact the acid catalyzed gels reesterify to a greater extent after gelling as compared to the base catalyzed gels and so contain more organic matter.

Thus by controlling the reaction condition (pH and the amount of water) one can vary the structure of the condensation products between linear chains, suitable for film and fiber forming, to highly cross linked polymers useful for powder preparation.

10.2.4.4 *Thin film deposition from a sol*

There are several ways in which a sol can be deposited to obtain a film. The two most common techniques are dip coating and spin coating. In dip coating, the substrate is vertically immersed in the sol and withdrawn at a slow rate (Fig. 10.16). A sol layer adheres to the substrate in which the precursors get concentrated as the drying proceeds. There is an 18 to 36 fold increase in concentration, bringing the precursors in close proximity; this leads to rapid gelation (Fig. 10.16 c).

During the withdrawal of the substrate, the adhering liquid experiences a viscous drag in the upward direction and a gravitation force in the downward direction. The viscous drag varies as ~ $\eta u/h$ where η is the viscosity of the liquid, u is the withdrawal rate and h is the thickness of the liquid/gel layer while the gravitational pull is ρgh. Equating the two, one gets

$$h = c_1(\eta u/\rho g)^{1/2} \text{ with } c_1 = 0.8 \text{ for Newtonian liquids.}$$

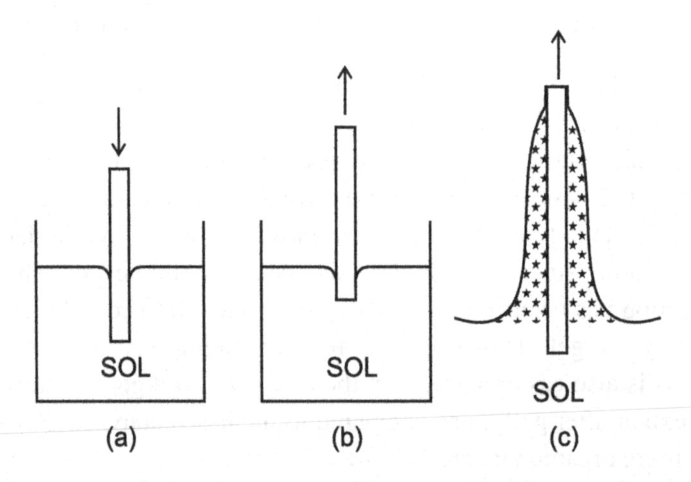

Fig.10.16. The process of dip coating (a) the substrate is vertically inserted in the sol and (b) withdrawn at a slow rate (c) As the substrate is withdrawn, the precursors get concentrated leading to rapid gelation.

However, the surface tension forces are also present and can not be neglected as u and η are low. Taking this into consideration results in the following expression for h, the film thickness

$$h = \frac{0.94(\eta u)^{2/3}}{\gamma_{LV}^{1/6}(\rho g)^{1/2}} \tag{10.1}$$

Thus the thickness increases as a power of the viscosity and the withdrawal rate. Experimentally, a log-log plot of the film thickness *vs.* withdrawal rate gives a straight line with a slope of between 0.47 to 0.64 depending on the structure of the sol *i.e.* the state of branching, particulate formation (Fig. 10.17). Note also that the viscosity constantly increases during the withdrawal step.

In spin coating, a few drops of the sol are placed on a rapidly rotating (few thousand rpm) substrate. The sol spreads on the substrate and forms a uniform coating. The thickness of the film depends on the rotation speed and other parameters as follows

$$h = \left(1 - \frac{\rho_A}{\rho_A^o}\right)\left(\frac{3\eta\, e}{2\rho_A^o \omega^2}\right) \tag{10.2}$$

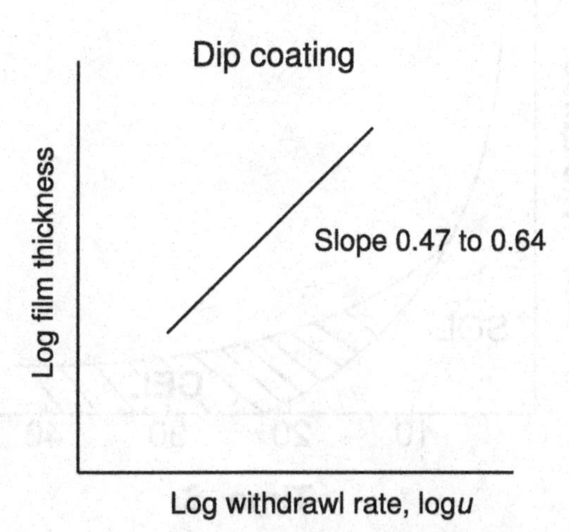

Fig. 10.17. A plot of the film thickness *vs.* the withdrawal speed in dip coating.

where ρ_A = mass of the volatile solvent per unit volume

ρ_A^o = initial value of ρ_A

ω = angular velocity of the substrate

e = evaporation rate.

In both the dip coating as well as the spin coating, thicker films can be produced by repeated coating and drying before the final sintering.

During the film formation, a gel forms within 30 to 40 seconds after deposition of the sol. Figure 10.18 shows the change in the film thickness with time as it undergoes the sol to gel transformation.

The porosity and structure of the dried film depends on the sol composition and the evaporation rate of the film after coating. The sol composition controls the amount of cross linking in the gel. The linear chains, obtained under the acidic conditions, can interpenetrate and are "transparent" while the highly branched clusters can not interpenetrate and are opaque. In the latter case the pore size is nearly the same size as the cluster size while in the former case very fine pores and very low porosity results. Furthermore, under the basic conditions with high water availability, a high condensation rate is obtained. The structure becomes

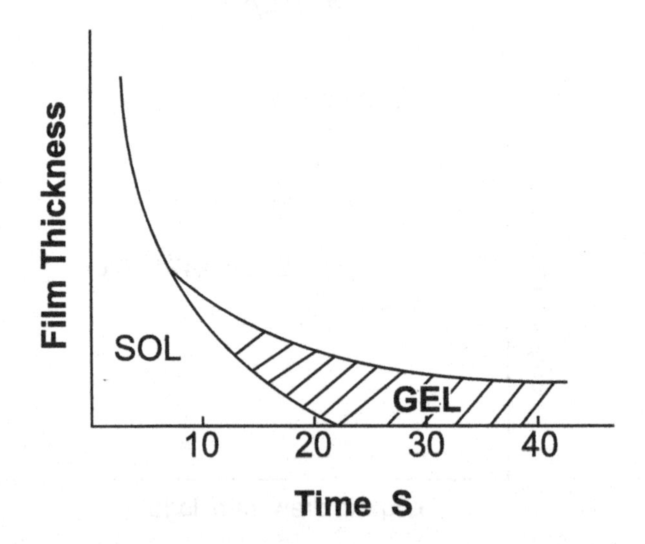

Fig. 10.18. Change in the film thickness during sol to gel transformation just after deposition.

rigid before it can be compacted by evaporation so that a more porous film results.

The evaporation rate also affects the film structure. With a low evaporation rate and high gelation rate the film gels before drying and traps a large quantity of liquid; the network becomes rigid with further gelation and can not shrink. This results in a porous film. If the amount of water used to prepare the sol is large and a large amount of base is used, then the gelation rate is high. In this case the gel becomes rigid before the evaporation compacts the structure and a more porous film results.

After drying, the film is heated at 500–700 ^{0}C to remove the organics and further densify the film and also to improve the adhesion between the film and the substrate.

10.2.4.5 *Cracking of gels and films*

An important consideration in the sol-gel films is the cracking that is likely to occur during the drying and sintering steps if appropriate precautions are not taken. There is a severe reduction in volume during sol to gel transformation and a further shrinkage occurs during the sintering of the film. This produces large stresses which may result in the cracking of the film.

Several processes occur during the drying of a gel which may lead to the generation of stresses [195]. Here we consider a simple model [196, 197].

Consider the case of a liquid in a capillary (Fig. 10.19 a). Due to the negative curvature of the meniscus, the pressure inside the liquid is smaller by an amount ΔP given by

$$\Delta P = \frac{2\gamma \cos\vartheta}{r}$$

Here, γ is the surface energy of the liquid – vapor interface, r is the capillary radius and θ is shown in Fig. 10.19 (a). The ΔP is negative as r is negative. A force trying to collapse the capillary is therefore exerted on the walls of the capillary as shown in Fig. 10.19 (b). As in the drying

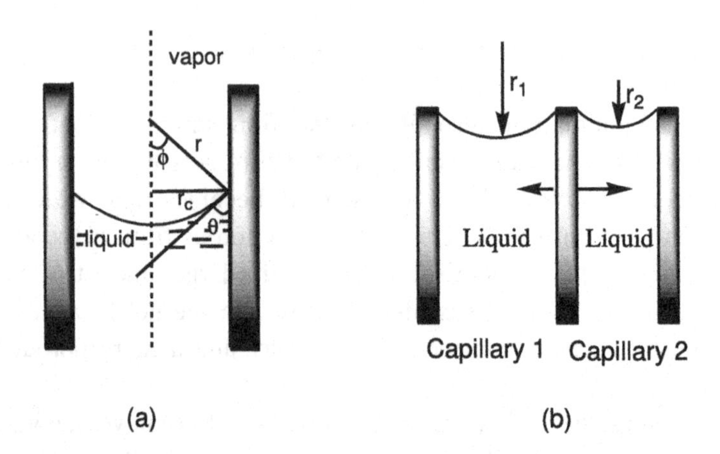

Fig. 10.19. (a) The meniscus in a capillary; a negative pressure exists in the liquid inside the capillary due to the curvature of the interface (b) adjacent capillaries with different diameters experience different forces on their walls resulting in a net force.

gel the radii of the different channels in which the liquid is trapped is varying, a net force is exerted on the gel walls which can result in its cracking (Fig. 10.19 b)

The cracking can be avoided by slower drying and limiting the dimension of the body *e.g.* thicker films are more likely to crack than the thinner films. This is because the substrate prevents the shrinkage of the film next to it. A stress gradient is set up in the film. If the film is thick, the stresses at the surface may reach a value high enough to cause cracking.

Another approach to avoid the cracking is to add small amounts of formamide ($HCONH_2$) or oxalic acid to the alkoxide sol. It is found that these additives, called "drying control chemical additives (DCCA)" significantly reduce the cracking tendency of the gel and even monolithic gels with very little shrinkage can be produced. The additives change the pore size distribution in the gel. This is believed to reduce the amount of shrinkage in the gel.

Monolithic gels can be dried with almost no shrinkage to produce what is known as an *aerogel* using supercritical drying. The process involves taking the gel around the triple point (the point at which the densities of the liquid and the vapor phase become equal, see Fig. 10.20)

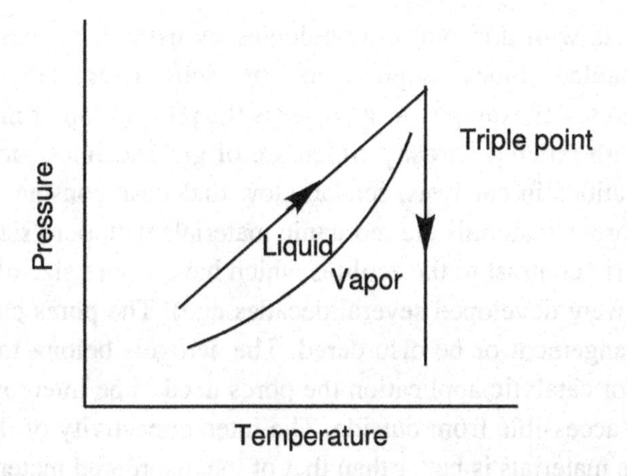

Fig. 10.20. The phase diagram showing the triple point where the densities of the liquid and the vapor are the same and the path followed in the supercritical drying.

from the liquid to the vapor region without crossing the boundary between the liquid and vapor regions.

The process can be illustrated by taking the example of a gel prepared from TEOS. The gel contains liquid, mostly ethyl alcohol, in the pores. It is called and *alcogel.* The alcogel is placed in an autoclave with excess alcohol and heated to 250 °C. The critical point for ethanol is 241 °C and 6.3 MPa. The pressure increases to about 10.5 MPa. Care is to be taken to not cross the liquid – vapor line in the phase diagram. The alcohol is then bled out at the rate of 10 ml min^{-1}. When the pressure reaches about 1 MPa, the autoclave is purged with argon to remove the remaining alcohol and cooled. An aerogel with only 2 to 3 vol. % shrinkage is obtained. The silica aerogels have a number of interesting properties including a very low density (~ 1 mg ml^{-1} as compared to the density of air, 1.2 mg ml^{-1}), very low thermal conductivity (~ 0.01 watt m^{-1}) and a high melting point of 1473 K.

10.2.4.6 *Preparation of mesoporous materials*

The sol-gel technique is widely used to prepare a wide range of nanomaterials, including particles, films, fibers, nanoporous materials, *etc.*, materials with different morphologies by using a template such as phase separated block copolymers or self assembled surfactant nanostructures. An example of the latter is the preparation of mesoporous materials with ordered porosity which are of great technological interest with applications in catalysis, sensors, low dielectric constant materials, *etc.* Mesoporous materials are inorganic materials with pore size between 2 to 50 nm (in contrast to the zeolites which have a pore size of < 1.5 nm and which were developed several decades ago). The pores can have an ordered arrangement or be disordered. The aerogels belong to the latter category. For catalytic application the pores need to be interconnected to make them accessible from outside. The interconnectivity of the ordered mesoporous materials is better than that of the disordered materials.

The scientists at Mobil [198, 199] and also in Japan [200] discovered a simple and novel method to prepare ordered mesoporous materials with ordered porosity. A surfactant is mixed with an alkoxide sol and allowed to gel. As will be discussed in Chapter 14, the surfactant molecules, because of the presence of two different types of components (a hydrophobic head and a hydrophilic tail) arrange themselves in various structures in the gel *e.g.* lamellar, simple cubic, hexagonal, *etc.* and act as a template for the gel. The surfactant molecules can be later removed by heating or by washing, leaving behind a mesoporous material with an ordered pore structure.

10.2.5 *Atomic Layer Deposition (ALD)*

The PVD and the CVD techniques of thin film deposition have relatively high rates of deposition and have been routinely used, *e.g.* in the manufacture of integrated circuits. However, reduction of the feature size of the electronic circuits to less than 65 nm requires deposition of films with thickness less than 5 nm with good thickness control. In contrast the minimum film thicknesses that can be deposited by CVD and PVD are about 10 nm and 20 nm respectively due to problems with thickness control and uniformity. Moreover, CVD can provide 100% coverage on features with aspect ratios no larger than 10:1 while with PVD only 50% coverage is obtained at these aspect ratios. The atomic layer deposition

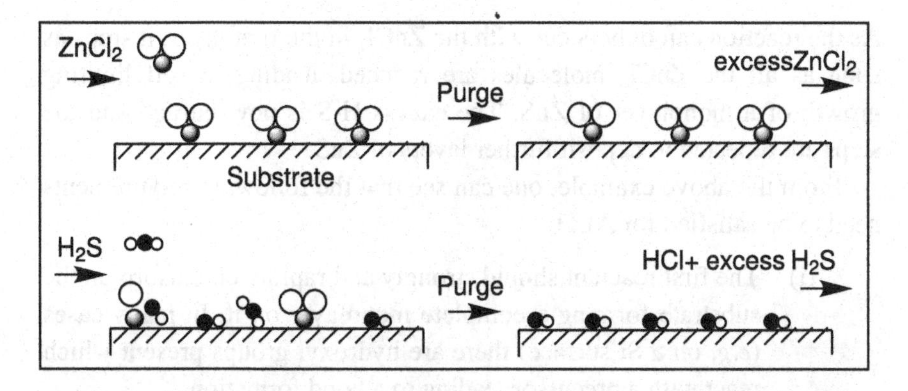

Fig. 10.21. Schematic showing the deposition of ZnS film using atomic layer deposition.

(ALD) technique provides a capability to cover features with large aspect ratios (~ 60:1). Such a capability makes it possible to fabricate a host of nanostructures in which the deposition of thin films on high aspect ratio features is required, *e.g.* formation of a nanotube by using a nanowire as a template for deposition followed by removal of the template by dissolution.

In the ALD technique, gaseous reactants are used to deposit one atomic layer at a time. An atomic layer is deposited by bond formation between the atoms of the previous layer and a reactant. An essential condition is that all the available surface sites must be completely used up in each step. After the deposition of a layer, the excess reactant is purged out from the deposition chamber and the process steps are repeated [191, 201, 202].

The ALD process can be illustrated using the example of the deposition of a thin film of ZnS using $ZnCl_2$ and H_2S precursors (Fig. 10.21). The process starts by the introduction of $ZnCl_2$ into the reaction chamber. The substrate should be such that $ZnCl_2$ readily chemisorbs on it. A monolayer of $ZnCl_2$ adsorbs on the substrate. The excess $ZnCl_2$ is purged and H_2S gas is introduced. A monolayer of ZnS forms due to the following reaction

$$ZnCl_2 + H_2S \rightarrow ZnS + 2HCl$$

As the reaction can only occur with the $ZnCl_2$ in the first layer, it stops as soon as all the $ZnCl_2$ molecules are reduced, leading to self limiting growth of a monolayer of ZnS. The excess H_2S is now purged and the steps are repeated to deposit further layers of ZnS.

From the above example, one can see that the following requirements need to be satisfied for ALD

(i) The first reactant should strongly and rapidly chemisorb on the substrate forming a complete monolayer on it. In many cases (*e.g.* on a Si surface) there are hydroxyl groups present which react with a precursor leading to a bond formation.
(ii) In order to limit the growth to only a monolayer, there should be no bonding between the reactant molecules themselves.
(iii) The second reactant should react aggressively with the exposed reactive groups of the first reactant and, in turn, expose its reactive groups for reaction with the first reactant.
(iv) None of the reactants should etch or dissolve the substrate or the film.

The rate of the ALD reactions using the gaseous precursors is maximum in a certain temperature range, usually from 200 °C to 400 °C. At lower temperatures, the thermally activated chemisorption step is slow while at too high a temperature the rate of reverse reaction (*i.e.* bond breakage) becomes significant and the bonding can not be sustained.

Some examples of the films that can be deposited and the precursors used are given in Table 10.2. It can be seen that both elemental and compound films can be deposited.

Two remarkable features of the ALD process are: (i) the film thickness can be precisely controlled by controlling the number of deposition cycles; thus films of thickness less than 5 nm can be deposited with good thickness control. (ii) Uniform coverage can be carried out on features having large aspect ratios (much larger than 50:1). One drawback to the ALD process is its slow rate of deposition, typically about 1 nm min^{-1}. These aspects are summarized in Fig. 10.22. Because of the capabilities of precise thickness control down to a few nm and step coverage at high aspect ratios, ALD is likely to replace CVD and PVD in the IC technology in applications such as deposition of the high dielectric

Table 10.2. Some precursors for the deposition of different films by ALD.

Film	Precursors	Temperature (°C)	Application	Ref
ZnO	Diethylzinc, H_2O Zin acetate	130–180	Optical devices	[203]
Ru	$RuCp_2$, H_2O	300	Interconnect	[204]
Al_2O_3	$Al(CH)_3$, H_2O or O_3		High-k dielectric	[205]
SiO_2	$SiCl_4$, H_2O		Dielectric	[206]
HfO_2	$Hf[N(CH_3)_2]_4$ (TDMAH), and H_2O; $HfCl_4$ or TEMAH, H_2O		High-k dielectric	[207, 208]
TiN	$TiCl_4$ or TiI_4, NH_3	350–400	Barrier	[209]
TiO_2	$TiCl_4$,H_2O	300	High-k dielectric	[210]
ZrO_2	$ZrCl_4$, H_2O	280–350	High-k Dielectric	[211]
W	WF_6, B_2H_6 or Si_2H_6	300–350		[212]

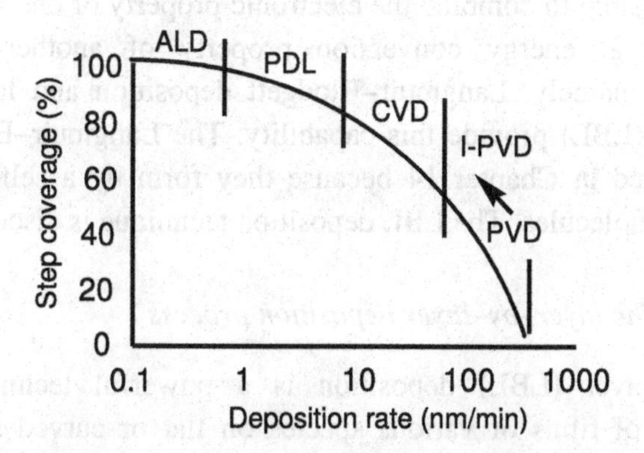

Fig. 10.22. Deposition rates and step coverage for various thin film deposition techniques (PDL = pulsed deposition layer – an ALD like process that deposits many layers at once, IPVD = ionized physical vapor deposition) [213] .

constant gate dielectric, DRAM capacitors and interconnect barrier layers. Furthermore, ALD has been demonstrated to be a viable technique to deposit on such geometries as nanowires and nanotubes. Thus, nanotubes of Al_2O_3 have been produced by atomic layer deposition of Al_2O_3 on an organic nanowire and subsequent dissolution of the nanowire [212] and for deposition of a dielectric layer on carbon nanotubes [214].

10.2.6 *Layer by layer (LBL) deposition*

10.2.6.1 *Introduction*

Multilayers of different organic molecules or multilayer composites consisting of organic and inorganic layers are of interest for several reasons. Although the organic molecules having a wide range of functionalities are available, to effectively use them in a nanostructure, some mechanical stability is necessary which can be imparted by making a composite consisting of the organic molecules and an inorganic component. Furthermore, it may also be possible to combine the functionalities of two different types of molecules if they can be arranged in layers with predetermined position and orientation. As an example, it may be possible to combine the electronic property of one layer with the photochemical energy conversion property of another layer. Two techniques, namely, Langmuir–Blodgett deposition and layer by layer deposition (LBL) provide this capability. The Langmuir–Blodgett films are discussed in Chapter 14 because they form by a self assembly of surfactant molecules. The LBL deposition technique is discussed below.

10.2.6.2 *The layer–by–layer deposition process*

Layer–by–layer (LBL) deposition is a powerful technique for the deposition of films of various species on flat or curved surfaces. The technique can be illustrated by considering the formation of a hollow nanosphere having walls with controllable and reversible permeability. Such a structure can be used for directed delivery of drugs or other species such as pesticides, herbicides, fragrances, dyes, *etc.*

A way to prepare the hollow nanospheres is as follows [215]. Polystyrene latex particles (640 nm) are first suspended in an aqueous solution of a cationic polyelectrolyte such as Poly (diallyldimethylammonium chloride) (PDAMAC). Typically the polyelectrolyte solution has a concentration of 1 mg ml^{-1} and also contains 0.5 M NaCl. The latex particles are sulfate stabilized and have, to start with, a negative charge on the surface. The positively charged PDAMAC forms a monolayer on the particle surface due to the electrostatic interaction (Fig. 10.23). Furthermore, the charge on the particles is also reversed.

The particles are now separated by centrifugation, rinsed with water and resuspended, this time in the solution of an anionic polyelectrolyte, poly (styrenesulfonate, sodium salt) (PSS). A monolayer of this electrolyte now forms on the particle and the charge on the surface again gets reversed. These steps are repeated to build several layers (~up to 60 or more) of the two polyelectrolytes. The latex particles are then

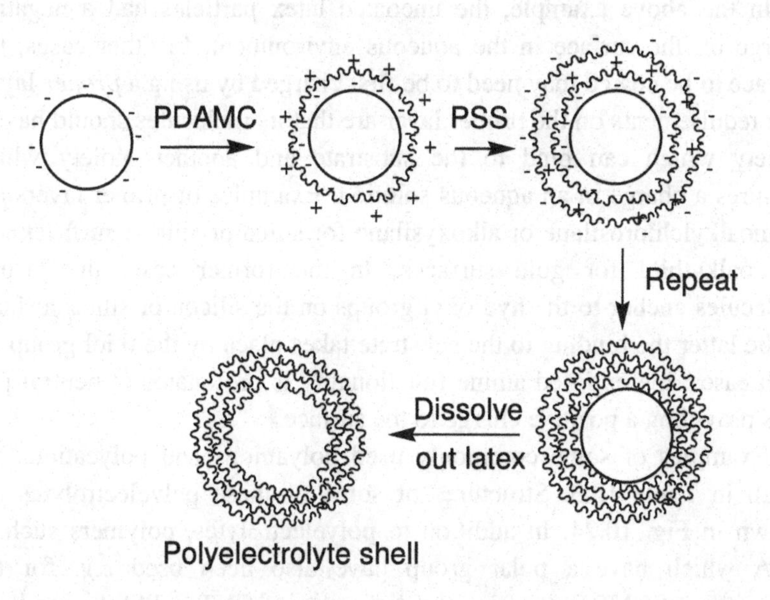

Fig. 10.23. Formation of a polyelectrolyte shell by sequential deposition of polyanions and polycations to produce multilayers by LBL technique.

dissolved out by treatment with a solvent, leaving behind a hollow shell having walls made of polyelectrolyte layers.

The conformation (and so the permeability) of the polyelectrolyte chains in the coating can be changed by changing the ionic strength (salt concentration) or the pH of the suspension as discussed later. This reversible change in the layer conformation can be used to load and subsequently release the molecules of a drug or other species. This has been demonstrated in model systems employing dye molecules [216, 217].

The polyelectrolyte chains in the different layers do not form a clean interface but are found to interpenetrate each other. This degree of interpenetration also depends on the chemical environment of the film.

One of the polyelectrolytes can be replaced by another charged species such as nanoparticles, nanosheets, proteins, polynucleic acids, viruses, polymetallic clusters, colloids, dyes, nanotubes and other nano objects making LBL a very versatile technique.

In the above example, the uncoated latex particles had a negative charge on the surface in the aqueous environment. In other cases, the surface to be coated may need to be first charged by using a *primer* layer. The requirements on the primer layer are that its molecules should have a moiety which can bind to the substrate and another moiety which acquires a charge in an aqueous solution. Examples of primer layers are aminoalkylchlorosilane or alkoxysilane for silica or silicon surfaces and aminoalkythiol for gold surfaces. In the former case, the primer molecules anchor to the hydroxyl groups on the silicon or silica surface; in the latter the binding to the substrate takes place by the thiol group. In both cases, the exposed amine functionality is protonated at neutral pH, thus providing a positive charge to the surface.

Examples of some commonly used polyanions and polycations are given in Table 10.3. Structures of some of these polyelectrolytes are shown in Fig. 10.24. In addition to polyelectrolytes, polymers such as PVA which have a polar group have also been used *e.g.* for the deposition of CNT films using the dispersion of CNT in PSS as one layer and PVA as another layer [218].

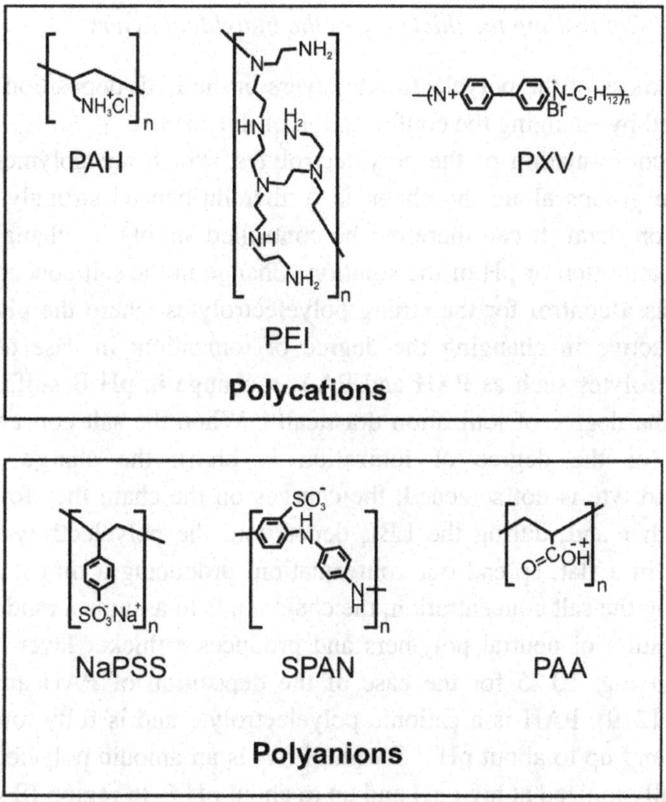

Fig. 10.24. Structures of some polyelectrolytes used in LBL deposition.

Table 10.3. Examples of polyelectrolytes used for coating in LBL deposition

Positively charge polyelectrolytes (Polycations)	Negatively charged polyelectrolytes (Polyanions)
1. Poly (diallyldimethylammonium chloride) (PDADMAC)	1. Poly (styrenesulfonate) (PSS)
2. Poly (allylaminehydrochloride) (PAH)	2. SPAN
3. Polyhexylviologen (PXV)	3. Polyacrylic acid (PAA)

10.2.6.3 *Controlling the thickness of the individual layers*

The thickness of the polyelectrolyte layers in the LBL deposition can be controlled by changing the conformation of the chains.

The conformation of the polyelectrolytes, which are polymers with ionizable groups along the chain, is further influenced strongly by the charges on them. It can therefore be controlled simply by changing the salt concentration or pH of the solution. Change in the salt concentration is used as a control for the strong polyelectrolytes where the pH is not very effective in changing the degree of ionization; in case of weak polyelectrolytes such as PAH and PAA, a change in pH is sufficient to change the degree of ionization drastically. When the salt concentration is low (or the degree of ionization is high), the charge on the polyelectrolyte is not screened; the charges on the chain therefore repel one another and, during the LBL deposition, the polyelectrolyte chain deposits in a flat, spread out conformation, producing a thin layer. On increasing the salt concentration, the chain tends to assume a random coil configuration of neutral polymers and produces a thicker layer. This is shown in Fig. 10.25 for the case of the deposition of PAH and PAA bilayers [219]. PAH is a cationic polyelectrolyte and is fully ionized at low pH and up to about pH 7.5 while PAA is an anionic polyelectrolyte and is fully ionized at high pH and up to about pH 6. In region III of the

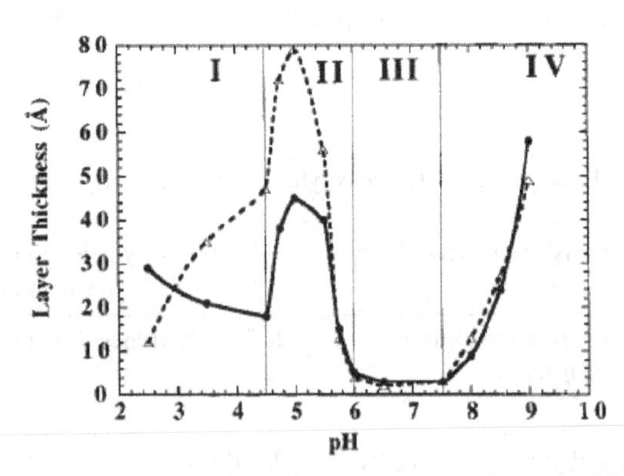

Fig. 10.25. Average incremental thickness contributed by PAA (solid line) and PAH (dashed line)adsorbed layers as a function of the solution pH (reproduced with permission from [219] copyright (2000) American Chemical Society).

figure, both the polyelectrolytes are fully ionized and each deposits in a thin layer of thickness ~ 3 A^0. In the regions II and IV where one of the layers is fully ionized while the other is nearly fully ionized, the thickness of the bilayer shoots up. This is not fully understood; it has been proposed that in this region the interpenetration of the chains is frustrated due to unequal charge density and so the molecules adopt a conformational arrangement that extends away from the surface; the chains may be still highly interpenetrated but they are organized in a different manner.

In region I, the PAH is fully ionized while the PAA is only partly ionized — its degree of ionization increases as the pH increases. The thickness of the PAA layer decreases as the charge density on it increases as explained earlier. On the other hand, more of the PAH is required to compensate the increasing charge on the PAA layer underneath so that the thickness of the depositing PAH layer increases with pH.

Thus the conformation (and so the permeability) of the polyelectrolyte chains in the coating can be changed by changing the ionic strength (salt concentration) or the pH of the suspension. As mentioned earlier, this reversible change in the layer conformation can be used to load and subsequently release the molecules of a drug or other species. This has been demonstrated in model systems employing dye molecules [216] [220].

The above example also shows that it is possible to obtain a subnanometer control on the film thickness in the LBL deposition.

10.2.6.4 *Incorporating nanoparticles in an LBL film*

A simple strategy to accomplish this consists of coating the nanoparticles by a polyelectrolyte. These coated nanoparticles can be then used in place of one of the polyelectrolytes in LBL deposition. An example is the preparation of LBL quantum dot–polyelectrolyte multilayer film from aqueous dispersion of anionic thioglycolic acid capped CdTe quantum dots and cationic PDADMAC [221].

It is straightforward to apply the LBL deposition technique to produce films with compositional gradients. For example, in the case of the quantum dot–polyelectrolyte film mentioned above, a film in which the size of the quantum dots increased gradually from bottom to top was actually prepared. As the band gap increases with a decreasing size of the quantum dot, the wavelength of the emitted luminescence decreases. This produces a rainbow effect when such a film, with a range of luminescent colors, is excited by a suitable radiation.

10.2.6.5 *Surface functionalization of nanoparticles*

The LBL technique can be very effectively used for surface functionalization of nanoparticles because of its capability of conformal coating. The several configurations that can be achieved are illustrated below where NP stands for nanoparticle and PE stands for polyelectrolyte.

(1) NP/PE1/PE2/PE1/PE2/…
(2) NP/PE/charged species/PE/ charged species/PE/..
(3) NP/PE1/PE2/PE1/PE2/…/charged species

In scheme 1, the coating consists of alternate layers of polyelectrolytes. This type of coating can be used to reverse the charge on the particle surface or to obtain a charged particle with an effectively larger diameter. In scheme 2, one of the polyelectrolytes is replaced by another functional charged species such as a dye molecule, protein, lipid, or nanoparticles, *etc.* The dye imparts fluorescence to the particles. The protein or lipid coated particles can serve as models for basic biophysical studies.

In case of alternate polyelectrolyte–nanoparticle coating, the polyelectrolyte acts as a glue to hold the nanoparticles. Such coatings if made on a micro-particle can impart to it the properties of the nanoparticles. Alternately, if the micro-particle is a latex particle then a high temperature treatment would burn it and sinter the nanoparticles to produce a hollow inorganic capsule. A thick double layer on the nanoparticle in the coating suspension would not allow the coating to be dense. The thickness of the double layer has to be reduced by adding an electrolyte to compress the double layer and allow the coating to be dense.

The LBL technique is a slow technique due to the time taken by each of the deposition steps (few seconds to hours) and the rinsing steps (few minutes). In addition, the rinsing step results in the production of a large amount of waste liquids. Efforts are therefore being made to make this process less time consuming and less wasteful. Techniques such as spraying [222, 223], spin-assisted LBL [224], "electrically driven" LBL [225], roll-to-roll process [226], and "exponential" growth [227] have been reported. Addition of a dewetting agent such as DMF (dimethylformamide) to the depositing solutions has been proposed [218] to do away with the rinse step. However, acceleration of the buildup often results in some loss of structural control or universality.

10.3 Multilayers for mechanical engineering applications

10.3.1 *Introduction*

An important application of the nanoscale thin films is in mechanical engineering as hard coatings on machining tools or as friction resistant (tribological) coatings. Thick (few microns) coatings of hard materials like TiC, TiN *etc.* have been in use for many years for this purpose. The maximum hardness for such coatings is about 20 GPa. In contrast, hardness as high as 60 GPa and more has been achieved using nanoscale thin films. The thin structures having such high hardness can be broadly divided into two categories.

(i) Multilayer structures: In this case the thin films are deposited as multilayers consisting of alternate layers of two different materials. Two different types of structures are produced depending on the choice of the film material. In one case the two film materials are chosen to be isostructural (*i.e.* having the same structure) but with different shear modulus (*e.g.* TiN and VN). In this case the epitaxial deposition occurs and a coherency is achieved across the interface between the layers Such structures are called *superlattices*. The lattice parameter mismatch generates some residual stresses in the layers. However, as discussed later, this residual stress is not the reason for the high hardness of these multilayers.

If the multilayers consist of materials with different crystal structures, the layers are polycrystalline. The enhancement in the mechanical properties is observed in this case also.

(ii) Nanocomposite films: Another important class of nanoscale films for mechanical engineering applications are the nanocomposite films in which nanocrystallites are embedded in a continuous matrix phase. Both the nanocrystalline phase and the matrix phase are simultaneously codeposited by choosing a suitable composition of the source.
The deposition of such films is a challenging problem because two different materials have to be deposited simultaneously and the deposition must be stable over a long period of time to achieve nm thick layers [228].

The mechanisms responsible for high hardness in these two categories of films are discussed below.

10.3.2 *The hardening mechanisms for superlattices*

It was mentioned above that the superlattices consisting of alternate layers of two materials with different shear moduli result in improved mechanical properties. An example of a superlattice structure is shown in Fig. 10.26 and a plot of the hardness of the superlattices consisting of alternate layers of TiN and VN deposited with (100) plane as the plane of the films is shown in Fig. 10.27. The multilayer period λ is the distance between the centers of the interfaces between the layers. It need not be equal to the thickness of the individual layers even if the two layers are of equal thickness because the interface may have a finite width. Figure 10.27 shows that the superlattice hardness reaches a value of about 5600 kg/mm^2 (~ 54.94 GPa, taking 1 kg/mm^2 = 9.81 MPa), higher by a factor of about 3 from the hardness of either VN or TiN. It also shows that the hardness increases first with the increasing multilayer period, reaches a peak value and then decreases as the period is further increased.

The reason for such high hardness in the superlattice structures is the inability of the dislocations to move by any large distances (see Appendix II). The number of dislocations within each layer is very small

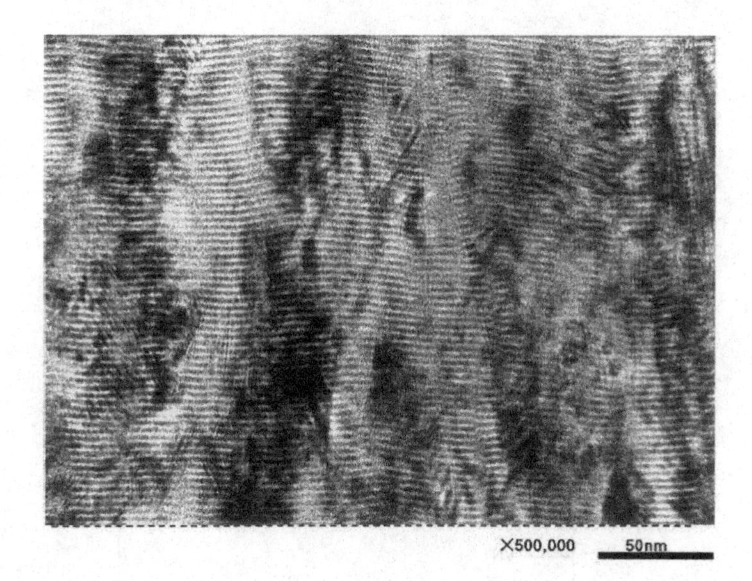

Fig. 10.26. Transmission electron microscope image from a TiN/AlN multilayer with a period Λ 3.0 nm; the dark layers are TiN and the bright layers are AlN (reproduced [229] copyright (1996) with permission of Elsevier).

and the stress required to move these dislocations or to nucleate fresh dislocations increases as the thickness of the layers decreases. Below a critical thickness, any significant dislocation motion can, therefore, take place only if the dislocations move across the interfaces. However, this also requires considerable energy if the dislocation is moving from a low modulus layer to a high modulus layer. These two situations are discussed below.

10.3.2.1 *Dislocation nucleation and multiplication within a layer*

A dislocation in a bulk crystalline solid experiences forces from several sources, such as other dislocations and point defects and a stress is required to overcome these forces and cause the dislocation to move. Even in a crystal with no force acting on the dislocation, a certain stress is required to move the dislocation because the motion of the dislocation

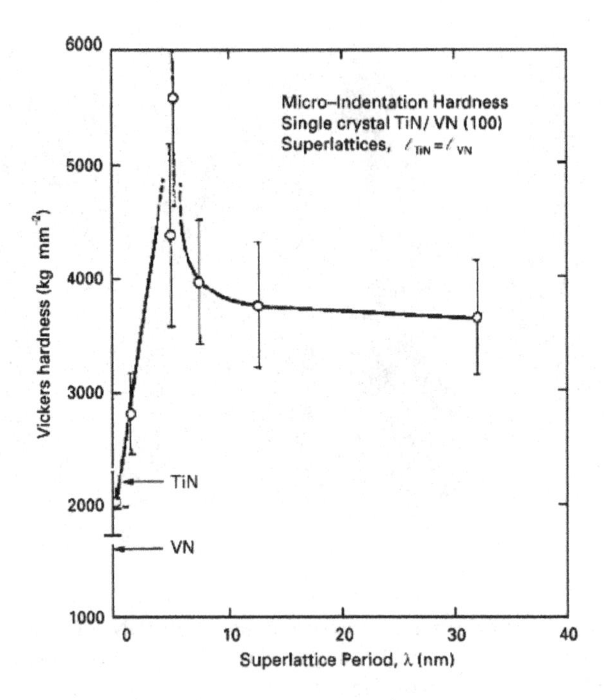

Fig. 10.27. Microindentation hardness vs. multilayer period for (100) oriented superlattices of TiN/VN; the layers of TiN and VN are of equal thickness and the total film thickness is 2.5 μm (reproduced with permission from [230], copyright (1987) American Institute of Physics).

changes the strain field and causes an increase in energy. All these effects are taken into account by calling the stress required to move the dislocation in the bulk solid as the *friction stress*, τ_0.

For a dislocation constrained to move within a thin layer, an additional stress is required to move the dislocation. This additional stress changes as $1/t$ where t is the thickness of the layer. The total stress to move the dislocation is given by

$$\tau = \tau_0 + \frac{MAG_1 b}{2\pi t}\ln\left(\frac{t}{b}\right)$$

Here, M is the orientation factor or the Taylor factor (see Appendix II), ($M = 3$ for fcc structure and $M = 2$ for bcc structure), $A = 1.21$ for mixed dislocation, G_l is the shear modulus of the layer and b is the Burgers vector of the dislocation [231].

The stress required to move a dislocation within a layer therefore increases as the layer thickness decreases (lnt varies slowly as compared to $1/t$). At sufficiently small layer thickness, no dislocations may exist within the layer. To estimate the magnitude of this critical thickness, one has to consider the equilibrium spacing between the dislocations lying on a slip plane. If τ is the resolved shear stress on the slip plane, then the average spacing between the dislocations is given by

$$\lambda = \frac{Gb}{3\tau}$$

This expression is arrived at by considering the case of a dislocation source generating dislocations which pile up against an obstacle such as a grain boundary. Each dislocation exerts a repulsive force on the other dislocations, making the dislocations spaced farther from each other as one moves from the obstacle to the source (Fig. AII.8, Appendix II). At equilibrium, the stress acting on the source is balanced by the back stress generated by the pile up of dislocations. The total stress in the surrounding region is relaxed by a total shear displacement of $\tau L/G$ where L is the length of the slip band. For this the number of dislocations needed is given by

$$n = \frac{\tau L}{Gb}$$

and so the average spacing between the dislocations is

$$\lambda = \frac{L}{n} = \frac{Gb}{\tau}$$

More refined calculations result in a factor of 3 in the denominator. The value of λ gives the critical thickness below which no dislocation may exist within a layer.

When there are no dislocations within a layer, the yielding may still occur by the generation or nucleation of fresh dislocations. Fresh dislocations may be generated by the operation of a Frank – Read source. The stress for this is given by $\tau = Gb/l$. Here, G is the shear modulus, b is the Burgers vector and l is the length of the dislocation line between the pinning points. In a multilayer structure, the interfaces on either side of a layer act as pinning locations for a dislocation line so that l can be taken equal to t the thickness of the layer. Therefore, the stress required to operate the dislocation source, and hence the yield stress, increases as the multilayer period decreases. As hardness is nothing but the resistance to plastic deformation, the hardness is expected to increase as the multilayer period decreases.

Expressions have also been derived for the nucleation of dislocations [232]. The stress required to nucleate dislocations is also found to vary as $1/t$.

In summary, the stress required to move a dislocation within a layer as well as the stress needed to nucleate new dislocations increases as $1/t$ according to the simple considerations used here.

10.3.2.2 *Dislocation movement across an interface*

In the previous section we have considered the yielding by motion of dislocations within a layer *i.e.* nearly in a plane parallel to the plane of the multilayers. However, there is also the possibility of the dislocations on an inclined plane crossing the interface into the adjacent layer. The movement of a dislocation from a low modulus layer to a higher modulus layer requires an expenditure of energy. This can be seen by considering the energy of a dislocation. A dislocation can be imagined to be constructed by shearing a solid across a partial cut followed by joining together the displaced parts as shown in Fig. 10.28. The action of joining together after the cut introduces stresses around the dislocation. The strain energy due to this can be calculated and is the energy of the dislocation. The strain energy per unit length of a screw dislocation is given by

$$E = \frac{Gb^2}{4\pi}\log(\frac{r_1}{r_0})$$

For an edge dislocation, the expression is

$$E = \frac{Gb^2}{4\pi(1-\nu)}\log(\frac{r_1}{r_0})$$

Here G is the shear modulus, b is the Burger's vector, ν is the Poisson's ratio, r_1 is the radius of the crystal and r_0 is the radius of the dislocation "core". The dislocation core is a region surrounding the dislocation line. r_0 is usually taken to be equal to b.

The energy of a dislocation line is, therefore, proportional to G, the shear modulus of the material in which the dislocation exists. Thus to

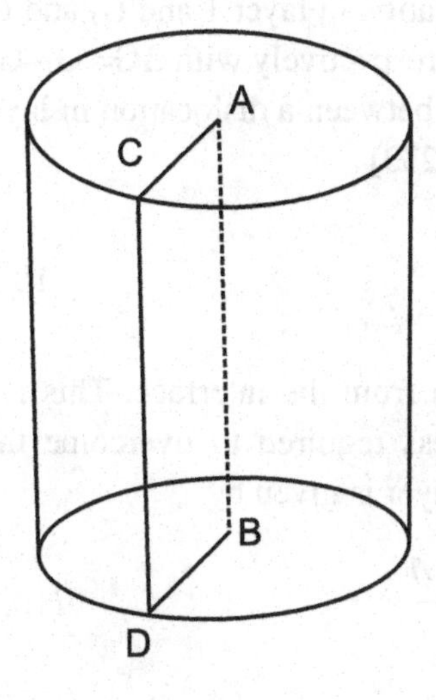

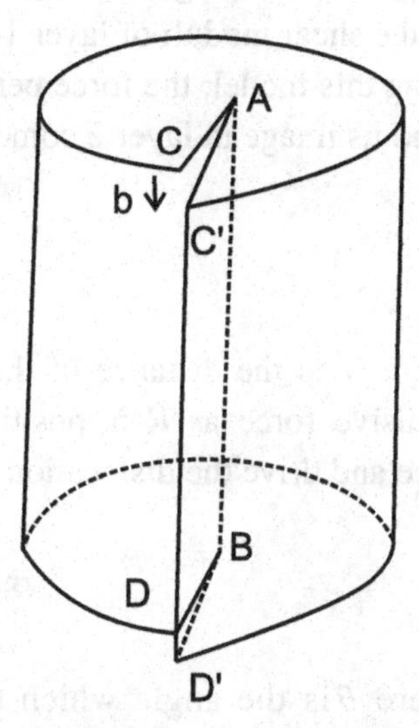

Fig. 10.28. Introducing a dislocation in a solid: (a) a cut is made in the solid along the rectangle (b) the two halves are sheared with respect to each other by an amount equal to b, such that the point C moves to C′ and D moves to D′ , and "welded" together. This action effectively creates a screw dislocation line AB with a Burger's vector equal to **b**.

make a dislocation in a low shear modulus layer cross the interface into a higher modulus layer would require an expenditure of energy. This energy has to be supplied by the external stress in order to cause the yielding of the bilayer structure by this mechanism. The stress required to move the dislocation across the interface is calculated using the method of *image forces*. As a dislocation in the lower modulus layer (layer 1) moves towards the interface, its stress field deforms the material of the second, higher modulus layer (layer 2). This exerts a repulsive force on the dislocation. The elasticity theory shows that this force is the same as would be exerted by a dislocation (called the image dislocation) located in the high modulus layer at an equal distance from the interface and having a Burger's vector given by

$$b' = \frac{G_2 - G_1}{G_2 + G_1} b = Rb \qquad (\text{with } R = \frac{G_2 - G_1}{G_2 + G_1})$$

where b is the Burger's vector of the dislocation in layer 1 and G_1 and G_2 are the shear moduli of layer 1 and layer 2 respectively with $\Delta G = G_1 - G_2$. Using this model, the force per unit length between a dislocation in layer 1 and its image in layer 2 comes out to be [233]

$$F = \frac{b^2 R G_1}{4\pi r} \tag{10.3}$$

where r is the distance of the dislocation from the interface. This is a repulsive force as R is positive. The stress required to overcome this force and drive the dislocation across the layer is given by

$$\sigma = \frac{bRG_1 \sin \vartheta}{4\pi r} \tag{10.4}$$

where θ is the angle which the slip plane makes with the interface. Considering the example of NbN (shear modulus G_1 = 142 GPa) and TiN (shear modulus G_2 = 192 GPa) and taking the smallest distance to which the dislocation can approach the interface as $2b$ and $sin\theta = 0.8$ we get $\sigma \approx G_1/200$ which is a very high stress, comparable to the theoretical strength of the low modulus layer. In his seminal paper, Koehler

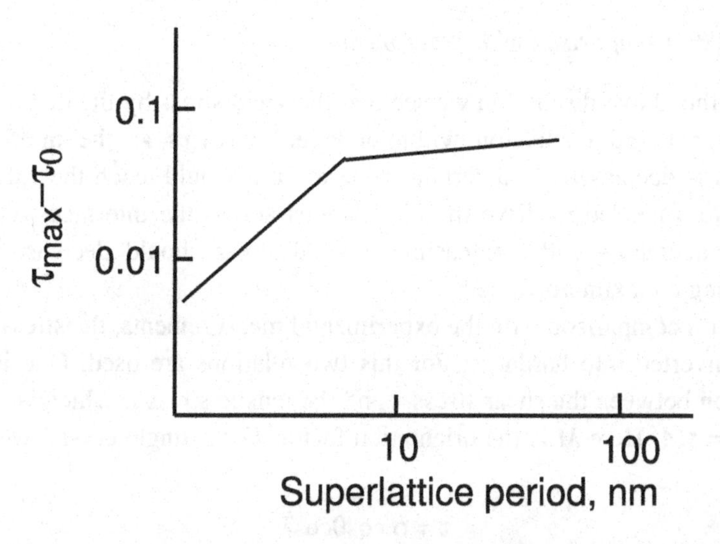

Fig 10.29. Schematic of the variation of the shear stress needed to move the dislocation across the interface from the low modulus layer into the high modulus layer; the plot is between $(\tau_{max} - \tau_0)$ normalized by ΔG and the superlattice period where τ_{max} is the stress required to drive the dislocation across the interface (i.e. the yield stress according to this mechanism) and τ_0 is the average stress required for slip in homogeneous material as defined in the text [232].

considered the example of Cu/Ni multilayers for which the stress comes out to be 1/100th of the shear modulus of Cu [233].

The expression in Eq. (10.4) provides only an upper limit value of the strength enhancement as it is based on the calculation of the stress on a dislocation near a single abrupt interface. In the multilayer structure, the interfaces may not be abrupt and may have a finite width. Also, the effect of the other interfaces in a multilayer structure has to be considered. The modulus in the interface region changes from G_1 to G_2 and so the average modulus is lower than G_2. This has the effect of reducing the stress needed to drive the dislocation across the interface. The net effect of the other interfaces is also to reduce this stress. The result is that the yield stress decreases from a high value at large multilayer spacing as this spacing is decreased (Fig. 10.29).

10.3.2.3 *Comparison with experiments*

From the above discussion we see that the yield stress by the dislocation movement and nucleation within a layer increases as the multilayer period is decreased. At a certain value of Λ, it would reach the value of the stress needed to drive the dislocations across the interface. As this stress decreases with decreasing Λ, yield stress should decrease after reaching a maximum value.

For a comparison with the experimental measurements, the stress is to be converted into hardness. For this two relations are used. One is the relation between the shear stress τ and the tensile stress σ which is given by $\sigma = \tau M$. Here M is the orientation factor. For a single crystal we saw that

$$\tau = \sigma \cos\theta \cos\lambda$$

or

$$M = \frac{1}{\cos\theta \cos\lambda}$$

Here θ and λ are the angles made by the slip plane normal and the slip direction with the loading direction. For a polycrystal, the orientation factor varies from grain to grain. Its value has been calculated to be 3.1 for the polycrystalline FCC materials [234] and 3.3 for the transition metal nitrides with NaCl structure [228].

The relation between the hardness and the yield strength for most metals is

$$\sigma_y = \frac{1}{3} H$$

Chu and Barnett [232] have compared the calculations with the experimental results. Their results are shown in Fig. 10.30. It can be seen that, in general, the trend expected from the theory is observed experimentally. It should be noted that several effects such as the thickness of the interface, effect of the coherency stresses, etc. are difficult to model quantitatively.

Note that in the multilayer structure made of isostructural (having same crystal structure such as Cu/Ni or TiN/VN) materials, the layers are coherent at the interface and so there is a strain in the region near the interface due to the lattice mismatch between the two layers. This introduces stresses in the structure when the layers are thin (see Chapter 4). The magnitude and the distribution of these stresses depend on the extent of the lattice mismatch. However, experimentally it is found that there is hardly any increase in the hardness if both the layers have nearly the same modulus but different lattice parameters, while the layers with nearly the same lattice parameter but with different moduli show a large increase in the hardness [235]. Hence the primary cause of increased hardness in the thin multilayer structures is the difference in the moduli of the two layers which restricts the motion of the dislocations across the interface.

In the system $V_{0.6}Nb_{0.4}N/NbN$, the constituent layers have nearly equal elastic moduli and a lattice mismatch of 3.5 % while the TiN/NbN have a large ΔG. The maximum hardness achieved in the case of $V_{0.6}Nb_{0.4}N/NbN$ multilayers was only higher by 4 GPa from the hardness expected from the rule of mixtures while in the case of the TiN/NbN multilayers an increase by about 30 GPa is observed.

10.3.3 *Mechanisms for nanocomposite coatings*

In a nanocomposite coating, nanocrystals about 3 to 10 nm in size are embedded in a matrix such that the matrix phase forms a continuous film around the nanocrystals with a thickness of about 1 to 3 nm *i.e.* the thickness of the interface between two nanocrystals is about 1 to 3 nm. The matrix phase is amorphous (*a*) in most cases, but can also be crystalline (*c*). Some of the various possible nanocrystalline/matrix phase combinations reported are TiN/*a*-Si_3N_4, TiN/*c*-BN, TiN/*a*-(TiB_2+TiB +B_2O_3), TiN/TiB_2, TiC/TiB_2, W_2N/a-Si_3N_4, VN/a-Si_3N_4, TiN/Ni, ZrN/Cu, ZrN/Y, TiAlN/AlN, CrN/Ni, Mo_2C/*a*-(carbon+Mo_2N), TiC/DLC (diamond like carbon), and WC/DLC [236]. Vickers hardness as high as 100 GPa has been achieved in these "superhard" coatings as compared to a hardness of 70 to 90 GPa for diamond and 48 GPa for BN, the only two intrinsically superhard materials known.

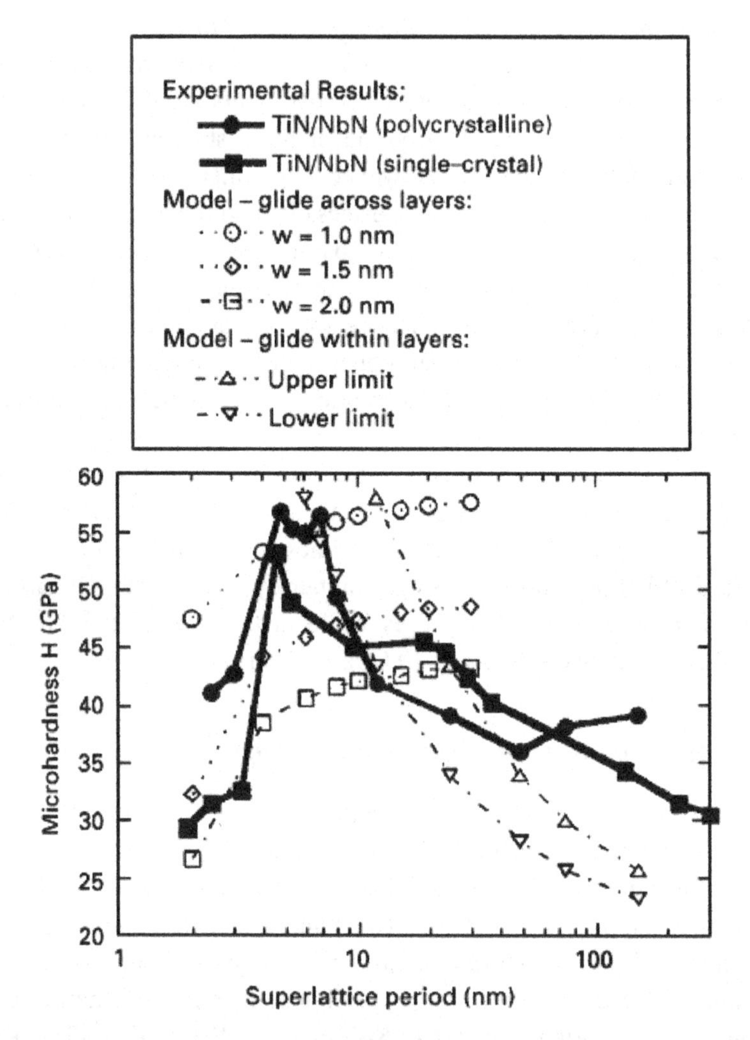

Fig. 10.30. The solid curves give the experimental variation of hardness with superlattice period for the TiN/NbN multilayers in the single crystalline and polycrystalline forms. The latter are formed when the multilayer period is increased; the dotted curves on the left (for low superlattice period) are predictions for the case when the yielding takes place by the movement of the dislocation across the interfaces as modified for a finite interface width; the dotted curves on the right (for higher superlattice period) are the predictions for yielding by motion of dislocation loops within a layer (marked lower limit) and for the nucleation of dislocations within a layer (marked upper limit) respectively (reproduced with permission from [232] copyright (1995) American Institute of Physics).

Such nanocomposite films are prepared by codeposition from a source which contains the ingredients for both the nanocrystals and the matrix phase by a process such as plasma enhanced CVD (P-CVD) [236].

The mechanisms leading to high hardness in such coatings are essentially the same as discussed for the multilayer coatings above. These are:

- Because of the small size of the nanocrystals, high stresses are required to move or nucleate dislocations within them as discussed above. The size of the nanocrystals should be in the lower nanometer range, close to the stability limit of the crystalline phase, so that they have no dislocations. A 3 nm grain size is found to be close to this minimum limit.
- Dislocations are not able to cross from the nanocrystal to the matrix phase due to a higher modulus of the matrix phase or due to the matrix phase being amorphous.
- In case of the polycrystalline solids, the inverse Hall Petch behavior is observed and the hardness decreases as the grain size is decreased below a critical size as the process of grain boundary sliding sets in,

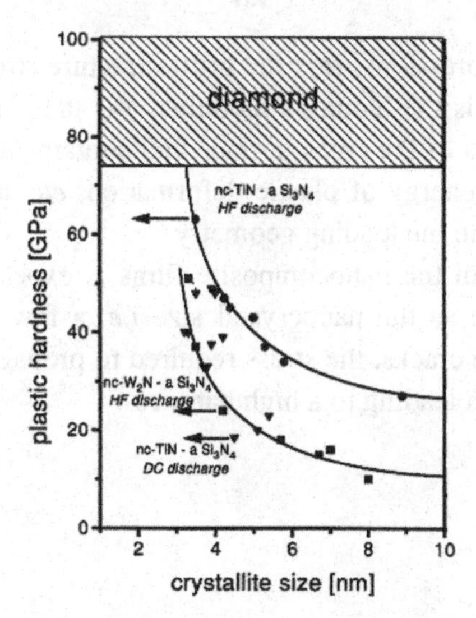

Fig.10.31. Variation of the hardness with the crystallite size in some nanocomposite coatings (reproduced from [237] copyright (1998) with permission of Elsevier).

as discussed in Chapter 11. In case of the nanocomposite coatings, the nanocrystal size is in the range of the inverse Hall Petch effect. However, due to the high resistance of the interfacial matrix region to grain boundary sliding, there is no inverse Hall Petch effect and the hardness of the composites is in fact found to be strongly increasing in this crystallite size range as shown in Fig. 10.31.

- If the yielding due to dislocation motion or grain boundary sliding is suppressed due to the mechanisms described above, failure can still occur by brittle fracture. However, the brittle fracture of such coatings also requires very high stresses. This is because the stress for brittle fracture scales as the square root of the crack length. This can be seen by considering the expressions for the fracture toughness of the brittle solids (see Appendix II):

$$K_c = Y\sigma(\pi a)^{1/2}$$

or alternately from the Griffith criterion which gives the brittle fracture stress in terms of the energy required for crack extension

$$\sigma = \left(\frac{E\gamma}{\pi a}\right)^{1/2}$$

In the above expressions, σ is the brittle fracture stress, a is the half crack length, E is the Young's modulus, γ is the energy required to create a unit area of the crack surface and contains terms such as the surface energy, energy of plastic deformation, *etc.* and Y is a factor which depends on the loading geometry.

The crack size in the nanocomposite films is expected to be of the same magnitude as the nanocrystal size *i.e.* a few nm. Due to the small size of the cracks, the stress required to propagate the cracks is exceedingly high leading to a high hardness.

Problems

(1) Micro devices usually contain a stack of layers. What considerations, other than the functional requirements come into play in selecting the materials of the layers in the stack and their relative positions?

(2) It is difficult to deposit a noble metal like gold or platinum on an oxide due to poor adhesion. How is this problem solved?

(3) How does the structure of a film deposited from the vapor phase depend on the mobility of the vapor phase atoms on the solid substrate? How can this mobility be changed?

(4) What is the special advantage of the evaporation process for thin film deposition as compared to other processes such as sputtering as far as nanotechnology is concerned.

(5) What is meant by epitaxial film growth? How is it achieved?

(6) Explain why, in the film formation from a TEOS sol, a film with fine pore size and low total porosity is formed under acidic conditions while a film with a large pore size and a large total porosity is likely to result under less acidic conditions and with larger amounts of water.

(7) Describe how you will prepare a nanosized hollow sphere with a shell of silica using LBL technique.

(8) Describe how you can use the LBL tecnique to produce an array of cantilevers consisting of polyelectrolyte-clay layers (see [238]).

(9) A poyelectrolyte with a degree of polymeriztion equal to 1000 has a chain expansion factor of 2.8 and a charactreristic ratio of 6.9 in its solution. The solution is used for LBL deposition. Assume that the polyelectrolyte essentially contains a linear C-C chain.

    (i) Assuming that the polyelectrolyte deposits in a layer of thickness equal to twice its radius of gyration calculated from its actual dimension, calculate the thickness of the layer in LBL deposition assuming that there is no interpenetration of the successive layers.

    (ii) A salt is added to solution changing the characteristic ratio of the electrolyte from 6.9 to 3.5. Find the new value of the layer thickness.

(10) Equation (10.3) gives the force experienced by a dislocation in the low modulus layer 1 and its image in layer 2. In a multilayer configuration, a dislocation would experience forces due to its images in all the other layers. Taking into account only the forces due to image dislocations in layers 2 on either side of layer 1, show that the net force is always positive and that the dislocation is repelled from the nearest interface.

(11) Calculate the critical thickness for the dislocation glide to take place under a shear stress of 1 GPA for a (i) copper layer (ii) nickel layer. The values of the various properties for the two metals are as follows

Metal	Young's modulus, E (GPa)	Poisson's ratio, ν	Burgers vector, b (nm)
Copper	110	0.34	0.181
Nickel	210	0.31	0.176

(Hint: Use $G = E/2(1+\nu)$ and the relation $\lambda = Gb/3\tau$ as given in the text).

Chapter 11

Bulk Nanostructured Materials

11.1 Introduction

Nanocrystalline metals, ceramics and nanocomposites have the potential to offer materials with extremely high strengths and other mechanical properties. The discovery of the carbon nanotubes, the material with the highest specific strength known today, has brought nearer to reality the dream of a space elevator for lifting cargo to distances of ~ 100,000 km (62,000 miles) into space, a distance of almost one fourth of the distance from earth to moon.

A polycrystalline solid consists of a number of single crystals joined together. Each crystal is called a *grain* and the boundaries between the different grains are called the *grain boundaries*. The hardness and the strength of a polycrystalline metal increase with decreasing grain size as (grain size)$^{-1/2}$ according to the well known Hall–Petch relation. However, there is a change in behavior when the grain size approaches a few nanometers. Such changes in the properties of nanocrystalline solids are discussed in this chapter. In Appendix II, an introduction to the mechanical properties of solids is given. The matter included there is a prerequisite to the material discussed in this chapter. Before describing the properties, the preparation of bulk nanocrystalline (nc) metals is briefly discussed.

11.2 Preparation of bulk nanocrystalline metals

The usual methods of refining (*i.e.* reducing) the grain size of metals such as cold working followed by recrystallization or seeding a melt of the metal with "seeds" which provide nucleation sites for solidification can at best lead to sizes close to a micron. Several alternate routes have been designed to produce nanocrystalline bulk metals. Most important of these are the following.

11.2.1 *Consolidation of fine metal powder by cold compaction*

Production of metal parts by compaction of its powder in a die and subsequent "sintering" of the compact by heating at a temperature much below the melting point is a well established process. The compaction step produces a porous compact with porosities which range between ~2 to ~10 %. Subsequent sintering removes most of the pores. However, the sintering step also leads to some grain growth.

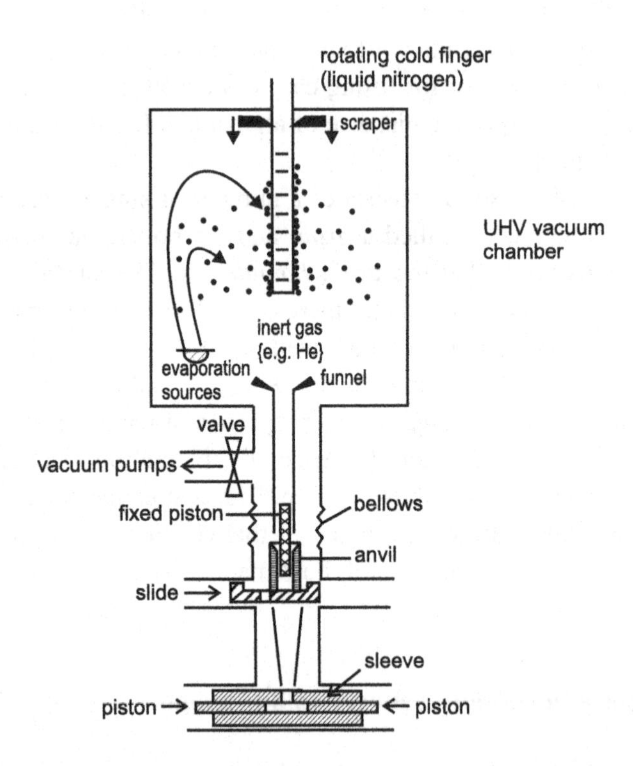

Fig. 11.1. A schematic of the apparatus used to prepare nanocrystalline metal disk by evaporation – condensation and cold compaction (reproduced with [239]copyright (1989) with permission of Elsevier).

In the production of nanocrystalline metals, the sintering step is avoided. The cold compaction itself leads to densities > 98 % because of the ability of metals to deform and fill the pores. Early studies on nanocrystalline metals were carried out on such compacts. The technique is not suitable for ceramics because the porosity in the cold pressed compacts can not be reduced to acceptable level due to the lack of the plastic deformation. Even in case of the metals, the porosity left behind after cold compaction is large enough to significantly affect the mechanical properties.

Production of nanometer sized metal powders is usually carried out by "evaporation–condensation". Metal vapors are produced by evaporation in a chamber evacuated to a vacuum of ~10^{-7} torr and back filled by an inert gas at a low pressure. The energy for evaporation can be provided by resistive heating, electron beam heating, laser, or other means. The vapors form clusters in the vapor phase and condense on a "cold finger" cooled to liquid nitrogen temperature. The powder is equiaxed with a grain size of a few nanometers and a narrow size distribution. A finer grain size is produced by using lower gas pressure, a lighter inert gas such as helium and by using low evaporation rate. Powders of oxides, nitrides and carbides can also be produced if the inert gas is replaced by oxygen, nitrogen or a carbonaceous gas respectively. The powder is scraped from the cold finger and fed into a die and compacted without exposing the powder to air in order to avoid its oxidation and contamination. A schematic of the apparatus is shown in Fig. 11.1.

11.2.2 *Electrodeposition*

Using high deposition rates and by the addition of suitable additives to form complexes in the bath it has been possible to form electrodeposited metals, alloys and composites with nanometer grain size. The high deposition rate is achieved if a current pulse is used rather than continuous current. The deposit has crystalline as well as amorphous patches in it. Plates having a thickness of a few mm formed by electrodeposition are commercially available.

11.2.3 *Crystallization of metallic glasses*

Metallic glasses are metals with an amorphous structure. They are usually produced by rapid cooling (> 10^6 degrees.sec^{-1}) of a melt of a metallic alloy with suitable composition. Such high cooling rates are achieved by "melt spinning" in which a stream of liquid metal impacts a rotating copper disc, spreads into a ribbon and is thrown off the disc as a metallic glass ribbon. Another method is to rapidly squeeze a drop of the melt between two metal blocks which yields a foil of the metallic glass. The rapid solidification does not provide enough time to the atoms to arrange themselves in the equilibrium crystalline structure. The earliest compositions to be made in metallic glass were based on Fe with small amounts of B or Si. At present this range has expanded considerably and even some pure metals can be produced in the glass form. Techniques such as electrodeposition, mechanical alloying and vapor deposition can also be used to produce metallic glasses.

A metallic glass can reduce its free energy by transforming to the stable crystalline phase. If this transformation is carried out at low temperatures, the resulting crystallite size is in the nanometer range. Thus a grain size between 5 and 8 nm is observed when a Ni–25 at % W alloy is annealed at 723 K for 24 h in vacuum [240]. The nanocrystalline materials are unstable at high temperatures because of the increased grain growth rate.

11.2.4 *Severe plastic deformation (SPD)*

Severe plastic deformation is also used to produce nanocrystalline metals. Ball milling of metal powders in a high energy ball mill is one way to produce severe plastic deformation of the particles. The metal powders weld, fracture and reweld during the ball milling. Nanometer sized grains are obtained after sufficient milling time. The final size obtained varies inversely as the melting point of the metal. If powders of two or more metals are used, the deformation causes alloy formation as well as nanocrystal formation. While this technique is useful to produce alloyed powders of otherwise difficult to produce alloys, it may not be suitable for producing bulk nanocrystalline parts, as the subsequent

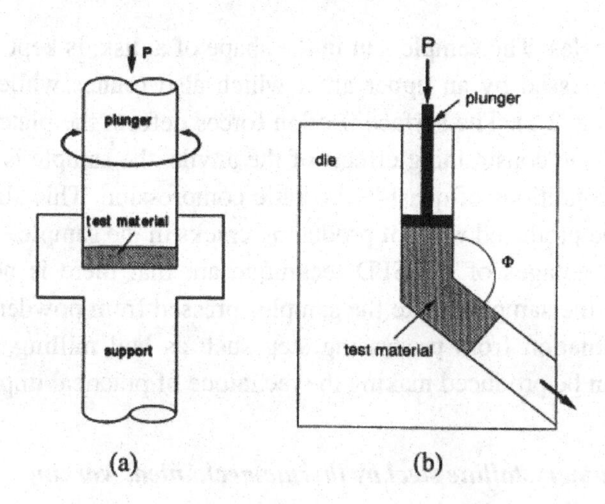

Fig. 11.2. Configuration of the dies used in (a) torsion straining under high pressure (b) Equal Channel Angular Pressing (ECAP) (reproduced from [241] copyright (2000) with permission of Elsevier).

consolidation step may not lead to high density unless a high temperature is used which can cause grain growth and destroy the nanocrystalline nature of the product.

Severe plastic deformation of bulk metals, for example by wire drawing or rolling results in the formation of a substructure of cells within grains. The cell boundaries are low angle boundaries and are not as stable as the high angle grain boundaries.

Several other techniques are also used to produce severe plastic deformation [241]. In one of these, called equal channel angular pressing (ECAP), a die containing a continuous channel which is bent nearly midway at about 90° is used (Fig. 11.2). The metal piece is first machined to fit the channel. It is inserted in the channel and pressed using a piston. The metal deforms and flows in to the other channel. As the sample dimensions remain the same, the process can be repeated many times causing a severe plastic deformation in the metal. Grain sizes produced in aluminum are about 200 nm, although sizes up to 50 nm can be produced.

Another technique used for producing high deformations is "torsion straining under high pressure". This method can be used to produce disk shaped samples. The sample, cut in the shape of a disk, is kept in a lower anvil and pressed by an upper anvil which also rotates while applying pressure (Fig. 2 a). The surface friction forces deform the plate by shear. Because of the constraining effects of the anvils, the sample is deformed under the conditions of quasi-hydrostatic compression. This allows large strains to be produced without producing cracks in the sample.

The advantages of the SPD technique are that there is no residual porosity in the sample unlike the samples pressed from powders, there is no contamination from processing step such as ball milling, and large samples can be produced making the technique of practical importance.

11.2.5 *Nanocrystalline steel by thermomechanical working*

The prospect of superior properties has motivated extensive research into the processing of nanocrystalline steel. Steel has a rich microstructure consisting of several stable and metastable phases which can be manipulated by heat treatment and mechanical working. This has led to the development of several thermomechanical processes to produce steel with nanometer grain size. While some success has been achieved in case of low carbon steels in which ferrite is the dominant phase, the material has a tendency to coarsen due to insufficient amount of carbide phase to pin the grain boundaries [242].

11.3 Mechanical properties of nanoscale metals

A very useful technique for the measurement of the mechanical properties of the nanoscale metals is nanoindentation. In the following, this technique is first described followed by a description of the mechanical properties of nanoscale metals.

11.3.1 *Hardness and nanoindentation*

Hardness can be defined as the resistance of a material to deformation. The usual method of measuring the hardness is to press an indenter into a sample. The hardness is then obtained by dividing the load by the area of the indent. Several different techniques such as Brinell, Vickers and Rockwell are used in the engineering practice. These techniques are distinguished mainly by the type of indenter used. These include a spherical indenter, a square based pyramid, a conical indenter with a rounded point, *etc.* [234].

The yield strength of a metal can be estimated from its hardness [243]. A relation which is found to be approximately true for most metals is

$$\sigma_y = \frac{1}{3} H$$

where H is the hardness in MPa and σ_y is the yield strength.

Nanoindentation is a useful technique to measure the mechanical properties of nanomaterials. In a nanoindentation experiment, a *nano*indenter is pressed into a sample and the applied force and the depth of penetration are continuously recorded. Such instruments are now commercially available. The indenter has a standard shape to facilitate the analysis. The most common indenter shapes used are pyramidal with square base (Vickers), pyramidal with triangular base (Berkovich and cube corner) and spherical. The geometries of the Vickers and the Berkovich indenters are shown in Fig. 11.3. To give an idea of the level of precision currently available, Table 11.1 gives the values of some parameters for a commercial instrument.

The force–indentation depth plot in a nanoindentation test is shown schematically in Fig. 11.4. Knowing the geometry of the indenter the contact area A at maximum load F_{max} can be calculated. The indentation hardness H_{IT} is then given by

$$H_{IT} = \frac{F_{\max}}{A}$$

Other properties which can be obtained from the indentation test include the modulus and the yield strength [244].

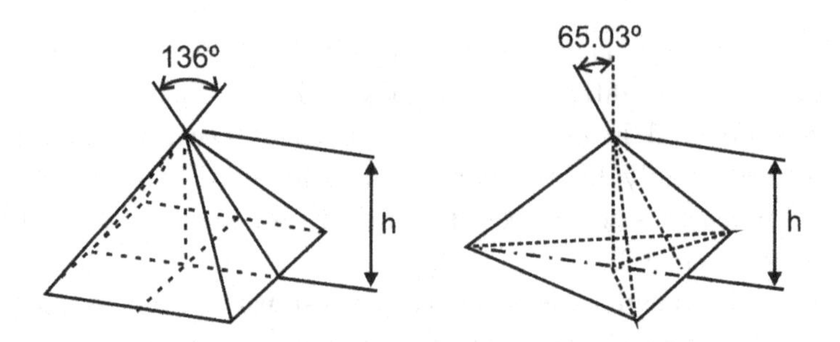

Fig. 11.3. Geometries of Vickers (left) and Berkovich (right) indenter (reproduced from [244] copyright (2000) with permission of Elsevier).

Table 11.1: Typical specifications of nanoindentation instruments

Displacement resolution	<0.01 nm
Total indenter travel	2 μm
Maximum indentation depth	> 500 μm
Maximum load	500 mN – 1 kg
Load resolution	50 nN

11.3.2 *Effect of nano-grain size on the concentration of defects*

Because of the influence of the defects on the mechanical properties, it is necessary to consider the effect on the density of defects as the grain size decreases to the nanometer size range [52]. At any temperature a crystal contains an equilibrium concentration of vacancies which is given by $f = exp(-Q/kT)$ where f is the atom fraction of vacancies and Q is the activation energy for vacancy formation. If the crystal size is reduced, a stage may be reached where the crystal is thermodynamically free of vacancies because even a single vacancy in it would cause the equilibrium concentration to exceed (see the example below).

Example 11.1. *Find the maximum size of a crystal of aluminum and of copper which can be free of vacancies at a temperature of 900K.*

Solution. If f is the equilibrium concentration of vacancies at 900 K in the crystal then a unit volume of the crystal, having n_0 atoms, will

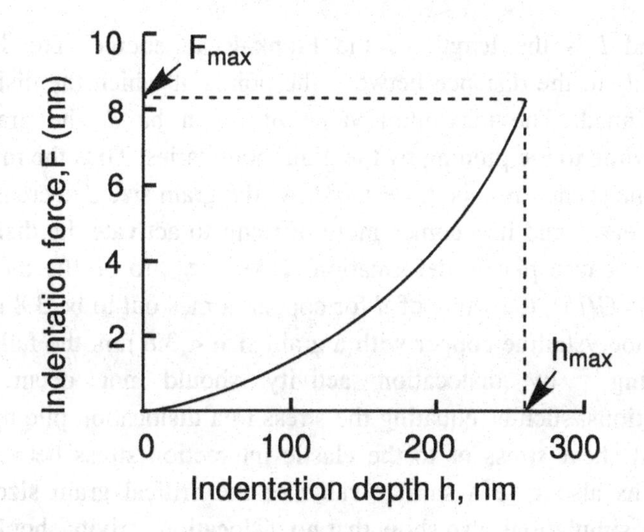

Fig. 11.4. Force–indentation depth plot in a nanoindentation test.

contain $n_0 f$ vacancies. The minimum volume for a vacancy to exist, therefore, is $(n_0 f)^{-1} = \pi d_{min}^{\ 3}/6$ where d_{min} is the size of a crystal with just one vacancy. Hence a crystal with size $< d_{min}$ will not contain any vacancy. This size is

$$d_{min} = [6/\pi n_0 f]^{1/3} = [6 \exp(Q/kT)/\pi n_0]^{1/3}$$

Substituting 0.66 eV and 1.29 eV for Q at 900 K for Al and Cu, one finds d_{min} for Al and Cu to be 6 and 86 nm respectively. □

A low or zero vacancy concentration implies that the processes such as the Herring–Nabarro creep which operate *via* lattice diffusion of vacancies will not operate in nanocrystalline solids with a grain size below the critical value. On the other hand, Coble creep, which causes deformation by diffusion along the grain boundaries, may be significant because of the high concentration of the grain boundaries and because of the ability of the grain boundaries to act both as a source and a sink for vacancies.

The shear stress required to operate a Frank–Read dislocation source is given by $\tau = Gb/l$ where G is the shear modulus, b is the Burger's vector and l is the length of the Frank–Read source. The length l corresponds to the distance between the points at which the dislocation line is pinned. The maximum value of l can be d, the grain size corresponding to the pinning by the grain boundaries. Thus the minimum value of the shear stress is $\tau_{min} = Gb/d$. As the grain size d decreases, this stress increases and it becomes more difficult to activate the dislocation source and cause plastic deformation. Taking τ_{min} to be the theoretical strength ($\approx G/15$), the value of d for copper comes out to be 3.8 nm [52] *i.e.* in nanocrystalline copper with a grain size < 3.8 nm, the failure due to yielding by dislocation activity should not occur. Other considerations, such as equating the stress in a dislocation pile up to the theoretical shear stress or to the elastic interaction stress between two dislocations also give a similar value of the critical grain size [245]. Computer simulations also show that no dislocation activity should occur in metals at a grain size below ~ 10 nm [246, 247]. Such a sample should therefore show yield strength close to the theoretical strength. However, it is found experimentally that the yield strength of a metal does increase as the grain size is reduced but then decreases as the grain size is further reduced below a critical grain size due to other mechanisms of plastic deformation coming into play.

11.3.3 *Effect of grain size on the yield strength: the inverse Hall–Petch effect*

In the early 1950's Hall [248] and Petch [249] empirically determined that the yield strength of metals increases as their grain size decreases according to the relation

$$\sigma = \sigma_o + kd^{-1/2} \tag{11.1}$$

Here σ_o and k are material constants and d is the grain size. Several explanations for the Hall–Petch effect have subsequently been proposed [245, 250-253]. According to these models the Hall–Petch behavior is associated with the resistance to the motion of dislocations provided by the grain boundaries and the stress concentration due to the resulting pile up of dislocations.

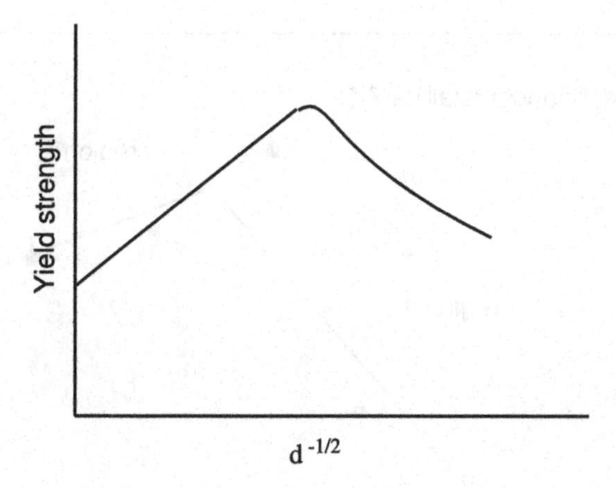

Fig. 11.5. Schematic showing the change in the yield strength with grain size in the nano grain size regime; note that in the inverse Hall – Petch region, the data does not usually follow a straight line behavior.

The Hall–Petch relation is found to be well obeyed for coarse grained (d > 1 μm) metals. Thus a plot of strength *vs.* $d^{-1/2}$ is found to be a straight line with a slope close to 1. For grain sizes below about 50 nm, the experiments show that the strength reaches a maximum and then decreases as the grain size is further reduced (Fig. 11.5). The reversal in the strength with decreasing grain size below a critical grain size is termed as *"inverse Hall– Petch effect"*. As the sample preparation techniques improved, the grain size d_c, at which the inverse Hall–Petch effect commenced was observed to move downward.

The reason for the larger critical grain size d_c in the earlier work was attributed to the poor quality of the samples due to voids, incomplete bonding between the grains, contamination *etc.* as most of these samples were prepared by consolidating the nanopowders. Nevertheless, the inverse Hall–Petch effect has been shown to be a real effect by measurements on carefully prepared samples. An example is shown in Fig. 11.6 which shows the plot of hardness against inverse square root grain size for carefully prepared ZrN coatings [254]. In this case the inverse Hall–Petch relation starts at about 16 nm. Many theoretical models also support the inverse Hall–Petch effect.

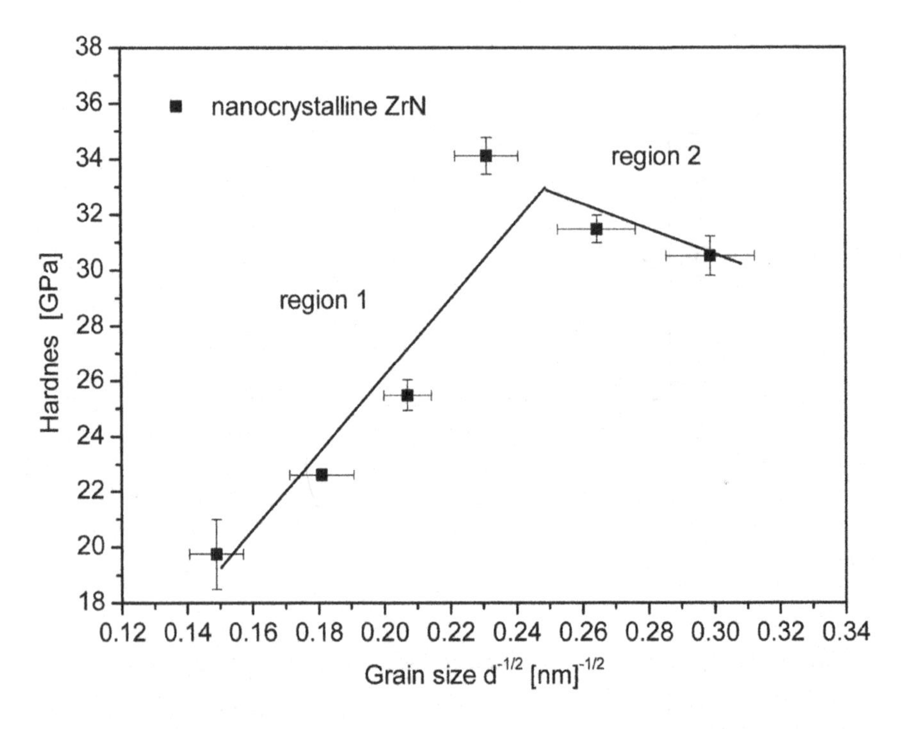

Fig. 11.6. Plot of the hardness of ZrN coatings vs. inverse square root grain size – inverse Hall – Petch relation is seen to start at a grain size ≈16 nm. (Reproduced from [254] copyright (2011) with permission of Elsevier).

Latest results for the FCC metals show that the Hall–Petch behavior persists at least down to 10 nm, the smallest grain size for which the data on well prepared samples are available [255, 256]. It should be noted that the hardness of the copper samples with average grain size of 10 nm is found to be 3 GPa which translates to a yield strength of 1 GPa (using the conversion: hardness ≈ 3 yield strength). This is more than an order of magnitude larger than the yield strength (~ 50 MPa) for coarse grained copper and illustrates the promise of nanocrystalline materials for achieving extraordinarily high strengths.

The yielding in the large (> 1 μm) grained metals occurs due to the onset of dislocation activity within the grains. Most common yielding mechanism is the activation of a Frank–Read source. When the grain size reaches the nm range, the stresses required to activate a dislocation source become very high. Computer simulations have also shown that no dislocation activity occurs within the grains below a critical size of ~ 10 nm in metals [246, 247]. Hence the dislocation mechanisms are no longer available for yielding and strain hardening below a grain size of ~ 10 nm. The reason for the break-down of the Hall–Petch relation below a critical grain size d_c is believed to be a transition of the deformation mechanism from the intra-grain dislocation glide to some grain boundary deformation mechanism.

One possible grain boundary deformation mechanism could be the Coble creep *i.e.* deformation produced due to atomic diffusion through the grain boundaries. In the Coble creep, the strain rate varies as d^{-3}, where d is the grain size. However, even in the nanocrystalline range, the d^{-3} dependence is not sufficient to produce the measured strain rates in many metals at room temperature. This mechanism is essentially important only at a high enough value of T/T_m (T_m = melting point) where the diffusion rate becomes significant or at very low strain rates where sufficient time is available for diffusion.

The grain boundary deformation mechanism which is believed to be dominant at low grain sizes in the nanocrystalline metals is the *grain boundary sliding*. It is not a pure grain boundary shear but a grain boundary shear accompanied by a mechanism such as dislocation glide or atom diffusion to accommodate the strain and to avoid the formation of cavities at the triple grain junctions. The mechanism can lead to large strains if the strain rate and the temperature are not too low.

In case the strain rate is high and the temperature is low, then the diffusion would not be sufficient to accommodate the mismatch strain accompanying the grain boundary sliding. The stresses would build up locally at the triple grain junctions, leading eventually to the formation of cracks, causing failure of the sample. The strength is now limited by this mechanism rather than by the Hall–Petch mechanism. This also appears to be the essential reason behind the low ductility of nc metals as discussed next.

Conrad and Narayan [245] have further developed this model in terms of thermally activated shear at the grain boundaries and have found that the experimental results from several metals agree with this model as shown in Fig. 11.7.

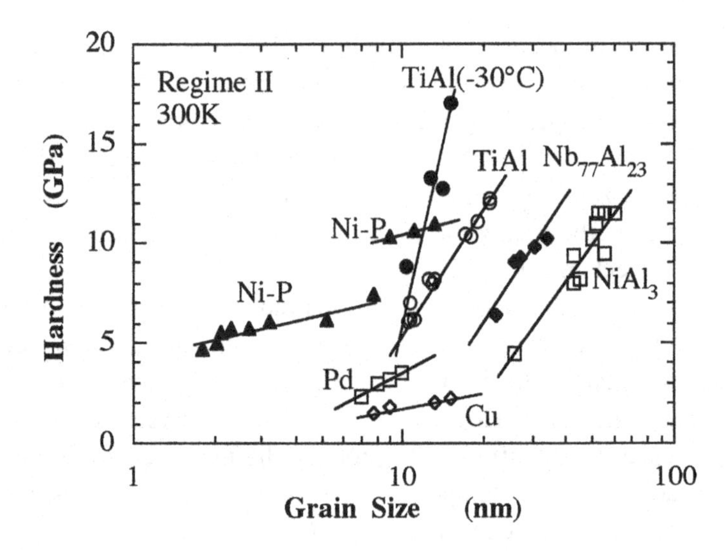

Fig. 11.7. Hardness *vs.* log d in the grain size softening region for several metals (reproduced from [245]copyright 2000 with permission of Elsevier).

11.3.4 *Ductility of the nanocrystalline metals*

Ductility is a key property of metals. It enables them to be shaped by deformation (*e.g.* by sheet metal deformation, rolling, forging, wire drawing) and prevents catastrophic failure in metal structures. While the ductility of metals with the usual grain sizes (few microns) is high, the ductility is found to be very limited when the grain size is in the nanocrystalline range. A measure of the ductility is the % elongation measured in a tensile test. For the coarse grained metals the % elongation ranges from 40 % to 70 %. In contrast, for the nc metals it is generally < 2 %.

In the nc metals, prepared by consolidation of powders or by similar techniques, some porosity is inevitably present. The stress concentration at these pores causes early failure before any significant elongation can occur. Such samples, therefore, have low ductility.

Even in the pore free nc metals the ductility is found to be low. This is due to their low values of the strain hardening exponent (n) and the strain rate sensitivity (m). As discussed in Appendix II, a sufficiently large strain hardening and/or strain rate sensitivity is needed to achieve large uniform strains and delay the onset of plastic instability.

The reasons for the poor ductility of the nc metals are not well understood at present. A key requirement for good ductility is a high strain hardening ability of the metal. In the coarse grained metals the ductility is limited by the onset of plastic instability or "necking" in the tensile test. In the region of the neck formation, the rate of strain hardening is not sufficient so that the metal keeps on deforming locally. This leads to the formation of voids which grow and join together, causing the metal to fail. For metals with a high rate of strain hardening, this phenomenon is delayed and a high ductility is obtained.

A measure of the strain hardening rate ability is the strain rate sensitivity '*m*' in the expression

$$\sigma = k\dot{\varepsilon}^{m}$$

where σ is the flow stress, $\dot{\varepsilon}$ is the strain rate and k is a constant. The strain rate sensitivity m is therefore

$$m = \frac{\delta \ln \sigma}{\delta \dot{\varepsilon}} \Big|_{\varepsilon,T}$$

If the value of m is positive and sufficiently large, the strain hardening rate is high, the neck formation is suppressed and high ductility results. Thus m=1 for superplastic forming (see below) and 0.5 for grain boundary sliding. For plastic deformation by dislocation glide, m is < 0.2.

The strain hardening occurs due to a build up of a high dislocation density within the grains. Such build up of dislocation density is thwarted in the nc metals because of the annihilation of dislocations and their absorption in the grain boundaries due to the proximity of the dislocations to one another and to the grain boundaries. This process is called *dynamic recovery.* Thus it is expected that the nc metals will have a low value of m and consequently a low ductility [257]. Table 11.2 below lists the values of m for some nanocrystalline metals.

One way to obtain high strength together with good ductility is to design a structure with a distribution of grain sizes extending from a few nm to 100 or more nm. The smaller grains provide the strength while the larger grains permit dislocation activity and strain hardening, thus contributing to the increased ductility. A few examples of this approach exist such as the work by Koch *et al.* in which the Zn samples were prepared by cryomilling for a controlled duration [258].

In addition to the plastic instability, another mechanism which limits the ductility of the nc metals is the growth of cracks initiating at preexisting voids. In the nc metals made by consolidation of powders, pores are present which act as the points from which the failure initiates. Such nc metals invariably have a low ductility. Figure 11.8 compares the tensile stress– strain plots for three cases. The conventional coarse grained sample (grain size 80 μm) has low strength but high ductility while the nanocrystalline sample prepared by conventional consolidation has a high strength but very low ductility. The ductility for the nanocrystalline copper increases dramatically if the sample is prepared so as to avoid the formation of pores — in pore free samples of Cu with a mean grain size of 23 nm, an elongation of 15.5 % was achieved with a yield strength of 800 MPa which is about 11 times the yield strength of coarse grained copper. However, it should be noted that the ductility obtained is still considerably lower than 70 % for the coarse grained copper.

In the above discussion we noted that the ductility of the nc metals is low under the usual testing conditions. However, at high temperatures or

Table 11.2 : Values of the strain rate sensitivity m for some metals [259]

Metal	Grain size (nm)	Condition	m
Fe	80		0.004
Fe	300		0.04
Fe	20	Ball milled	0.006
Ti	260		0.01

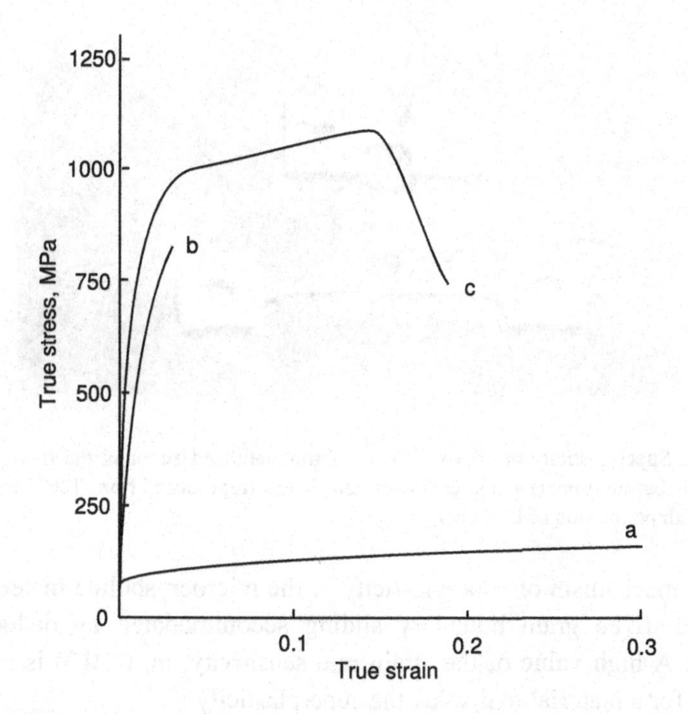

Fig. 11.8. Stress–strain plots of three copper samples; (a) coarse grained sample with a grain size of 80 μm (b) nc sample (grain size 23 nm) was prepared by consolidation of powder (c) nc sample (grain size 26 nm) prepared by in-situ consolidation using mechanical milling [260].

at low enough strain rates, the nc metals can display very high ductility and even show the phenomenon of *superplasticity* where they deform, without necking, to elongations of a few hundred percent or more. The phenomenon of superplasticity was in fact noted first in the microcrystalline materials [261]. Several *micro*crystalline materials in which the grain size is equiaxed and less than about 10 μm and which do not undergo significant grain growth during high temperature deformation show superplasticity at temperatures greater than about 0.4 of the melting point, T_m. Superplasticity has been observed in metals, alloys as well as in ceramics and composites. Figure 11.9 shows an example of superplasticity in a ceramic composite.

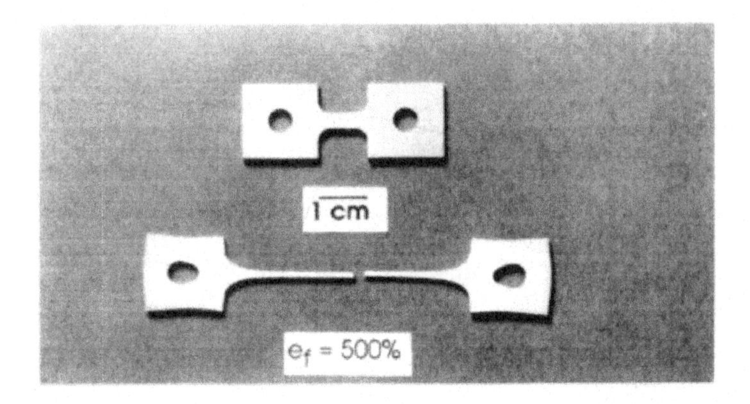

Fig. 11.9. Superplasticity in 20 vol% Al_2O_3/Yttria stabilized zirconia; the figure shows the sample before (upper) and after (lower) tensile test (reproduced from [262] copyright (1989) with permission of Elsevier).

The mechanism of superplasticity in the microcrystalline materials is believed to be grain boundary sliding accommodated by dislocation motion. A high value of the strain rate sensitivity, m, (≥ 0.3) is usually needed for a material to display the superplasticity.

As discussed earlier, in the nc metals, the deformation mechanism is believed to change to grain boundary sliding below a critical grain size, d_c.

The strains produced by the grain boundary sliding, if not accommodated, lead to the formation of cracks at the triple grain junctions which can grow and cause the failure of the metal. The accommodation of the strains in the *micro*crystalline materials can occur by dislocation mechanisms. However, in the nc metals such mechanisms may no longer be sufficiently extensive to accommodate the strain. At high temperatures, the diffusion of atoms, if sufficiently rapid, can also accommodate this strain. Due to the large density of the grain boundaries in the nc metals, the grain boundary diffusion is a plausible mechanism for the strain accommodation during the superplastic flow. Thus a low strain rate, a high temperature and a small grain size are needed to achieve superplastic deformation.

The dependence of the strain rate in the superplastic deformation is given by the following relation [263]

$$\dot{\varepsilon} \propto D d^{-p} \sigma^{\frac{1}{m}}$$

Here D is the diffusion coefficient, d is the grain size and σ is the applied stress, p is the grain size dependence exponent and m is the strain rate sensitivity. For microcrystalline superplasticity, $p \geq 2$ and $m \geq 0.3$. If the grain size d is reduced to the nanocrystalline regime, then the same strain rate can be achieved at either a lower temperature (through the dependence of D) or/and at low stresses *i.e.* it becomes favorable for the superplasticity to occur (see Problem 9).

Problems

(1) A tensile test specimen is strained at a rate of $\dot{\varepsilon} = 10^{-3}\ s^{-1}$. Show that it requires approximately 400 s to elongate the sample by 50 %. (Hint: First convert the engineering strain to true strain).

(2) In the tensile test, necking or the plastic instability sets in at the maximum value of the load, *P*. Using the conditions that at necking $dP = 0$, and the condition of constancy of volume ($LA = V =$ constant), show that the strain hardening exponent

$$n = \frac{\delta \ln \sigma}{\delta \varepsilon}\bigg|_{\dot{\varepsilon},T}$$

is equal to the true strain ε at necking.

(3) Stress required to multiply the dislocations by the Frank–Read mechanism in aluminum (G = 27.6 GPa) is found to be 10 MPa. Find the largest distance possible between the points pinning the dislocations. The Burger's vector is 3×10^{-10} m.

(4) Calculate the strain energy density in a low carbon steel sample loaded to its elastic limit of 500 MPa. Young's modulus of this steel is 210 GPa.

(5) Particles of a hard second phase of average size 0.8 micron are uniformly dispersed in a metal having a lattice parameter of 3×10^{-10}m and a shear modulus of 1000 GPa. Assuming that the strengthening takes place by bowing of the dislocations between the particles, *and the particles can be assumed to be located in a cubic network,* calculate the approximate yield stress of the material. The volume fraction of the particles is 0.1

(6) Find the minimum approximate size of a copper crystal containing not more than 5 vacancies at 900 K. Activation energy for vacancy formation in Cu is 0.66 eV.

(7) A simple cubic crystal at 0° K has a dislocation density of $10^{9}\ m^{-2}$. What concentration of vacancies would be created if the dislocations climb down on the average by 10 A°. The lattice parameter of the crystal is 2 A°.

(8) If the yield strength of a gold wire of grain size 5 μm is 118 MPa, and that of a wire with grain size 3 μm is 135 MPa, what would be the yield strength of a wire with a grain size of 1 μm. Assume that the Hall–Petch relation is followed.

(9) A microcrystalline alloy with a grain size of 10 μm deforms superplastically at 773 K. If the grain size is reduced to 50 nm, find the temperature at which the same strain rate can be achieved. Given: activation energy for diffusion = 150 $kJmol^{-1}$ and it can be assumed to remain unchanged in the temperature range involved; strain rate sensitivity m = 0.5 and the inverse grain size dependence parameter p =3.

Chapter 12

Polymer Nanocomposites

12.1 Introduction

A composite is a material made from two chemically distinct materials which maintain their identity and have a distinct interface between them in the finished structure. A well known example of a composite material is a polymer reinforced with glass fibers. Here the glass fibers are the stronger phase and serve to impart improved mechanical properties to the polymer. Wood is a natural composite made of cellulose fibers and lignin. The continuous phase, *e.g.* the polymer in the fiber reinforced composites or the lignin in the wood, is called the matrix phase and the discontinuous phase (*e.g.* the glass fibers or the cellulose fibers) is called the reinforcement.

Composites can be made using polymers, metals or ceramics as the matrix phase and fibers and particles of different shapes as reinforcements. Because of their extraordinary mechanical properties, the composites using carbon nanotubes are of great interest since the properties of the composite are a strong function of the properties of the reinforcement. Another important type of nanocomposite is the polymer–exfoliated clay composite in which the nanosized clay platelets act as the reinforcement. These two types of nanocomposites, with polymers as the matrix, will be discussed in this chapter with emphasis on their mechanical properties. A brief introduction to the mechanical properties of composites will be first given.

A major concern in the preparation of nanocomposites is to achieve a good dispersion of the reinforcing phase in the matrix. The magnitude of the problem can be appreciated by considering the changes in the number of particles and the interface area between the particles and the matrix as the size of the particles decreases. If we consider a composite consisting

Table 12.1: Number of particles and total interface area in 1 cm^3 of a composite containing 30 vol. % particles of different sizes.

Particle size	No. of particles per cm^3	Interface area per cm^3	Particle separation
10 μm	5.8×10^8	0.18 m^2	12 μm
1 μm	5.8×10^{11}	1.82 m^2	1.2 μm
10 nm	5.8×10^{17}	181.8 m^2	12 nm
1 nm	5.8×10^{20}	1818 m^2	1.2 nm

of 30 vol. % of spherical particles in a matrix, and assume the particles to be located at the corners of a cube, then the number of particles and the interface area in 1 cm^3 of the composite is as given in Table 12.1. This Table shows that there is an enormous increase in the interface area as the particle size reduces to the nanometer range; the separation between the particles also reduces drastically, becoming comparable to the diameter of a polymer chain (~ 1 nm). The calculations are for a high volume fraction of the particles (30%). This also shows why it is extremely difficult to obtain nanocomposites with well dispersed particles when the volume fraction is high. The volume fraction in well dispersed nanocomposite systems is usually less than 5%.

12.2 Introduction to the mechanical properties of composites

To discuss the mechanical properties of composites, it is illustrative to start with unidirectional, continuous fiber composites. In such a composite, the fibers are all aligned in the same direction and are continuous along the whole length of the specimen (Fig. 12.1 a). Thin sheet, called a lamina, of such a composite can be made by winding a continuous fiber (*e.g.* that of glass or carbon) soaked in a liquid polymer (*e.g.* epoxy resin) around a flat template. Many such laminas can subsequently be pressed together to yield a thick laminate. Another type of composite is the one in which the fibers are long and aligned but not

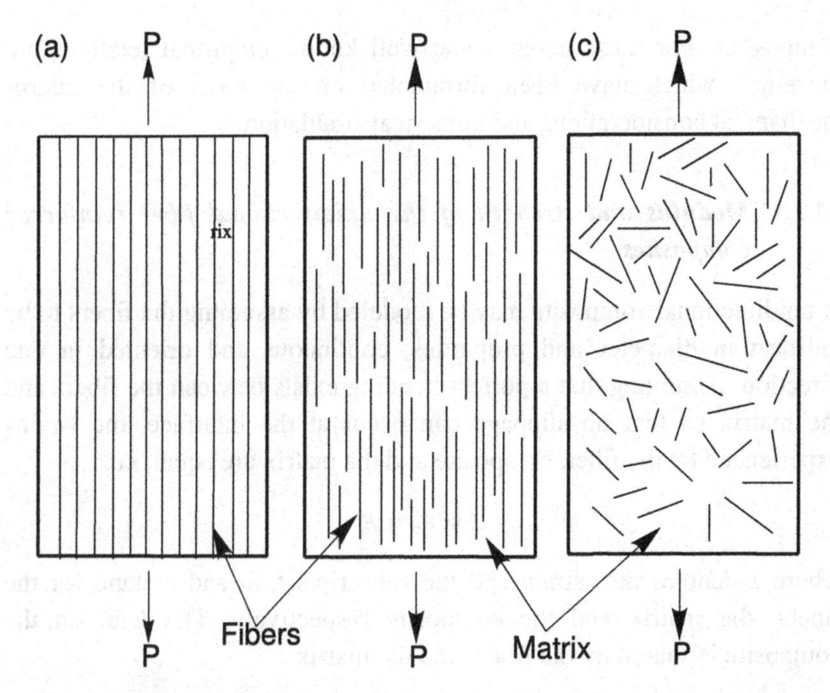

Fig. 12.1. Schematic drawing of (a) unidirectional, continuous fiber reinforced composite (b) oriented long fiber composite (c) random short fiber composite; P is the applied load.

long enough to extend along the whole length of the composite (Fig. 12.1 b). A third case is the one in which the fibers are short and randomly aligned (Fig. 12.1 c). Particle reinforced composites are a special case of this type of composites in which the short fibers are replaced by particles of any shape (spheroid, plate, discs, *etc.*).

The unidirectional fiber composites and the oriented fiber composites are anisotropic *i.e.* their properties are different along different directions. They have high values of modulus and strength along the fiber direction as compared to a transverse direction. The random short fiber composites on the other hand tend to be isotropic in the plane of the composite because of the random orientation of the fibers in this plane.

In the following, the expressions for the modulus and the strength of the unidirectional composites are first derived in terms of the values of the corresponding quantities for the fibers and the matrix. Such derivation is not easy in case of the short fiber or particle reinforced composites. For these cases, some well known empirical relations are presented which have been formulated on the basis of the micro-mechanical considerations and numerical simulation.

12.2.1 *Modulus and strength of the unidirectional fiber reinforced composites*

A unidirectional composite may be modeled by assuming the fibers to be uniform in diameter and properties, continuous and oriented in one direction. Assuming that a perfect bonding exists between the fibers and the matrix so that no slippage can occur at the interface, the strains experienced by the fiber, composite and the matrix are equal, *i.e.*

$$\varepsilon_f = \varepsilon_m = \varepsilon_c$$

where ε denotes the strain and the subscripts f, m and c stand for the fibers, the matrix and the composite respectively. The load on the composite is shared by the fibers and the matrix

$$P_c = P_f + P_m$$

or

$$\sigma_c A_c = \sigma_f A_f + \sigma_m A_m$$

where σ stands for stress and A_c, A_f and A_m are the total cross sectional areas of the composite, the fibers and the matrix respectively, with $A_c = A_f + A_m$. Multiplying both sides by the length of the composite and dividing by the total volume of the composite, we get

$$\sigma_c = \sigma_f V_f + \sigma_m V_m \qquad (12.1)$$

where V_f and V_m are the volume fractions of the fibers and the matrix respectively ($V_f + V_m = 1$).

Equation (12.1) also gives the strength of the composite if σ_f and σ_m are replaced by the strengths of the fibers and the matrix respectively.

Dividing both sides of Eq. (12.1) by strain (which is same in the composite, the fibers and the matrix), we get

$$E_c = E_f V_f + E_m V_m \tag{12.2}$$

Here, E_c, E_f and E_m are the Young's modulus of the composite, the fibers and the matrix respectively in the longitudinal direction *i.e.* in the direction of the fibers. Equation (12.2) shows that the modulus of the composite is given by a simple rule of mixtures.

The fraction of the total load carried by the fibers is given by

$$\frac{P_f}{P_c} = \frac{P_f}{P_f + P_m} = \frac{\frac{P_f}{P_m}}{1 + \frac{P_f}{P_m}} \tag{12.3}$$

Also

$$\frac{P_f}{P_m} = \frac{\sigma_f A_f}{\sigma_m A_m} = \frac{\sigma_f V_f}{\sigma_m V_m} = \frac{E_f}{E_m} \cdot \frac{V_f}{V_m}$$

Substituting in (12.3), we get

$$\frac{P_f}{P_c} = \frac{\frac{P_f}{P_m}}{1 + \frac{P_f}{P_m}} = \frac{E_f V_f}{E_m V_m + E_f V_f} = \frac{\frac{E_f}{E_m}}{\frac{V_m}{V_f} + \frac{E_f}{E_m}} \tag{12.4}$$

Taking a value of 0.05 (5 %) as the volume fraction of the fibers, and 1.2 GPa, 25 GPa and 1000 GPa as the moduli of the polymer matrix, the glass fibers and the carbon nanotubes respectively, the fraction of load carried comes out to be 0.52 for the glass fibers and 0.98 for the carbon nanotubes. This shows the potential of the carbon nanotubes in the composites.

The properties of the unidirectional composites in the transverse direction are much less attractive. When a composite is loaded in the transverse direction, the total elongation of the composite is the sum of the fiber elongation and the matrix elongation (Fig. 12.2)

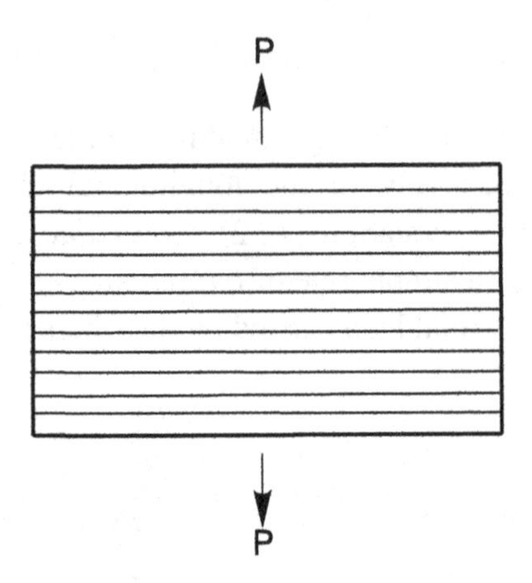

Fig. 12.2. Transverse loading of a unidirectional fiber reinforced composite.

$$\delta_c = \delta_f + \delta_m$$

where δ denotes elongation. From this, it is easy to show that

$$\varepsilon_c = \varepsilon_f V_f + \varepsilon_m V_m$$

or

$$\frac{\sigma_c}{E_c} = \frac{\sigma_f}{E_f} V_f + \frac{\sigma_m}{E_m} V_m$$

In case of the transverse loading, the fibers and the matrix experience the same stress so that

$$\frac{1}{E_c} = \frac{V_f}{E_f} + \frac{V_m}{E_m} \tag{12.5}$$

Equations (12.2) and (12.5) are plotted in Fig. 12.3. It can be seen that the fibers are much more effective in enhancing the longitudinal modulus.

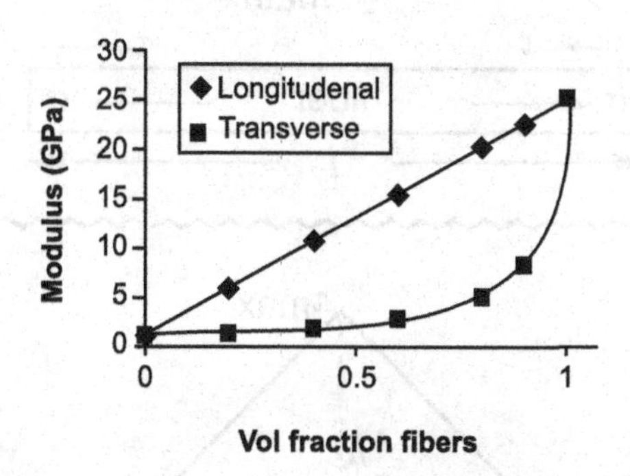

Fig. 12.3. Variation of the longitudinal and the transverse modulus of a uniaxial fiber composite; values used in the calculation are E_m=1.2 GPa, E_f = 25 GPa.

12.2.2 *Short fiber and particle reinforced composites*

12.2.2.1 *Importance of the aspect ratio*

When the fibers are shorter than the specimen length, the stress transferred to the fibers depends on the fiber length or the aspect ratio, *l/d* where *l* is the length and *d* is the diameter of the fiber. This consideration is important in the short fiber composites, the carbon nanotube composites as well as the composites reinforced with particles having an aspect ratio greater than one. To understand the importance of the aspect ratio, consider a cylindrical fiber embedded in a matrix. Assume that there is no slippage between the fiber and the matrix upon loading. The strain in the fiber and the matrix is therefore the same. The stress is transferred from the matrix to the fiber via the matrix–fiber interface. If the fiber and the matrix were free, the former would experience a much lower strain due to its higher modulus. To keep the strain equal, a shear stress is set up at the interface. This shear stress gradually builds up from zero at the tips of the fiber to a maximum value at the middle of the fiber as shown in Fig. 12.4. This is balanced by the stress produced in the fiber.

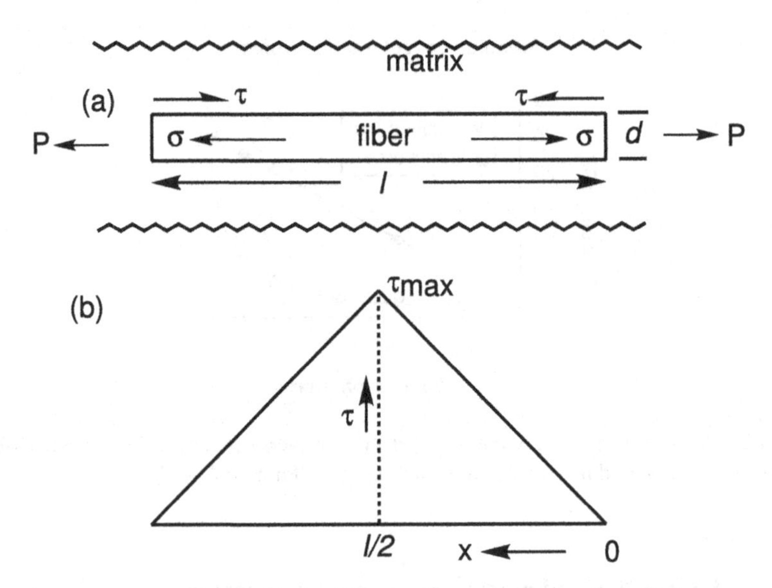

Fig. 12.4. (a) A shear stress is developed at the fiber– matrix interface (b) the shear stress increases from zero at the tips of the fiber to a maximum value at the middle.

Let σ be the uniform stress in the fiber and τ the shear stress at the interface at any point x from the tip of the fiber. Only one half of the fiber need be considered because of the symmetry of the problem. Taking a force balance we get

$$\pi \frac{d^2}{4}\sigma = \int_0^{l/2} \pi d\tau dx = \int_0^{l/2} \pi d(2\tau_{max}\frac{x}{l})dx; \text{ (using } \frac{\tau}{x} = \frac{\tau_{max}}{l/2})$$

This gives

$$\frac{l}{d} = aspect\ ratio = \frac{\sigma}{2\tau_{max}} \tag{12.6}$$

The maximum stress that the fiber can stand is its breaking strength, σ_f. Therefore, for full stress transfer, the aspect ratio should not be less than $(\sigma_f/2\tau_{max})$. If the aspect ratio is less than this value, the stress transferred to the fiber will be smaller. This is illustrated in Fig. 12.5 which gives the calculated longitudinal stress in the fiber and the shear stress at the fiber–matrix interface along the length of the fiber for a model system of a single fiber embedded in a matrix [264]. It can be seen that the stress transferred to the fiber increases with the aspect ratio.

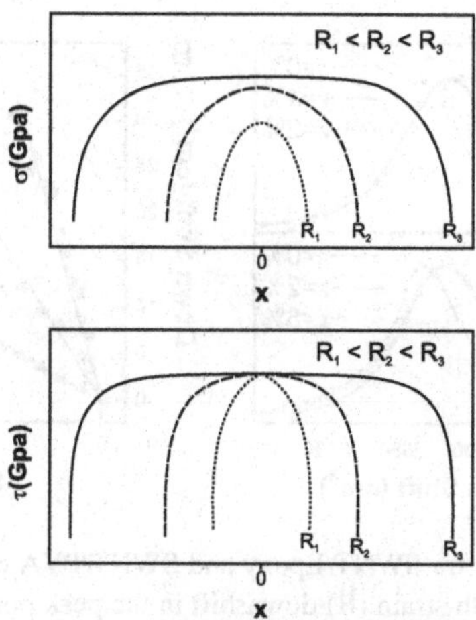

Fig. 12.5. (a) Distribution of the longitudinal stress along the fiber length for three different aspect ratios, R_1, R_2 and R_3 with $R_3 > R_2 > R_1$; note that the stress in the fiber increases and extends to longer lengths as the aspect ratio increases (b) distribution of the shear stress in the fiber – matrix interface for different aspect ratios [264].

The extent of the stress transferred to a carbon nanotube can be measured using Raman spectroscopy. It is known that the position of the Raman *G'* peak from a nanotube shifts to lower wave numbers when a strain is applied to the tube (Fig. 12.6). On the average there is a downshift of 37.5 cm^{-1} when a 1% axial strain is applied to an individual SWNT [265]. Using this knowledge and by measuring the shift in the Raman peak, the stress in the tube can be monitored (see Example 12.1).

Example 12.1. *In a polymer–SWNT composite, there is a shift in the G' Raman band by 4.8 cm^{-1} when the composite is strained by 1 % and a shift by 15 cm^{-1} when the composite fails by fracture. It is known that there is a downshift in the position of the G' band of SWNT by 37.5 cm^{-1} on straining the tube by 1 %.*

(i) What is the strain in the SWNT when the composite is strained by 1%?

(ii) What is the strain in SWNT at the fracture point?

Comment on the strength of the polymer – CNT interface.

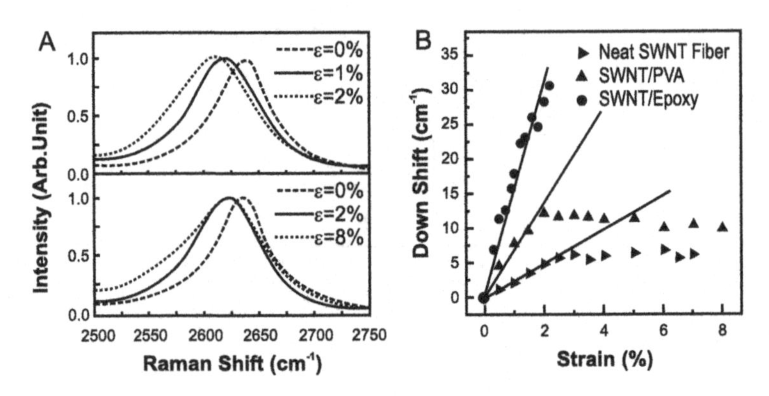

Fig. 12.6. (A) Raman spectra SWNT/Epoxy and SWNT/PVA composite fibers showing the shift in the G′ band with strain (B) downshift in the peak position with strain for neat SWNT fiber and the two composite fibers (Adapted with permission from [266] Copyright (2009) American Chemical Society).

Solution.

(i) Strain in the SWNT when there is a 1% strain in the composite = (4.8/37.5)% = 0.128% or the SWNT has a strain which is only 12.8 % of the strain in the composite.

(ii) Strain in the SWNT when the composite fractures = 15/37.5 = 0.4 %.

These calculations show that the strength of the interface is low as a result of which only a small fraction of the strain is transferred to the SWNT. □

12.2.2.2 *Modulus and strength of short fiber composites*

As is clear from the above discussion, the short fibers carry the load less effectively than the long fibers. To derive an expression for the modulus of the composite in terms of the moduli of the fibers and the matrix in this case is not easy. One way is to use the expression (12.2) derived for the uniaxial composites by replacing the modulus of the fibers by an effective modulus $\eta_l E_f$ where η_l is called the length efficiency factor [267, 268]. The modulus of a composite with *oriented* short fibers is then obtained from Eq. (12.2) as

$$E_c = \eta_l E_f v_f + E_m v_m \tag{12.7}$$

The factor η_l is given by

$$\eta_l = 1 - \frac{\tanh(a.l/d)}{a.l/d} \tag{12.8}$$

with
$$a = \sqrt{\frac{-3E_m}{2E_f \ln V_f}}$$

The value of η_l increases as the aspect ratio increases and approaches 1 for an aspect ratio of ~10.

In a similar manner, the strength of the *oriented* short fiber composites can be obtained from Eq. (12.1)

$$\sigma_f = \eta_s \sigma_f V_f + \sigma_m V_m \tag{12.9}$$

Here η_s is called the strength efficiency factor and is given by $\eta_s = (1 - l_c/2l)$ for $l > l_c$. In this case, the fibers can be stressed to failure.

For $l < l_c$, the stress in the fibers does not reach their breaking stress and the failure instead occurs by the matrix failure. In this case the strength is controlled by the interface strength τ as follows

$$\sigma_f = \frac{\tau\, l}{d}\, \sigma_f V_f + \sigma_m V_m \tag{12.10}$$

The above discussion has been for the oriented fibers. When the fibers are not aligned, the modulus is given by incorporating another factor, called the *orientation efficiency factor* in Eq. (12.7) [267]

$$E_c = \eta_o \eta_l E_f v_f + E_m v_m \tag{12.11}$$

The value of the orientation efficiency factor η_o is 1 for aligned fibers, 3/8 for fibers aligned in a plane and 1/5 for randomly oriented fibers.

12.2.2.3 *The Halpin–Tsai and Mori –Tanaka formulations*

In the previous sections we have presented the relations for the modulus and strength of the fiber reinforced composites which are relevant for the carbon nanotube composites. In the next section we will consider the polymer–clay composites in which the clay platelets act as the reinforcement. Several models have been proposed to predict the moduli of the particle or platelet reinforced composites. These models are also applicable to the fiber reinforced composites. They are based on micromechanical models which consider the stress distribution in the reinforcement and the surrounding matrix and extend it to the whole composite, to arrive at an average stress and strain. Most of these micromechanical models require considerable numerical simulation which has to be carried out separately for each composite system. The empirical models, instead, give simple relations which can be used to get useful predictions.

Halpin and Tsai, taking into consideration the results of the micromechanical models, have proposed a set of empirical equations which have become popular for predicting the moduli of the composites [269, 270]. The equations contain an unknown parameter ζ, whose value is decided on the basis of the available experimental data (Table 12.2). These equations have proved very useful for a range of composites containing short fibers, particles or platelets as reinforcement. The equations are as follows:

$$M_c = M_m\left(\frac{1+\zeta\mu V_f}{1-\mu V_f}\right) \text{ with } \mu = \frac{\frac{M_f}{M_m}-1}{\frac{M_f}{M_m}+\zeta} \qquad (12.12)$$

Here M is a relevant mechanical property, v is the volume fraction, ζ is a parameter which depends on the property and the geometry of the filler particles and the subscripts c, m and f refer to the composite, the matrix and the reinforcement respectively. The mechanical properties that are usually predicted using these equations are the longitudinal and the transverse tensile moduli and the shear modulus. The Poisson's ratio for

Table 12.2 : Values of the Halpin-Tsai parameter ζ for different filler geometries (L = length of fiber, rod, plate or whisker, D = diameter of particle, rod, fiber or whisker, W = width of plate, T = thickness of plate) (adapted from [270]).

Filler geometry	Longitudenal Modulus, E_{11}	Transverse Modulus, E_{22}	Shear Modulus G_{12}
Spherical particles	2	2	1
Oriented short fibers	$2(\frac{L}{D})$	2	1
Oriented plates	$2(\frac{L}{T})$	$2(\frac{L}{W})$	$(\frac{L+W}{2T})^{1.73}$
Oriented whiskers	$2(\frac{L}{D})$	2	$(\frac{L}{D})^{1.73}$

all geometries is given by the simple rule of mixture $\nu_c = \nu_f V_f + \nu_m V_m$ where ν stands for the Poisson's ratio and V stands for the volume fraction. The values of the parameter ζ are given in Table 12.2 for the various geometries of the reinforcement when the reinforcement is preferentially oriented in one direction (for reinforcements with aspect ratio > 1) [270]. The limiting values of ζ are 0 and ∞. For these limiting values, the Halpin–Tsai equations reduce to the reciprocal rule of mixture

$$\frac{1}{M_C} = \frac{V_f}{M_f} + \frac{V_m}{M_m}$$

and the simple rule of mixture ($M_c = V_f M_f + V_m M_m$) respectively. Note that these are the same as the relations derived for the transverse and the longitudinal moduli of a unidirectional fiber reinforced composite.

Example 12.2. *Calculate and plot the longitudinal and the transverse moduli of an oriented short glass fiber reinforced polypropylene as a function of the volume fraction of the glass fibers from 0 to 60 vol. % using the Halpin–Tsai equation. Given: modulus of polypropylene = 1.2 GPa, modulus of glass fibers = 75 GPa; fiber aspect ratio = 180.*

Solution. From Table 12.2, the values of ζ for calculating the longitudinal and the transverse moduli are 2 x 180 = 360 and 2 respectively.

Using Eq. (12.12), the values of the parameter μ for the two cases are

$$\mu_{\text{longitudenal}} = \frac{\frac{M_f}{M_m} - 1}{\frac{M_f}{M_m} + \zeta} = \frac{\frac{75}{1.2} - 1}{\frac{75}{1.2} + 360} = 0.145562$$

$$\mu_{\text{transverse}} = \frac{\frac{M_f}{M_m} - 1}{\frac{M_f}{M_m} + \zeta} = \frac{\frac{75}{1.2} - 1}{\frac{75}{1.2} + 2} = 0.9535$$

substituting in Eq. (12.12) one gets,

$$M_{longitudenal} = M_m \left(\frac{1 + \zeta\mu v_f}{1 - \mu v_f}\right) = 1.2 \times \left(\frac{1 + 360 \times 0.145562 v_f}{1 - 0.145562 v_f}\right)$$

$$= 1.2 \text{ x} \frac{1 + 52.402 \times v_f}{1 - 0.145562 \times v_f}$$

and

$$M_{transverse} = M_m \left(\frac{1 + \zeta\mu v_f}{1 - \mu v_f}\right) = 1.2 \times \left(\frac{1 + 2 \times 0.9535 v_f}{1 - 0.9535 v_f}\right)$$

$$= 1.2 \times \frac{1+1.907\times v_f}{1-0.9535\times v_f}$$

The values of the longitudinal and the transverse modulus have been calculated for different volume fractions of the fibers and plotted in Fig 12.7. In Table 12.2, the values of ζ have been given for the cases when the fibers or plates are fully oriented. This produces a large difference in the values of the longitudinal and transverse modulus. In practice, the orientation is not perfect — there may be some preferred orientation in a particular direction depending on the fabrication technique used or, in the extreme case, the orientation may be nearly random. The result is that the values predicted by the Halpin–Tsai equations lie somewhere between the longitudinal and transverse cases in Fig. 12.7.

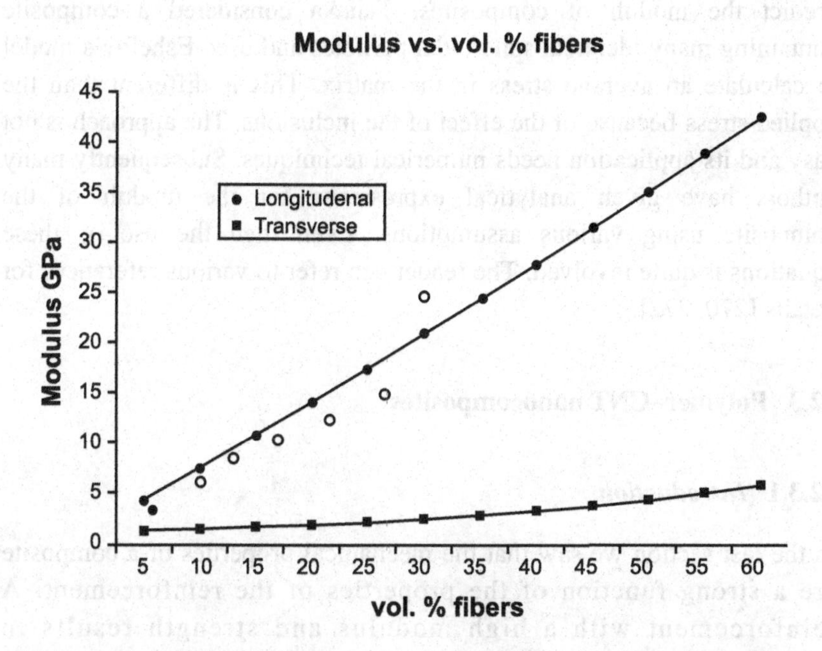

Fig.12.7. Variation of the longitudinal and the transverse modulus of a short glass fiber polymer composite according to the Halpin–Tsai equation (lines); the circles show how the experimental data usually lie.

When the fibers or the reinforcement particles are oriented randomly, the modulus of the composite can be expressed as a weighted average of the longitudinal and transverse moduli

$$M_c = f_1 M_{11} + f_2 M_{22}$$

where M_{11} and M_{22} are the calculated longitudinal and transverse moduli and f_1 and f_2 are the weighing factors with $f_1 + f_2 = 1$. The values of f_1 and f_2 are usually taken to be 3/8 and 5/8 respectively [271].

When the fibers assume a preferred orientation due to processing effects or some other reason, the weighing factors can be back calculated from the measured modulus and can serve as a measure of the preferred orientation of the particles.

The Halpin–Tsai equations are found to agree with the experimental data to moderate volume fractions (about 0.4).

The Mori–Tanaka average stress theory is another approach used to predict the moduli of composites. Tanaka considered a composite containing many identical spheroidal particles and used Eshelby's model to calculate an average stress in the matrix. This is different than the applied stress because of the effect of the inclusions. The approach is not easy and its application needs numerical techniques. Subsequently many authors have given analytical expressions for the moduli of the composite using various assumptions. Even then the use of these equations is quite involved. The reader can refer to various references for details [270, 272].

12.3 Polymer–CNT nanocomposites

12.3.1 *Introduction*

In the last section we saw that the mechanical properties of a composite are a strong function of the properties of the reinforcement. A reinforcement with a high modulus and strength results in correspondingly high values of these properties in the composite. From this point of view, the carbon nanotubes with a Young's modulus of 600 to 1000 GPa, strength ranging from 35 to 110 GPa and a density as low

Table 12.3 Some nanoscale fillers used for polymers [272].

Filler	Properties enhanced or imparted
Exfoliated clay	modulus, flame resistance, barrier property, making different polymers compatible in a polymer blend
Carbon nanotubes	Mechanical properties, electrical conductivity.
Graphene	Mechanical properties, electrical conductivity.
CdSe, CdTe	Charge transport
ZnO	UV absorption
Silver nanoparticles	Antibacterial
Silica	Rheology control

as 1.3 gcm^{-3} can be said to be the ultimate reinforcement material. Much effort is therefore being devoted to the study of polymer, metal and ceramic based CNT composites. The polymer–CNT composites are of special interest because of their low density for applications such as aerospace and automobile where weight saving is a very important concern.

In addition to the CNT, several other nanofillers have been used to prepare nanocomposites. Most important amongst these are the composites of polymer with exfoliated clay particles. Other fillers of interest are carbon nanofibers, graphene, and metal nanoparticles. Some of the fillers and the corresponding property enhancements are listed in Table 12.3.

12.3.2 *Some important considerations in CNT–composites*

To fully exploit the excellent mechanical properties of the CNT in the composites, some of the factors that need to be considered are as follows. These considerations, also apply to the composites in general.

12.3.2.1 *Aspect ratio of the CNT*

Both the Halpin–Tsai and Mori–Tanaka theories show the importance of the aspect ratio and the modulus of the reinforcing phase [272]. The polymer–CNT composites show similar behavior. An increase in the aspect ratio leads to an increase in the modulus of the composite and this effect is greatly amplified when the modulus of the filler is increased. This shows why it is important to have a high aspect ratio of the carbon nanotubes and to obtain complete exfoliation of the clay platelets in the case of the polymer–clay nanocomposites discussed later.

12.3.2.2 *Shear strength of the matrix–CNT interface*

In order to produce a good bonding between the CNT and the matrix, the latter must wet the CNT. The equilibrium contact angle between two solids is a measure of the wetting, a low contact angle signifying better wetting. The contact angle depends on the surface energy of the various interfaces (Chapter 2). If γ represents the specific surface energy and the subscripts *s1*, *s2* and V denote the solid 1, solid 2 and the vapor phase (air) respectively, then

$$\cos\theta = (\gamma_{SL,V} - \gamma_{S1,S2}) / \gamma_{S2,V}$$

For a low value of θ, the surface energy of the matrix phase, γ_{S2V}, should be low. It has been proposed on the basis of experiments, that this value should be between 100 to 200 mNm^{-1} for good wetting. Thus aluminum with $\gamma = 840$ mNm^{-1} does not wet the CNT while oxides such as V_2O_5 ($\gamma = 80$ mNm^{-1}) readily wet the CNT.

The wetting behavior, and hence the polymers–CNT interface strength, can be drastically changed by a surface treatment such as oxidation, acid treatment or functionalization by attaching suitable groups of atoms to the surface of the CNT [273, 274]. However, the process of functionalization also introduces some defects.

Figure 12.8 shows the results of a model calculation on a composite consisting of a single nanotube embedded in a matrix [264]. The figure shows that the modulus is higher for the hard interface; however, the effect of the aspect ratio on the modulus is much stronger than that of the interface strength.

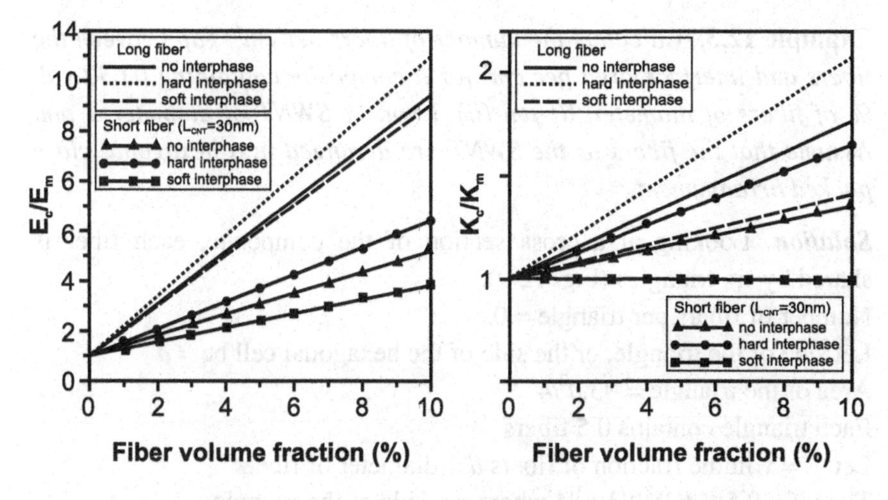

Fig.12.8. Results of calculations for a model consisting of a single nanotube embedded in a matrix. A hard interface results in a stiffer composite. Note that the effect of the aspect ratio is more pronounced than the effect of the interface. The soft and hard interfaces were modeled by taking the ratio of the moduli of the CNT and the matrix to be 10 and 0.3 respectively. For the short fibers, the aspect ratio was 30 while for the long fibers the results remained unchanged beyond an aspect ratio of ~ 200. (left) effective Yong's modulus (right) effective bulk modulus (Reprinted from [264] with permission from Elsevier).

12.3.2.3 *Dispersion of the nanotubes in the matrix*

For maximum effectiveness, the nanotubes must be uniformly dispersed in the composite so that they exist as individual tubes with only the polymer surrounding them. In this condition, the stress is transferred to the nanotubes most efficiently and uniformly throughout the composite. However, the nanotubes tend to agglomerate and form clumps when introduced into a matrix due to Van der Waals and other forces, especially when their volume fraction increases beyond ~ 2 vol. %. Recall from the earlier discussion that the number of particles in a unit volume increases rapidly and the separation between them drastically reduces as the size of the particles is reduced. It can be shown (see Example 12.3) that in a composite containing 3 vol. % of SWNT of diameter 1 nm, there are ~ 4 x 10^{12} nanotubes per cm^3 and the separation between the tubes is about 4.5 nm only.

Example 12.3. *Calculate the number of fibers per cm^3, gap between the fibers and interface area per cm^3 for a composite containing (i) 30 vol. % of fibers of diameter 10 μm (ii) 3 vol. % SWNT of diameter 1 nm. Assume that the fibers or the SWNT are arranged in a hexagonal close packed arrangement.*

Solution. Looking at a cross section of the composite, each fiber is shared by six triangles (Fig. 12.9).
Number of fibers per triangle = 0.5
Let side of the triangle, or the side of the hexagonal cell be = a
Area of the triangle = $\sqrt{3}.a^2/4$
Each triangle contains 0.5 fibers
Let f = volume fraction of fibers d = diameter of fibers
Then $f = 0.5\pi(d/2)^2/\sqrt{3}.a^2/4$ where a = side of the triangle.
This gives "a".
Spacing between the fibers, surface to surface = $a - d$
(i) Calculation of the number of fibers per cm^3:
There is 0.5 fiber to an area of $\sqrt{3}.a^2/4$. Then in 1 cm^2 cross sectional area there are $2/\sqrt{3}.a^2$ fibers. (a in cm)
(ii) Interface area per cm^3
In $\sqrt{3}.a^2/4 \times l$ volume, interface area = $0.5 \times 2\,\pi r l$
In 1 cm^3 volume, interface area = $\pi r/\sqrt{3}.a^2/4 = 4\pi r/\sqrt{3}.a^2$

The following results are obtained using the above relations

Fiber diameter	*10 μm*	*1 nm*
Volume fraction	0.3	0.03
Side of cell, a	17.4 μm	5.5 nm
Gap between fibers	7.4 μm	4.5 nm
Number cm^{-3}	3.8×10^5	3.8×10^{12}
Interface area cm^{-3}	0.12 m^2	120 m^2

Note that for the nanotubes the gap between the nanotubes is of the same order as the polymer chain size. This makes it difficult to infiltrate the matrix phase in between the nanotubes. □

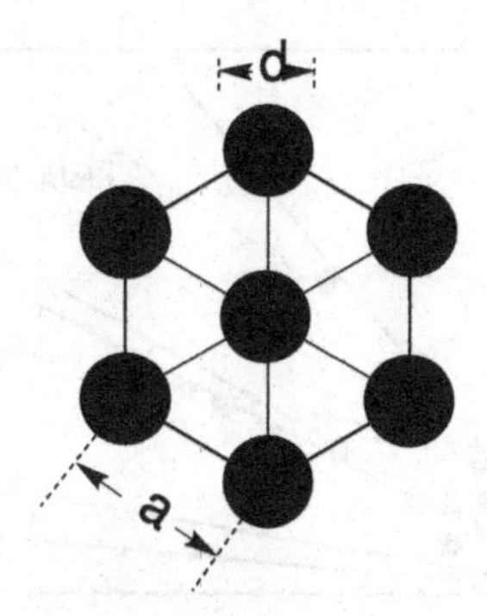

Fig. 12.9 Assumed arrangement of the fibers or the nanotubes

It therefore becomes increasingly difficult for the polymer molecules to penetrate the space between the nanotubes as their concentration increases and there is increasing tendency for the nanotubes to clump together *i.e.* to aggregate. Thus in general it is difficult to achieve good dispersion beyond 5 vol. % CNT and the properties improve less rapidly or not at all if the vol. % is increased beyond this level. This is shown for the aluminum–CNT composites in Fig. 12.10 where it can be seen that the modulus increases rapidly at first and then at a reducing rate as more and more CNT are added to the composite beyond about 4 vol. %. Use of suitable surface treatments, additives and dispersion techniques is required to achieve a uniform dispersion of the CNT in the matrix.

12.3.2.4 *Waviness of the nanotubes*

The carbon nanotubes in a composite are not straight but are curved or can be said to have "waviness". The waviness is a consequence of the low stiffness of the tubes due to their large aspect ratio. This waviness can be quantified in terms of a parameter called the curl parameter λ which is the ratio of the actual length of the tube to its curled length (Fig. 12.11 a) [150]. The modulus of the composite reduces as the curl parameter increases; however, the composite maintains the high modulus to larger strains (Fig. 12.11 b). The stretching and aligning of the nanotubes during the processing of the composites should lead to a higher modulus.

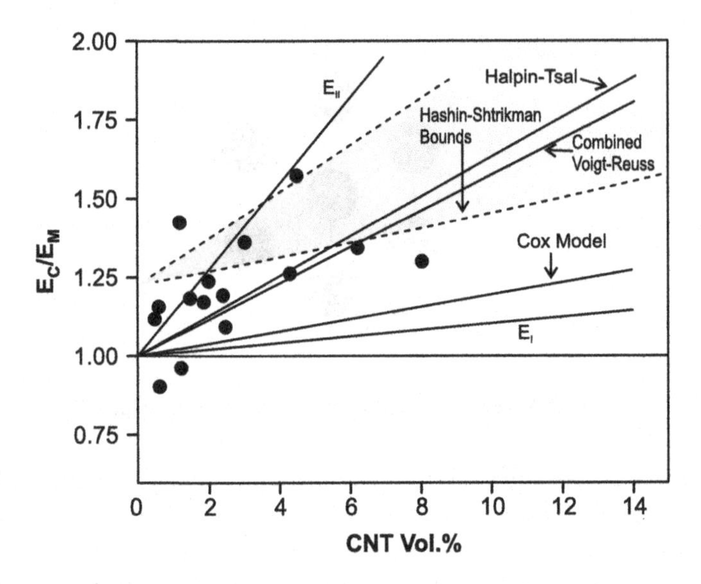

Fig.12.10. Change in the modulus of the composite relative to that of the matrix with increasing CNT content for an aluminum–CNT composite; the predictions of the various models are also shown; the modulus increases less rapidly after 4 vol. % due to the aggregation of the carbon nanotubes (reprinted from [275] with permission from Elsevier).

12.3.3 *Processing and properties of the polymer–CNT composites*

Several methods are used to prepare the CNT–polymer composites [267, 273, 276]. In the *solution processing* method, the CNT and the polymer are mixed in a solvent followed by evaporation of the solvent to yield a composite film. Good dispersion of CNT can be obtained by effective agitation of the solution. In *melt processing*, the CNT are mixed in a molten thermoplastic polymer and the shaping of the mix is carried out by the usual polymer processing techniques such as injection molding, compression molding or extrusion. The method has the advantage that it can be readily adapted to the existing polymer processing practices. However, it needs to be further developed as at present it is difficult to attain good dispersion of the CNT and there is a greater chance of the CNT getting shortened and damaged. The melt processing can also be

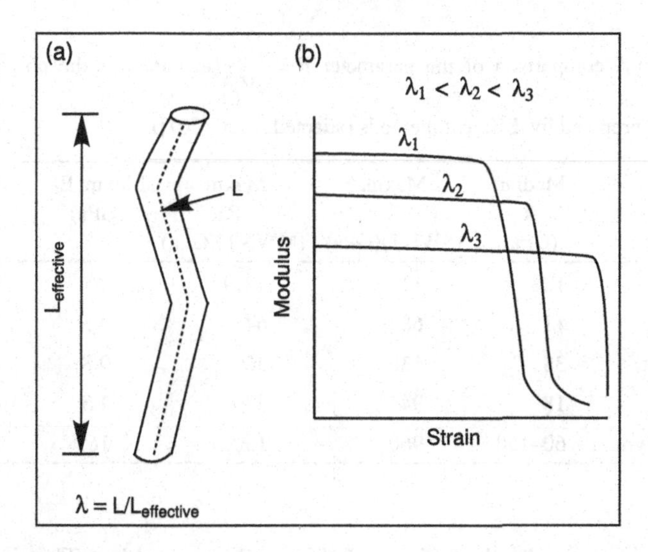

Fig. 12.11. Effect of the waviness of the carbon nanotubes on the modulus of the composite (a) the waviness is quantified by the curl parameter λ which is the ratio of the total length of the tube to its curled length (b) the modulus decreases as the curl factor increases.

used to prepare the composite fibers by extrusion followed by fiber drawing or other techniques. Another method, that of *in-situ polymerization*, is useful for making composites with polymers which are insoluble or thermally unstable; it can also be used for grafting polymer molecules onto the nanotubes to make them more compatible with a polymer. In this technique, the fibers are mixed with the monomers and the polymerization is carried out. The method can be used with any polymer.

The studies so far carried out on polymer–CNT composites have found significant enhancement in modulus and strength. However, the quantum of improvement is a strong function of the method used to prepare the composite. This is understandable because different methods result in different extents of dispersion of the nanotubes in the composite. Although, other factors such as type of the nanotubes used, their functionalization and whether the composite is in the form of bulk, fiber or film also have significant influence, the overwhelming influence of

Table 9.4 : A comparison of the parameter $R = \frac{dE_c}{dV_f}$ (in GPa) for the polymer–CNT composites prepared by different methods (adapted from [267]).

Method	Median R (GPa)	Maxm. R (SWNT)(GPa)	Maxm. R (MWNT)(GPa)	Maxm. E_C (GPa)	Maxm. σ (MPa)
Solution	128	112	1244	7	348
Melt	11	68	64	4.5	80
Melt (fiber)	38	530	36	9.8	1032
Epoxy	18	94	330	4.5	41
In-situ polym.	60–150	960	150	167	4200

the dispersion masks the effects of these other variables. This is clearly brought out by a comparison of the quantity dE_c/dV_f, the initial rate of increase of the modulus with the volume fraction of the CNT. Table 12.4 gives the median and the maximum values of this quantity for the composites processed by different methods. The data for the maximum values for the composites containing SWNT and MWNT is given separately. Taking the median value as a measure of the strengthening effect, it is seen that the solution processing and in situ polymerization lead almost to the same values of the median, dE_c/dV_f. This is not surprising because the processing in all these cases starts in a liquid in which it is relatively easy to achieve a good dispersion of the CNT. The value for the melt processed composites is the minimum, emphasizing that the melt processing is not able to achieve a good dispersion of the nanotubes.

The maximum values given in Table 12.4 should be taken only as an indication of what can be achieved and not as a comparison of different methods because of the limited data that has been used.

The polymer –CNT composites are yet to match the level of the mechanical properties which is routinely achieved in the polymer–carbon fiber composites, because of the difficulty of introducing a high volume fraction of CNT and poor dispersion. Better mechanical properties of the composites are expected if the dispersion of the nanotubes in the matrix is made uniform and the interfacial strength is optimized. Composites of CNT with metallic and ceramic matrices also hold a good potential.

12.4 Polymer–clay nanocomposites

Interest in the polymer–clay nanocomposites was triggered by the results reported by the Toyota Research Laboratories in 1989 on nylon-6–exfoliated clay composites. These nanocomposites have the advantage of inexpensive, easily handled filler as compared to the composites with other fillers such as carbon nanotubes. In the following we describe this important class of nanocomposites.

12.4.1 *Filled polymers vs. the polymer nanocomposites*

Mixtures of polymers with finely divided substances such as carbon black, minerals and short fibers have been in use for a long time. Such mixtures are called *filled polymers*. The preparation and application of the filled polymers is very well established with a worldwide consumption of more than 50 million tons of fillers in 2007. Incorporation of filler in a polymer or a rubber is used to make it more useful by changing its physical properties such as density, conductivity or flame resistance or the mechanical properties such as modulus, strength or viscosity.

The size range of the carbon black is 10 to 500 nm while the size range of mineral fillers is usually between 100 to 10000 nm. The short fibers typically have a diameter of 10 μm and lengths ranging from 1 to 10 mm, giving an aspect ratio (length/diameter) ranging from 100 to 1000. A comparison in terms of the surface area of some of the fillers used in the filled composites is given in Table 12.5. The specific surface area of the various filler ranges from 2 to 240 m^2g^{-1} except for the fine carbon black where it can reach 500 m^2g^{-1}. In comparison, the surface area of fully dispersed montmorillonite clay used in polymer–clay nanocomposites, as described later, is about 750 m^2g^{-1}.

Table 12.5. Specific surface area of some of the fillers used in the filled polymers [270].

Filler	Specific surface area (m^2/gm)
Clay kaolinites	10 – 30
Talc	2 – 12
Mica	19 – 140
Silica (synthetic)	100 – 240
Calcium carbonate (synthetic)	3 – 120
Carbon black, different grades	6 – 500

According to our definition of a nanostructure having one of its dimensions less than 100 nm, many of the carbon black filled polymers would qualify as nanocomposites. However, the term is more commonly restricted to the case when the filler or the reinforcing phase has at least one dimension in the range of 1 to 10 nm. In this case the dimensions of the filler are of the order of a characteristic dimension (such as the radius of gyration) of the matrix polymer phase. However, whether this leads to any special effects (the so called "nano effects"), especially in case of the mechanical properties, is uncertain. All the changes in the properties can usually be accounted for by the theories used for microcomposites.

While the filled polymers have been on the scene for a long time, the interest in the polymer nanocomposites really took off only in 1989 when the researchers at the Toyota Central Research Laboratories reported a nylon-6–clay nanocomposite that was utilized in a timing belt cover on the Toyota Camry [277, 278]. Although the polymer nanocomposites are still mostly in the research stage with small applications beginning to appear, their potential is huge, with a multibillion dollar market being forecast within the next decade.

As mentioned earlier, the clays have the advantage of being a naturally occurring inexpensive bulk raw material. They have been used for decades in the ceramic industry for making ceramic products such as pottery, tiles, sanitary ware, *etc.* as well as a filler for polymers to enhance properties such as strength, elastic modulus and flame resistance.

In the filled composites, the clay exists as micron sized particles. As discussed later, a basic structural unit in a clay particle is a platelet with a thickness of about 1 nm and a lateral dimension of 100 to 200 nm. A number of these platelets stacked one above the other form the next unit, called a *tactoid,* and several tactoids assemble to form a particle. The polymer nanocomposites differ from the filled polymers by the fact that in the former the clay is dispersed as nano sized individual platelets rather than micron sized particles. This results in a significant increase in the strength and modulus *without degrading the impact resistance*, makes the polymer less permeable to undesirable molecular species, improves its flame resistance while simultaneously retaining the optical transparency. And all this is accomplished at a much lower loading of the polymer by clay, typically less than 5 wt % as compared to more than 20 wt % for the filled polymers.

12.4.2 *Clays used in the nanocomposites*

Clays are naturally occurring alumino-silicate minerals having a layered structure. The clays commonly used in polymer nanocomposites belong to the structural family known as 2:1 phyllosilicates in which each layer consists of three sheets of atoms (Fig. 12.12). The central sheet consists of Al or Mg surrounded by six oxygens in an octahedral coordination. It is sandwiched between two sheets of tetrahedrally coordinated silicon (each silicon surrounded by four oxygens at the corners of a tetrahedron, with silicon in the center of the tetrahedron) so that some of the oxygens belong to both the sheets as shown in Fig 12.12. These three sheets

Table 12.6. Chemical formulae of the commonly used 2:1 phyllosilicates [279].

2:1 Phyllosilicate	Chemical formula
Montmorillonite	$(Al_{2-x}M_xMg_x)(Si_2O_5)_2(OH)_2$
Hectorite	$(Mg_{3-x}M_xLi_x)(Si_2O_5)_2(OH)_2$
Saponite	$M_xMg_3\,(Si_{4-x}Al_x)O_{10}(OH)_2$
	($x = 0.25$ to 0.65; M= monovalent cation)

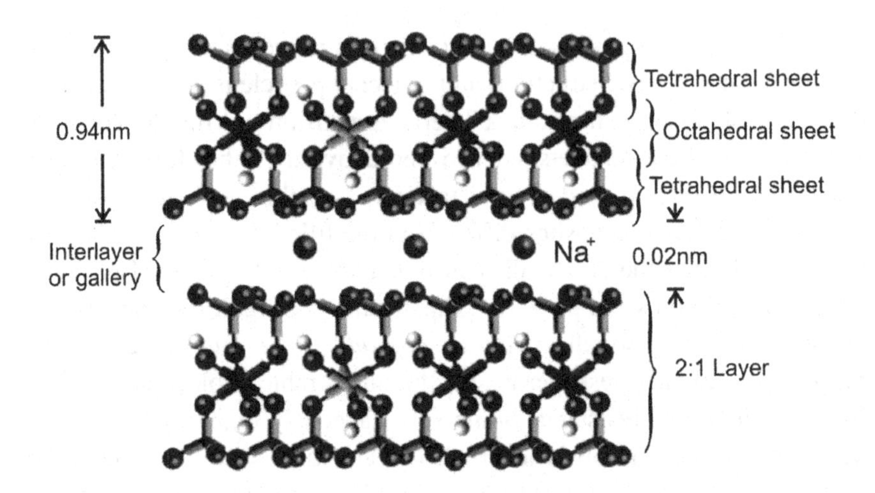

Fig.12.12. Structure of sodium montmorillonite platelets; each platelet consists of a octahedral sheet sandwiched between two tetrahedral sheets. The platelets are separated by interlayer gallery (Adapted, courtesy Southern Clay Products, Gonazales).

together constitute a *platelet*. The platelets organize themselves into stacks called *tactoids*, the bonding between the platelets being by the van der Waals force. The space between the platelets, or the van der Waal's gap is called the interlayer *gallery*. Several tactoids agglomerate to form a pyrophillite clay particle.

If in the original 2:1 phyllosilicate structure, some of cations in the central octahedral sheets are replaced by lower valence cations (*e.g.* Al^{3+} replaced by Mg^{2+} or Fe^{2+} or, Mg^{2+} replaced by K^+) a deficiency of positive charge is created which is balanced by the attachment of cations from the geological environment of the clay. The cation is normally Na^+, K^+, Ca^{++} or Mg^{++}. These cations are located in the galleries of the particle. In sodium montmorillonite (Na-MMT), the layered silicate most commonly used in the polymer nanocomposites, Al^{3+} is replaced by Mg^{2+} and the charge is balanced by attachment of Na^+ ions (Fig. 12.12). The other layered silicates commonly used in the polymer nanocomposites are hectorite and saponite. Their chemical formulae are given in Table 12.6.

The negative charge on clay, before its neutralization by cations, is known as the *cation exchange capacity* (CEC) of the clay. This is also equal to the combined charge of the counterbalancing ions. It is expressed in *meq* (milliequivalent)g^{-1}. The negative charge attracts the positive ions, hence the name "cation exchange capacity". It ranges from 0.9 to 1.2 meq.g^{-1} in montmorillonites from different sources [280].

In the presence of water, the cations get hydrated and cause the galleries to expand. Hydrate formation, and consequent swelling, is easier for the monovalent alkali cations as compared to 2^+ and higher cations. For separation of the clay into platelets, this swelling is a desirable step. In the clays with a 2^+ or higher valent cations, an exchange step first needs to be carried out to exchange the cations with the monovalent cations. Hydration of the cations leads to the swelling of the clay and eventual separation of a clay particle into individual platelets. This process is called *exfoliation.* In water, the platelets can be fully exfoliated.

Example 12.4. *The composition of a clay has been determined to be given by the formula*

$$Na_{0.33}K_{0.01}Ca_{0.02}(Mg_{0.26}Fe_{0.23}Al_{1.51})(Al_{0.12}Si_{3.88})O_{10}(OH)_2$$

Find its cation exchange capacity.

Solution. From the formula, it can be seen that in the octahedral layers, the aluminum ions have been partly replaced by Mg and Fe ions while in the tetrahedral layers Si has been partly replaced by aluminum ions. The counterbalancing charges are provided by the sodium, potassium and calcium ions.

The total equivalents of the exchangeable cations (Na^+, K^+ and Ca^{2+}) are: 0.33 x 1 + 0.01 x 1 + 0.02 x 2 = 0.38

The formula weight of the clay is calculated to be: 374.9

The cation exchange capacity therefore is = 0.38/374.9 = 1.01 x 10^{-3} eq.g^{-1} or 1.01 meq.g^{-1}. □

The x-ray *d* spacing of a completely dry sodium MMT is 0.96 nm while the thickness of each platelet is 0.94 nm and the gallery height is

0.02 nm (Fig. 12.12) [272]. The lateral dimension of the platelets vary because the process of their formation involves growth from solution and the different platelets grow to a different extent. Figure 12.13 shows a carefully measured distribution of the quantity $\sqrt{A}/t$ in a commonly used Na-MMT where A is the platelet area and t is the platelet thickness [281]. Since t is approximately 1 nm, the figure implies that the most probable lateral dimension is in the range 100–200 nm.

The lateral dimension of the platelets is often erroneously quoted to be of the order of several microns. This may happen because in a tactoid, the platelets are not stacked with perfect registry but are staggered as shown schematically in Fig. 12.14. When the platelets in a tactoid are dispersed, many overlapping platelets may appear to be a single platelet in the microscope image unless care is taken to achieve a full dispersion.

To achieve the best properties in the clay–polymer nanocomposites, the full exfoliation of the clay platelets is required. As mentioned earlier, the clays having alkali ions as the exchange cations are highly hydrophilic and get fully exfoliated in aqueous solutions. Effective nanocomposites can, therefore, be easily formed with polymers which are water soluble such as polyvinyl alcohol, polyethylene oxides, latex,

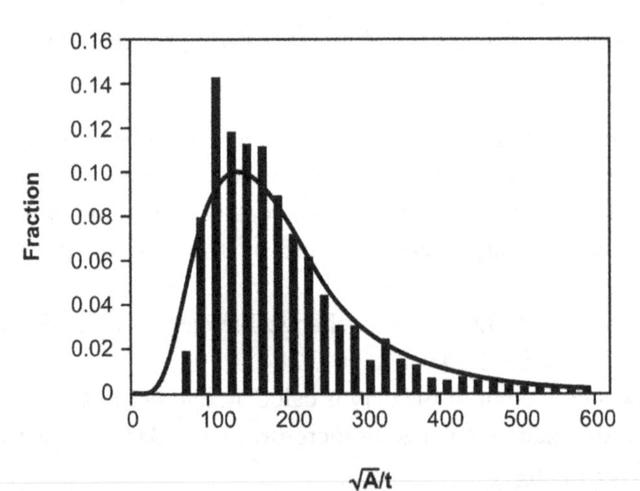

Fig. 12.13. Distribution of the aspect ratio as measured by the quantity $\sqrt{A}/t$ (where A is the platelet area and t is the platelet thickness) for native sodium montmorillonite platelets (Reprinted from [281, 282] with permission of the American Chemical Society).

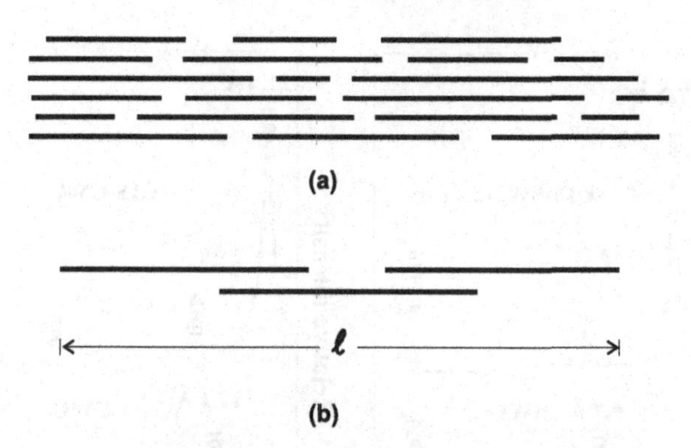

Fig. 12.14. (a) Schematic representation of the arrangement of platelets in a tactoid (b) several overlapping platelets may be confused for a single platelet of much larger lateral dimension.

polyvinylpyrrolidone, *etc.* Full exfoliation can also be achieved in systems based on polyamide, polyimide and epoxy resins because of the high polarity of these polymers.

In order to disperse the clay in the hydrophobic polymers, it is necessary to modify the clay so that it becomes hydrophobic. This is accomplished by treating the clay with cationic surfactants such as alkylammonium or alkylphosphonium derived from organic onium salts such as quaternary ammonium salts *e.g.* $RNH^{3+}Cl^{-}$ where R is an organic chain (see chapter 13). An example is octadecylammonium chloride $CH_3(CH_2)_{17}NH^{3+}Cl^{-}$. The cations in the galleries of the clay are replaced by the onium ion (*i.e.* $CH_3(CH_2)_{17}NH^{3+}$ in the case of octadecylammonium chloride). The exposed alkyl tail of the onium ion makes the clay hydrophobic and the height of the galleries is increased. The resulting product is called *organoclay*. Figure 12.15 shows schematically the x-ray scans of two clays before and after the treatment with a surfactant. The increase in the gallery spacing is indicated by the shift in the peaks to lower angles.

The organic tail can be modified to contain functional groups which would react with the polymer or act as initiator for polymerization of a prepolymer. The organophilic nature of the organoclay makes it possible for the polymer to diffuse into the galleries and eventually exfoliate the clay.

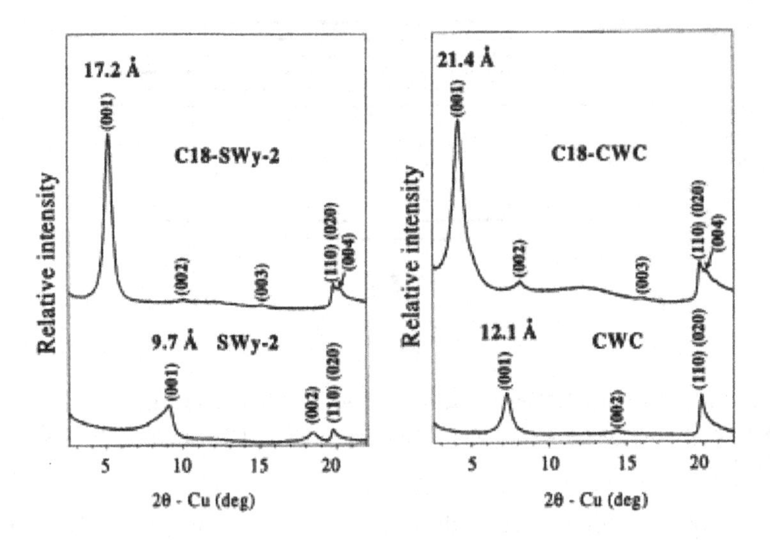

Fig. 12.15. XRD patterns of the SWy-2 (left) and the CWC (right) clays before (bottom) and after (top) alkylammonium treatment (reproduced from [283] copyright (2001) with permission from Elsevier).

12.4.3 *Preparation of the polymer–exfoliated clay nano-composites*

Several methods have been reported for the formation of polymer–layered silicate nanocomposites. An ideal composite structure is one in which complete separation (exfoliation) and dispersion of the clay platelets in the polymer matrix is achieved. The extent of this dispersion may vary depending on the affinity between the clay and the polymer and the process used. The morphologies achieved can be broadly classified into three categories: (i) immiscible (ii) intercalated and (iii) exfoliated. In the immiscible composites, the clay exists more or less as the original micron sized particles. This is the case for the filled polymers (conventional or the microcomposites). In the intercalated composites the polymer chains enter (intercalate) the galleries of the clay tactoids while in the exfoliated composites the platelets are completed separated and dispersed in the polymer matrix. These different cases are depicted in Fig. 12.16 [282].

The extent of polymer intercalation into the clay can be monitored by wide angle x-ray diffraction. The x-ray scans obtained from the three morphologies are also shown schematically in Fig. 12.16. In the case of the microcomposite, the x-ray scan is the same as that for the clay as there is no change in the *d* spacing of the clay. In case of the intercalated composite, the *d* spacing increases due to the entry of the polymer chains between the platelets while in the exfoliated clay the platelets are dispersed randomly and no x-ray peak is observed.

The four principal methods for the preparation of polymer–layered silicate composites are the following.

(1) In situ template synthesis [284]: In this method the clay particles are synthesized using the sol gel method in a water solution of the polymer. This technique has been successful with composites of hectorite type clay minerals.

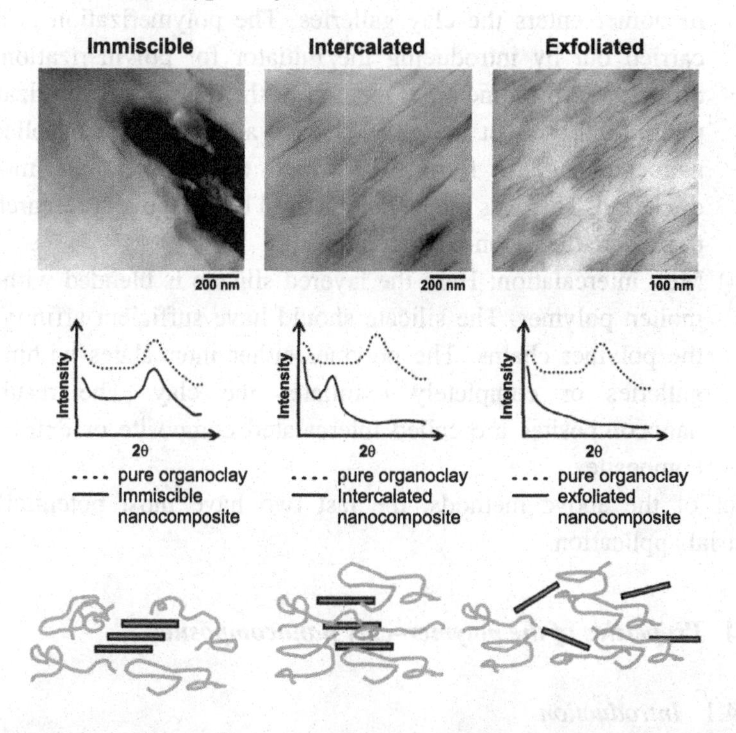

Fig.12.16. Schematic representation of the relative arrangement of the clay platelets and the polymer chains in the immiscible, intercalated and the exfoliated composites together with the TEM micrographs and the schematics of the wide angle x-ray scans; reprinted from [282] copyright (2008) with permission from Elsevier).

(2) Intercalation of polymer from solution: Intercalation refers to insertion of the polymer chains in the gallery spaces of the clay. In this method, the clay is first exfoliated in a solvent in which the polymer is also soluble. Then the polymer is added to the solution. Intercalation of the polymer chains within the clay layers takes place. The solvent is then removed. The sheets reassemble with polymer chains between them forming the nanocomposite.
The major problem with this method is the requirement of handling large volumes of solvent.
(3) In situ intercalation polymerization: In this technique, the monomer solution is added to the clay. The clay swells and the monomer enters the clay galleries. The polymerization is then carried out by introducing the initiator for polymerization by diffusion or by some other method. If the rates of polymerization within and without the gallery space are properly controlled, a nanocomposite is formed in which the platelets are mostly dispersed. This was the technique used by the Toyota researchers to prepare the nylon-6 nanocomposite.
(4) Melt intercalation: Here the layered silicate is blended with the molten polymer. The silicate should have sufficient affinity for the polymer chains. The polymer either intercalates within the galleries or completely exfoliates the clay. The resulting nanocomposites are called intercalated composite or exfoliated composite.

Out of the above methods, the last two have most potential for industrial application.

12.4.4 *Properties of the polymer–clay nanocomposites*

12.4.4.1 *Introduction*

The polymer–clay nanocomposites show a remarkable enhancement in several properties as compared to the matrix polymer. Even more remarkably, such enhancements are displayed at a very low clay content which is a fraction of what is needed to achieve the same properties in the filled polymers. This results in considerable weight saving and also helps to maintain the transparency and toughness which degrade at a high loading as used in the filled polymers.

The major interest in using polymer–clay nanocomposites is the enhancement in the mechanical properties. However, several other properties also show remarkable changes. Properties such as flame resistance, reduced permeability to moisture or other undesirable agents (barrier properties), ability to blend two difficult to blend polymers, *etc.* are also significantly increased in the polymer–clay nanocomposites. In the following we focus on the mechanical and barrier properties of such composites.

12.4.4.2 *Mechanical properties*

The introduction of exfoliated clay into a polymer is found to result in a very significant increase in the modulus with only a few vol. % of the

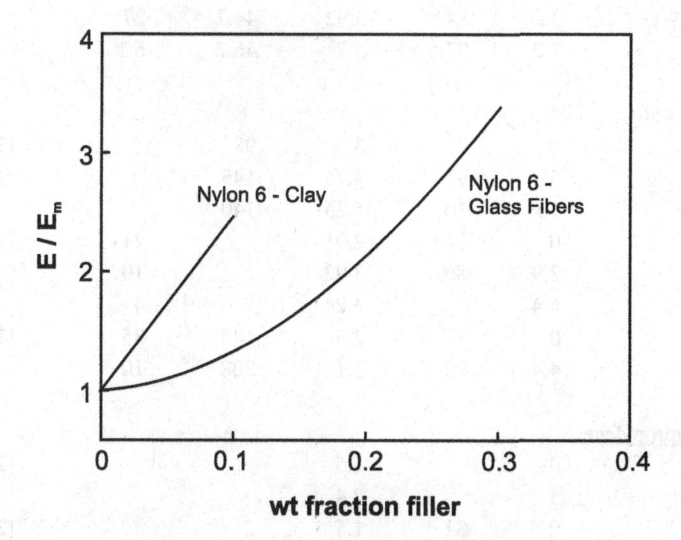

Fig. 12.17. Changes in the modulus of a nanocomposite and a filled polymer with nylon-6 as the matrix; the reinforcements are organically modified montmorillonite in the nanocomposite and glass fibers in the filled polymer [272].

Table 12.7: Mechanical properties of some polymer–clay nanocomposites ([284, 285]).

Polymer	Clay wt%	Tensile strength (MPa)	Tensile modulus (GPa)	Impact strength (J/m)	Elongation to break %	Ref
(i) Nylon 6						
(a)	0	69	1.1	6.2		[286]
	4.2	107	2.1	6.1		
(b)	0	68.6	1.11			[287]
	4.7	97.2	1.87			
	5.3	97.3	2.04			
(c)	0	68	1.08	50	75	[288]
	1.6	82	1.38	44		
	1.8	82	1.41	42		
(d) Low Mol. Wt.	0	69.1	2.82	36	232	[289]
	3.2	78.9	3.65	32.3	12	
	6.4	83.6	4.92	32	2.4	
(e) Med. Mol. Wt.	0	70.2	2.71	39.3	269	
	3.1	85.6	3.66	38.3	81	
	7.1	95.2	5.61	39.3	2.5	
(f) High Mol. Wt.	0	69.7	2.75	43.9	129	
	3.2	84.9	3.92	44.7	27	
	7.2	97.6	5.7	46.2	6.1	
(ii) Nylon 66						
(a)	0	77	3	98		[290]
	5	97	4.75	145		
	10	107	5.25	140		
(b)	0	72.6	2.91		211	[291]
	2.9	80.4	3.92		10	
	4.4		4.24		4	
(c)	0	79	2.5	145	35	[292]
	4.4	90	3.1	208	10	
(iii) Polypropylene						
(a)	0		1.5			[293]
	5		2.4			
(b)	0	31	1.5	2		[294]
	3	39	2.1	3.4		

(Contd.)

Table 12.7 (Contd.)

(c)	0	37.9			1350	[295]
	2	39.4			36	
(iv) Polyethylene						
(a) HDPE	0		1.02			[296]
	0.9		1.06			
	1.8		1.25			
	2.8		1.38			
	4.0		1.36			
(b)	0		0.183			[297]
	3.5		0.258			
(c)	0	28.9	0.26		945	[298]
	1.5	30.2	0.31		860	
(d)	0	26.1	0.8			[299]
	5.4	33.1	1.67			
(e)	0		0.19		>400	[300]
	4.6		0.48		>400	
(f)	0	21.1	0.649		773	[301]
	6	27.4	0.753		168	
(g)	0	22				[302]
	5	25				
	10	27				
	15	28				
(v) Poly(ethylene terephthalate) (PET)						
(a)	0	46	2.21		3	[303]
	3	71	4.1		3	
(b)	0	32	2.55		370	[304]
	3.5	47	3.35		<5%	
(c)	0	49	1.3			[305]
	2	64	1.7			
(d)	0	5.9	142			
	5	6.2				
	10	6.5				
	21.5	8.3				

Contd.

Table 12.7 (Contd.)

(vi) Poly(butylenes terephthalate) (PBT)				
(a)	0	41	1.37	[306]
	3	60	1.76	
	5	40	1.86	
(b)	0	1.16		[307]
	3	1.25		
	5	1.35		
(vii) Polyurethane (PU)				
(a) soft PU	0	45		[308]
	3	31		
	7	21		
(b) Hard PU	0	58		[308]
	3	44		
	7	34		

clay. Figure 12.17 compares the moduli of nylon-6 composites made with nanoclay and with short glass fibers [272]. The improvement can be by as much as a factor of ten in case of rubbers and low modulus polymers as illustrated by Fig 12.18. The rate of increase in the modulus is steeper in case of the nanocomposite; consequently much smaller amounts of the reinforcement are needed as compared to the microcomposites or the filled polymers.

A summary of some of the experimental data on the polymer–clay nanocomposites is given in Table 12.7.

Attempts have been made to fit the data on the polymer–clay nanocomposites to the Halpin–Tsai equations. The agreement of the Halpin–Tsai equations or the Mori–Tanaka theory with the experiment is, in general, not very good. This is due to several factors such as (i) the theories assume the filler particles to be all oriented in the same direction which is not the case in the experiments (ii) the particles have a distribution of aspect ratios (iii) the bonding between the filler and the matrix is not perfect as assumed in the theories (iv) the filler does not have the same modulus in all directions (v) in case of a polymer matrix, the modulus is not isotropic due to polymer chain orientations and the presence of crystalline regions in the polymer (vi) the particles may be

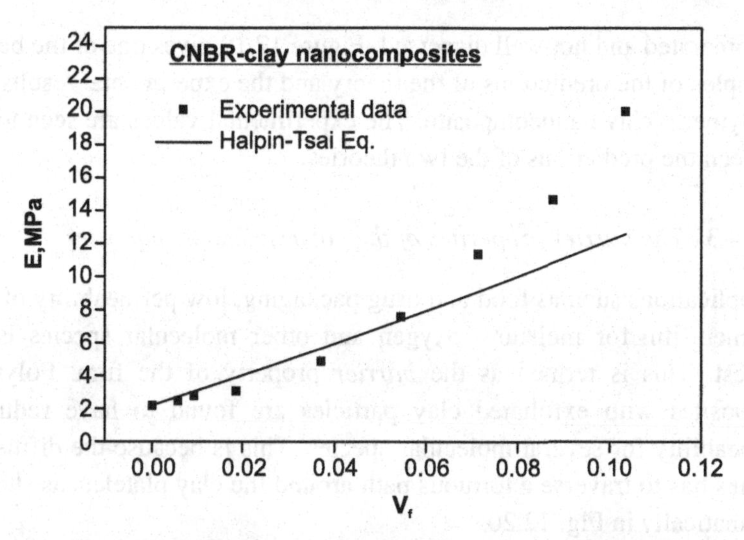

Fig. 12.18. Change in the modulus of a rubber (CNBR)–clay nanocomposite with vol % of clay. The line gives the predictions of the Halpin–Tsai equation (adapted from [309] copyright (2004) with permission from Elsevier).

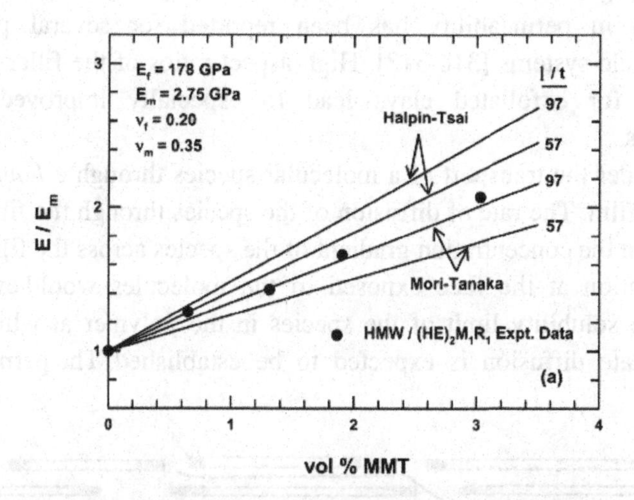

Fig. 12.19. Comparison of the relative modulus, E/E_m ,with the predictions of the Halpin –Tsai equations and the Mori–Tanaka theory for a high molecular weight (HMW) nylon-6 (average number molecular weight = 29,300)– montmorillonite organoclay composite; the lines are drawn for two values of the aspect ratio, 97 for complete exfoliation and 57 determined using a statistical image analysis procedure; 0.2 and 0.35 are the Poisson's ratios for the reinforcement and the matrix phase (reproduced from [272] copyright (2003) with permission from Elsevier).

agglomerated and not well dispersed. Figure 12.19 gives one of the better examples of the predictions of the theory and the experimental results for a polymer – clay nanocomposite. The experimental values are seen to lie between the predictions of the two theories.

12.4.4.3 *The barrier properties of the polymer nanocomposites*

In applications such as food and drug packaging, low permeability of the polymer film for moisture, oxygen and other molecular species is of interest. This is termed as the *barrier* property of the film. Polymer composites with exfoliated clay particles are found to have reduced permeability for several molecular species. This is because the diffusing species has to traverse a tortuous path around the clay platelets as shown schematically in Fig. 12.20.

Figure 12.21 shows the effect on the permeability of various gases in a butyl rubber/vermiculite nanocomposite as a function of vermiculite content. Significant reduction in the permeability can be seen. Similar reduction in permeability has been reported for several polymer–nanoparticle systems [310-313]. High aspect ratios of the filler particles such as for exfoliated clays lead to especially improved barrier properties.

Consider the transport of a molecular species through a *homogenous* polymer film. The rate of diffusion of the species through the film would depend on the concentration gradient of the species across the film. The concentration at the face exposed to the molecules would eventually reach the solubility limit of the species in the polymer at which point steady state diffusion is expected to be established. The permeability

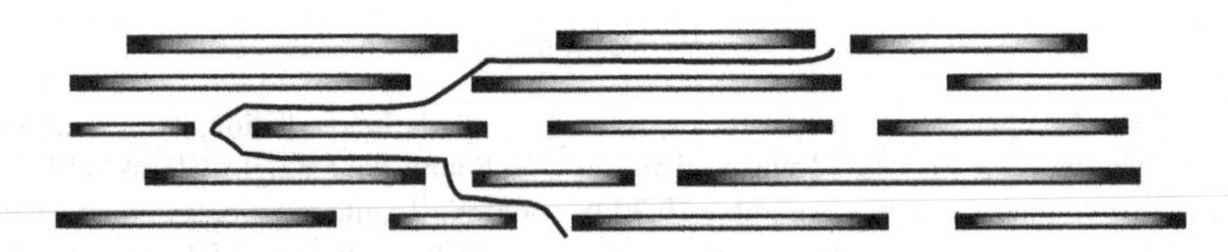

Fig. 12.20. The presence of nanoparticles in the polymer matrix imposes a tortuous path on the diffusing species.

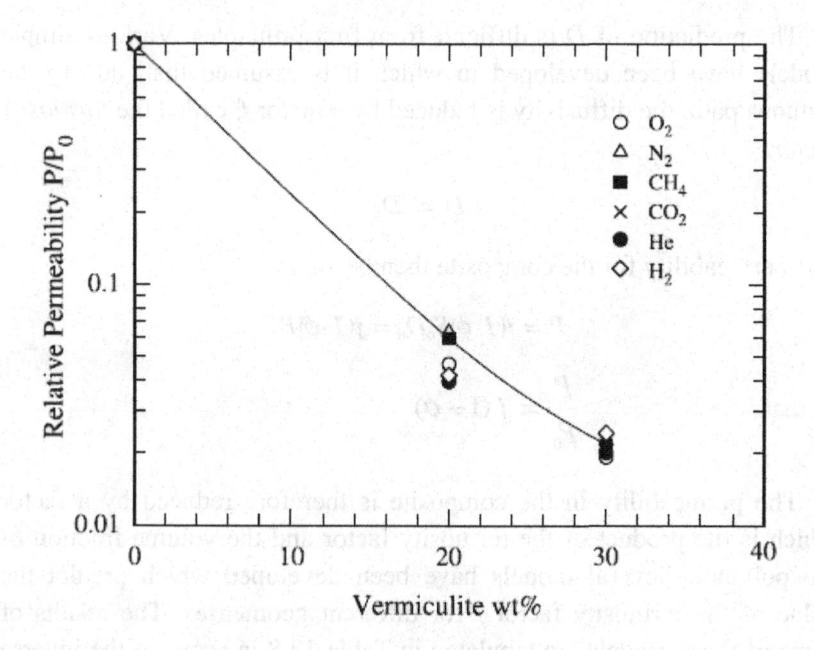

Fig. 12.21. Relative permeability of various gases *vs.* vermiculite content in a butyl rubber/vermiculite nanocomposites at 30 °C and 2 atm. pressure (reprinted from [314] copyright (2006) with permission from Elsevier).

coefficient P_0 is defined as the product of S_o, the solubility of the penetrant in the polymer and its diffusivity, D_o

$$P_o = S_o D_o$$

If it is assumed that the presence of the particles does not affect the matrix (*i.e.* there are no nano effects), and the filler does not absorb the penetrant, then the effective solubility for the composite can be taken as $S_o(1-\phi)$, where ϕ is the volume fraction of the particles. The permeability for the composite is then given by

$$P = S_o\,(1-\phi)D$$

where D is the diffusivity in the composite.

The prediction of D is difficult from first principles. Various simple models have been developed in which it is assumed that, due to the tortuous path, the diffusivity is reduced by a factor f, called the *tortuosity factor*,

$$D = fD_o$$

The permeability for the composite then becomes

$$P = f(1-\phi)S_oD_o = f(1-\phi)P_o$$

so that
$$\frac{P}{P_0} = f(1-\phi)$$

The permeability in the composite is therefore reduced by a factor which is the product of the tortuosity factor and the volume fraction of the polymer. Several models have been developed which predict the value of the tortuosity factor f for different geometries. The results of some of these models are tabulated in Table 12.8 in terms of the inverse of the tortuosity factor f, given by

$$f^{-1} = (\frac{P_o}{P})(1-\varphi)$$

Table 9.8: Models for predicting the barrier properties of platelet filled nanocomposites [282]

(i) Ribbon (w = width, t = thickness, a = aspect ratio = w/t)

(a) Nielsen model [315] : $(P_o/P)(1-\phi) = 1 + a\phi/2$
(b) Cussler (regular array) [316] : $(P_o/P)(1-\phi) = 1 + (a\phi/2)^2$
(c) Cussler (random array) [316] : $(P_o/P)(1-\phi) = (1 + (a\phi)/3)^2$

(ii) Disk (d = diameter, t = thickness, $a = d/t$)

(a) Gusev and Lusti [317] : $(P_o/P)(1-\phi) = \exp(a\phi/3.47)^{0.71}$
(b) Fredrickson and Bicerano[318] : $(P_o/P)(1-\phi) = (1+\chi + 0.1245\chi^2)/(2+\chi)^2$
where $\chi = a\phi/2\ln(a/2)$
(c) Bhardwaj [319] : : $(P_o/P)(1-\phi) = 1 + 0.667(S+0.5)$
where S = orientation factor with $1/2 \leq S \leq 1$)

However, the models are at present not sufficiently well developed for making predictions with any significant level of accuracy. A major difficulty is that usually the precise value of the aspect ratio is not known [314]. Even for model systems with reasonably well known aspect ratios, the predictions of the different models vary widely.

12.4.5 *The nanoeffects*

The term "nano effect" here refers to a new synergistic effect arising in the polymer nanocomposite due to the small size of the particles which is not expected on the basis of the macroscopic theories. Such effects can be expected to arise when the size of the particle becomes comparable to some characteristic dimension of the matrix phase, such as the radius of gyration of the polymer chain.

Two effects which can be attributed to the incorporation of nanoparticles in a polymer matrix are a change in the glass transition temperature (T_g) and a change in the crystallization behavior of the polymer. It is found that the *Tg* of a polymer can either be raised or suppressed by the addition of nanoparticles depending on the type of polymer. The increase in the T_g results due to a reduced chain mobility due to the presence of the nanoparticles while a decrease can occur due to different reasons. In PMMA loaded with silica nanoparticles, a decrease in the T_g is observed and is attributed to the poor wetting of silica particles by PMMA. This causes an increase in the free volume which leads to a decrease in the T_g.

Another effect which is known to occur in the composites is the change in the crystallization behavior of the polymer. The nanoparticles can act either as nucleation centers for crystallization or, at higher loading, hinder the crystallization by reducing the chain mobility.

The changes in the T_g are small, generally less than 10° C. If the change in the T_g due to the nanoparticles results in a change in a property such as the modulus, then the effect can justifiably be called a nano effect. Same is true for the changes in properties due to changed crystallinity of the polymer. However, these changes, if present, are smaller than the uncertainty associated with the predictions of the microcomposite theories. While discussing the mechanical properties of the polymer nanocomposites we saw that the experimental results fall in the ranges predicted by the composite theories such as the Halpin–Tsai equations or the Mori–Tanaka theory. Due to the several assumptions in the theories which are not usually true in practice, it is not surprising that the agreement between the experimental results and the theory is not very precise. Nevertheless the results obtained are on the lines predicted by the composite theories and there is no need to invoke any so called "nano effect". However, in several other areas, nano effects have to be invoked to explain the observed properties such as changes in the aging behavior of the thin films of polymers and increase in the free volume in a polymer by the introduction of nanoparticles due to disruption of packing of rigid chain polymers which causes an increase in the permeability [320].

Problems

(1) Find out the volume fraction of the MWNT of diameter 10 nm at which each tube will be separated from the other on the average by (i) 20 nm (ii) 5 nm. On the basis of your results, in which case, SWNT or MWNT, would it be easier to achieve high volume fractions of CNT in the composite.

(2) Show that the Halpin–Tsai equation reduces to the rule of mixtures for the longitudinal modulus of a unidirectional fiber composite when $\zeta = \infty$ and to the transverse modulus of such a composite when $\zeta = 0$.

(3) (a) Calculate the longitudinal modulus predicted by the Halpin–Tsai equations for a rubber– clay nanocomposite containing 2, 4, 6 and 8 vol % of clay. Given: modulus of clay = 170 GPa, modulus of rubber = 0.64 MPa, aspect ratio of clay platelets = 26

(b) The measured modulus values of the above composite for clay volume fractions of 2, 4, 6 and 8 % are 1.05, 1.55, 1.75 and 2.08 MPa respectively. Compare the values predicted in (a) with the measured values and suggest the reasons for the discrepancy.

(4) Show that the critical length of a nanotube in a composite for the stress to reach the fracture strength of a nanotube is given by

$$l_c = \frac{\sigma_f d_o}{2\tau_{\max}}\left[1 - \frac{d_i^2}{d_o^2}\right]$$

where d_i and d_o are the inner and the outer diameters of the tube respectively, σ_f is the strength of the tube and τ_{max} is the maximum interfacial shear stress at fracture.

(5) Calculate the critical length l_c for maximum stress transfer for the composites prepared using (i) SWNT of fracture strength σ_f = 50 GPa and outer diameter 1 nm and inner diameter 0.6 nm (ii) arc MWNT of fracture strength σ_f = 50 GPa, outer diameter 2 nm and inner diameter 0.7 nm and (iii) CVD MWNT of fracture strength σ_f = 10 GPa, outer diameter 5 nm and inner diameter 1.8 nm. Assume interface shear strength of 50 MPa in each case.

(6) In Fig.12.13 the most probable value of the quantity $\sqrt{A/t}$ is about 150. Calculate the specific surface area of the platelets (in m^2g^{-1}). Take the thickness of a platelet to be 1 nm and the density of montmorillonite as 2.83 gcm^{-3}. Compare your result with the specific surface area for (i) carbon black (ii) carbon nanotubes (c) a typical white filler for filled composites (*e.g.* kaolinite clay).

(7) The composition of a clay is given by: $Na_{0.51}K_{0.03}Ca_{0.03}(Mg_{0.36}Fe_{0.14}Al_{1.46})(Al_{0.13}Si_{3.87})O_{10}(OH)_2$

Show that the cation exchange capacity of this clay is 1.59 meq/gm.

(8) The quantity $\dfrac{dE_C}{dV_f}$ for low volume fractions of the reinforcing phase is a measure of the efficiency of the reinforcement. Calculate this quantity for a CNT–HDPE (high density polyethylene) composite assuming that the composite follows the Halpin–Tsai equation as modified for randomly oriented fibers for the following volume fractions of the CNT : 0.1 %, 0.2 %, 0.3 % and 0.4 %. Take the Young's modulus of the CNT and the HDPE to be 1000 and 1 GPa respectively and the aspect ratio of the CNT as 100.

Chapter 13

Molecules for Nanotechnology: Polymers, Biopolymers, Dendrimers and Surfactants

In this chapter some important classes of molecules widely used in nanotechnology are briefly discussed. These are polymers and biopolymers, surfactants and dendrimers. Most of the material may be already familiar to many of the readers. Nonetheless it is thought that it should be useful to have this information readily available for refreshing the reader's memory as well as to serve as a reference.

13.1 Polymers

There are numerous applications in nanotechnology where polymers play an important role. Some examples are (i) use of polymers as key matrix materials for lithographic resists in micro and nanotechnology and in stamps as well as substrates and inks for micro contact printing (ii) exploitation of the self organization of diblock copolymers to prepare novel nanostructures (iii) self assembly of π conjugated polymers to form conducting nanowires (iv) in drug delivery and as implants and scaffolds for tissue engineering (v) for steric stabilization of nanoparticle suspensions (vi) coatings to provide protection and functionality (vii) nanoparticle–polymer composites (viii) novel applications of polymer nanoparticles.

The application of polymers in nanotechnology has expanded rapidly and a new field of *polymer nanotechnology* is emerging. In this chapter we provide a brief introduction to some essential aspects of the polymers.

13.1.1 *Some common polymers*

Polymers are formed by linking together a large number of "monomers". The repeat unit of a polymer is either a monomer or the monomer with a few atoms missing from it which separate out as small molecules (*e.g.* H_2O) during the polymerization process. In the former case the polymerization is said to be *addition polymerization* or *chain polymerization* while in the latter case it is called *condensation* or *stepwise* polymerization. Examples of some common polymers synthesized by the two techniques are given in Table 13. 1.

13.1.2 *Properties of polymers*

13.1.2.1 *Molecular weight*

Since the polymerization reactions yield chains having a distribution of lengths, a polymer has to be characterized by an average molecular weight (molar mass). The kind of the average molecular weight obtained depends on the experimental method employed. Thus the number average molecular weight $\bar{M}_n$ is obtained by the osmotic pressure method.

$$\bar{M}_n = \frac{\sum (N_i m_i) m_i}{\sum N_i m_i}$$

Here N_i is the number of mers in the ith chain and m_i is the molecular weight of each mer. Modern gel permeation chromatographs directly give the distribution of molecular weights in the polymer.

Most of the synthetic polymers have a backbone made of C-C bonds. The carbon atoms in the polymers are hybridized. A saturated carbon atom with four hydrogen atoms or other substituents forms a tetrahedron with an H–C–H angle of 109° 28′. In a chain, the C–C bond can therefore assume any of the three possible sp^3 orientations. Furthermore, in a molten polymer or a polymer solution, bond rotation angles can have additional small variations due to thermal oscillations. The result is that a

Table 13.1. Examples of polymers

(a) Polymers synthesized using chain polymerization

Monomer		*Polymer*	
Ethylene	$CH_2=CH_2$	Polyethylene	$-[CH_2-CH_2]_n-$
Propylene	$CH(CH_3)=CH_2$	Polypropylene	$-[CH(CH_3)-CH_2]_n-$
Vinylchoride	$CH(Cl)=CH_2$	Polyvinylchloride	$-[CH(Cl)-CH_2]_n-$
Styrene	$CH(C_6H_5)=CH_2$	Polystyrene	$-[CH(C_6H_5)-CH_2]_n-$
Methyl-methacrylate	$C(CH_3)(COOCH_3)=CH_2$	Polymethyl methacrylate	$-[C(CH_3)(COOCH_3)-CH_2]_n-$
Acrylic acid	$CH(COOH)=CH_2$	Polyacrylic acid	$-[CH(COOH)-CH]_n-$
Tetra-fluoroethylene	$CF_2=CF_2$	Polytetra-fluoroethylene	$-[CF_2-CF_2]_n-$

Contd.

(b) Polymers synthesized using stepwise polymerization

Inter-unit linkage	Polymer	
$-\overset{\overset{\displaystyle O}{\|\|}}{C}-C$	$HO\left[(CH_2)_m COO\right]_n H$	Polyester
$-\overset{\overset{\displaystyle O}{\|\|}}{C}-NH-$	$H\left[NH(CH_2)_m CO\right]_n OH$	Polyamide
$-Si(R)_2-O-Si(R)_2-$	$HO\left[Si(CH_3)_2-O\right]_y H$	Polysiloxane

long polymer chain is randomly coiled with large number of possible conformations.

13.1.2.2 *Length of a polymer chain*

It is important to have an estimate of the actual volume that a polymer chain occupies. If a polymer chain is stretched out from end to end, its length is called the *contour length*. For a chain with n bonds of length l (Fig 13.1) the contour length is simply nl. However, as mentioned above,

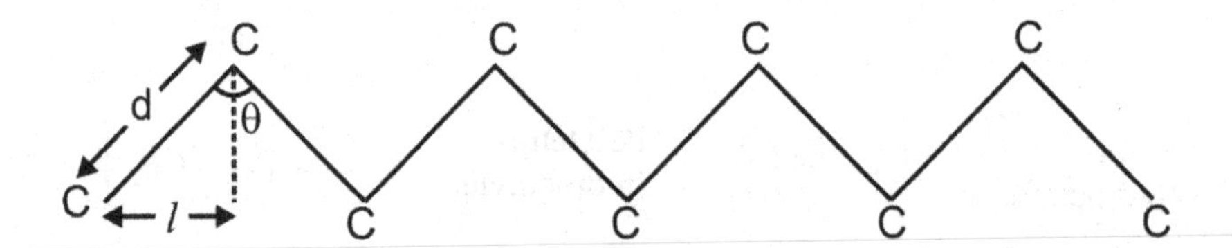

Fig. 13.1. The contour length of a chain with n bonds of length l each is the stretched out length nl.

the polymer chains are actually coiled, the extent of coiling depending on the temperature or , in case of a solution, the quality of the solvent. There are several models to estimate the polymer chain length.

In the *freely jointed chain model* the chain is assumed to consist of volume-less links which can rotate freely in space. This is equivalent to a random walk problem in which each link is equivalent to a step in the random walk. The "mean square end to end distance" of the chain in this case is given by

$$\langle r^2 \rangle = nl^2$$

Another measure of the length of a polymer chain is the *root mean square (r.m.s.) radius of gyration,* $\langle R_g^2 \rangle^{1/2}$ which is a measure of average distance of chain segment from the center of mass of the coil. For a freely jointed chain, the radius of gyration for a linear polymer is given by:

$$\langle R_g^2 \rangle^{1/2} = \left[\frac{<r^2>}{6} \right]^{1/2}$$

For extended (branched) chain, $R_g^2 \leq \dfrac{r^2}{6}$

The freely jointed chain model does not take into account several restrictions to completely free rotation. Thus the bond angle θ can not vary continuously but is fixed. There are other restrictions to free rotation due to steric hindrance or potential energy barriers. Effect of these short range interactions (*i.e.* a fixed bond angle, steric restrictions and potential energy barriers) is to expand the end to end mean square chain length over that arrived at from the freely jointed chain model. The actual chain length in the absence of long range interaction (to be discussed below) is termed as *unperturbed chain length* r_0. The ratio $< r_0^2 >$ to the mean square end to end distance, nl^2, is called the *characteristic ratio,* κ. Its values range between 5 and 10.

In the freely jointed chain model, volume–less links are assumed and two or more segments of the chain can "overlap". However, in real chains, each segment exists within a volume from which the other segments are excluded. Practically this "excluded volume" effect also leads to a further expansion of the chain by a factor α due to this long range interaction *i.e.* $<r^2>^{1/2} = \alpha < r_o^2 >^{1/2} = \kappa^{1/2}.\alpha <nl^2>^{1/2}$. The value of α is large in a good solvent. A solvent for which $\alpha = 1$ at a given temperature is called the *theta* (θ) solvent and the corresponding temperature is called the *Flory temperature*, θ. At θ temperature, the excluded volume is zero.

Example 13.1. *Polystyrene of molecular weight 468,000 is dissolved in toluene. Calculate (a) the contour length (b) the random flight end to end distance (c) the random flight radius of gyration (d) unperturbed end to end distance and (e) the actual end to end distance. Given: C-C bond length = 0.154 nm, C-C bond angle = 109.5°, chain expansion factor* α *= 2.8; characteristic ratio = 8.*

Solution. Molecular weight of the monomeric unit –[CH_2 – CHC_6H_5]- is 104.

Number of monomeric units in the polymer = 468,000/104 = 4500

There are two C–C bonds along the chain length in each monomeric unit so that n = 4500 x 2 = 9000.

Referring to Fig. 13.1, l = $d \sin\theta/2$ = 0.1257 nm

(i) Contour length = nl = 0.1257 x 9000 = 1131.3 nm

(ii) The random flight end to end distance, r, is given by $\sqrt{<r^2>} = \sqrt{(nl^2)}$ = 11.92 nm

(iii) Random flight radius of gyration is given by $\sqrt{<(R_g^2)>} = \sqrt{(<r^2>/6)}$ = $11.92/\sqrt{6}$ = 4.87 nm

(iv) Unperturbed end to end distance, r_0, is given by $<r_0^2> = 8\ nl^2$ = 8 x 9000 x $(0.1257)^2$ = 1137.6 nm so that $\sqrt{<r_0^2>}$ = 33.7 nm

(v) Actual end to end distance = α x $\sqrt{<r_0^2>}$ = 2.8 x 33.7 = 94.4 nm □

13.1.2.3 *Bonding within and between polymer chains: thermoplastics, thermosets and elastometers*

How are the polymers chain held together in solid polymers? We know that the C–C bond along the chain is a strong covalent bond. However, the bonding of a polymer chain to chains surrounding it can be by weak bonds (Van der Waal's bond or a hydrogen bond) or through strong covalent bonds (Fig. 13.2). The strength of the C–C bond and some other primary bonds as well as the van der Waal's bond and the hydrogen bond are given in Table 13.2.

The weak bonds such as the Van der Waals are always present. Below the glass transition temperature *Tg*, the thermal energy is not sufficient to disrupt them and the polymer behaves as a solid with high elastic modulus. The polymers having weak bonds between the chains are called thermoplastics. Examples of thermoplastics are nylon 6-6, vinyl polymers, acrylics, styrenes, polyolefins, polycarbonates, polyamides, *etc.* The thermoplastics can be brought into a viscous liquid form by heating to a temperature high enough such that the weak bonds between the chains become ineffective.

In some polymers, primary bonds between the chains are deliberately formed by subjecting the polymer to a "curing" operation (Fig 13.2 b). Such polymers are called *thermosets.* Thus reacting phenol with formaldehyde at a high temperature in the presence of a catalyst results in the formation of phenolic resin. Strong primary bonds, called cross links, are next formed (after the resin has been molded in the desired shape) by reacting with additional formaldehyde at high temperature in

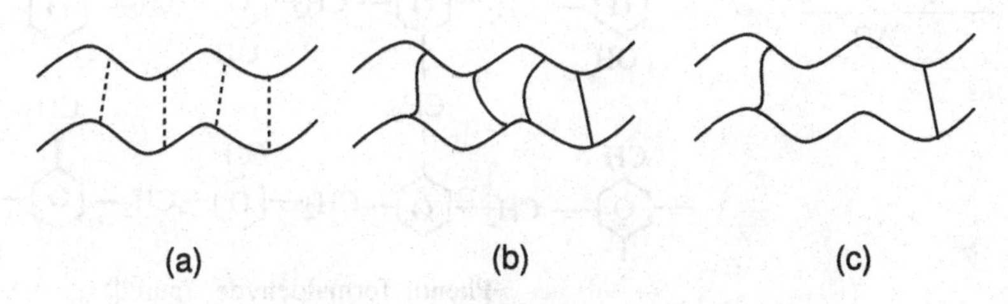

Fig. 13.2. Bonding between polymer chains can be by (a) weak bonds shown by dotted lines or (b) and (c) by strong primary bonds shown by solid lines; in a rubber the number of bonds per unit chain length is small (c).

Table 13.2. Bond strengths.

Primary bond	Dissociation energy kcal/ mole
C–C	83
C=C	146
C–H	99
C–N	73
C–O	80
C=O	179
C≡N	213
C–Cl	81
Hydrogen bond	1–3
van der Waals bond	0.5–1

the presence of a second catalyst (Fig. 13.3). The product is called phenol formaldehyde or bakelite. It is strong and has electrical and thermal resistance and finds wide applications in electrical fixtures. Other examples of thermosets are epoxy resins and rubbers. In *rubbers* or *elastometers*, the number of cross links per unit length (cross link

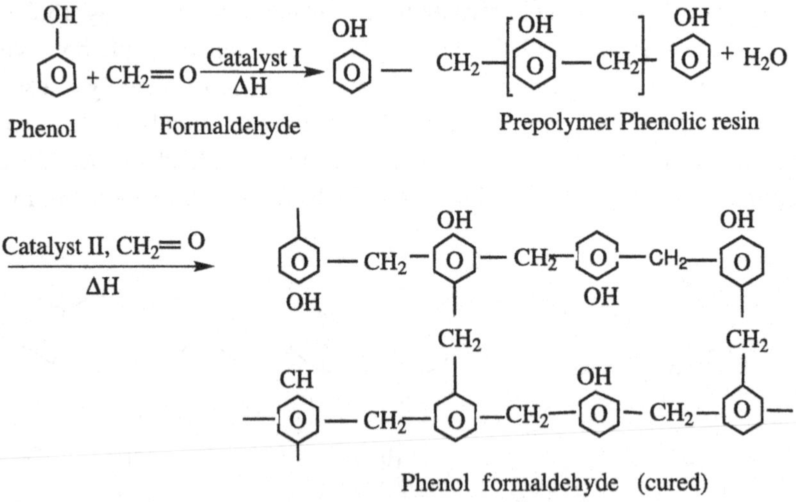

Phenol + $CH_2{=}O$ —Catalyst I, ΔH→ Prepolymer Phenolic resin + H_2O

Phenol Formaldehyde Prepolymer Phenolic resin

—Catalyst II, $CH_2{=}O$, ΔH→ Phenol formaldehyde (cured)

Fig.13.3. Formation of phenol formaldehyde.

density) is kept low so that the chains can uncoil between the cross links (Fig. 13.2 c).

13.1.2.4 *Glass transition temperature*

On cooling from a molten state some polymers form an amorphous solid while others form mixtures of amorphous and crystalline solids. The extent of the crystalline fraction depends on the polymer and the rate of cooling. The following discussion is applicable to the fully amorphous polymers. The crystallinity of polymers is dealt with later.

Consider a polymer in the molten state. If a shear stress is applied to a melt of polymer, the polymer chains are able to uncoil and move and slide past each other; the polymer thereby undergoes plastic (permanent) deformation or viscous flow.

At slightly lower temperatures, the ability of the chains to slide past each other decreases but the chains can still uncoil rapidly (in ~ 10^{-6} s). The uncoiling of the chain produces large elongations (Fig 13.4) and the polymer behaves as a rubber. Earlier we had mentioned that in rubbers the chains are cross linked, the cross link density being small. In amorphous polymers without cross links the chain "entanglements" act as temporary cross links so that the polymer displays rubber like behavior in a certain temperature range.

At very low temperatures the chain uncoiling time is very large (~10^{10} s) so that no deformation due to chain uncoiling is seen; the only deformation is due to stretching of the polymer (*i.e.* stretching of the C–C bonds). In this temperature range, the polymer behaves as a rigid glass with a very high modulus.

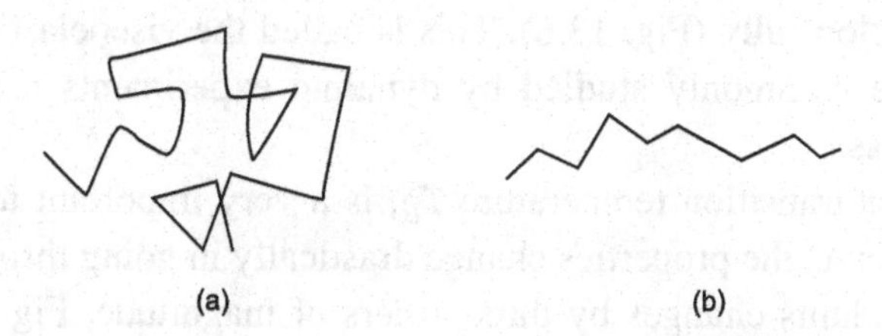

Fig.13.4 (a) An unstressed coil (b) bonds in the chain tend to align along the direction of stress upon loading.

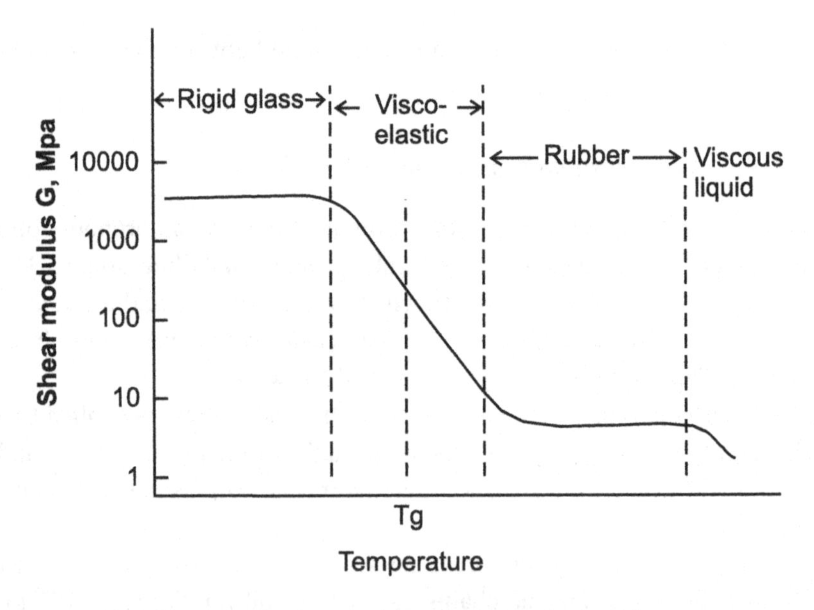

Fig. 13.5. Change in the shear modulus of a thermoplastic with temperature.

The response of an amorphous polymer to an external shear stress over a wide range of temperature is shown in Fig. 13.5. The shear modulus typically varies from 7 MPa in the *rubbery* region to 7 GPa at low temperatures where the polymer is a rigid glass. The point of inflexion in the curve corresponds to *Tg*, the glass transition temperature. The *Tg* for polymers varies from -123 °C for some rubbers to 100 °C for polystyrene and 300 °C or more for the highly cross linked polymers. The uncoiling time of the chains around the *Tg* is of the order of 10 s and the polymer shows a viscoelastic behavior where the strain takes some time to develop fully (Fig. 13.6). This is called the viscoelastic response and is more commonly studied by dynamic experiments using a time varying stress.

The glass transition temperature *Tg*, is a very important temperature for a polymer as the properties change drastically in going through the *Tg* (*e.g.* the modulus changes by three orders of magnitude, Fig 13.5). The glass transition temperature is most conveniently determined using a differential scanning calorimeter (DSC).

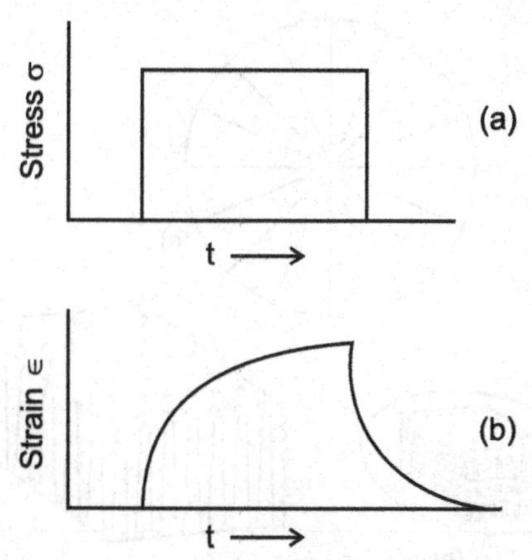

Fig. 13.6. Response of an amorphous polymer to a stress pulse in the viscoelastic region (a) the stress pulse (b) the corresponding strain response. The small instantaneous response during loading and unloading is due to bond stretching while the slow response is due to chain uncoiling.

All those factors which increase the flexibility of polymer chain decrease the *Tg*. These include: less bulky side groups, less compact arrangement of atoms in the side groups, side groups placed symmetrically on the chain, low polarity, less crystalline fraction, *etc.* Compounds called *plasticizers* are sometimes added to polymers to reduce the glass transition temperature.

13.1.3 *Crystalline polymers*

X-ray diffraction and other evidence shows that many polymers are partly crystalline, having both crystalline and amorphous fractions. The x–ray pattern typically consists of sharp peaks from the crystalline regions superimposed on a broad peak from the amorphous regions. A sharp peak in the DSC curve due to melting of the crystalline region is also a sign of crystallinity in the polymers.

Polymer fibers tend to have a high crystalline fraction because during the drawing process the chains get extended and can easily organize into a crystalline structure. Single crystals of polymers, about 10 nm thick and few microns long can be grown from very dilute solutions. The polymer chain in these crystals lies in the thickness direction and repeatedly folds upon itself.

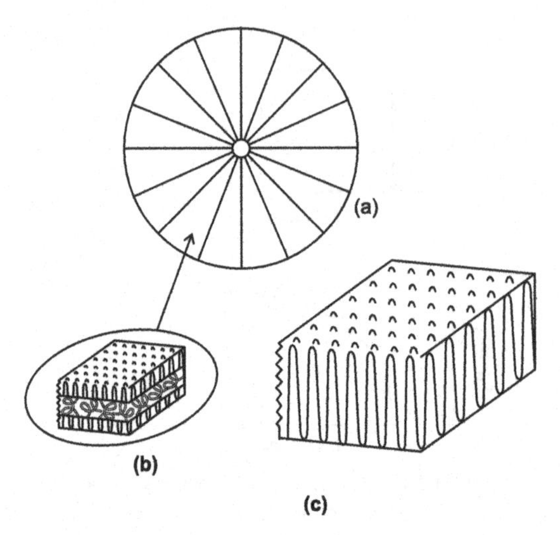

Fig. 13.7. Morphology of crystalline polymers: (a) crystalline regions appear as spherulites in optical microscope (b) the region between the "spokes" of the spherulite is made up of lamellar stacks with alternate layers of crystalline lamella and amorphous layer (c) lamella or a single crystal, is 5 – 10 nm thick and a few microns wide; the polymer chains fold back upon themselves to form a three dimensional ordered arrangement.

Crystalline polymers can also form from melt. In this case the polymer exists as a mixture of crystalline and amorphous regions. At least three different levels of structures in this case can be identified. The smallest unit is a lamella which is like a single crystal, with 5 to 10 nm thickness. The lamellae are stacked over one another, each lamella separated by an amorphous layer. A polymer chain can leave a lamella, enter the amorphous region and reenter the same or a different lamella. The lamellar stacks are further organized into superstructure morphologies. Most common of these is a spherulite which is spherical in shape and is made up of fibrils growing from its centre. The fibrils are formed from lamellar stacks (Fig. 13.7).

13.1.4 *Polymer solutions: Phase separation*

In a polymer solution, the chains adopt a conformation which is determined by the interaction between the chain segments and the solvent. In a good solvent this interaction energy is negative and the chain expands to maximize the segment–solvent contacts while in a poor solvent the chain contracts and clumps together.

Phase separation at a nanoscale in block copolymers is important in nanotechnology as the phase separated polymers can be used as templates to prepare different nanostructures as discussed later. Phase separation, however, is a general phenomenon and occurs more commonly in the polymer solutions. As we noted earlier, the mixing of a polymer and a solvent or of two polymers can occur if it is accompanied by a decrease in the free energy.

$$\Delta G_m = G_{AB} - (G_A + G_B) \quad < O$$

Here A and B denote the polymer and the solvent respectively. If ΔG_m is negative for all compositions at a certain temperature, then the system is completely miscible — the polymer dissolves in the solvent at all concentrations. As the temperature is reduced, a maximum and two minima usually occur in the ΔG_M *vs.* concentration plot below a certain temperature called the upper critical temperature UCT (Fig. 13.8).

At a temperature below UCT, such as T_3, a common tangent can be drawn to the two minima, the point of tangency corresponding to the compositions Φ_1, Φ_2. For any composition between Φ_1 and Φ_2, the free energy is lowest if the mixture separates into two phases of compositions Φ_1 and Φ_2 in appropriate ratio. This separation of a homogeneous solution into two phases can occur either by spinodal decomposition or by nucleation and growth depending on the overall composition. In the lower diagram, (Fig.13.8) two curves are drawn; one is the locus of the points of tangency and is called the *binodal curve* and the other, inner curve, is the locus of the two points of inflexion (between the points of tangency) of the ΔG_M plots at different temperatures and is called the *spinodal curve*. The curvature of the free energy curve ($\delta^2 G/\delta\phi^2$) is negative within the spinodal. For any point B within the spinodal, an arbitrarily small fluctuation in composition leads to a decrease in the free energy (Fig. 13.9 a). The system therefore easily phase separates into two phases differing only by a small amount in composition; the two phases

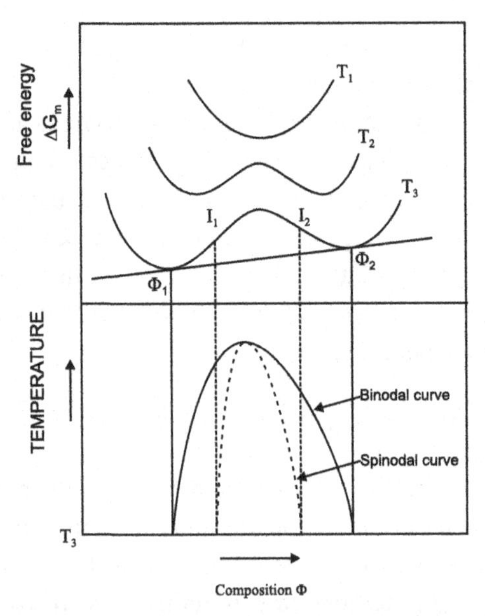

Fig. 13.8. (Top) Free energy *vs.* composition at three different temperatures (T1 > T2 > T3); at T3, the common tangent touches the curve at compositions Φ_1 and Φ_2 ; I_1 and I_2 are the inflexion points at which $(\delta^2 G/\delta\phi^2)_{T,P} = 0$; (bottom) the locus of the points of tangency is the ***binodal*** curve and the locus of the inflexion points gives the ***spinodal*** curve. Within the spinodal curve, the phase transformation proceeds by spinodal decomposition while for compositions between the binodal and the spinodal the phase transformation is by nucleation and growth.

are spread throughout the whole volume forming a bicontinuous structure. This process is called *spinodal* decomposition. With time, the composition difference between the phases may increase producing a droplet structure (Fig. 13.10).

When $(\delta^2 G/\delta\phi^2) > 0$, *i.e.* for a composition lying between the two curves (spinodal and the binodal) in Fig.15 (b), a fluctuation in composition is accompanied by an increase in free energy and so a phase separation of the spinodal type can not occur. The phase separation then occurs by nucleation and growth and eventually the system consists of two phases distributed uniformly in the whole volume with distinct boundaries. In case of the spinodal decomposition, the phases initially have a small difference in composition but large spatial variation. Given enough

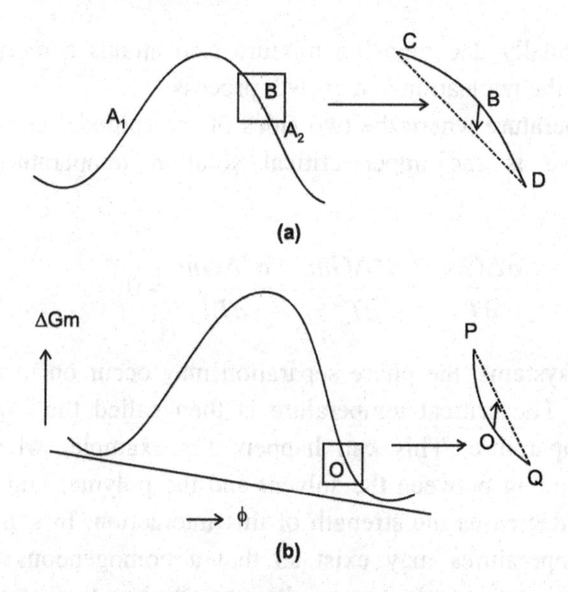

Fig. 13.9. (a) For any point B, any small fluctuation in composition leads to a decrease in the free energy as indicated by the figure on the right; the solution spontaneously phase separates into phases of composition C and D; A_1 and A_2 are the inflexion points. (b) For a composition O lying outside the inflexion points, a fluctuation in composition leads to an increases in free energy as shown by the arrow; this energy barrier has to be overcome before phase separation can occur into phases of compositions corresponding to the two minima in the free energy curve.

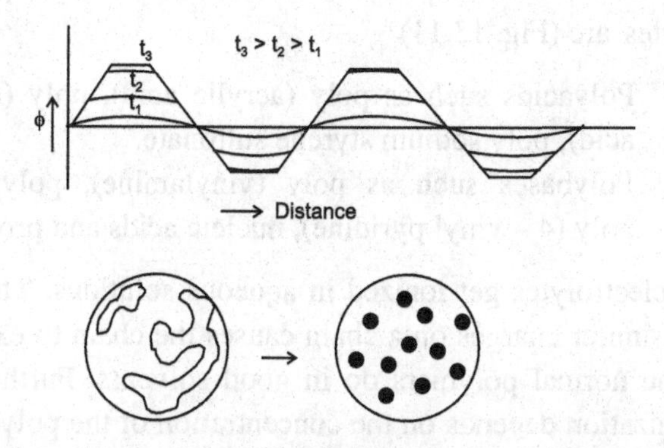

Fig. 13.10. (Top) In the spinodal decomposition a phase with initially uniform composition separates into two phases which initially ($t = t_1$) differ only slightly in composition; with time the composition fluctuation sharpens; (bottom) eventually an initially bicontinuous structure transforms to a droplet structure.

time, a spinodally decomposing mixture also attains a morphology as presented by the nucleation and growth process.

The temperature where the two sides of the spinodal curve meet the binodal curve is the upper critical solution temperature. At this temperature,

$$\frac{\partial \Delta Gm}{\partial T} = \frac{\partial^2 \Delta Gm}{\partial T^2} = \frac{\partial^3 \Delta Gm}{\partial T^3} = 0$$

In some systems, the phase separation may occur on increasing the temperature. The critical temperature is then called the lower critical solution temperature. This can happen, for example, when there is hydrogen bonding between the solvent and the polymer and raising the temperature decreases the strength of this interaction. In some systems, both the temperatures may exist so that a homogeneous solution is possible only in a temperature range between the two temperatures.

13.2 Polyelectrolytes

Polyelectrolytes are polymers with ionizable groups along their chains. One example of their use in nanotechnology is for the preparation of thin films by layer–by–layer deposition (LBL). Some examples of the polyelectrolytes are (Fig. 13.11)

(i) Polyacids such as poly (acrylic acid), poly (methacrylic acid), poly sodium styrene sulfonate.

(ii) Polybases such as poly (vinylamine), polyphosphates, poly (4 – vinyl pyridine), nucleic acids and proteins.

The polyelectrolytes get ionized in aqueous solutions. The repulsion between the similar charges on a chain causes the chain to expand much more than the normal polymers do in good solvents. Furthermore, the degree of ionization depends on the concentration of the polymer as well as that of any added salts. Due to these factors, the properties of polyelectrolytes such as viscosity and the intensity of light scattering behave quite differently as compared to the normal polymers. The polyelectrolytes behave as a normal polymer in non ionizing solvent as

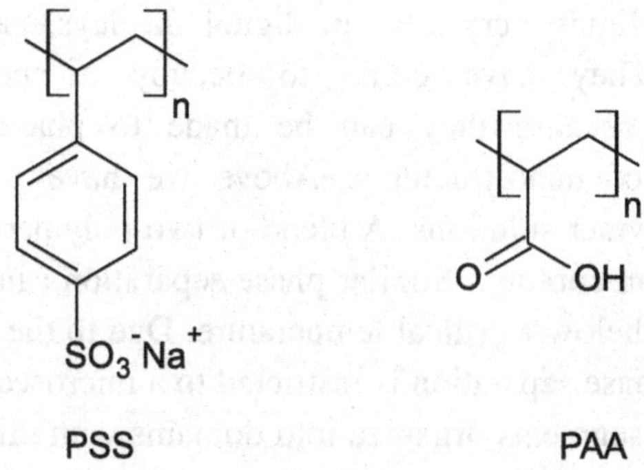

Fig. 13.11. Chemical structures of two synthetic polyelectrolytes. To the left is *poly(sodium styrene sulfonate)* (*PSS*), and to the right is *poly(acrylic acid)* (*PAA*). Both are negatively charged polyelectrolytes when dissociated. *PSS* is a 'strong' polyelectrolyte (fully charged in solution), whereas *PAA* is 'weak' (partially charged).

well as in ionizing solvents containing large amount of salt so that the double layer (see Chapter 7) around the poly electrolyte chain is completely screened.

13.3 Copolymers

13.3.1 *Types of copolymers*

Copolymers are obtained by simultaneous polymerization of two or more monomers. Copolymers can be classified as random, alternating, block and graft copolymers, depending on the sequence of the two monomers A and B in the polymer chain (Fig 13.12). In the graft copolymer, the blocks of one monomer are grafted onto a backbone of the second monomer type.

Copolymers are useful because they combine the properties of the two homopolymers. Thus polystyrene is a low cost material and clear but brittle while rubber is quite flexible. Copolymerizing the styrene and the butadiene monomers yields butadiene styrene copolymer which has improved mechanical properties.

13.3.2 *Phase separation in block copolymers*

Block copolymers find applications as thermoplastics for shoe soles, car tires as well as liquid crystals in digital displays, carriers for drug delivery, *etc*. They have come to occupy a special place in nanotechnology because they can be made to phase separate into different kinds of nanostructures. Above we have discussed phase separation in polymer solutions. A blend of two polymers nearly always phase separates on cooling. Similar phase separation can occur in block copolymers also below a critical temperature. Due to the connectivity of the blocks, the phase separation is restricted to a microscale. Usually, the individual block segments organize into domains with dimensions in the range 10 to 100 nm. The process is termed microphase separation. The two blocks separate out and can form a variety of ordered nanostructures. These nanostructures can be classified as lamellar, gyroid, hexagonal close packed cylinders and body centered cubic (Fig. 13.13). On heating, the disordered phase is recovered.

These structures are thermodynamically stable and can be shown on a phase diagram (Fig. 13.14). In the phase diagram the temperature is replaced by χN where N is the degree of polymerization and χ is the Flory-Huggins parameter which varies with temperature as follows:

(a) ABAABABBABABBAABAB.....

(b) ABABAB.......... ABAB.........

(c) AAABB....BBBAAA.....AAABB....BBAA.....AABB.....

(d) A A A A A A A A A A A A A

```
        B          B          B
        B          B          B
        B          B          B
  :          :          :
```

Fig. 13.12. Sequence of monomers A and B in (a) random (b) alternating (c) block and (d) graft copolymers.

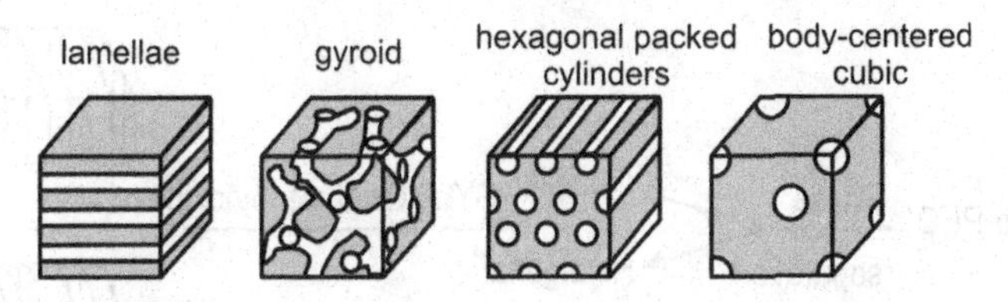

Fig. 13.13. The ordered nanostructures formed due to phase separation in block copolymers.

$$\chi = A + \frac{B}{T}$$

Here A and B are empirical constants. The Flory-Huggins parameter is a measure of the extent to which the energy is lowered by replacing the polymer–polymer contacts by polymer–solvent contacts [97].

The micro phase separation in the block copolymers has been successfully used to create useful nanostructures. As an example polymer inorganic nano composites can be formed by impregnating one of the blocks with a metal alkoxide precursor which later forms the inorganic phase by sol-gel reaction (hydrolysis and condensation). Templin *et al.*

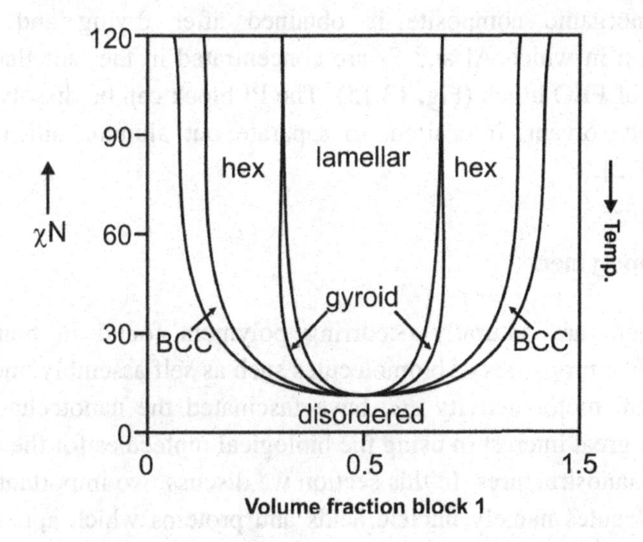

Fig. 13.14. Phase diagram for a diblock copolymer showing the regions for the various ordered structures.

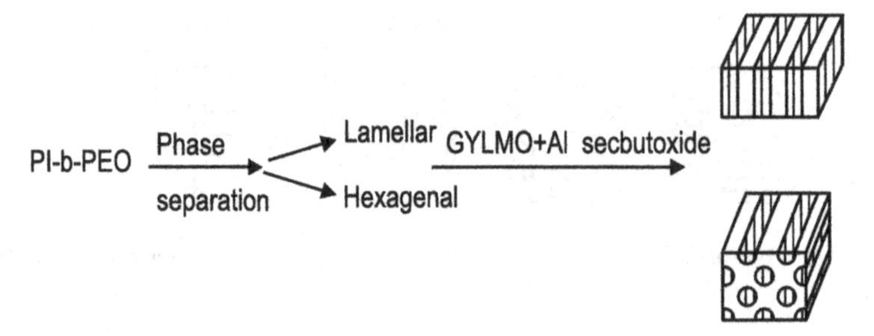

Fig. 13.15. Organic–inorganic composites by exploiting the phase separation in block copolymers.

[321] used poly (isoprene)–b–poly (ethylene oxide) (PI–PEO) block copolymer to prepare a composite containing aluminum silicate. A prehydrolyzed mixture of two metal alkoxides, (3-glycidyloxypropyl-) trimethoxysilane, $(CH_3O)_3Si(CH_2)_3OCH_2CHCH_2$-GLYMO), and aluminum *sec*-butoxide, $Al(OBu)_3$, is added to the phase separated block copolymer. The block copolymer composition is chosen to either yield a lamellar or a hexagonal structure after phase separation in which the PEO forms lamellae or rods respectively. The hydrolysis products of the alkoxides go selectively to the PEO block due to hydrogen bonding. An organic–inorganic composite is obtained after drying and solvent evaporation in which Al and Si are concentrated in the lamellae or the cylinders of PEO block (Fig. 13.15). The PI block can be dissolved with an organic solvent, if desired, to separate out alumina silicate nano objects [322].

13.4 Biopolymers

Biopolymers are naturally occurring polymers found in plants and animals. The properties of biomolecules such as self assembly, molecular recognition, motor activity *etc.* have fascinated the nanotechnologists. There is a great interest in using the biological molecules for the creation of useful nanostructures. In this section we discuss two important groups of biomolecules namely nucleic acids and proteins which appear to be most promising for applications in nanotechnology.

13.4.1 *Proteins*

Proteins are made of amino acids. Amino acids contain an amino (NH_2) group and a carboxylic group (COOH) connected to a central C atom called α carbon. A side chain *R* is also attached to the α carbon (Fig. 13.16 a). Different amino acids have different side chains. There are many different amino acids but only 20 of them are most commonly found in proteins. Out of these about eight have *R* groups that are ionizable or have polar side chains. The rest have non polar hydrocarbon side groups. Examples of amino acids are given in Fig.13.17. The amino group of one amino acid combines with the carboxylic group of the other to form a peptide linkage (Fig. 13.16 b). A protein molecule is formed by linking together a large number of amino acid units. When the number of peptide linkages is small (molar mass of the chain ≤ 10000 g mol^{-1}), the molecule is called a peptide or a polypeptide while the larger chains (having a molecular weight ~ 10^6) are called proteins. In contrast to proteins, the polypeptides do not have the tertiary or higher structures

(a)

$$H^+ \; H_2N - \underset{H}{\overset{R}{C}} - COO^-$$

(b)

$$H^+ \; H_2N - \underset{H}{\overset{R_1}{C}} - \underset{O}{\overset{}{C}} - \overset{H}{N} - \underset{R_2}{\overset{H}{C}} - C \begin{matrix} {}^{\nearrow O} \\ {}_{\searrow O^-} \end{matrix}$$

Peptide bond

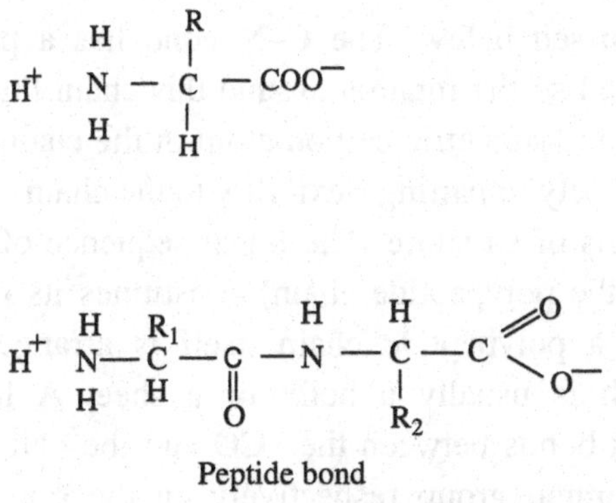

or

$$N^+H_3\text{–Peptidelink–}\underset{H}{\overset{R_2}{C}}\text{–Peptidelink–}\underset{H}{\overset{R_3}{C}}\text{–Peptidelink} \ldots\ldots \underset{H}{\overset{Rn}{C}}\text{–COOH}$$

Fig. 13.16. (a) An amino acid has an amino (NH_2) and a carboxylic group joined to a central C atom which also has a side chain *R*. (b) two amino acids link via a peptide bond formed between an amino group and a carboxyl.

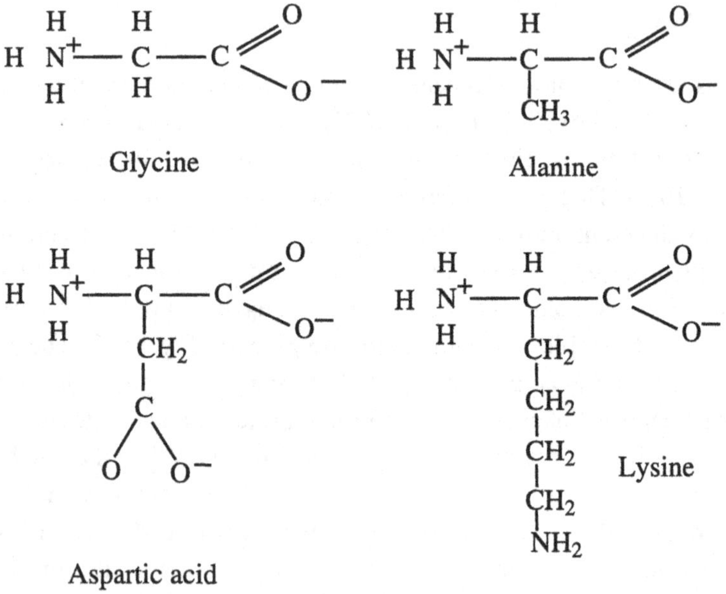

Glycine

Alanine

Aspartic acid

Lysine

Fig. 13.17. Examples of amino acids. Glycine and alanine are nonpolar, aspartic acid is acidic while lysine is a basic amino acid.

found in proteins as discussed below. The C–N bond has a partially double bonded character and so the rotation around this chain linkage is limited. The linkages to the asymmetric carbon atom in the chain are by single bonds; these rotate freely imparting flexibility to the chain.

A protein has four levels of structure. The linear sequence of amino acids within a chain (*i.e.* the polypeptide chain) constitutes its primary structure. In the protein, a polypeptide chain itself is arranged in a secondary structure which is usually a helix or a sheet. A helix is stabilized by the hydrogen bonds between the –CO and the –NH group on the carbonyl and the imino group respectively on the same chain while a sheet forms by such bonds between two adjacent chains. The helices or the sheets further fold so as to place the hydrophobic side groups in the interior of the chain, away from the external aqueous environment. This is the tertiary structure of a single peptide chain. Note that this structure forms only in proteins and the smaller polypetides do not form this structure. Many peptide chains then associate together to give the protein its quaternary structure (Fig. 13.18).

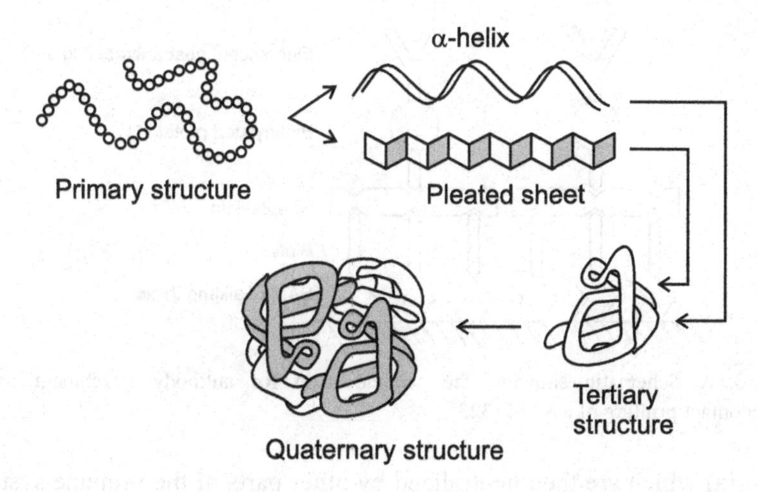

Fig. 13.18. Four levels of protein structure.

The characteristic shapes in which the proteins are folded are critical to their functioning. By such folding, critical bonding sites are formed which allow other molecules to bond to the proteins. The bonding of proteins to DNA, for example, is responsible for the packaging of about 2 meters of double stranded DNA in the nucleus of a cell which is only 2 μm across. As another example, recognition of a double stranded DNA by a protein occurs by its ability to insert an α helix of the protein in the major groove in DNA. The specificity of a protein towards other proteins, receptors or molecules is due both to the shape factors and the non covalent interactions such as hydrogen bonding, ionic bonding or hydrophobic interaction. Such interactions can be used for self assembly. The most exploited ligand–receptor system for nano assembly is that of avidin and biotin. Avidin is a protein found in eggs and has four bonding sites to biotin, a ligand which is a vitamin. Functionalization of two surfaces with avidin and biotin respectively can enable them to assemble in predesigned manner (Fig. 13.19).

Enzymes are soluble proteins that catalyze the biological reactions. *Antibodies* (also known as immunogloblins, abbreviated as Ig) are gamma globulin proteins that bind to specific antigens (*e.g.* virus,

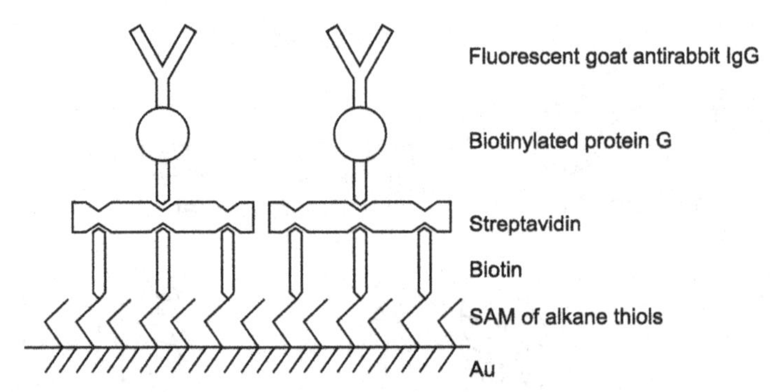

Fig 13.19. Schematic showing the scheme used for antibody attachment using microcontact printing of a SAM [323].

bacteria) which are then neutralized by other parts of the immune system of the body. *Motor proteins* (myosin, kinesin, dynesin) are nanometer sized molecular motors that move organelles (constituents of the cell), the whole cell and entire muscles.

For exploiting the proteins for self assembly of structures, it will be necessary to bind proteins to micro fabricated surfaces, while preserving their folded shape and hence the functionality. This is not easy because proteins readily denature when taken out of the aqueous environment. Nevertheless in several cases it has been done. The strategy used is to functionalize one end of the protein with chemical groups that can bond to a particular surface, which itself may have to be functionalized with a complimentary group. Lahiri *et al.* [323] first deposited a SAM (self assembled monolayer) of alkane thiols on a Au layer. Next biotin was printed on SAM by micro contact printing of 5 μm squares. A layer of streptavidin followed by biotinylated protein G bound to fluorescent goat anti rabbit 1gG protein was next deposited (Fig. 13.19). Fluorescence revealed that the protein has been patterned within the 5 μm squares.

13.4.2 *Nucleic acids*

Deoxyribonucleic acid (DNA) and ribonucleic acid (RNA) are polymers having molecular mass ranging from 10^4 to 10^9. They are found in the nucleus and cytoplasm of the cell.

13.4.2.1 *Deoxyribonucleic acid (DNA)*

DNA is of especial importance in nanotechnology because of the ability of its two complementary single strands to join together by hydrogen bonds. The double stranded helical structure of DNA is the store house of the genetic information of an organism. This information resides in the sequence in which four bases (adenine (A), cytosine(C), guanine (G) and thymine (T)) are arranged on the backbone of sugar molecules and phosphate ions. The bases cytosine, thymine (and uracil in RNA) are pyrimidine bases while adenine and guanine are purine bases.

The repeating unit in each strand of DNA molecule consists of a sugar, a phosphate and a base. This repeating unit is called a *nucleotide.* (The part containing the base and the sugar only, without the phosphate linkage, is called *nucleoside*). The sugar molecule in the DNA is deoxyribose (Fig. 13.20 a). A base (A, T, G or C) is attached to the 1′ carbon of the deoxyribose. The 5' hydroxyl of each deoxyribose unit is replaced by a phosphate that is attached to the 3' carbon of the deoxyribose in the preceding unit.

The bases on the two strands of DNA are bonded through hydrogen bonds such that adenine bonds only to thymine and cytosine bonds to guanine. This complementary nature of bonding between the bases is the key to the use of DNA strands for self assembly. This complimentary bonding property of the bases is shown schematically in Fig. 13.21. The joining together of two single stranded (SS) DNA strands is called *hybridization*. The two strands run antiparallel (one from 3′ to 5′ and the other from 5′ to 3′) to each other forming a double helix. This is the secondary structure of DNA.

The individual base pairs in DNA are spaced about 0.34 nm apart. One turn of the double helix contains about ten and a half base pairs. The inside diameter of the double helix is 1.1 nm and the outside diameter is 2 nm. In an imaginary cylinder just enclosing the double helix, there are two groove shaped empty spaces for every turn of the helix. The larger of these is called the major groove and the smaller one, the minor groove. These grooves are the sites where the peptide chains with complementary shape fit in.

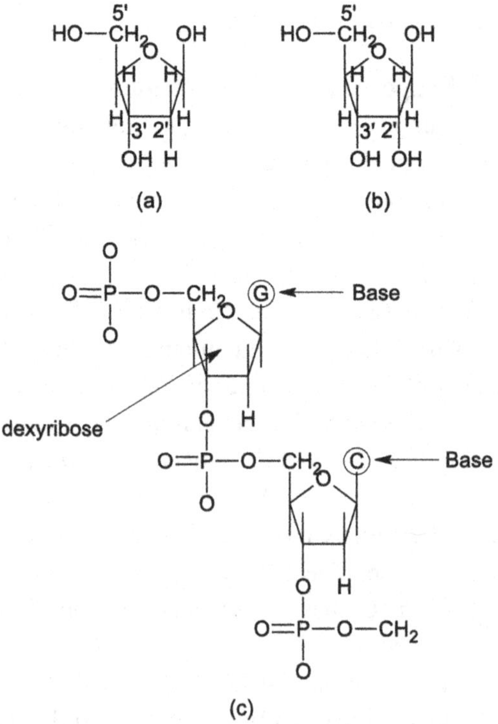

Fig. 13.20. (a) deoxyribose molecule with 2', 3' and 5' carbon atoms marked (b) the ribose molecule which is the sugar in RNA (c) a part of a DNA strand.

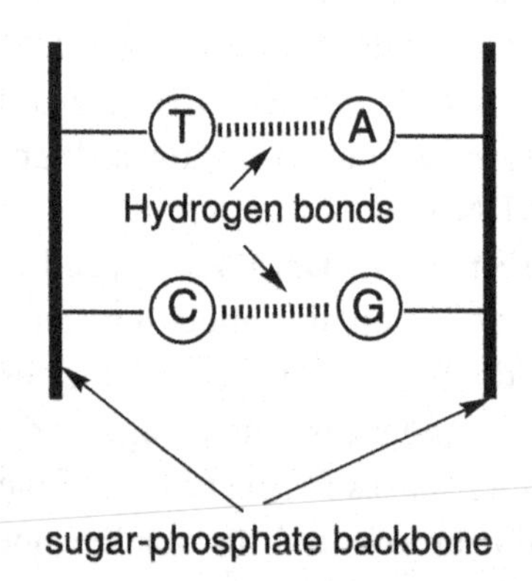

Fig. 13.21. Schematic showing the bonding of the bases thymine and cytocine on one strand with adenine and guanine on the other strand of a DNA molecule.

On heating above a temperature, called the melting temperature, the two strands can separate or denature. On cooling they can diffuse, come close and rejoin; this is called *rehybridization* or recapturing. The phosphate ion in the DNA backbone is negatively charged so that positive ions in the solution are needed to counter the repulsion between the phosphate ions on the two strands. The melting temperature thus depends on the ionic strength in the surrounding medium as well as the G-C content in the strands. DNA molecules with known sequence of bases are now available commercially.

13.4.2.2 *Ribonucleic acid (RNA)*

RNA has a structure similar to DNA except for a few differences. Firstly, unlike DNA it has only a single strand. Secondly the sugar in it is D–ribose in which there is a hydroxyl group attached to the 2′ carbon (Fig. 13.20 b). The presence of this hydroxyl group makes the RNA less flexible than DNA. Thirdly, out of the four bases of DNA, in RNA the thymine is replaced by another base, uracil, while the remaining three bases are the same.

RNA exists in three major forms. These are *transfer* RNA (*t*RNA), *ribosomal* RNA (*rRNA*) and *messenger* RNA (*m*RNA). The base sequence of all the RNA is replicated from the DNA in the cell. The tRNA is the smallest in size. Its function is to transport amino acids to ribosomes, which are the site for the production of proteins. For every amino acid, there is at least one tRNA which binds specifically to it. Thus there are many different species of tRNA.

The ribosomal RNA, rRNA, combines with proteins to form the ribosomes. In a ribosome there are 60 to 65 % by weight of rRNA and the remaining are proteins. The rRNA are of several kinds and are of variable size. The ribosome is the machine where the synthesis of the proteins takes place.

The mRNA are also of variable length. They are responsible for directing the aminoacids to follow the specific sequence in a growing protein chain.

Unlike DNA, the RNAs do not form double helix. Instead, the complementary base sequences in the different regions of the same chain pair up and form double helices. Thus parts of RNA are folded up into double helices while the non hydrogen bonded portions of the chain form loops. The helical regions can pack against each other *via* additional tertiary interactions to form a tertiary structure.

The structural features of the biomolecules discussed above make them potential building blocks for the bottom up fabrication of nanodevices. The property of the DNA and RNA to specifically bond to complementary sequence of bases is the easiest to exploit for self assembly. The structural features of proteins also make them well suited for such applications. The diversity of function and structure of RNA make it potentially much more powerful building block for a host of nanodevices including synthetic molecular machines to perform a variety of functions [324, 325].

13.5 Surfactants

13.5.1 *Introduction*

The term surfactant is derived from *surface active reagent*. These are molecules which change the surface energy of the liquid–air, liquid–liquid or liquid–solid interface by absorbing at the interface. As an example, if we add oil to a beaker containing water, we find that the two do not mix; in fact, the oil does not even spread on water uniformly but floats on it as a compact liquid lens. This is because the surface energy of the interface between the oil and water is high — by forming a layer the interface area is minimized. However, if a small amount of sodium dodecyl sulfate (SDS) is added and the mixture is agitated, then the oil is dispersed in the form of small droplets in water, forming a oil–in–water emulsion.

Molecules such as SDS have two parts; one part is attracted towards water and the other part is repelled by water but attracted towards the oil. The former is called the *hydrophilic* (water liking) part and the latter is

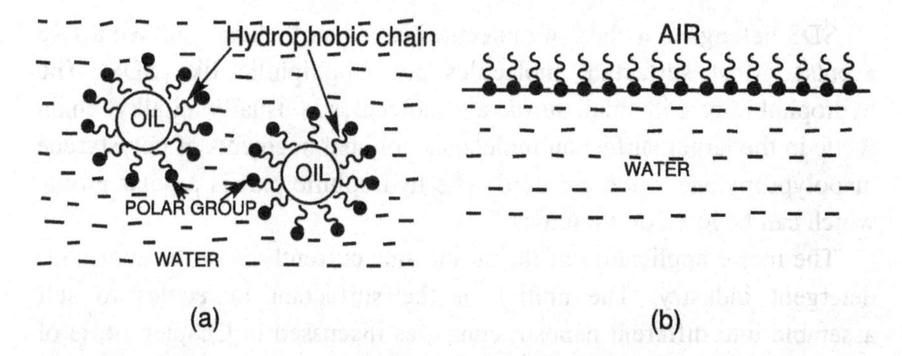

Fig. 13.22. (a) Surfactant molecules at the oil water interface with polar group toward water (b) Surfactant molecules segregate to the water–air interface with the polar group pointing towards water and the hydrophobic tail pointing towards the air.

called *hydrophobic* (water hating) part. (The other terms used to denote affinity of a molecule to oil or water or to other solvent are listed in Table 13.3).The molecule as a whole is called an *amphiphile* because it has liking to both. In the emulsion, the molecule attaches to the water–oil interface with the hydrophilic part oriented towards water and the hydrophobic part oriented towards the oil droplet (Fig. 13.22 a). The net effect is a drastic lowering of the oil–water interfacial energy. If the surfactant is mixed only with water, it tends to go to the surface of water with the hydrophobic part pointing up toward the air (Fig. 13.22 b).

Table 13.3. Terms used to denote affinity or lack of affinity to water, oil or a solvent.

Term	Meaning
Amphipathic, amphiphile	A single molecule containing both kinds (*e. g.* water loving and oil loving) moieties.
Amphiphilic	Having affinity to both water and oil
Hydrophilic	Having affinity for water
Hydrophobic	No affinity for water
Lipophilic	Having affinity for oil
Lyophilic	Having affinity for solvent
Lyophobic	No affinity for solvent

SDS belongs to a class of molecules called surfactants (surface active agents). Most surfactant molecules are amphiphilic like SDS. The hydrophobic part in small surfactant molecules is usually an alkyl chain while in the larger surfactant molecules polymer segments of polystyrene or polypropylene oxide are used. The hydrophilic part is a polar group, which can be ionic or nonionic.

The major application of the surfactants currently is in the enormous detergent industry. The ability of the surfactant molecules to self assemble into different nanostructures, as discussed in Chapter 14, is of special interest in nanotechnology. One of the earliest applications of these self assembled surfactant nanostructures was in the synthesis of mesoporous silica by Mobil scientists in 1992 [198, 199]. Micelles, one of the nanostructures formed by the surfactant molecules, can be used as nanoreactors within which the desired reactions can be carried out. Synthesis of nanoparticles within the micelles is an example of this. Furthermore, the tendency of the surfactant molecules to form a monolayer on the surface of a liquid has been exploited for the deposition of the monolayers and multilayers of surfactants as well as other molecules, nanoparticles, fullerenes, *etc.* on solid substrates by the Langmuir–Blodgett technique discussed in Chapter 14.

13.5.2 *Types of surfactants*

As mentioned above, the amphiphiles have two parts. One part (lyophilic) likes the solvent while the other part (lyophobic) is repelled by it. While there are surfactants which do not have such distinct parts, a strong surfactant action is obtained if the surfactant has a lyophilic and a lyophobic part. Some of the hydrophilic and hydrophobic moieties which can be combined to produce a surfactant are given below.

(a) Hydrophilic groups

(i) Ionic: Carboxylate ($–CO_2H^-$), sulfate ($–OSO_3^-$),sulfonate ($– SO_3^-$), quaternary ammonium ($R_3\,N^+$) where R can be H or an alkyl group *e.g.* CH_3.

(ii) Non ionic hydrophilic groups: Fatty acid ($-CO_2H$), alcohol ($-CR_2OH$), ether (–COC–), polyethylene Oxide ($H-[OCH_2CH_2]_n-OH$).

(b) Hydrophobic or lipophilic groups: Straight or branched alkyl chains (C_8 to C_{18}), polystyrene oxide $H\,[OCH\,(C_6H_5CH_2]_nOH)$, polypropylene oxide $(H\,[OCH\,(CH_3)\,CH_2]_n\,OH$, polysiloxane $H\,[OSi\,(CH_3)_2]n\,OH$.

In the following are given some examples of the surfactants with different hydrophilic and hydrophobic groups. A surfactant with an ionic hydrophilic group is called either anionic or cationic depending on the polar group. An anionic surfactant can be represented in Fig. 13.23 (a).

Examples of anionic surfactants are sodium dodecyl sulfate (SDS) $C_{12}H_{25}SO_4Na$ and sodium dodecyl benzene sulfonate, $C_{12}H_{25}C_6H_5SO_3Na$.

A simple cationic surfactant is shown in Fig. 13.23 (b). The hydrogen in NH_3 can be replaced by one or more methyl or other groups, *e.g.* as in Fig. 13.23 (c). As the head group is based on a quaternary ammonium compound, these surfactants are called alkyl *quats*. Two well known examples of alkyl quats are cetyltrimethyl ammonium chloride, CTAC

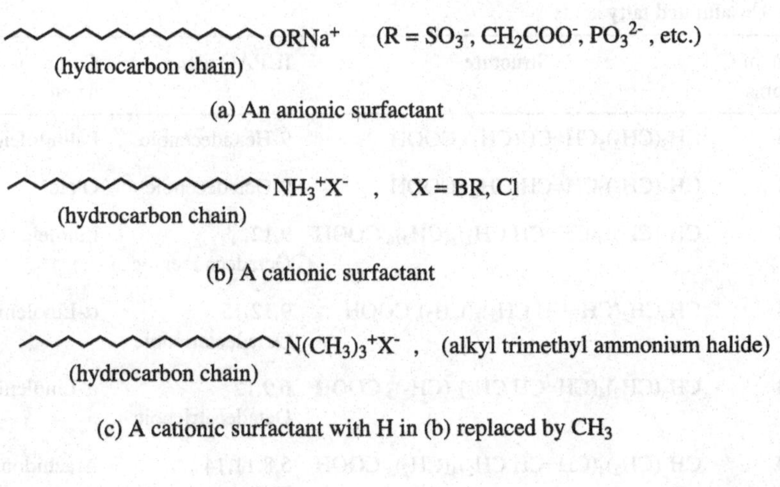

(a) An anionic surfactant

(b) A cationic surfactant

(c) A cationic surfactant with H in (b) replaced by CH_3

Fig. 13.23. Anionic and cationic surfactants.

($C_{16}H_{33}(CH_3)_3{}^+NCl^-$) and cetyltrimethyl ammonium bromide (CTAB). An example of a nonionic surfactant with polyethylene oxide as the hydrophilic group and a hydrocarbon chain as the hydrophobic part is dodecyl hexaethylene glycol monoester, $C_{12}–H_{25}–(CH_2–CH_2–O)_6–OH$ denoted as $C_{12}E_6$.

Table 13.4. IUPAC and common names of some fatty acids.

(a) Saturated fatty acids $CH_3\ (CH_2)_n\ COOH$

Number of carbon atoms	n	IUPAC name	Common name
12	10	Dodecanoic	Lauric
14	12	Tetradecanoic	Myristic
16	14	Hexadecanoic	Palmitic
17	15	Heptadecanoic	Magaric
18	16	Octadecanoic	Stearic
20	18	Eicosanoic	Arachidic
22	20	Docosanoic	Behenic
24	22	Tetracosanoic	Lignoceric

(b) Unsaturated fatty acids

No. of C atoms	Structure	IUPAC name	Common name
16	$CH_3(CH_2)_5CH{=}CH(CH_2)_7COOH$	9-Hexadecenoic	Palmitoleic
18	$CH_3(CH_2)_7CH{=}CH(CH_2)_7COOH$	9-Octadecenoic	Oleic
18	$CH_3(CH_2)_4(CH{=}CH\ CH_2)_2(CH_2)_6\ COOH$	9,12 Octadecadoenoic	Linoleic
18	$CH_3CH_2(CH{=}CH\ CH_2)_3(CH_2)_6COOH$	9,12,15 Octadecatrienoic	α-Linolenic
18	$CH_3(CH_2)_4(CH{=}CH\ CH_2)_3(CH_2)_3\ COOH$	6,9,12 Octadecatrienoic	α-Linolenic
20	$CH_3(CH_2)_4(CH{=}CH\ CH_2)_4(CH_2)_3\ COOH$	5,8,11,14 Eicosatitraenoic	Arachidonic

The fatty acids are widely used as surfactants as they were the earliest ones to be investigated. It is helpful to be familiar with both the IUPAC and common names of the fatty acids. Some of these are listed in Table 13.4.

There are also surfactants which contain both the positive and the negative charges in the head group. These surfactants are called *zwitterionic* and are used in cosmetic products since they do not irritate the skin or the eyes. The positive charge is provided usually by the ammonium group while the negative charge can come from carboxylate (COO^-), SO_3^-, *etc.* An example of a zwitterionic surfactant is 3 dimethyl dodecyl propane sulphonate (betaines), $C_{12}H_{25}\,N^+\,(CH_3)_2\,CH_2CH_2SO_3^-$.

Amphoteric surfactants are those which can behave as cationic, zwitter ionic or anionic depending on the pH, being cationic at low pH and anionic at high pH.

An important class of surfactants are lipids which are molecules of biological origin. They are essentially fatty acid salts but are rarely found in pure form. Usually they contain two or more alkyl chains as in phospholipids (Fig. 13.24). Examples of phospholipids are phosphatidylethanolamine (R = $CH_2CH_2NH_3^+$) and phosphatidylcholine or lecithin (R = $CH_2CH_2N(CH_3)_3^+$).

Several polymers also act as surfactants. A well known example is polyvinyl alcohol which is partially hydrolyzed polyvinyl acetate (Fig. 13.25).

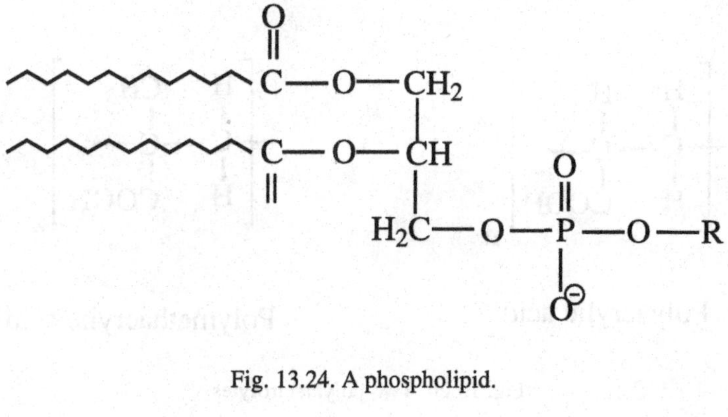

O
‖
C—O—CH₂
C—O—CH
‖
H₂C—O—P(=O)(O⁻)—O—R

Fig. 13.24. A phospholipid.

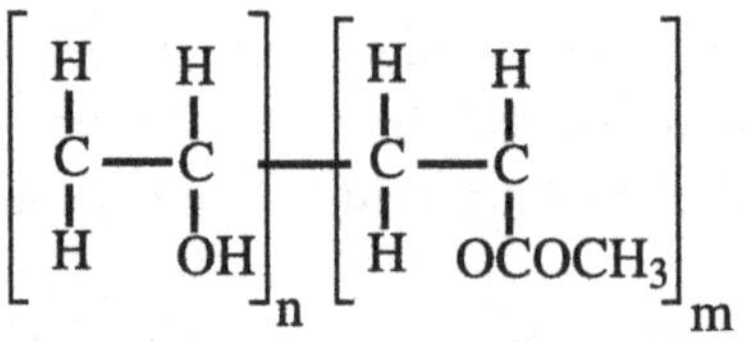

Fig. 13.25. Partially hydrolyzed polyvinyl acetate.

Another important class of surfactants is polyelectrolytes such as polyacrylic acid, polymethacrylic acid (Fig. 13.26).

13.5.3 *Absorption of surfactant molecules in a solution*

13.5.3.1 *Introduction*

Adsorption of surfactant molecules in a liquid and their location and arrangement in the liquid have applications in areas such as the formation of the Langmuir–Blodgett films and the surfactant nanostructures. An indication that these features of a surfactant solution depend on the concentration is provided by the changes in surface tension. The change in surface tension with concentration of a solute, c_2, can be represented by the curves in Fig. 13.27. For a surfactant (curve1) and an organic solute (curve 2) the slope is negative while for an electrolyte the slope is positive. The change in surface tension is related

Polyacrylic acid

Polymethacrylic acid

Fig. 13.26. Two polyelectrolytes

to the adsorption of the solute on the surface. In case of the surfactant and the organic solute, there is a positive adsorption on the surface while for an electrolyte there is depletion because the ions prefer to be hydrated and moving them to surface requires energy to partially dehydrate them. This causes a depletion of the electrolyte from the surface.

For a surfactant solution (curve 1, Fig. 13.27) the surface tension decreases linearly with concentration 'c_2' when c_2 is small

$$\gamma = \gamma_0 - kc_2 \tag{13.1}$$

Although not shown in Fig. 13.27, at higher concentrations, γ varies approximately linearly with lnc_2 instead of c_2 up to a certain concentration called the critical micelle concentration, (CMC) beyond which the surface tension does not change further with an increase in the concentration of the surfactant. At CMC the location and arrangement of the surfactant molecules in solution changes drastically. It is at this concentration and beyond that the surfactant molecules assemble to form a variety of nanostructures.

The arrangement and location which the surfactant molecules adopt in a solution, depends strongly on their concentration. At low

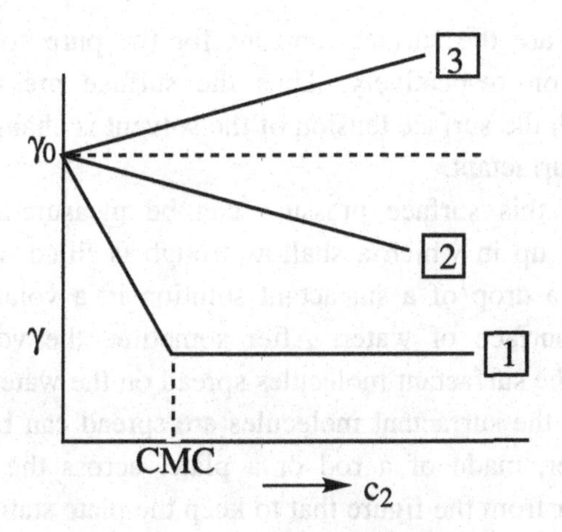

Fig. 13.27. Change in surface tension of a water solution with concentration of the solute for (1) an amphiphile (2) a simple organic solute (3) an electrolyte.

concentration, the surfactant mostly segregates to the solution–air interface and forms a monolayer (one molecule thick layer) on the surface. The surfactant molecules in this concentration range behave as a two dimensional gas as discussed later. As the concentration is increased, the concentration of the surfactant molecules at the surface also increases and they form an increasingly compact layer. With the increase in concentration, the molecules in the compact layer assume different arrangements due to lateral forces between them. A most general sequence of phase formation is as follows

$$G \rightarrow E \rightarrow C$$

Here G denotes a gas like phase at low concentrations, E is an extended phase at intermediate concentration and C is a compact phase at high concentration. The surface area per molecule is highest for the G phase and lowest for the E phase. In actual experiments, a surfactant may show many more phases depending on the type of surfactant.

The common way to investigate the phase formation is to measure a quantity called surface pressure π *vs.* the concentration at a constant temperature. The surface pressure is defined as

$$\pi = \gamma_0 - \gamma$$

where γ_0 and γ are the surface tensions for the pure solvent and the surfactant solution respectively. Thus the surface pressure π is the amount by which the surface tension of the solvent is changed due to the addition of the surfactant.

To see how this surface pressure can be measured, consider an experimental set up in which a shallow trough is filled up to the brim with water and a drop of a surfactant solution in a volatile solvent is placed on the surface of water. After sometime the volatile solvent evaporates and the surfactant molecules spread on the water surface. The area over which the surfactant molecules are spread can be changed by moving a barrier, made of a rod or a plate, across the surface (Fig. 13.28). It is clear from the figure that to keep the plate stationary, a force per unit length equal to π has to be maintained on it to overcome the force due to difference in the surface tension on the two sides. As the

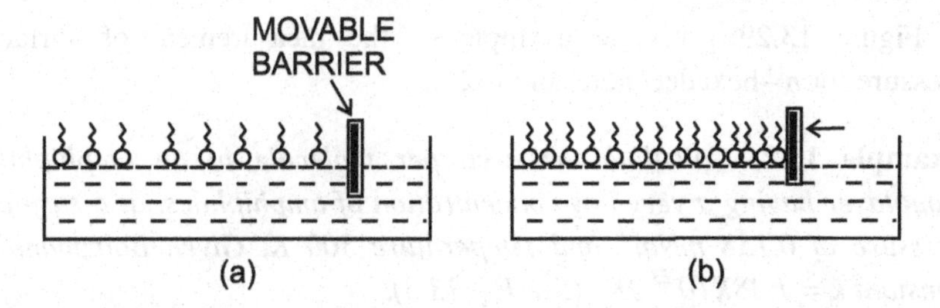

Fig. 13.28. The surfactant layer on the surface is compressed by moving the barrier (a). The force experienced by the barrier is the difference between the surface tension on the two sides of it (b).

layer is compressed by moving the barrier, the surfactant molecules arrange themselves into increasingly compact arrangement as described earlier. Eventually a close packed, compact monolayer is obtained beyond which the surface pressure increases rapidly [97]. Beyond the CMC, this layer cannot be compressed further. The excess surfactant molecules then self assemble into well ordered nanostructures inside the solution. These self assembled nanostructures have been put to various uses in nanotechnology. These are described later in Chapter 14.

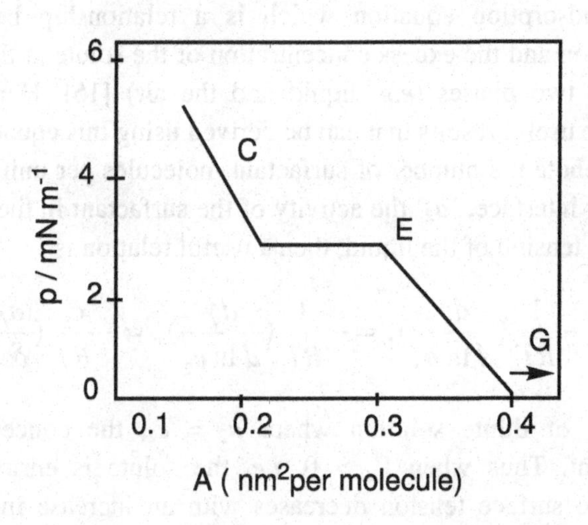

Fig. 13.29. Change in surface pressure with composition of the layer for n-hexadecanoic acid.

Figure 13.29 gives an example of the measurement of surface pressure for n –hexadecanoic acid [326].

Example 13.2. *Calculate the area per molecule in an amphiphile monolayer having a very low concentration of amphiphiles, at a surface pressure of 0.138 mNm^{-1} and temperature 300 K. Given Boltzmann's constant k = 1.38X10^{-23} JK^{-1} (See Eq. 13.3).*

Solution. At low concentration the monolayer behaves as a two dimensional gas for which $\pi A = n_S\ RT$ or $\dfrac{\pi A}{n_S X\ N_A} = kT$ or

$$\frac{A}{n_S \times N_A} = area\,/\,molecule = \frac{kT}{\pi} = \frac{1.38\times10^{-23}\times300}{0.138\times10^{-3}} = 30nm^2$$ □

13.5.3.2 *The Gibbs absorption equation*

As was mentioned above, at low concentrations the surfactant mostly segregates to the surface of the liquid. The relation between the amount of solute absorbed on the surface and the surface tension is provided by the Gibbs adsorption equation which is a relationship between the surface tension and the excess concentration of the solute at the interface between the two phases (*e.g.* liquid and the air) [15]. Here we only discuss some useful results that can be derived using this equation.

Let Γ_2 denote the number of surfactant molecules per unit area at the liquid–vapor interface, 'a_2' the activity of the surfactant in the liquid and γ the surface tension of the liquid, then a useful relation is

$$\Gamma_2 = -\frac{1}{RT}\left(\frac{d\gamma}{d\ln a_2}\right)_T = -\frac{1}{RT}\left(\frac{d\gamma}{d\ln c_2}\right)_T = -\frac{c_2}{RT}\left(\frac{\delta\gamma}{\delta c_2}\right)_T \quad (13.2)$$

in the limit of dilute solution where $a_2 = c_2$, the concentration of the surfactant. Thus when $\Gamma_2 > 0$, *i.e.* the solute is enriched at the interface, the surface tension decreases with an increase in the solute concentration.

As Γ_2 denotes the number of surfactant molecules per unit area at the liquid–vapor interface, reciprocal of Γ_2 gives the average area occupied by one molecule of the solute. Just before the CMC, the γ *vs.* $\log c_2$ plot is linear so that the slope at this concentration gives the value of Γ_2. Beyond the CMC, the surface tension does not change because there is no change in the chemical potential with concentration and the excess surfactant goes inside the solution.

As noted earlier, the surface pressure π is defined as the reduction in γ due to the formation of the surfactant monolayer

$$\pi = \gamma_0 - \gamma$$

Using Eq. (13.1) we get

$$\pi = kc_2$$

Thus at low concentrations, the surface tension varies linearly with concentration.

Differentiating we get,

$$d\gamma/dc_2 = -k$$

Then using Eq. (13.2),

$$\pi = \Gamma_2 RT$$

and , using

$$\Gamma_2 = \frac{n_2^S}{A},$$

one gets

$$\pi A = n_2^S RT \tag{13.3}$$

where n_2^S is the (excess) moles of the surfactant molecules in the surface layer and A is the total area of the surface layer. Comparing this with the ideal gas equation at low pressures PV = nRT, we see that Eq. (13.3) can be thought of as an equation for a bidimensional gas formed by the surfactant monolayer.

Example 13.3. *The plot of γ vs $\log_{10} c_2$ for the dodecyl ether of hexaethylene oxide at three temperatures is shown in Fig. 13.30. Calculate the surface excess and the area per molecule for each temperature.*

Solution. For 25^0, the slope is –16.7 mNm^{-1} at the low concentration side of the flat region, just below the CMC. For this region

$$\Gamma_2 = -\frac{1}{RT}\left(\frac{d\gamma}{d\ln c_2}\right) = -\frac{1}{2.303RT}\left(\frac{d\gamma}{d\ln_{10} c_2}\right)$$

$$= -\frac{1}{2.303\times8.314\times298}\left(-16.7\times10^{-3}\right)$$

$$= 2.93\times10^{-6}\, mole.m^{-2}$$

Therefore, area per molecule

$$= \frac{1}{2.93\times10^{-6}\times6.02\times10^{23}} = 0.56\; nm^2$$

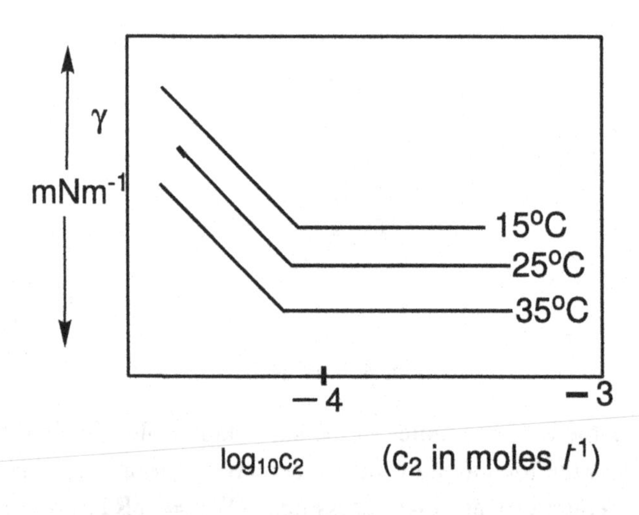

Fig. 13.30. A Variation of the surface tension of dodecyl ether of hexaehylene oxide with concentration at three different temperatures [327].

As mentioned above, at the CMC, the surfactant molecules begin to self assemble into various nanostructure in the bulk of the solution. □

13.6 Dendrimers

13.6.1 *Synthesis*

Dendrimers are large molecules having a core that repeatedly branches outwards [328-330] . The word is derived from dendron, meaning a tree in Greek. A dendrimer can be grown by starting from the core and then growing the branches outwards, or first synthesizing the branches and then joining them to the core; the former is called the *divergent* method while the latter is called the *convergent* method. The branches are also called *dendrons*. A core molecule must have multiple reaction sites that are identical. The divergent method is illustrated in Fig. 13.31.

As an example of a dendrimer, consider the polyamidoamines (PAMAMs), the first dendrimers to be synthesized. In these, the ammonia is used as the core molecule which is first reacted with methyl

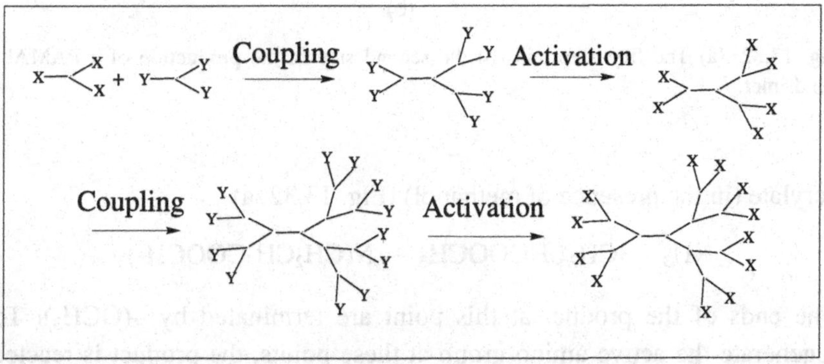

Fig. 13.31. Divergent method of dendrimers growth: "X" are the active sites on the core at which the monomer unit joins. The outer ends of the monomer (Y) are inactive to start with. They have to be activated to place active sites (X) at their outer ends so that the next layer of monomers can attach.

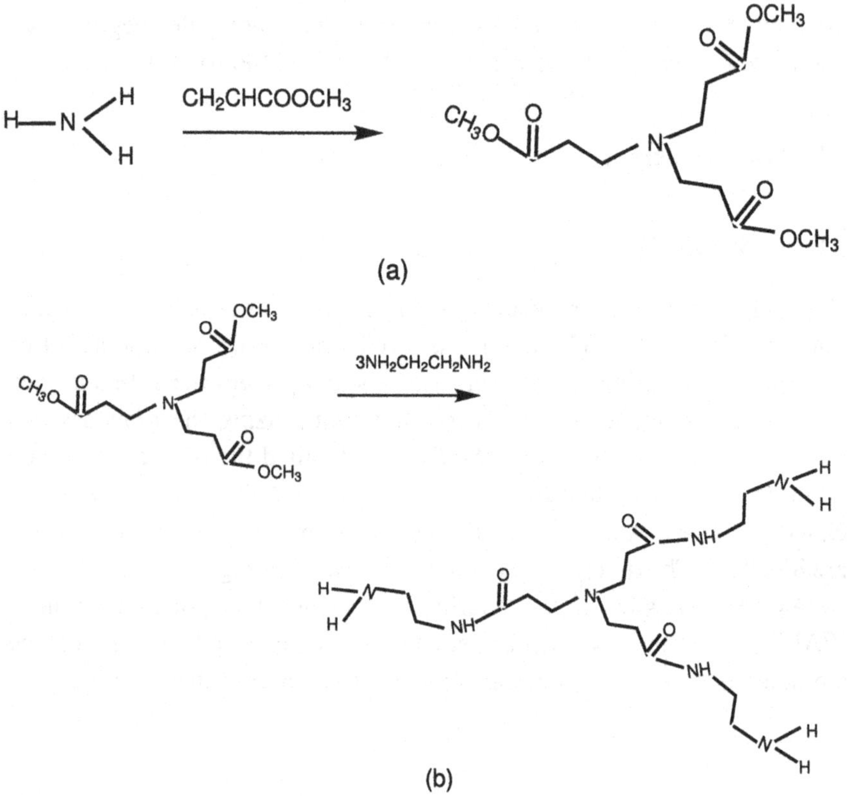

Fig. 13.32. (a) The first step and (b) the second step in the production of a PAMAM dendrimer.

acrylate (in the presence of methanol) (Fig. 13.32 a)

$$NH_3 + 3CH_2CHCOOCH_3 \rightarrow N(CH_2CH_2COOCH_3)_3$$

The ends of the product at this point are terminated by $-(OCH_3)$. To regenerate the active amino group at these points, the product is reacted with ethelenediamine (Fig. 31.32 b)

$$N(CH_2CH_2COOCH_3)_3 + 3NH_2CH_2CH_2CH_2NH_2$$
$$\rightarrow N(CH_2CH_2CONH\ CH_2CH_2NH_2)_3 + 3CH_3OH$$

This complete reaction sequence results in the completion of the first *generation* of the dendrimers. Each repetition of this two step sequence produces a fresh generation. The number of reactive surface sites is doubled with every generation while the mass increases more than twice. The generation after the first reaction is called the half generation; here the terminal groups are acidic. After the second reaction, the full generation is completed and the surface of the molecule consists of multiple amines. A third generation dendrimer is shown in Fig. 13.33.

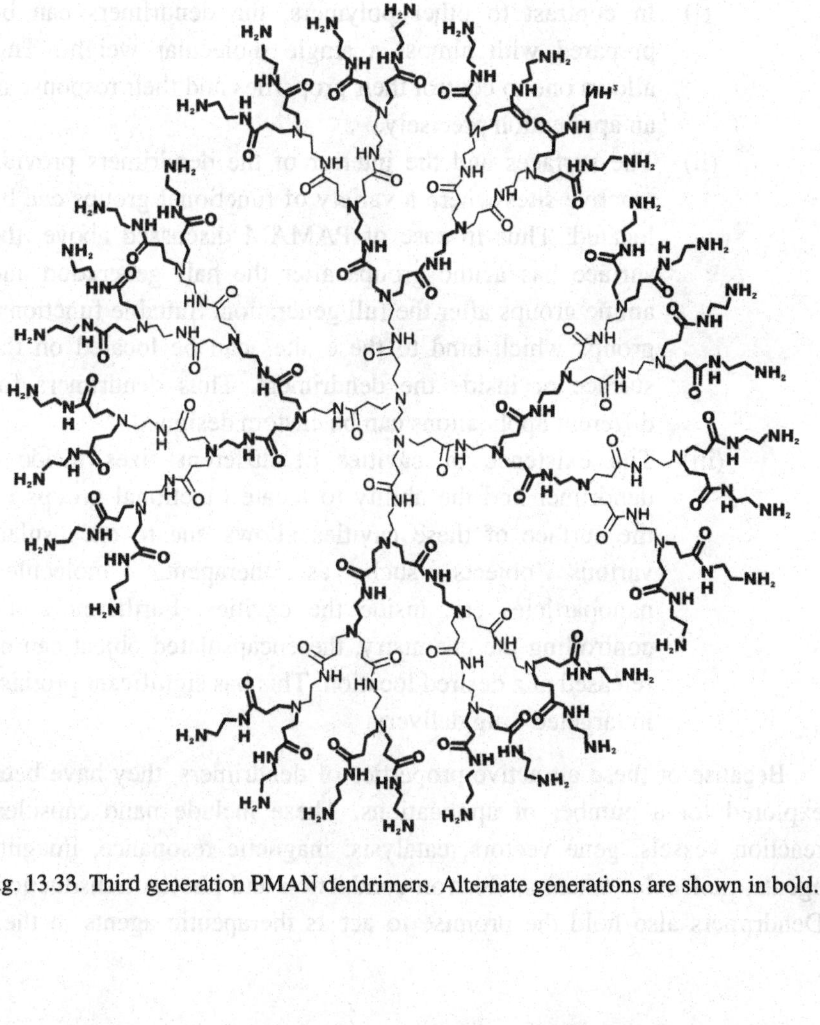

Fig. 13.33. Third generation PMAN dendrimers. Alternate generations are shown in bold.

Up to generation three (G3), the dendrimers are small, without any specific three dimensional structure. With higher generations they begin to become spherical and assume a distinct three dimensional structure. Beyond G5, they are highly structured spheres. The diameter of a G9 dendrimer is typically about 10 nm. The number of primary functional amine groups in a PAMAM dendrimer of generation n is given by 2^{n+2}.

13.6.2 *Properties and applications*

The dendrimers are distinguished by the following characteristics.

(i) In contrast to other polymers, the dendrimers can be prepared with almost a single molecular weight. This allows one to control their properties and their response in an application precisely.

(ii) The surfaces and the interior of the dendrimers provide reactive sites where a variety of functional groups can be located. Thus in case of PAMAM discussed above, the surface has acidic groups after the half generation and amine groups after the full generation. Suitable functional groups which bind to these sites can be located on the surface or inside the dendrimers. Thus dendrimers for different applications can be custom designed.

(iii) The existence of cavities of different sizes inside a dendrimer and the ability to locate functional groups on the surface of these cavities allows one to encapsulate various objects such as therapeutic molecules, nanoparticles etc. inside the cavities. Furthermore, by controlling the chemistry, the encapsulated object can be released at a desired location. This has significant promise in targeted drug delivery.

Because of these attractive properties of dendrimers, they have been explored for a number of applications. These include nano capsules, reaction vessels, gene vectors, catalysis, magnetic resonance, imaging agents, molecular antenna, electron conduction and photon transduction. Dendrimers also hold the promise to act as therapeutic agents in their own right just like many other macromolecules such as recombinant proteins, liposomes, new synthetic peptides and recombinant protein–polymers.

Many types of dendrimers are now commercially available. Some of these are listed below:

(i) PAMAM dendrimers based on either an ethylenediamene or ammonia core and amino groups on the surface, manufactured by Dendritech (USA).
(ii) Poly (propylene imine) dendrimers sold under the trade name Astrmol by DSM (Netherlands).
(iii) Therapeutic dendrimers manufactured by Starpharma (Australia).

Problems

(1) For polystyrene of molecular weight 416,000 dissolved in toluene, the chain expansion factor α is 3.2 and the characteristic ratio is 10.0. Calculate (a) the contour length (b) the random flight end to end distance (c) the random flight radius of gyration (d) unperturbed end to end distance and (e) the actual end to end distance. Given: C-C bond length = 0.154 nm, C-C bond angle = 109.5°.

(2) Assuming that the molecules of polystyrene in Problem 1 pack together with a packing factor of 0.6 (i.e. void space of 40 %), calculate the maximum number of molecules that can form a nanoparticle (diameter $\leq$ 100 nm).

(3) (a) The T_g's for three polymers are -100 C, 20 C and 100 C. Which polymer would be most suitable to make a microlithography stamp and why?

(b) From amongst each row of the Table below, which polymer is likely to have a higher T_g in each case and why?

$(CH_2\text{-}CHCH_3)_n$ (Polypropylene)	$[CH_2\text{-}C(CH_3)2]_n$ (Polyisobutylene)
$(CH_2\text{-}CH_2)_n$ (Polyethylene)	$(CH_2\text{-}CHCl)_n$ (PVC)
$(CH_2\text{-}CHCH_2CH_2CH_2CH_3)_n$	$[(CH_2\text{-}C(CH_3)_3]_n$
(Polyvinyl n butyl ether)	(Polyvinyl t butyl ether)

(4) The polymers polystyrene and poly L-lactide form a block copolymer. The polystyrene component can be leached out after hydrolysis. Suggest a method for preparing a structure made of SiO_2 which is in the form of nanoscale interconnected gyroidal network. Explain the principles involved. (see [331]).

(5) A surfactant is needed to stabilize nanoparticles in an *organic solvent*. From amongst the groups below, give three pairs of groups you would chose to attach to the ends of the surfactant chain and specify which end is to be attached to the particle and which end would be exposed to the solvent?
Polystyrene, polyacrylic acid, polyvinyl acetate, PMMA, polyethylene oxide, PVA.

(6) What type of bonds are involved in the formation of the primary, the secondary, the tertiary and the quaternary structures of proteins? Discuss briefly.

(7) Give an example from the literature where the proteins have been used in the formation of nanostructures.

(8) Why should the proteins containing a very large number of side groups be more stable in an aqueous solution as compared to those containing mostly polar groups? *(Hint: On unfolding of the molecule, the polar groups can quickly form hydrogen bonds with water, with little change in the free energy).*

(9) Polypeptides have been used to produce what is called a "leucine zipper". Describe and explain. (see [332]).

Chapter 14

Self Assembly and Self Organization

14.1 Introduction

The manufacture of microelectronic components such as a computer chip is at present carried out by the so called "top down" method in which a thin film is first deposited and parts of it are then etched away to yield the desired pattern. The feature size that can be produced in this method has exponentially decreased over the years as discussed in Chapter 1. The current technology is expected to reach its limit of about 10 nm by the year 2016. New strategies are needed to continue further with the miniaturization of the components.

One of the approaches being investigated to assemble nanosized structures is the self assembly process. In this process the aim is to assemble a structure starting from some building blocks. A building block can be a single molecule, a preassembled set of molecules, a larger object such as a colloidal particle or even a millimeter sized object. In contrast to the strong covalent or other primary bonds that are used in the synthesis of a molecule, it is the much weaker secondary bonds that are used in the self assembly process. Although individually weak, their large number imparts a reasonable rigidity and strength to the structure. The weak bonds permit self correction of any defects during the assembly process because such bonds can be easily broken and remade to create a structure with the lowest free energy. The self assembly process thus operates under the "thermodynamic control". This again contrasts with the case of a chemical reaction in which the product formed is not necessarily the most stable but nevertheless is favored because the reaction leading to it is the fastest *i.e.* a chemical reaction operates under the "kinetic control".

In the following we first describe briefly the various secondary interactions that are made use of in the self assembly. This is followed by a discussion of some important examples.

14.2 Interactions important in self assembly

As mentioned above, it is the weak interactions which are made use of in the self assembly because of the capability of self correction. While the primary interactions such as the ionic and the covalent bonds have energies of several hundred $kJ.mol^{-1}$, the secondary interactions seldom exceed 100 $kJ.mol^{-1}$. The various interactions important in self assembly are described below. A summary of these is given in Table 14.1.

14.2.1 *Ionic and electrostatic interactions*

Electrostatic bonds can be of the type ion–ion, ion–dipole and dipole–dipole. The dipole–dipole interactions are usually grouped under the van der Waal's forces, described in Chapter 7.

Molecular ions and polyions are of particular interest in nanotechnology rather than the ionized atoms as in salts. Polyionic macro molecules are those in which a number of similar functional groups are ionized. These polyions can bond electrostatically with small ions to form complex molecular aggregates. These polyions can also interact with oppositely charged polyions to form ultra thin layers as in the layer-by-layer (LBL) assembly.

In the ion-dipole interaction, an ion is attracted to the oppositely charged end of the dipole on a polar molecule. The interaction energy is directly proportional to the charge on the ion (q), the dipole moment of the dipole (μ) and inversely proportional to the distance (d) between the ion and the dipole

$$E \propto \frac{q\mu}{d^2}$$

An example of this interaction is the attraction of ions of a salt such as Na^+ and Cl^- to the water molecules in an aqueous solution of NaCl.

Table 14.1 A summary of the interactions important in the self–assembly.

Type	Description/example	Energy, kJmol^{-1}
1. Electrostatic		
(i) Ion–ion;	Macromolecules with ionized functional groups interact with small ions	> 190
(ii) Ion–dipole	Salt ions (e.g. Na^+ and Cl^-) with dipoles on water molecules in aqueous solution of the salt.	40 to 120
2. π interactions	The electron rich interior of a benzene ring is attracted to the electron poor exterior of another benzene or other aromatic ring	10 to 20
3. Hydrogen bond	A–H --- B where A and B are both highly electronegative	5 to 50
4. Van der Waal's interaction		
(i) dipole–dipole		5 to 40
(ii) Fluctuating dipoles (dispersion or London force)	Condensation of inert gas molecules at low temps; formation of colloidal crystals	< 5
5. Coordinate bond	Bond between metal and a small molecule (ligand) : the molecule donates a pair of electrons to the metal	20-380
6. Dative bond	Special case of the coordinate bond in which the bond is between the main group–main group elements or between metal–metal.	20 to 380
7. Hydrophobic effect		

The electrostatic interactions are weaker than the covalent bonds and are useful where adequate bond strength together with motion and structural rearrangement are required.

14.2.2 *The π–interactions*

Although the π bonds are also part of the covalent interactions, they deserve special mention because of their importance in the formation of nanostructures by the molecular assembly. They play an important role in the stacking (one above the other) of bases in a DNA molecule, in determining the shapes of the proteins, in the packing of molecular crystals, in molecular electronics and in other areas. The π bonds in a molecule form a region in which the electrons are delocalized. As an example, in an aromatic molecule such as benzene, the *p* orbital perpendicular to the plane of the molecule overlaps with the *p* orbital of the adjacent carbon atoms to form π bonds. These bonds are not localized but form two continuous doughnut shaped electron clouds, one on each side of the molecular plane. These electron clouds can form non-covalent bonds with a system with a net positive charge region. Some examples of these interactions are shown in Fig. 14.1 [333].

As in an aromatic ring the π electron clouds exist below and above the molecular plane, the stacking interactions such as those depicted in Fig. 14.1 can occur repeatedly, over long distances, providing considerable electrostatic stability and alignment. Further, the interactions involved in such structures can be adjusted by hetero atom substitution as in the bases purenes and pyrimidenes in DNA. Moreover, the π interactions can also be manipulated by changing the solvent. These features are very useful for nanoscale design. However, an unpredictable stacking sequence and large size of the aromatic rings as compared to the space required for the more direct hydrogen bond or metal–ligand coordination bond pose some limitations to design of nanostructures by the π interactions.

As mentioned above, in an aromatic ring, the electron cloud exists below and above the molecular plane. If several of these aromatic rings are joined in a long chain (Fig. 14.2 a) then the electron clouds of

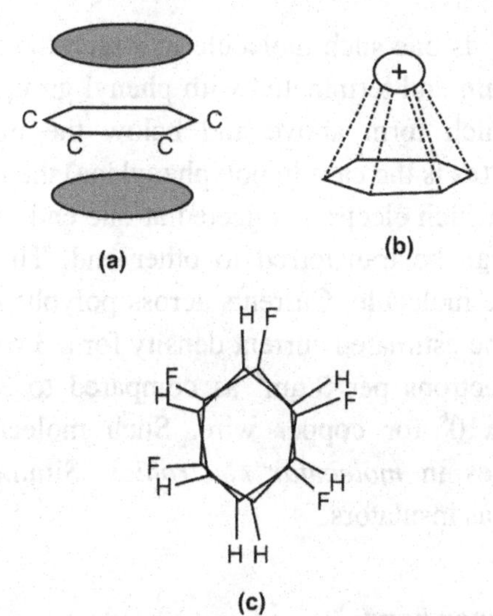

Fig. 14.1. (a) Overlapping of π orbitals on the adjacent carbon atoms in benzene leads to the formation of doughnut shaped electron clouds above and below the plane of the molecule (b) π-cation interaction (c) Stacking of benzene and perfluorobenzene; in benzene the electron density is highest on the π system while the σ system of carbons with peripheral hydrogen have a net positive charge. In perfluorobenzene, the F atoms have the highest electron density which interacts with the net positive charge on the hydrogens in benzene leading to stacking.

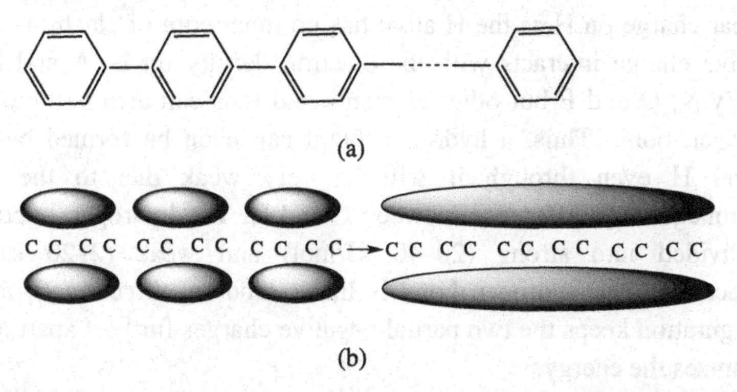

Fig. 14.2. (a). Polyphenylene, a molecule consisting of a linear chain of aromatic rings (b) in such a molecule the individual electron clouds of the aromatic rings combine to form continuous electron distribution above and below the plane to lower the energy.

individual molecules join together to form a continuous electron distribution above and below the molecule (Fig.14.2 b).

Polyphenylene is one such molecule in which the phenylene groups are joined in a chain and terminated with phenyl groups at both ends. If the π orbitals which form above and below the molecule are only partially occupied (as is the case in polyphenylene) then they can form channels through which electrons injected at one end of the molecule by applying a bias can be transferred to other end. This imparts certain conductivity to the molecule. Currents across polyphenylene wires have been measured. The estimated current density for a 3 ring polyphenylene wire is 4×10^{12} electrons per $S.nm^2$ as compared to 2×10^{11} for carbon nanotubes and 2×10^6 for copper wire. Such molecule are therefore promising as wires in *molecular electronics*. Similarly the aliphatic molecules can act as insulators.

14.2.3 *The hydrogen bond*

Hydrogen bond plays a very important role in nanotechnology because of its directionally and reliability. A hydrogen bond arises from a link of the type A–H...B. Here A and B both are highly electronegative elements and B possesses a lone pair of electrons. The pull of electron of hydrogen towards A forms a polar bond between A and H and exposes the positive nuclear charge on H as the H atom has no inner core of electrons. This positive charge interacts with the electron density on B. A and B are usually N, O and F but other elements and ions can also participate in hydrogen bond. Thus, a hydrogen bond can even be formed between C and H even through it will be very weak due to the small electronegativity difference between C and H. The hydrogen bonds can be divided into strong (20–40 kJ/mol) and weak (2–20 kJ/mol) interactions. The hydrogen bond is linear (and so, directional), as this configuration keeps the two partial negative charges furthest apart and so minimizes the energy.

A familiar example of the hydrogen bonds is in the base pairing in DNA (thymine (T) with adenine (A) and cytosine (C) with guanine (G)). The bases, consisting of small aromatic heterocycles have strongly bonding functional groups incorporated in them.

As was mentioned in the Introduction to this chapter, the building blocks for self assembly range from molecules to larger, even millimeter sized object. Supramolecular chemistry can be thought of as a sub area of self assembly which concerns itself with assembling of larger structures form molecular building blocks (MBB). The molecular building block may be a single molecule or a number of molecules which are covalently bonded but have functionalities on the outside which enables them to joine with other MBB by hydrogen bonding or other electrostatic interactions. The hydrogen bond is extensively used in supramolecular chemistry because of its several advantages: (1) Hydrogen bonds are self assembling and self directing; an example is the spontaneous formation of protein secondary structure in aqueous media. (2) There are a number of functional groups such as OH, C=O, N–H, C=N, COOH, NH_2, NOO– which can be incorporated in the covalent framework to provide hydrogen bond. (3) Hydrogen bonds occupy very little space as compared to, for example, π interactions which require large area. (4) Multiple hydrogen bonds can be incorporated in a structure to provide orientational specificity.

The hydrogen bond can be used to form large structures such as molecular crystals and extended arrays of massive molecules. Fig.14.3 below shows, as an example, a portion of the hydrogen bonding network in the peptide β sheets.

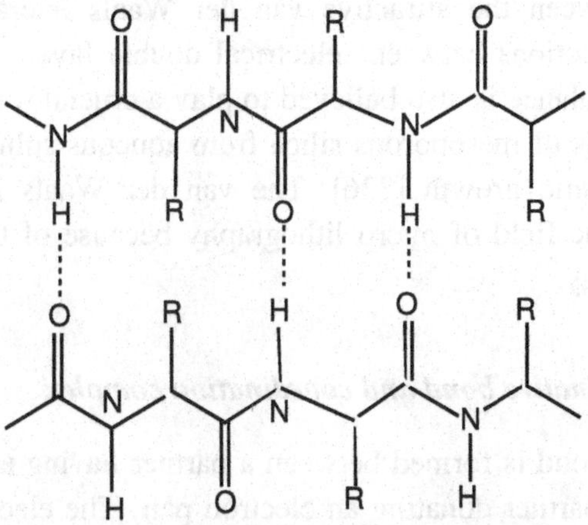

Fig. 14.3. Hydrogen bonds can be used to stabilize massive structures as shown here by the stabilization of peptide sheets by hydrogen bonds.

A limitation of the hydrogen bond is its relatively high strength. This makes it difficult to correct any errors in the assembly. Another problem often encountered, especially in an aqueous medium, is that there may be many other interactions which have comparable strength and may occur depending on the conformation of the molecules. Then the desired hydrogen bonding may become difficult to predict or control.

14.2.4 *van der Waals interactions*

The Van der Waals interactions have already been discussed in Chapter 7. They play an important role during the fabrication and manipulation of nanostructures in many cases. The hydrophobic interactions (discussed below) are dependent on the van der Waals interactions. The hydrophobic interactions are crucial in protein folding, in the holding together of the cell membranes and in other biological structures. Such biological structures serve as models for the fabrication of useful nanostructures by supramolecular chemistry. As another example, the lateral van der Waals interactions between nanowires (or nanotubes) cause them to grow in a self organized parallel aligned fashion *e.g.* growth of parallel aligned wurtzite ZnO nanowires on sapphire substrate [334] and growth of parallel array of carbon nanotubes on silicon. Formation of colloidal crystals from their suspensions occurs because of a balance between the attractive van der Waals interactions and the repulsive interactions between electrical double layers or steric forces [97] ; such a balance is also believed to play a crucial role in the growth of curved forms of mesoporous silica from aqueous solutions [335] and in the biomimetic growth [336]. The van der Waals interactions are important in the field of micro lithography because of their role in the resist technology.

14.2.5 *Coordinative bond and coordination complex*

A coordinate bond is formed between a partner having an empty orbital and the other partner donating an electron pair. The electron donors are denoted as Lewis bases and the electron acceptors as Lewis acids. The electron acceptors are usually the metal atoms or metal ions while the electron donors are small molecules (ligands). Metals are known to coordinate with up to 12 ligands though usually the number varies from 4 to 8. The metal ligand compounds are known as complexes because of the presence of a large number of coordinating ligands. The bonding between the metal and the ligand can be by a single bond which is analogous to a main group σ bond (Fig. 14.4 a); it can be by coordination of metal by more than one pair of electrons (Fig.14.4 b) when the bond is termed as a chelating ligand bond or it can be by metal ligand π bond Fig. 14.4 (c). An example of (c) is ferrocene in which Fe (II) is coordinated by two five member aromatic cyclo-pentadienyl rings ($[C_5H_5]^-$). In such systems, the rotation about the metal and ring center axis can takes place easily, making them useful in the design of molecular rotors.

The strength of the complex bonds covers a wide range between weak interactions (*e.g.* dipole interactions) and the strong covalent bonds. This adjustable bonding strength can be used to yield adjustable lifetimes of molecules. This is useful in the creation of nanostructures using the supramolecular architecture.

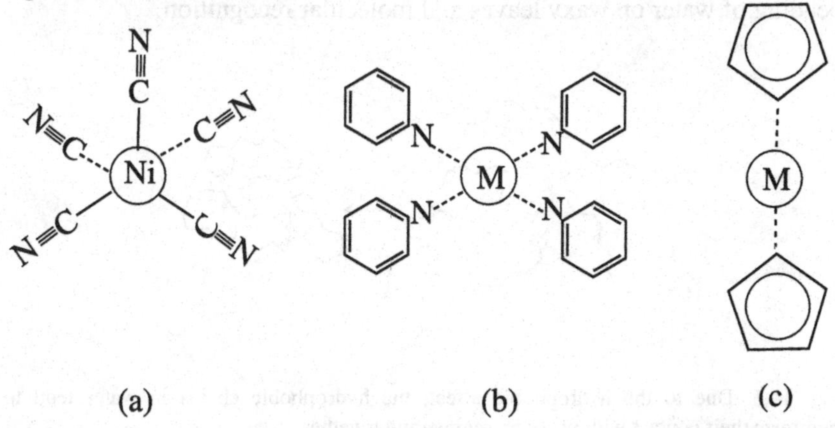

Fig. 14.4. (a) A Ni atom coordinated to five ligands by single bonds (b) metal M is coordinated to two chelating ligands by two bonds each (c) Metal–ligand π bond.

14.2.6 *Dative bond*

When the coordinate bond is limited to main group–main group elements or to metal– metal interaction, it is called a dative bond. Such bond is most common between group III and group V atoms where the group III element has an empty orbital and group V element has at least one lone pair available for bonding. A well known example of a dative bond is the bond between boron and nitrogen. The strength of the dative bond is usually between 50 to 85 kJ/mol and can be adjusted by changing the donor and acceptor elements. The dative bond is directional like the covalent bond but has lower strength, making it useful for the fabrication of nanostructures.

14.2.7 *The hydrophobic effect*

In an aqueous medium, hydrophobic substances, such as oil, segregate together, expelling water from between them. Examples of the manifestation of this effect are in the self organization of amphiphile molecules into micelles and membranes, in the folding of proteins, beading of water on waxy leaves and molecular recognition.

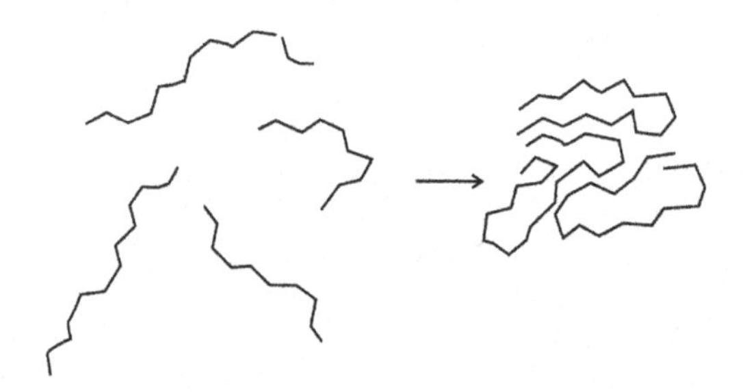

Fig. 14.5. Due to the hydrophobic effect, the hydrophobic chains in water tend to minimize their contact with water by segregating together.

The hydrophobic effect is an entropic effect. Molecular dynamics simulations of water show that a water molecule is surrounded on the average by 4 other nearest neighbor water molecules. Also the water molecule is hydrogen bonded to two out of these four water molecules at any instant. Thus this system of one water molecule surrounded by four other molecules has a number of configuration states available to it. A hydrophobic surface, exposed to water, is unable to form hydrogen bonds with the water molecules. The number of configurational states available to water are therefore drastically reduced. This leads to a decrease in entropy. The layer of the water molecules around the hydrophobic surface is, therefore, organized in an "ordered layer" with much less freedom to translate and rotate. If another hydrophobic entity closes in on the previous hydrophobic surface and replaces this ordered water layer (Fig. 14.5), there is an increase in entropy and a decrease in the free energy of the system. This decrease in the free energy due to increase in the entropy is the driving force for the hydrophobic effect.

14.3 Examples of self assembled systems

14.3.1 *Self assembled monolayers (SAM)*

Two important methods for the preparation of metal nanoparticles are evaporation–condensation (see Chapter 11) and precipitation from a salt solution. The most important consideration in any preparation scheme is that the particles need to be covered with a layer of some molecules as soon as they are produced. This is necessary to prevent the aggregation of the particles. The molecular layer does this by introducing a steric or electrostatic repulsion between the particles. Such particles are usually referred to as monolayer protected clusters (MPC)..

The most used stabilizing molecules for noble metals such as gold and silver are alkane thiols, $C_nH_{2n+1}SH$. The sulfur atom in these molecules forms a strong covalent bond with the metal with an energy of 40–45 Kcal/mol. The exact mechanism of this bond formation is not yet established. The various mechanisms which have been suggested are the following:

(1) Oxide on Au is reduced by thiol

$$Au_nAu_2O + 2RSH \rightarrow Au_n(Au{-}SR)_2 + H_2O$$

(2) Hydrogen evolution

$$Au_n–Au_2 + 2RSH \rightarrow Au_n(Au\text{-}SR)_2 + H_2$$

(3) Redox process in the presence of a disulfide

$$Au_n–(Au)_2 + R–S–S–R \rightarrow Au_n(Au–SR)_2$$

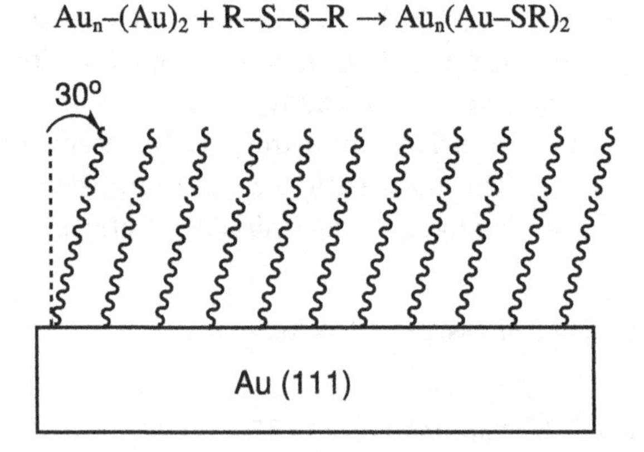

Fig. 14.6. Self assembled alkanethiol chains on gold surface; the S atoms at one end of the chains bond with the gold surface and arrange themselves in a hexagonal close packed arrangement on the (111) gold surface while the alkyl chains are inclined at an angle of about 30° from the surface normal to maximize the inter chain van der Waal's interaction.

The interesting property of the thiols is that they form a monolayer on the particle surface; furthermore, this monolayer reorganizes to a dense hexagonal packing by van der Waals bonding. The alkane chains tilt at 30° from the surface to maximize the van der Waals interaction (Fig. 14.6). The monolayers are called Self Assembled Monolayers or SAMs.

14.3.2 *Self assembly of amphiphiles into nanostructures*

14.3.2.1 *The critical micelle concentration (CMC)*

A well known example of self assembly due to the hydrophobic effect is the organization of amphiphiles into various structures. In Chapter 13, the amphiphile molecules which are commonly used as surfactants were described. Such molecules have a hydrophilic head and a hydrophobic tail. If a drop of a surfactant solution is placed on the surface of water, the molecules form a monolayer on the surface with their polar head pointing towards the water and the hydrophobic tail pointing away from it. If the concentration of the surfactant molecules is increased, the layer becomes more compact. This is accompanied by a decrease in the surface tension. Beyond a critical concentration, called the critical micelle concentration (CMC), the surface tension becomes constant indicating that the excess molecules are now going into solution inside the water. It is found that the surfactant molecules self assemble into nanostructures, called micelles, in the bulk of the solution. The type of nanostructure formed depends on the surfactant and its concentration.

The CMC is defined as the concentration at which there is a sharp increase in the number of micelles. This increase is accompanied by a change in many properties of the solution and any of these can be used to locate the CMC. Apart from the surface tension (which is the property most commonly used to determine the CMC) properties like osmotic pressure, self diffusion of surfactant, turbidity *etc.* show a sharp change at CMC. The exact value of CMC will depend on the technique used to measure it. Thus CMC is not a thermodynamic phase transition but only a phenomenologically defined quantity.

14.3.2.2 *Spherical micelle — the simplest amphiphile nanostructure*

The simplest of the nanostructures forming at CMC is a spherical micelle. The other structures are described later. In a spherical micelle, (Fig. 14.7) the hydrophobic tails of the amphiphile point towards the center of the micelle and the polar head groups face outwards forming a corona around the hydrophobic tails. The polar head groups thus are exposed to water while the hydrophobic tails are shielded from water.

The driving force for micellization is the hydrophobic effect. Around the tail of a surfactant molecule in solution, the water molecules have an ordered arrangement due to hydrophobicity of the tail. When the surfactant molecule becomes part of a micelle, this water is expelled outside the micelle where its order no longer persists. There is thus an increase in entropy. There is some reduction in entropy due to the ordering of surfactant molecules in a micelle; also there is an increase in enthalpy due to demixing of water and surfactant. However, the last two contributions are small so that there is a net decrease in free energy accompanying the micelle formation.

14.3.2.3 *Other nanostructures — relation to the packing factor*

In addition to the spherical micelle, the molecules can form aggregates having a variety of structures. A simple way to express the dependence of the nanostructure formed on the nature of the molecule is to consider the packing factor of the molecules in a structure. This can be illustrated

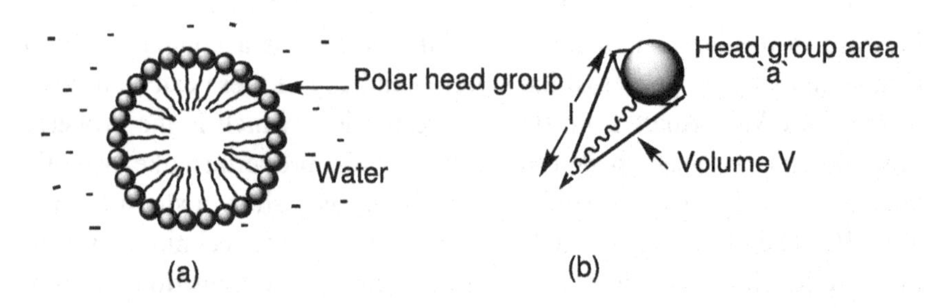

Fig.14.7. Schematic arrangement of surfactant in a spherical micelle (a) polar head groups are on the outside, exposed to water while the hydrophobic tails point towards the center, away from water (b) a spherical micelle can be looked at as a packing of cones.

by taking the example of a spherical micelle. A spherical micelle can be considered as a packing of cones with the head group at the base of the cone and the lipophilic tail pointing towards the apex of the cone (Fig. 14.7 b). The area occupied by a head group, `$a$` is optimized to balance the tendency of the lipophilic tails to attract each other (tending to reduce `$a$`) and the repulsion between the ionic head groups. The upper limit on the length ℓ of the cone is the length of the fully extended hydrocarbon chain. Let the number of surfactant molecules in a micelle be n_a, the aggregation number. Then for a spherical micelle,

$$n_a = \frac{\frac{4}{3}\pi R_s^3}{V} \tag{14.1}$$

where R_s is the radius of the micelle and V is the volume per molecule. The value of n_a is also obtained by dividing the surface area of the micelle by `$a$`

$$n_a = \frac{4\pi R_s^2}{a} \tag{14.2}$$

Combing Eqs. (14.1) and (14.2) one gets

$$\frac{V}{aR_s} = \frac{1}{3}$$

As $R_s \leq \ell$, the length of fully extended chain, hence

$$\frac{V}{a\ell} \leq \frac{1}{3}$$

The quantity V/al is called the packing factor p. Thus p has to be less than or equal to 1/3 for a spherical micelle. For large values of p, the micelles assume different shapes as shown in Fig. 14.8.

Estimates of the values of ℓ and V for an alkyl chain containing n_c carbon atoms, can be obtained from the following equations [337, 338]

$$\ell = 0.154 + 0.127\, n_C \text{ (nm)} \tag{14.3}$$

and

$$V = 0.027\,(n_C + n_{Me})\ (\text{nm}^3) \tag{14.4}$$

where n_{Me} is 1 for single chain amphiphiles and 2 for double chain amphiphiles.

Example 14.1. *Calculate the area of the head group of a phospholipid with the structure given in Fig. 14.9 assuming that the packing factor is the highest value possible for the nanostructure that forms (see Fig. 14.8).*

Solution. From Fig. 14.8 we find that the double chained lipids form planar bilayers or lamellae with $V/al = 1$.

In this case n_c is the average of the number of carbon atoms in the two chains
Number of carbon atoms, n_c = [15 (palmitate) + 18 (oleate)]/2 = 16.5
Number of chains $n_m = 2$
Using equations (14.3) and (14.4), we get
$l = 2.2495$, V = 0.4995 which gives area of the head group $a = V/l =$ 0.222 nm^2. □

14.3.3 *Langmuir–Blodgett (L–B) films*

Consider the following experiment. A hydrophilic substrate such as glass is dipped in water contained in a trough. A few drops of an amphiphile in a volatile solvent are now spread on the water surface. The solvent evaporates, leaving a monolayer of amphiphiles on the water surface. The monolayer is compressed by moving a barrier so that a close packed monolayer of amphiphile is obtained. The substrate is now withdrawn at a controlled rate. This causes a monolayer of the amphiphile to be transferred to both sides of the substrate with the polar groups attached to the substrate (Fig. 14.10).

This technique of film deposition is called the Langmuir–Blodgett) (L–B) technique and the deposited film is called the L–B film. The film thickness can be increased by repeated deposition of monolayers [339, 340].

Packing factor and shape	Structure	Example
$V/a\ell < \frac{1}{3}$ 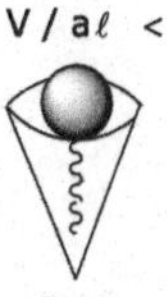 Cone	Spherical micelles	Surfactants with large 'a':SDS in low salt solution
$\frac{1}{3} < \frac{V}{a\ell} < \frac{1}{2}$  Truncated	Cylindrical micelles	Smaller head group area: nonionic lipids, CTB in high salt solution
$\frac{1}{2} < \frac{V}{a\ell} < 1$  Trancated Cone	Flexible bilayer, Vesicles	Double chained lipids with high head group areas
$\frac{V}{a\ell} \approx 1$ 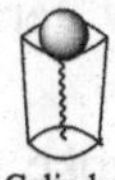 Cylinder	Planar bilayers, lamellae	Double chained lipids with small head group areas
$\frac{V}{a\ell} > 1$ 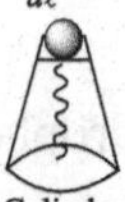 Cylinder	Inverse micelles	Nonionic lipids

Fig. 14.8. Relation of the shape of a surfactant aggregate to the packing parameter [338].

Thus in the above experiment, if the substrate is again dipped in the trough, a second monolayer is deposited over the first monolayer with the hydrophobic tails of the two layers pointing towards one another. Pulling out the substrate causes a third layer to be deposited with its

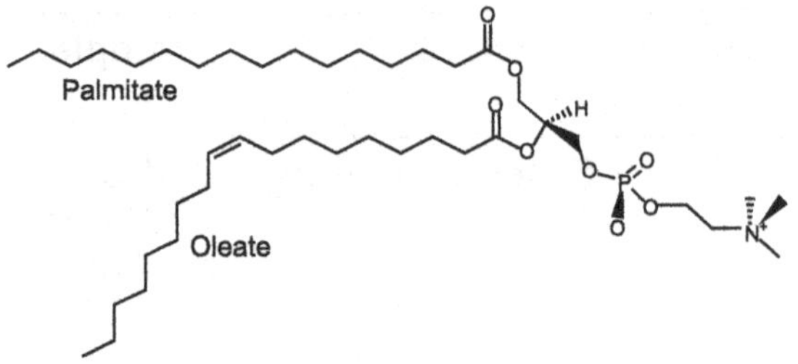

Fig. 14.9. A phospholipid.

polar groups pointing towards the substrate. The process can be repeated to give a film consisting of several monolayers.

The L–B films can be made more robust by first treating the substrate to create on it some moieties which would form a covalent bond with the surfactant molecule *e.g.* OH groups on the substrate would bond to the alkoxysilanes producing a strongly bound, close packed and well ordered film [341]. Depending on the substrate–amphiphile combination, the monolayer may be transferred to the substrate during both dipping and pulling out of the substrate (down stroke and upstroke, respectively) or during one of the strokes only. The films are accordingly termed differently.

- X type film: Monolayer transferred only during the down stroke.
- Y type film: Monolayer transferred both during the down stroke and the upstroke.
- Z type film: Monolayer transferred only during the upstroke.

The relative arrangements of the polar head groups and hydrophobic tails for a hydrophilic substrate are shown in Fig. 14.11.

By selecting suitable surfactant–substrate combinations, it has been possible to design L–B films for various applications. By incorporating objects such as nanoparticles, fullerene, nanowires etc. in the L–B films even greater versatility in applications has resulted.

The Langmuir–Blodgett films have potential applications in diverse areas. These include photovoltaic cells, electron beam microlithography, two dimensional magnetic arrays, integrated optics, electronic displays, biological membranes, molecular electronics, sensors *etc.* The L–B

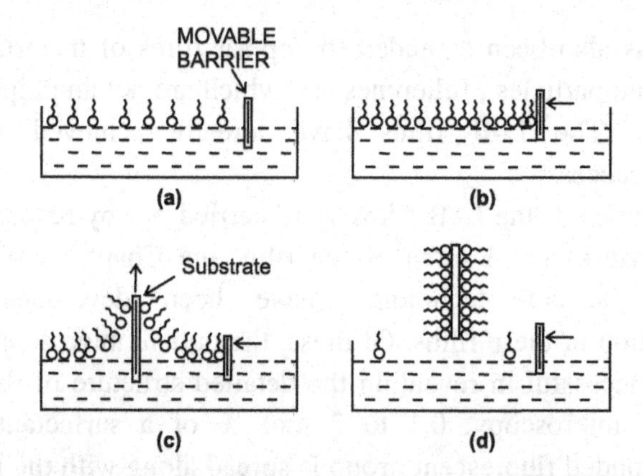

Fig. 14.10. Deposition of L-B film. (a) A solution of an amphiphile is spread on the water surface (b) the solvent evaporates; the amphiphile monolayer is compacted by moving the barrier (c) the substrate is withdrawn; the film pressure π is held constant by adjusting the barrier position using a feedback circuit (d) a monolayer is transferred to each side of the substrate with polar groups in contact with the hydrophilic substrate

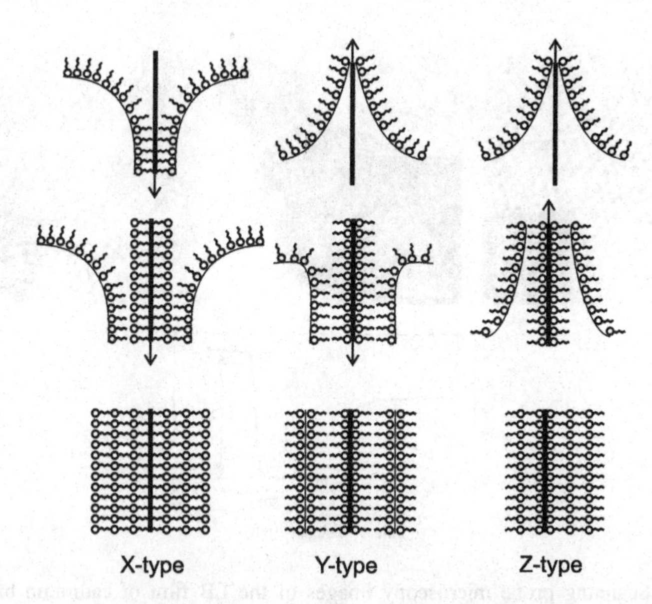

Fig.14.11. Relative arrangements of head groups and hydrophobic tails in the three types of L-B films. The monolayer deposition takes place only during the down stroke in the X type film, during the down stroke as well as the upstroke in the Y type film and only during the upstroke in the Z type film.

technique has also been extended to deposit films of materials such as polymers, nanoparticles , fullerenes *etc.* which are not amphiphiles in the usual sense. The L–B films have already achieved widespread commercial success.

Initial studies of the L–B films were carried out by recording the π (surface pressure) *vs.* A (area of the film, see Chapter 13) isotherms. With time, several techniques have been developed for the characterization of these films. Of these, the optical microscopy methods have been successful in revealing the detailed structure of the films. In fluorescence microscopy, 0.1 to 2 mol % of a surfactant having a covalently bonded fluorescent group is spread along with the monolayer. This probe surfactant molecule is so chosen that it partitions to a particular phase which therefore appears bright [342].

Scanning probe microscopy has also been applied to observe the molecular structure of LB films. An example of an SPM image is shown in Fig 14.12.

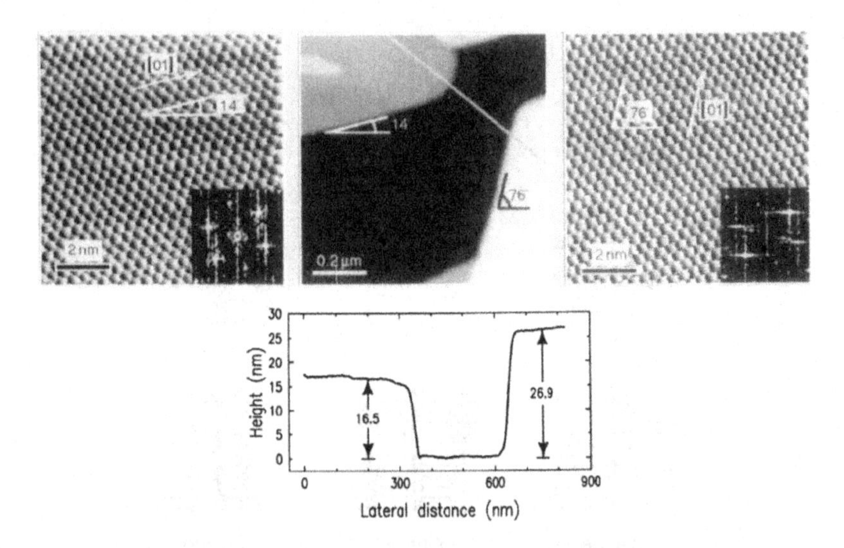

Fig.14.12. Scanning probe microscopy images of the LB film of cadmium branchiate transferred into mica. The image in the center is a 1µm X 1µm showing parts of three–bilayer (right of image); the high resolution images of the two regions are shown in the left image (three–bilayer) and right (five–bilayer) respectively. (Reproduced with permission from [343] copyright (1992) American Chemical Society).

14.3.3.1 *Deposition of nanoparticles films using the L–B technique*

The L–B technique can be used to deposit films of nanoparticles fullerenes, nanowires, *etc.* The process essentially consists of capping the nanoparticles (or other objects) with a suitable agent such as alkanethiols (for noble metal particles), fatty acids, *etc.*, dissolving in a suitable liquid which is volatile (*e.g.* chloroform) and spreading the solution over another liquid to form a monolayer which is then transferred to a substrate. Examples of metal nanoparticle films prepared in this manner are platinum nanoparticles capped with polyvinylpyrrolidone dispersed in chloroform and spread on air–water interface [344], fatty acid capped silver nanoparticles dispersed in toluene and spread on TEM grids [345] and Au_{55} cluster capped by triphenylphosphine spread from freshly prepared solution in methylene chloride [346]. Metal nanoparticles can also be prepared below or within the L–B film [347] [348].

The films of semiconductor nanoparticles can also be prepared in a similar way [349, 350]. Instead of direct spreading of nanoparticles to form a monolayer on a sub-phase (*i.e.* the underlying liquid phase), nanoparticles have also been prepared under or within the monolayers of amphiphiles and subsequently deposited on substrates.

L–B films of various fullerenes (of which the most common are C_{60} and C_{70}, see Chapter 8) have also been deposited by the L–B Technique by forming a monolayer of C_{60} dispersed in benzene spread on water and transferring the film on silver coated quartz substrates .

14.3.4 *Liquid crystals*

Liquid crystals are another important example of self assembly in which the ordering of the constituent units can occur due to their anisotropic shape combined with electrostatic, steric or dispersive (London) interactions. The constituent units in this case are molecules with anisotropic shapes like rods or discs; the molecules capable of forming liquid crystals are called *mesogens*. The most important characteristic of liquid crystals is that their molecules have a long range *orientational* order *i.e.* they are preferentially oriented along a common axis, called the *director,* over macroscopic dimensions. Some liquid crystals also have a long range translational order in one or two dimensions. About 1% of all organic molecules undergo the transitions from a solid crystal to a liquid crystal phase before transforming to the liquid phase.

The self assembly of amphiphile molecules into various structures discussed earlier is another example of liquid crystals. In this case the liquid crystals are called *lyotropic* liquid crystals. As the concentration of the amphiphiles in the solution is increased, the molecules first form *micelles* which, with a further increase in concentration, assemble into different ordered structures such as two dimensional sheets stacked one over the other or three dimensional structures such as a cube.

Liquid crystals are being very actively investigated for nanotechnological applications. Self assembling properties of liquid crystals have been demonstrated to lead to spontaneous alignment of rod like structures such as carbon nanotubes, semiconducting nanorods and conjugating polymer chains in the direction of the nematic director [351, 352].

Thermotropic liquid crystals have shown great promise in controlling the optical properties of plasmonic nanostructures. By applying an electric field to the liquid crystal phase held near the plasmonic nanoparticles, the refractive index can be changed leading to a spectral shift of the surface plasmon resonance [353].

Tuning of the photonic band gap of a colloidal crystal can be carried out by infiltrating the void space in it by a liquid crystal and using the electric field to change the refractive index of the liquid crystal phase. An inverse colloidal crystal, *e.g.* the silica inverse colloid crystal, is best suited for this application. The spherical void spaces can be filled by the liquid crystal phase resulting in a system of liquid crystal droplets dispersed in a matrix of silica. The director in these droplets is oriented in tangential orientation such that the drops effectively have two poles which make them anisotropic. When no field is applied, the drops are oriented at random and have an isotropic refractive index; when a field is applied, the director tends to get oriented along the field and an anisotropy develops. This causes a change in the optical band gap.

14.3.5 *Dendrimer self assembly*

In Chapter 13 we discussed the structure and some important features of dendrimers. The outstanding characteristics of the dendrimers, as noted there, are that they can be prepared in different sizes, with a very uniform size distribution and a variety of functional groups can be precisely located in their interior and at the periphery. These characteristics have generated great interest in using the dendrimers as fundamental, nanometer sized building blocks for the construction of devices with applications in many fields such as biomedical, nanoelectronics, advanced material, nanocatlysis, *etc.*

One striking application of the uniform size of dendrimers is their use in the assembly of nanoparticles with controlled inter-particle spacing.

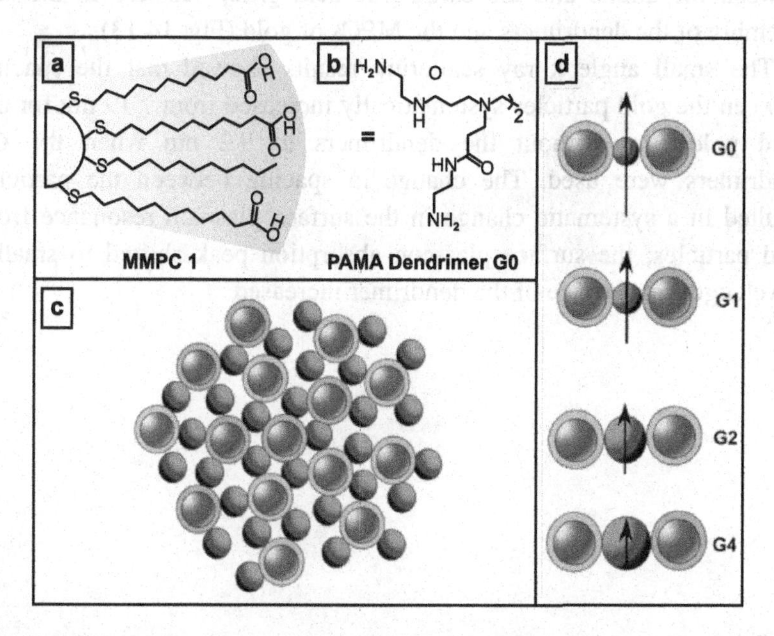

Fig. 14.13. (a) Thiol coated gold nanoparticles with carboxylic acid termination (b) PAMAM dendrimer with amine terminal group, generation zero (c) schematic showing the assembly of the dendrimer and the nanoparticles (d) the dipole-dipole interaction (length of the arrow) between the gold particles decreases as the spacing between the particles increases due to higher generation denedrimer spacers (reproduced with permission from [354] copyright (2005) American Chemical Society).

Such assemblies are useful in many applications. For example, the surface plasmon resonance from metal nanoparticles depends on many factors including the refractive index of the material surrounding the particle, the core size and the particle–particle interaction due to dipolar coupling. Of these, the particle–particle interaction depends strongly on the inter particle distance. This distance can be changed, for example, by applying a coating of uniform thickness on all the particles as was done by Ung *et al* [355] by coating the gold particles by silica. However, the dendrimers provide a more versatile method to control the inter-particle spacing. Srivastava *et al.* [354] added an excess of a solution in methyl alcohol of PAMAM dendrimers with amine terminated surface groups to a solution of thiol monolayer coated gold nanoparticles with carboxylic acid terminal group on the thiol layer. The electrostatic interaction between the amine and the carboxylic acid group resulted in the self assembly of the dendrimers and the MPCs of gold (Fig. 14.13).

The small angle x ray scattering results showed that the spacing between the gold particles systematically increased from 7.19 nm for the solid gold film without the dendrimers to 9.2 nm when the *G4* dendrimers were used. The change in spacing between the particles resulted in a systematic change in the surface plasmon resonance from gold particles; the surface plasmon absorption peak shifted to smaller wavelengths as the size of the dendrimer increased.

Problems

(1) What is meant by the statement that "the self assembly process is carried out under thermodynamic control"?

(2) Why are the directional bonds much more useful for self assembly than the non–directional bonds?

(3) Calculate the area of the head group for the following amphiphile molecules given that the packing factor is the highest value possible for the type of nanostructures they form: (i) SDS (ii) CTAB (iii) dicetyl dimethylammonium chloride.

A.1. Appendix 1: Description of crystal structure

A crystal structure can be thought of as a regular, periodic arrangement of points in space (Fig. A1.1) with an atom or a group of atoms located at each point. The periodic arrangement of points is called the lattice and the atom or the group of atoms placed on each point is called the "basis". It was shown by Bravais that there are only 14 arrangements of points in space possible such that every point has surroundings identical to that of every other point in the array. These 14 arrangements constitute the space lattices or Bravais lattices. Starting from any point in the space lattice, every other point can be reached by translation by a multiple of three fundamental translation vectors. Thus in the two dimensional square lattice shown in Fig. A1.2, these vectors are **a** and **b**.

A unit cell is the smallest unit which generates the space lattice when repeated in space indefinitely. In Fig. A1.2, the unit cell is any square *pqrs* obtained by joining the four lattice points.

Corresponding to each of the Bravais lattices a crystal structure can be obtained by combining it with a basis. The resulting crystal structures are grouped into seven crystal systems. Here we consider only the simplest of them, the cubic system. It includes the simple cubic, body centered cubic (BCC) and the face centered cubic (FCC) space lattices respectively. In the following discussion we assume that the basis consists of only a single atom. Then in the cubic system, atoms are placed at each corner of a cubic cell, in the bcc system there are atoms at

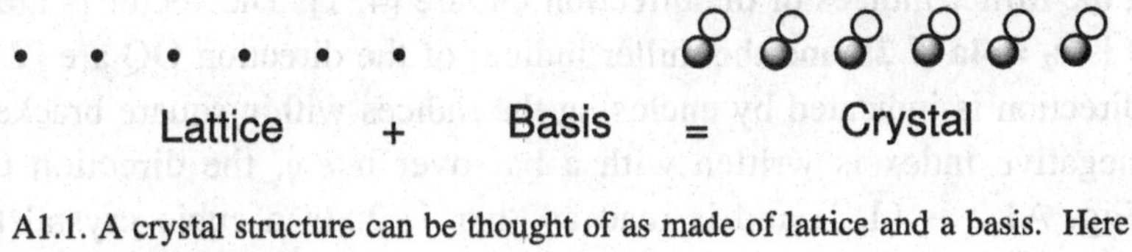

Lattice + Basis = Crystal

Fig. A1.1. A crystal structure can be thought of as made of lattice and a basis. Here the basis consists of two atoms.

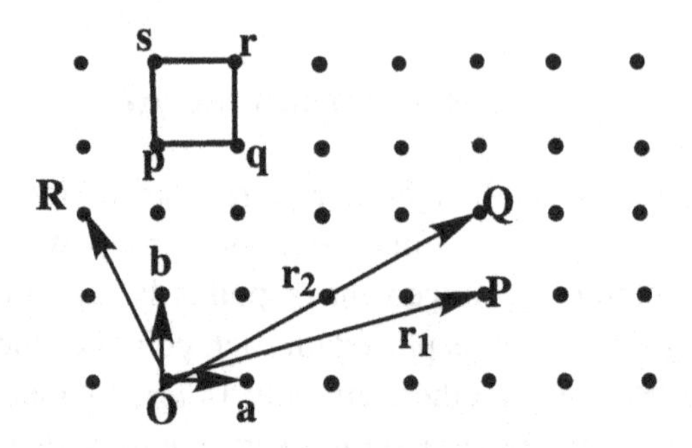

Fig. A1.2. A square lattice is shown; any point on the lattice can be reached from any other point by translating through a multiple of vectors **a** and **b**. Every point in the lattice has identical surroundings.

each cube corner and in the centre of the cell while in the FCC system there are atoms at the cube corners and the center of each face of the cube (Fig. A1.3). Since each corner of a cube is shared by eight other cubes, the eight atoms on the corners of a cube contribute 8x1/8 = 1 atom to the unit cell. In the BCC structure the number of atoms per unit cell is 2 while in the FCC it is (8x1/8 + 6x1/2) = 4 as each of the face centered atom is shared by two unit cells.

It is important to be able to specify directions and planes in a crystal. This is done using Miller indices. In a crystal, a valid direction is always along the vector joining one lattice point, which can be taken as origin, to another lattice point. If the fundamental translation vectors are **a**, **b**, and **c** then the vector corresponding to a direction can be written as $\mathbf{r} = n_1\mathbf{a} + n_2\mathbf{b} + n_3\mathbf{c}$. The Miller indices of the corresponding directions are then (n_1,n_2,n_3) provided that (n_1,n_2,n_3) are the smallest possible integers *e.g.* if $n_1 = 2m_1$, $n_2 = 2m_2$ and $n_3 = 2m_3$ then the Miller indices would be (m_1,m_2,m_3). For illustration, consider again the square lattice in Fig. A1.2. The vector $\mathbf{r}_1$ along OP is

$$\mathbf{r}_1 = 4\mathbf{a} + \mathbf{b}$$

and the Miller indices of the direction OP are [4, 1]. The vector $\mathbf{r}_2$ along OQ is $\mathbf{r}_2 = 4\mathbf{a} + 2\mathbf{b}$ and the miller indices of the direction OQ are [2,1]. A direction is indicated by enclosing the indices within square brackets. A negative index is written with a bar over it *e.g.* the direction OR in Fig. A1.2 is $(\bar{1},2)$ and is read as "bar 1, 2. In a cubic crystal the

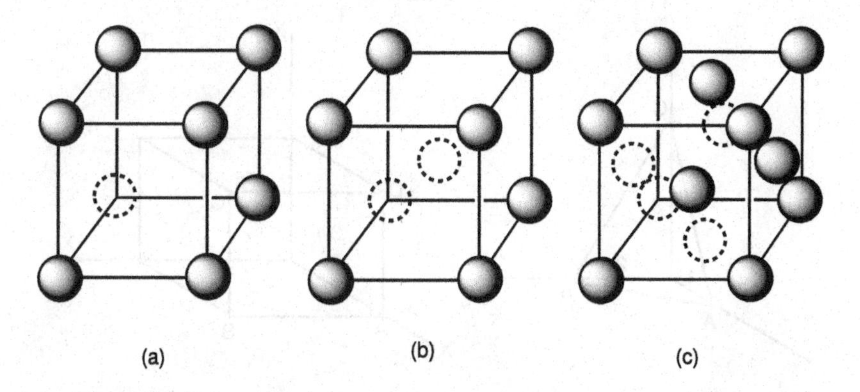

Fig. A1.3. Unit cells of (a) simple cubic (b) body centered cubic (bcc) and (iii) face centered cubic (fcc) structures; the dotted circles represent the atoms hidden from the view; for clarity the atoms are shown with a small size – they actually have radius equal to half the cube edge in (a), one fourth of the body diagonal in (b) and one fourth of the face diagonal in (c).

direction along the edges can be [100], [010], [001] as well as $[\bar{1}00]$, $[0\bar{1}0]$ and $[00\bar{1}]$. A family of directions such as this are denoted by enclosing one of the directions in triangular brackets *i.e.* <100>. Table A1.1 gives some of the family of directions in a cubic crystal and the number of directions in each family.

Table A1.1: Some of the family of directions in a cubic crystal.

Miller Indices	Direction	Members of the family
<100>	Cube edge	[100], [010], [001]
		$[\bar{1}00],[0\bar{1}0],[00\bar{1}]$
<110>	Face diagonal	[110], [101], [011] $[\bar{1}10],[\bar{1}01],[0\bar{1}1]$
		$[\bar{1}\bar{1}0],[\bar{1}0\bar{1}],[0\bar{1}\bar{1}],[1\bar{1}0],[10\bar{1}],[01\bar{1}]$
<111>	Body diagonall	$[111],[1\bar{1}\bar{1}],[\bar{1}1\bar{1}]$, $[11\bar{1}]$
		$[\bar{1}\bar{1}\bar{1}],[\bar{1}11],[1\bar{1}1]$, $[\bar{1}\bar{1}1]$

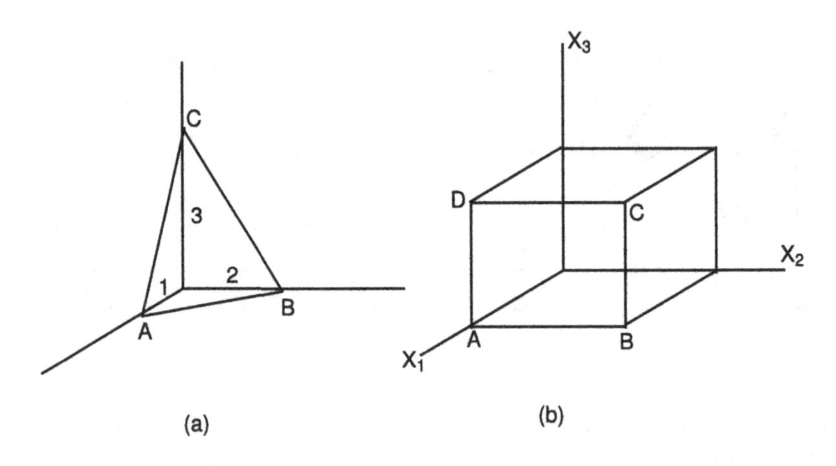

Fig. A1.4. (a) A plane ABC in a cubic crystal having intercepts 1, 2 and 3 on the axes has Miller indices (632) (b)The plane ABCD having Miller indices (100) has intercept 1 on the X_1 axis and is parallel to both OX_2 and OX_3 axes.

Note that in Table A1.1, the directions given in the second row in each case point 180° from that in the first row and are obtained by multiplying by −1 the indices in the first row. As one can arbitrarily chose the crystal orientation, the number of distinct directions in each case is just half of the number given in the last column *e.g.* 3, 6 and 4 along the cube edge, face diagonal and body diagonal respectively.

Miller indices of a crystal plane are defined as follows. Consider the plane ABC in Fig. A1.4. The Miller indices of this plane are obtained as follows:

(i) Find the intercepts OA, OB, OC of the plane with the axes in terms of the lattice constants. In this case these are 1,2,3.
(ii) Take the reciprocals of these numbers which gives 1, ½, 1/3.
(iii) Convert into smallest integers in the same ratio. This gives (6, 3, 2).
(iv) The numbers, included in parentheses (632) are the Miller indices of the plane ABC.

A value of 0 for any of the Miller indices indicates that the plane is parallel to the corresponding axis. Thus the plane (100) in the cubic lattice is the plane ABCD parallel to both the X_2 and X_3 axes (Fig. A1.4b).

For cubic crystals, a plane is perpendicular to a direction having the same indices as the plane *e.g.* [111] is perpendicular to (111). It follows that a plane containing any two directions can be obtained by taking a cross product of the two directions. Thus the plane containing the directions [111] and $[1\bar{2}3]$ is $(5\bar{2}\bar{3})$ as

$$(\mathbf{i}+\mathbf{j}+\mathbf{k}) \times (\mathbf{i}-2\mathbf{j}+3\mathbf{k}) = 5\mathbf{i}-2\mathbf{j}-3\mathbf{k}$$

For further information on this topic the reader can refer to texts like Kittel or Callister [72, 356].

A.2. Appendix 2: Review of the mechanical properties of crystalline solids

A2.1 Stress

The stress at a point is defined with reference to a plane passing through the point. The direction of the stress component and the plane on which it is acting are denoted by two subscripts. The first one gives the direction of the normal to the plane and the second one gives the direction in which the stress component is acting. If the normal to the plane is taken to be along the Z axis, then the normal stress is σ_{zz} and the two shear stresses are σ_{zx} and σ_{zy} respectively. Often, the numerals 1, 2 and 3 are used to represent the directions x, y and z and accordingly the above stresses would be written as σ_{33}, σ_{31} and σ_{32} respectively, or in general as σ_{ij}. If i = j then the stress is a normal stress, otherwise it is a shear stress. It can be shown, using the strain energy arguments, that $\sigma_{ij} = \sigma_{ji}$.

The stress at a point is totally defined by σ_{ij} with i and j assuming the values from 1 to 3. As $\sigma_{ij} = \sigma_{ji}$, only six of these nine numbers are independent. Actually, these nine numbers form a symmetric tensor of rank 2. Further information on tensor representation of stress can be found in the excellent text by Nye [357]. The nine components are written as an array as follows

$$\begin{bmatrix} \sigma_{11} & \sigma_{12} & \sigma_{13} \\ \sigma_{21} & \sigma_{22} & \sigma_{23} \\ \sigma_{31} & \sigma_{32} & \sigma_{33} \end{bmatrix}$$

Consider a small cube element around point P in a stressed body. The axes of the cube are taken to be parallel to the coordinate axes. In general, each face of the cube experiences both normal and shear stresses as shown in Fig. A2.1(a). It is always possible to reorient the cube so that only normal stresses act on any face, the shear stresses being zero for this orientation (Fig. A2.1b). The normal stresses along the three coordinate axes are denoted as σ_1. σ_2 and σ_3 respectively and are called the *principal stresses* while the axes along which the elementary cube is now oriented

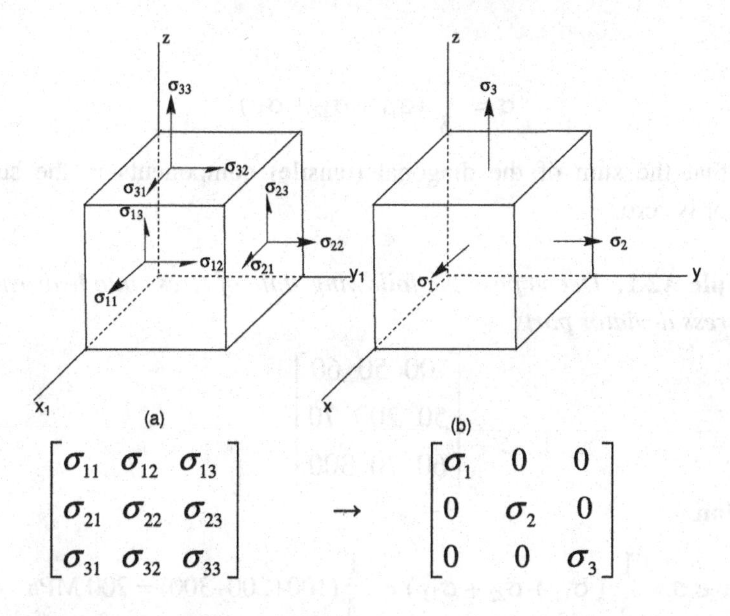

Fig. A2.1. (a) A cube element with both shear and normal stresses acting on each face (b) the cube oriented along the principal axes; in this orientation, only the normal stresses act on all faces, the shear stresses being zero.

(or, equivalently, the directions of the normals to the faces of the cube) are called the *principal directions*.

A body is said to be subjected to a hydrostatic stress if the stress at every point on its surface is a normal stress — at the faces of an elementary cube in the body only the normal stresses act and, furthermore, they are all equal *i.e.* $\sigma_1 = \sigma_2 = \sigma_3 = \sigma$.

An arbitrary state of stress can be considered to consist of two parts — a hydrostatic part and a pure shear part. The pure shear part is called the *stress deviator*.

$$\begin{bmatrix} \sigma_{11} & \sigma_{12} & \sigma_{13} \\ \sigma_{21} & \sigma_{22} & \sigma_{23} \\ \sigma_{31} & \sigma_{32} & \sigma_{33} \end{bmatrix} = \begin{bmatrix} \sigma & 0 & 0 \\ 0 & \sigma & 0 \\ 0 & 0 & \sigma \end{bmatrix} + \begin{bmatrix} (\sigma_{11}-\sigma) & \sigma_{12} & \sigma_{13} \\ \sigma_{21} & (\sigma_{22}-\sigma) & \sigma_{23} \\ \sigma_{31} & \sigma_{32} & (\sigma_{33}-\sigma) \end{bmatrix} \quad \text{(A2.1)}$$

hydrostatic stress deviator

with

$$\sigma = \frac{1}{3}(\sigma_{11} + \sigma_{22} + \sigma_{33})$$

Note that the sum of the diagonal (tensile) components in the stress deviator is zero.

Example A2.1. *Decompose the following state of stress into hydrostatic and stress deviator parts*

$$\begin{bmatrix} 100 & 50 & 60 \\ 50 & 200 & 70 \\ 60 & 70 & 300 \end{bmatrix}$$

Solution.

We have $\sigma = \frac{1}{3}(\sigma_{11} + \sigma_{22} + \sigma_{33}) = \frac{1}{3}(100+200+300) = 200$ MPa

The hydrostatic stress and the stress deviator are therefore as given below.

$$\begin{bmatrix} 100 & 50 & 60 \\ 50 & 200 & 70 \\ 60 & 70 & 300 \end{bmatrix} = \begin{bmatrix} 200 & 0 & 0 \\ 0 & 200 & 0 \\ 0 & 0 & 200 \end{bmatrix} \text{MPa} + \begin{bmatrix} -100 & 50 & 60 \\ 50 & 0 & 70 \\ 60 & 70 & 100 \end{bmatrix} \text{MPa} \quad \square$$

A2.2 Strain

The *tensile strain* is defined as the *change in length per unit length, $\Delta l/l_0$* where Δl is the change in length of an element of initial length l_0. Thus if in an elemental parallelepiped the sides of lengths x,y and z become x $+\Delta x$, y + Δy and z + Δz after the deformation then the tensile strains are

$$e_{11} = \Delta x/x,\ e_{22} = \Delta y/y \text{ and } e_{33} = \Delta z/z$$

To define the shear strains, consider a rectangular element having lengths of sides *dx* and *dy* which after deformation assumes the shape shown in Fig. A2.2. The total change in angle between the sides *dx* and *dy* is the shear strain γ and is given by

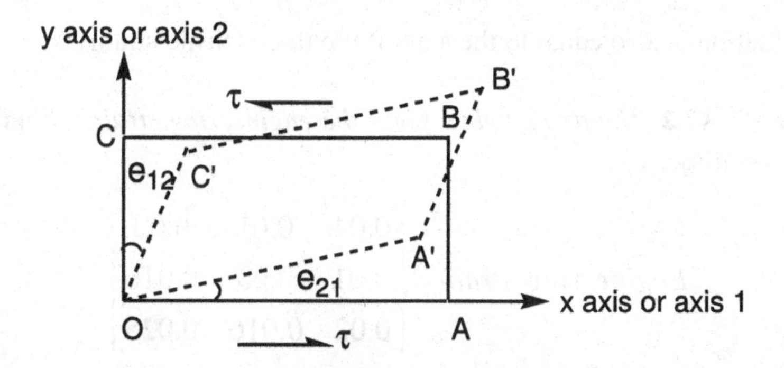

Fig. A2.2. A rectangular element OABC in the body deforms by the shear stress τ; after deformation the two sides rotate towards each other by angles e_{21} and e_{12}. The total change in angle between the sides is the shear strain $\gamma = e_{21} + e_{12}$.

$$\gamma_{12} = e_{12} + e_{21},$$

for small deformations. The strains γ_{23} ($= \gamma_{32}$) and γ_{31} (γ_{13}) can be similarly defined.

The tensile strains and the shear strains defined above are called engineering strains to distinguish them from the tensorial strains. The strain at a point in a body can also be defined by the nine components of a symmetric second rank tensor just like the stress. However, unlike the case of the stress, the engineering and tensor quantities are not the same in case of the strains. To fulfill the requirements of a second rank tensor, the components of strain tensor are related to the engineering strains as follows:

(i) The engineering and the tensor tensile strains are equal

$$e_{11} = \varepsilon_{11},\ e_{22} = \varepsilon_{22},\ e_{33} = \varepsilon_{33} \quad \text{or} \quad e_{ii} = \varepsilon_{ii}$$

(ii) The tensor shear strains are half of the corresponding engineering shear strains

$$\varepsilon_{12} = \tfrac{1}{2}\gamma_{12}\ \textit{etc.} \text{ or } \varepsilon_{ij} = \tfrac{1}{2}\gamma_{ij} \text{ for } i \neq j$$

The volumetric strain, or dilation, is defined as the change in volume per unit volume due to a hydrostatic stress

Dilation $\Delta = \delta V/V$

The dilation is also equal to the sum of the three tensile strains $\Delta = e_{11} + e_{22} + e_{33}$

Example A2.2. *The array below gives the engineering strains. Find the tensor strains.*

$$Engineering\ strain = \begin{bmatrix} 0.04 & 0.01 & 0.02 \\ 0.01 & 0.03 & 0.016 \\ 0.02 & 0.016 & 0.025 \end{bmatrix}$$

Solution.: Using the relation $\varepsilon_{ij} = \frac{1}{2}\gamma_{ij}$ for $i \neq j$, we get the tensor strains as given below:

$$Tensor\ strain = \begin{bmatrix} 0.04 & 0.005 & 0.01 \\ 0.005 & 0.03 & 0.008 \\ 0.01 & 0.008 & 0.025 \end{bmatrix}$$ □

A2.3 The elastic moduli

Application of a stress causes a deformation in a solid. Up to a certain limit, called the *elastic limit*, the strain produced is linearly related to the stress and is recovered upon the removal of the stress, *i.e.* $S = Me$ where S and e denote the stress and the strain respectively and M is the modulus. The modulus M is called Young's modulus E when the stress and strain are tensile and shear modulus G when a shear strain is produced by a shear stress. The Bulk modulus relates the hydrostatic stress to the dilation. For an isotropic solid, the various moduli are related so that there are only two independent moduli.

A tensile force applied in x direction produces an elongation in that direction. In addition, it also produces contractions in the transverse y and z directions. Experimentally, the transverse strain is related to the longitudinal strain by a constant called the Poisson ratio, ν

$$e_{22} = e_{33} = -\nu e_{11} = -\nu \frac{S_{11}}{E}$$

For most metals, the value of ν is about 0.33.

As mentioned above, for an isotropic solid the various moduli are related, only two being independent. One of the important relations is the one relating the Young's modulus, the shear modulus and the Poisson's ratio

$$G = \frac{E}{2(1+\nu)} \tag{A2.2}$$

A2.4 The elastic strain energy

When a body is deformed elastically, the work done in its deformation is stored as elastic strain energy. Consider the extension of a rod of unit area of cross section under a load. The load *vs.* elongation plot is a straight line up to the elastic limit. Work done in extending the rod by an extension Δl is

W = Strain energy stored =

$$\int_0^{\Delta l} P dl = \int_0^{\Delta l} \frac{PA_0}{A_0} \frac{l_0 dl}{l_0} = V \int_0^{e_0} S de = V \int_0^{e_0} E e de = VE \frac{e_0^2}{2} = V \frac{S_0^2}{2E}$$

Or the strain energy per unit volume

$$U_0 = \frac{S_0^2}{2E} = \frac{e_0^2 E}{2} \tag{A2.3}$$

Here P is the load applied at any time during the experiment, A_0 and l_0 are the initial area of cross section and the initial length respectively, V is the volume of the rod, e_0 is the final strain and S_0 is the final stress.

In a similar way, the strain energy per unit volume for the shear deformation is

$$U_0 = \frac{\tau^2}{2G} = \frac{\gamma^2 G}{2}$$

where τ and γ are the shear stress and the shear strain. For a general state of stress, the strain energy is obtained by adding together the contributions from the various stresses as follows:

$$U_0 = \frac{1}{2}(\sigma_{11}\varepsilon_{11} + \sigma_{11}\varepsilon_{11} + \sigma_{11}\varepsilon_{11} + \tau_{12}\gamma_{12} + \tau_{23}\gamma_{23} + \tau_{31}\gamma_{31}) \quad \text{(A2.4)}$$

and in the tensor notation

$$U_0 = \frac{1}{2}\sigma_{ij}\varepsilon_{ij}$$

Example A2.3. *A cylindrical rod of diameter 'd' and length 'l' of a material having a Young's modulus 'E' is subjected to a tensile stress σ along its axis. What is the total strain energy stored in the rod?*

Solution. The strain energy per unit volume

$$U_0 = \frac{S_0^2}{2E} = \frac{\sigma^2}{2E}$$

Therefore the total strain energy stored in the rod

$$= \text{volume x } U_0 = \frac{\pi d^2 l \sigma_0^2}{8E}$$

□

The strain energy plays an important role in many phenomena and processes. One example of this is the growth of the self assembled quantum dots by epitaxial deposition discussed in Chapter 4.

A2.5 The tensile test

A basic test to determine the mechanical properties is the tensile test. The test yields several numbers which are a measure of the various mechanical properties.

The test consists in pulling a sample in a tensile testing machine while measuring the applied load and the resulting elongation. The plot

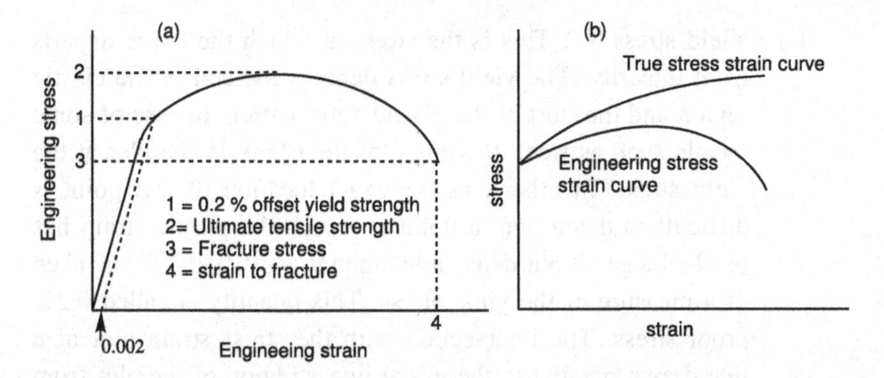

Fig. A2.3. (a) the engineering stress–engineering strain plot for a ductile metal obtained from a tensile test (b) comparison of the engineering and the true curves.

of load *vs.* elongation can be converted to an "engineering stress *S*" *vs.* "engineering strain *e* " plot by using the relations

$$S = \frac{P}{A_0} \text{ and } e = \frac{l - l_0}{l_0}$$

where P is the applied load, A_0 is the initial area of cross section of the gage length, l_0 is the gage length and l is the length of the gage section at any instant during the test. A typical stress–strain plot for a ductile metal such as copper is shown in Fig. A2.3 a. The plot shows an initial region of elastic deformation with a constant slope followed by the plastic region. In the latter region the stress required to deform the sample increases continuously. This phenomenon is called *work hardening* or *strain hardening*. The stress reaches a maximum and then decreases. The sample elongates uniformly up to the maximum. Beyond this point, the sample thins down ("necks") at some point in the middle and eventually breaks.

If the sample is unloaded before the limit of the linear region is reached, the sample regains its original shape and size; beyond the elastic limit the deformation is permanent and is called plastic deformation. The following quantities characterizing the material are obtained from the tensile test.

(i) Yield stress S_y : This is the stress at which the curve departs from linearity. The yield stress denotes the end of the elastic region and the start of the plastic deformation. In case of some metals such as steel, there is a distinct kink in the plot at the yield stress. In others, as the exact location of this point is difficult to determine, a point at which the plastic strain has reached a small but detectable amount, usually 0.2 %, is taken as a measure of the yield stress. This quantity is called 0.2% proof stress. The intersection with the stress strain plot of a line drawn parallel to the initial linear region of the plot from 0.2 % strain on the x axis gives the 0.2 % proof stress.

(ii) Ultimate tensile strength, S_u: It is the stress corresponding to the maximum in the plot. The sample deforms uniformly up to this point and beyond it undergoes "plastic instability" resulting in the formation of a neck

(iii) "Percent elongation" and the "percent reduction in area": Let l_{max} be the length of the gage section and A its cross section area in the failed region after the test, then

$$\text{Percent elongation} = \frac{l_{\max} - l_0}{l_0} \times 100$$

$$\text{Percent reduction in area} = \frac{A_0 - A}{A_0} \times 100$$

Both these quantities are measures of the *ductility* which is the capacity of the material to undergo plastic deformation before failure.

(iv) Young's modulus: Ideally, the slope of the initial linear region is the Young's modulus of the material. However, since the total elastic deformation (i.e. deformation in the linear region up to the yield point) is very small (usually < 1%), it is difficult to get an accurate value of the slope.

In defining the engineering stress, the initial area A_0 is used. During the tensile test, the area of cross section changes. To take this into account, a true stress is defined as the load divided by the minimum cross sectional area at that instant

$$\sigma = P/A$$

The true strain increment during the test is defined as

$$d\varepsilon = dl/l_0$$

so that the true strain at any instant is

$$\varepsilon = \int_{l_0}^{l} \frac{dl}{l_0} = \ln \frac{l}{l_0}$$

It can be shown that the engineering and the true quantities are related as follows in the region of uniform deformation, before the necking.

$$\sigma = s\,(1+e) \text{ and } \varepsilon = \ln(1+e) \qquad \text{(A2.5)}$$

A comparison of the engineering stress–engineering strain curve and the true stress–true strain curve is shown in Fig. A2.3 (b). This true stress –true strain plot is called the *flow curve.* Note that the flow curve does not have a maximum *i.e.* the material continues to strain harden up to fracture though with a decreasing rate.

Ignoring the initial elastic region, which is very small, the flow curve can be represented by the expression

$$\sigma = K\,\varepsilon^n \qquad \text{(A2.6)}$$

K is called the strength coefficient and n is the *strain hardening exponent*. It is given by the following relation

$$n = \left| \frac{\delta \ln \sigma}{\delta \varepsilon} \right|_{\dot{\varepsilon},T} \qquad \text{(A2.7)}$$

It can be easily shown that n corresponds to true strain at necking under uniaxial loading (see Problem 2, Chapter 11).

Usually in the tensile test, small changes in the rate of straining, $\dot{\varepsilon}$, do not have much effect. However, if the strain rate is changed by many orders of magnitude, or if the material is very strain rate sensitive, the stress–strain curve may change. The effect of strain rate is expressed by another power relation:

$$\sigma \sim \dot{\varepsilon}^{m}$$

so that

$$\sigma = A\dot{\varepsilon}^{m}\varepsilon^{n} \quad \text{(A2.8)}$$

The index m is called the *strain rate sensitivity*. It is given by the expression

$$m = \frac{\delta \ln \sigma}{\delta \dot{\varepsilon}}\Big|_{\varepsilon,T} \quad \text{(A2.9)}$$

The value of m for metals is usually ~ 0.2 while n is high for ductile metals like copper ($n \approx 0.5$) and small for metals like steel ($n \approx 0.15$). Materials with large values of n and m, can be given a large plastic strain before failure. It has been shown by Hart that the plastic instability or necking occurs when [358]

$$\frac{n}{\sigma} - 1 + m \leq 0$$

When n, m or both are high, the necking is delayed to high stresses. The reason for this is that as soon as the sample starts to neck at some point, the stress required to deform at that point increases rapidly due to high values of m and n , so the deformation rate falls down. As discussed in Chapter 11, an issue with the nanocrystalline metals is their low ductility.

A high value of m promotes a longer region of uniform elongation and delays the necking. High value of m (0.4–0.9) together with a fine grain size can result in the phenomenon of superplasticity under some conditions (low strain rate and high temperature) in which the sample elongates uniformly to very high strains which can be as large as 1000 % or more. Superplasticity has been observed in metals such as aluminum alloys as well as in ceramics, such as zirconia. As discussed in Chapter 11, one possible reason proposed for the low ductility of the nanocrystalline metals is their low *m*.

A2.6 The theoretical yield stress

Early experiments with single crystals of metals showed that lines or a band of lines appear on the surface of the crystal at the onset of the plastic deformation. These lines and bands were thought to appear due to relative slipping of the crystal planes, one over the other, and were termed slip lines and slip bands. The process is called *slip*. The stress required to cause such slipping of one crystal plane with respect to other is called the theoretical yield stress. It can be easily calculated and comes out to be ~ $G/2\pi$ where G is the shear modulus [234]. More refined calculations show that the theoretical strength is about $G/30$. However, the actual yield strength of metals is found to be lower by more than an order of magnitude. Thus the shear modulus of aluminum is about 30 GPa. The theoretical yield stress is therefore about 3 GPa but the experimentally determined yield stress is only about 40 MPa.

The discrepancy between the theoretical and the experimental yield stress arises because the slip does not take place by a simple shear of planes; instead it occurs by the movement of dislocations. As described in Chapter 2, the dislocations are line defects. The simplest dislocation to visualize is an edge dislocation which can be thought of as the edge of an extra partial plane inserted in the crystal as shown in Fig. A2.4(a). The atoms at the edge of this extra half plane can be moved by applying a small stress so that they form bonds with the atoms of the adjoining plane; this effectively results in the shifting of the dislocation by an atom spacing as shown in Fig. A2.4 (b). Thus the movement of the dislocation can be affected by a much lower stress than the theoretical yield stress.

When a dislocation moves to the surface of the crystal, a step of height equal to the Burgers vector **b** of the dislocation is created on the surface. Movement of a large number of dislocations produces observable plastic strain. The stress required to move a dislocation in the crystal lattice is called the *Peierl's stress* and is given by [234, 359]

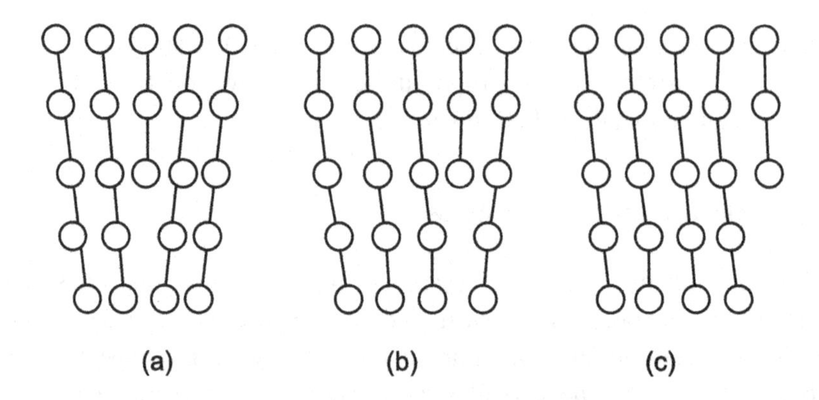

Fig. A2.4 (a) Movement of an edge dislocation due to an applied stress (a) initial position (b) after movement by b (c) after the dislocation has moved to the surface.

$$\tau_P = \frac{2G}{1-\nu} \exp^{-\frac{2\pi d}{b(1-\nu)}} \quad \text{(A2.10)}$$

Taking $d \approx b$ and $\nu = 0.33$, one gets $\tau_P = 2 \times 10^{-4}$ G which is much closer to the experimental yield stress (d is the interplanar spacing).

A2.7 The critical resolved stress

Equation A2.10 shows that the stress to move a dislocation depends on the interplanar spacing *d* and the lattice parameter (through the Burgers vector, **b**). Because of this, the Peierl's stress is lowest for a set of planes and directions. Generally such slip planes are the planes of greatest atomic density and the slip directions are the closest packed directions in the slip plane. These planes and directions are the planes and directions of easy slip. In the FCC crystals the slip planes are {111} and the slip directions are <110>. A slip plane together with a slip direction is called a *slip system*. For the FCC crystals, there are four distinct slip planes

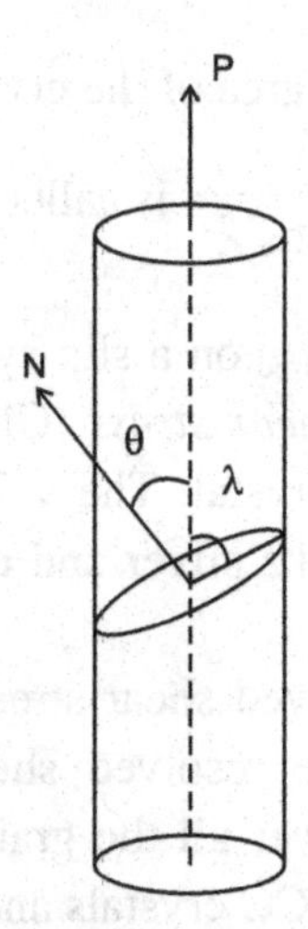

Fig. A2.5. A force F is applied to a cylindrical single crystal with a cross sectional area A; the resolved shear stress on a plane inclined at an angle θ to the crystal axis in a direction in this plane which is inclined at an angle λ to the crystal axis is (F/A)cosθcosλ.

e.g. (111), $(11\bar{1})$, $(1\bar{1}1)$ and $(\bar{1}11)$, each such plane with three <110> directions so that there are 12 distinct slip systems for the FCC crystals. In the BCC crystals, the slip planes can be {110}, {112} or {123}, all with <111> slip directions because the difference in the Peierl's stress for the different slip planes is small. In the HCP crystals the slip system is (0001) $<11\bar{2}0>$. The number of slip systems are 48 and 3 for the BCC and the HCP crystals respectively.

The stress required to produce slip by operation of a slip system depends on the Peierl's stress as well as the factors such as the concentration of dislocations, point defects and impurity atoms in the crystal. The stress acting on the slip system depends on the orientation of the crystal with respect to the loading axis. If the loading axis is inclined at an angle θ to the slip plane and at an angle λ to the slip direction (Fig. A2.5), then the resolved stress on the slip plane in the slip direction when a force P is applied along the crystal axis is

$$\tau_{resolved} = \frac{P}{A}\cos\theta\cos\lambda = \sigma\cos\theta\cos\lambda = \frac{\sigma}{M}$$

where A is the cross sectional area of the crystal, $P/A = \sigma$ is the applied tensile stress and $M = \dfrac{1}{\cos\theta \cos\lambda}$ is called the orientation factor.

The slip occurs when $\tau_{resolved}$ on a slip system reaches a critical value called the *critical resolved shear stress* (CRSS). This can be viewed as the yield stress for a single crystal. The value of CRSS can range from less than 1 MPa for metals like silver and copper to more than 90 MPa for 99.9 % pure titanium.

In a polycrystal, the resolved shear stress varies from grain to grain and to calculate the average resolved shear stress, the value of the orientation factor averaged over all the grains, $\overline{M}$, is needed. The best estimates of $\overline{M}$ are 3.1 for FCC crystals and 2 for BCC crystals [234].

A2.8 Dislocation multiplication and strain hardening

The plastic deformation in metals predominantly occurs by the motion (glide) of dislocations on the slip planes. Other, less important, mechanisms of plastic deformation are twinning and creep by diffusion of vacancies; the latter becomes important only at temperatures at which the diffusion rate is significant while the amount of deformation produced by twinning is small as compared to that by dislocation glide.

The *density of dislocations* is an important parameter in the mechanical properties of crystalline solids. It is expressed as the number of dislocations threading a unit area of the crystal, or, alternately, as the total length of dislocation lines in a unit volume of the crystal. The unit is same (number per m^2) in both the cases.

Not all the dislocations are able to glide when a stress is applied. The dislocations which glide on the slip plane and produce plastic deformation are called "glide dislocations" while the dislocations which are inclined to the slip plane and thread it are called the "forest dislocations". It is easy to calculate the strain produced due to the glide of dislocations. If n dislocations, each with a Burger's vector b, move out of a crystal of dimension d, the shear strain produced is ~ (nb/d). If the dislocations move with a velocity v, the shear strain rate is given by

$$\dot{\gamma} = \rho b v \tag{A2.11}$$

where ρ is the density of the glide dislocations and v is the dislocation velocity.

A2.8.1 Dislocation multiplication

As mentioned above, as the crystal deforms, the dislocations move out of the crystal and produce shear strain. The density of dislocations should therefore decrease with increasing deformation; instead it increases rapidly because of certain mechanisms of *dislocation multiplication.*

Emission of dislocations from the grain boundaries upon application of a stress is one mechanism leading to the increase in the dislocation density. The disordered structure of the grain boundaries permits them to behave both as a source as well as a sink for the dislocations and point defects. Figure A2.6 shows the emission of dislocations from grain

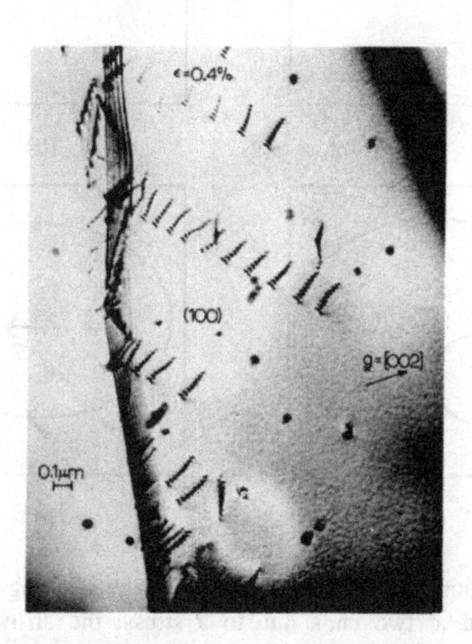

Fig. A2.6. Bright field electron transmission image of dislocation (emission) profiles associated with grain boundary ledge sources in stainless steel 304 after plastic straining to 0.4%. (reproduced from [360] copyright (1981) with permission of Elsevier).

boundary in a 304 stainless steel (steel containing 18 % Cr, 8 % Ni) after plastic straining to 0.4 % [360].

Another important mechanism of dislocation multiplication is the *Frank–Read mechanism.* The stress required to bow a dislocation segment pinned at both ends to an arc of radius R is given by

$$\tau = \frac{Gb}{2R} \tag{A2.12}$$

The maximum value of this quantity is reached when $R = l/2$ and l is the length of the pinned segment. If the applied stress is more than this, the dislocation bends further, becomes unstable and forms nearly a loop

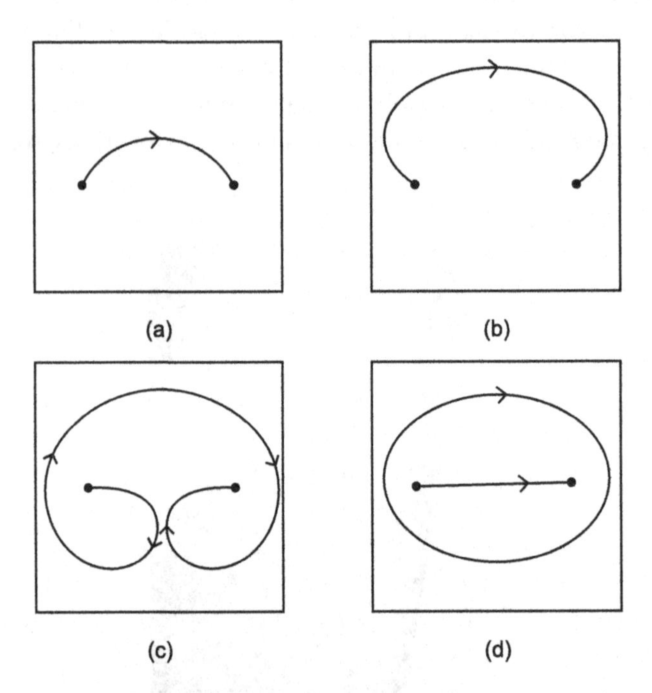

Fig. A2.7. The formation of a dislocation loop by the bending of an initially straight dislocation line pinned at two ends due to a stress; the arrow denotes the positive direction of the dislocation line; (a) the dislocation line is bent to a radius R > 2l (b) the dislocation line is bent beyond R = 2l to a radius R < 2l (c) two segments with opposite signs annihilate leading to the formation of (d) a complete dislocation loop and the dislocation line.

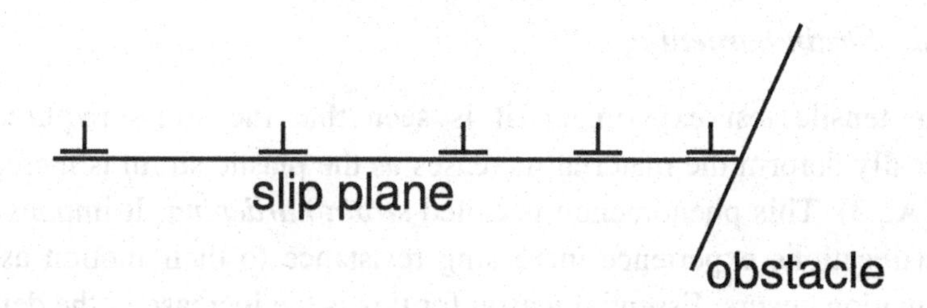

Fig. A2.8. Formation of a pile up of dislocation loops against an obstacle such as a grain boundary.

around the original dislocation line. When the segments of this near loop come in contact, they annihilate each other because they have different signs. This produces a dislocation loop as well as the original dislocation line as shown in Fig. A2.7.

The whole process can repeat many times to produce many dislocation loops. The first loop is stopped by the grain boundary or some other obstacle. The subsequent loops experience a repulsive force due to the first and the succeeding loops and assume an equilibrium separation as shown schematically in Fig. A2.8. When the cumulative repulsive force becomes larger than the applied shear stress, the further formation of loops stops. The arrangement of dislocations all lying on a slip plane, one behind the other as shown in Fig. A2.8 is called a *dislocation pile up*. The back stresses developed due to the pile up stop the Frank Read source.

Example A2.4. *Stress required to multiply the dislocations by the Frank Read mechanism in aluminum (G = 27.6 GPa) is found to be 10 MPa. Find the largest distance possible between the points pinning the dislocations. The Burger's vector is* 3×10^{-10} *m.*

Solution: Given $\sigma = 10$ MPa $= 2Gb/l = 2 \times 27.6$ GPa $\times 3.10^{-10}/l$
(the tensile stress is taken to be twice the shear stress)
which gives $l = 1.66$ μm □

A2.8.2 *Strain hardening*

In the tensile test experiment, it is seen that the stress required to plastically deform the material increases as the plastic strain is increased (Fig. A2.3). This phenomenon is called *strain hardening*. It implies that the dislocations experience increasing resistance to their motion as the deformation occurs. Essential reason for this is the increase in the density of dislocations which accompanies the plastic deformation. Two important sources for this resistance are:

(i) Dislocations piling up against a barrier: A pile up exerts a back stress on the dislocations behind it making it necessary to increase the stress to continue the glide of the dislocations.
(ii) Cutting through the forest dislocations: As a dislocation glides on a slip plane, it has to cut through the forest dislocations threading the slip plane. The cutting process may lead to production of jogs in the glide dislocation which makes the gliding difficult due to the jog not lying on the slip plane. The equation relating the flow stress to the dislocation density is

$$\tau = \tau_0 + \alpha G b \rho^{1/2} \tag{A2.13}$$

where τ_0 is the stress needed to move the dislocation in the absence of other dislocations, α is a numerical constant which varies from 0.3 to 0.6 for the fcc and bcc metals, G is the shear modulus and ρ is the dislocation density.

A2.9 Dispersion hardening

Incorporation of the particles of a second phase in a metal results in a dramatic increase in its yield strength. It requires additional stress to move the dislocations through these particles, resulting in an increase in the yield strength. Depending on the constitution and the size of the particles, the dislocations may either "extrude" between the particles or "cut through" them. The former occurs when the particles are hard and large while the latter mechanism operates for softer and smaller particles.

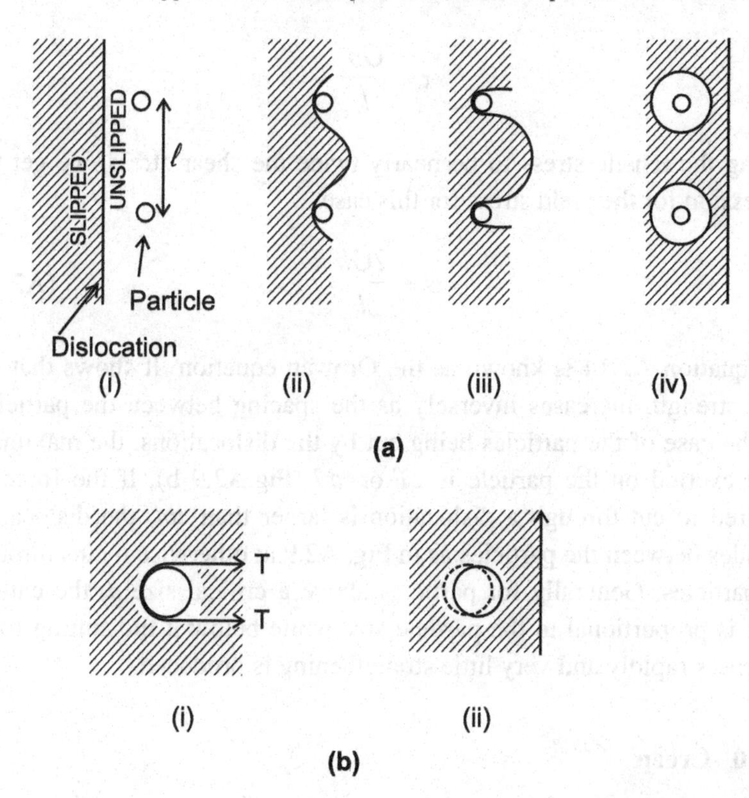

Fig. A2 9. (a) Extrusion of a dislocation between two second phase particles (b) cutting of particles by a dislocation.

The two mechanisms are shown schematically in Fig. A2. 9 (a) and (b). In (a) the maximum force is reached when the dislocation bends into a semicircle of diameter l, the distance between the particles. The force exerted by a shear stress per unit projected length of the dislocation loop is τb, where τ is the shear stress acting on the dislocation and b is its Burger's vector. In Fig. A2.9 (a iii) this force is opposed by $2T$ where T is the line tension of the dislocation ($T = ½\ Gb^2$, where G is the shear modulus). Equating the two we get

$$\tau = \frac{Gb}{l}$$

Taking the tensile stress to be nearly twice the shear stress, we get the expression for the yield stress for this case

$$\sigma = \frac{2Gb}{l} \tag{A2.14}$$

Equation A2.14 is known as the Orowan equation. It shows that the yield strength increases inversely as the spacing between the particles. For the case of the particles being cut by the dislocations, the maximum force exerted on the particle is $2T$ or τbl (Fig.A2.9 b). If the force F_c required to cut through a dislocation is larger than τbl, the dislocation extrudes between the particles as in Fig. A2.9 a; otherwise it cuts through the particles. Generally for particles above a critical size d, the cutting force is proportional to the particle size while below it the cutting force decreases rapidly and very little strengthening is produced.

A2.10 Creep

A wire of a low melting point metal like tin, loaded to a stress *below* its yield strength, undergoes appreciable increase in length with *time* at room temperature. The time dependent deformation of materials is called creep. Its rate becomes significant when the temperature is a significant fraction of the melting point.

The most important mechanism of deformation in creep is the dislocation glide, as it is at room temperature. The other mechanisms are creep due to diffusion (called diffusional creep) and grain boundary sliding. Usually a particular mechanism dominates in a range of stress and temperature for a given microstructure of the material. The diffusional creep becomes more important at high temperatures or at low stresses while the grain boundary sliding is appreciable at high temperatures.

The difference between the dislocation glide at low temperatures and at high temperatures is that while at a low temperature, the dislocations,

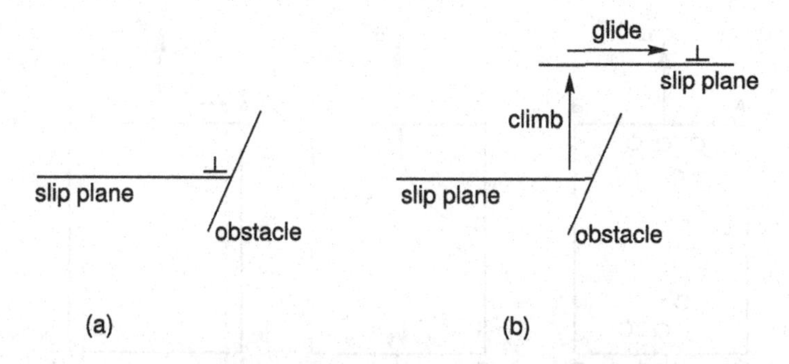

Fig. A2.10. Climb of an edge dislocation enables the dislocation to glide on another slip plane (a) dislocation held at a barrier (b) dislocation has climbed and glided on a parallel slip plane.

after gliding for some distance, are held up at barriers so that further deformation ceases, at a higher temperature, an edge dislocation can escape the barrier by climbing up and then continue gliding on a parallel slip plane, producing further deformation (Fig. A2.10).

Climb of an edge dislocation refers to its motion in a direction perpendicular to the slip plane due to the atoms from the lower edge of the extra half plane diffusing away.

The activation energy for creep deformation is found to be very close to that for self diffusion in a pure metal or the diffusion of a solute in an alloy. This confirms the climb mechanism of dislocation escape being responsible for the creep deformation.

As was mentioned above, the diffusional creep is another mechanism for creep. At low stresses and high temperatures the creep is known to occur predominantly by diffusion of atoms. In a stress gradient, a gradient in the concentration of vacancies is set up in a grain. This is because a tensile stress σ lowers the energy for the formation of vacancies by σb^3 where b is the Burger's vector and b^3 is taken to be nearly equal to the atomic volume. As a result, the equilibrium concentration of vacancies is higher under a face experiencing a tensile stress as compared to one under a compressive stress or under a lower tensile stress. There is thus a flux of vacancies from the former to the latter, or a flow of atoms in the opposite direction, leading to a mass redistribution and so a deformation as shown in Fig. A2.11. The creep

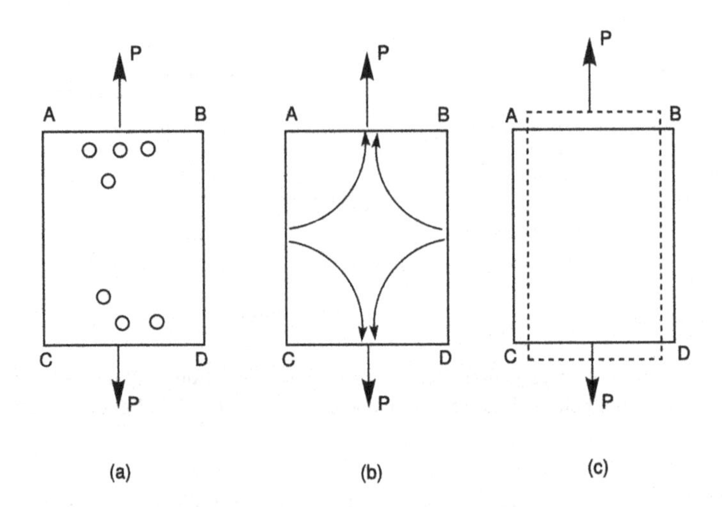

Fig. A2.11. (a) The concentration of vacancies is higher under the faces AB and CD of a grain due to these faces experiencing a larger tensile stress than the faces AC and BD (ii) there is a flux of vacancies from the faces AB and CD toward the faces AC and BD, or a flux of atoms in the opposite direction as shown by arrows (iii) the mass transfer leads to a change in shape i.e. deformation.

deformation here occurs due to diffusion of vacancies by lattice diffusion and is known as *Herring–Nabarro* creep.

The creep rate in this case is directly proportional to the applied stress σ and the diffusion coefficient D and inversely proportional to the square of the grain size d, *i.e.*

$$\dot{\varepsilon} \propto \frac{\sigma D}{d^2}$$

Under certain conditions, the diffusion of vacancies occurs through the grain boundaries. Such creep is called the *Coble creep.* In Coble creep, the creep rate varies inversely as the cube of the grain size (instead of square of the grain size, as in the H- N creep), *i.e.*

$$\dot{\varepsilon} \propto \frac{\sigma D}{d^3}$$

In nanocrystalline metals and ceramics, the grain boundary area per unit volume is very high. The Coble creep is therefore an important mechanism in this case.

A phenomenon which also occurs under creep conditions (high temperature and/or low strain rate) is *grain boundary sliding*. This can be observed by scribing a line on a polished and etched surface across a grain boundary and observing it after subjecting the sample to creep deformation. It is found that there is an offset where the line crosses the grain boundary (Fig. A2.12).

It is also seen that the offset varies from line to line along the grain boundary, indicating that it is not the pure sliding of one grain with respect to another. In many cases, the dislocation glide is the accommodating mechanism for grain boundary sliding. Another strain accommodating mechanism which can be dominant under certain conditions is the diffusion of atoms.

A phenomenological equation for creep rate which is found to fit the different mechanisms by using different values of the parameters is as follows

$$\dot{\varepsilon}_{ss} = A\frac{DGb}{kT}\left(\frac{b}{d}\right)^s\left(\frac{\sigma}{G}\right)^n \qquad \text{(A2.15)}$$

with $D = D_0 \exp(-Q/RT)$.Here D is the appropriate diffusion coefficient, G is the shear modulus, T is the temperature, b is the Burger's vector, d

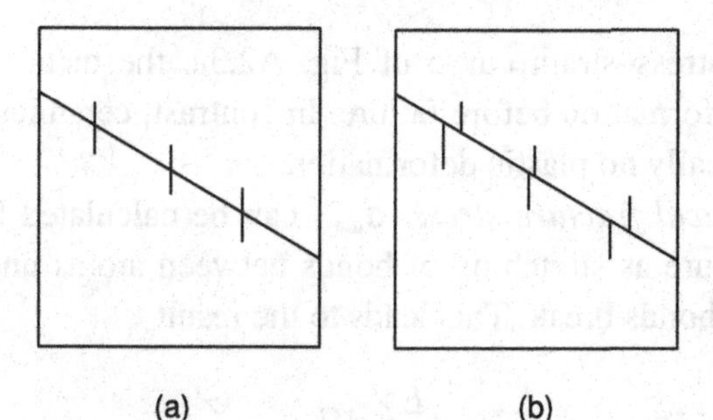

Fig. A2.12. Illustration of grain boundary sliding (a) lines scribed across a grain boundary (b) offset observed after subjecting the sample to creep; note that the offsets are unequal.

Table A2.1. Experimentally determined values of the exponents in the creep equation for different mechanisms.

Creep mechanism	s	n
(a) Dislocation creep		
(i) Glide, climb controlled	0	4–5
(ii) Glide, glide controlled	0	3
(iii) Climb without glide	0	3
(ii) Diffusion creep		
(i) Herring – Nabarro	2	1
(ii) Coble	3	1
(iii) Grain boundary sliding		
(i) With accommodation by slip	2	2
(ii) With liquid at grain boundary	3	1

is the grain size, σ is the applies stress and s and n ($=1/m$, where m is the strain rate sensitivity) are exponents whose values depend on the dominant mechanism. Table A2.1 gives the experimentally determined values of s and n for the various mechanisms.

It should be noted that while the dislocation creep has no dependence on the grain size, the diffusion creep and grain boundary sliding depend strongly on the grain size as indicated by large value of the exponent s.

A2.11 Brittle fracture

In the tensile stress–strain curve of Fig. A2.3a, the metal undergoes a large plastic deformation before failure. In contrast, ceramics and glasses fail with practically no plastic deformation.

The *theoretical fracture stress,* σ_{max}, can be calculated by modeling the brittle fracture as stretching of bonds between atoms under a tensile stress until the bonds break. This leads to the result

$$\sigma_{max} = \left(\frac{E\gamma}{a_0}\right)^{1/2} \approx \frac{E}{10}$$

(by substituting the approximate values of γ the surface energy and a_0 the lattice parameter for the solid).

The theoretical fracture stress should therefore be about one tenth of the Young's modulus. However, the experimental value of the fracture stress is found to be lower by a factor of 10 or more.

The discrepancy between the theoretical and experimental fracture stress was explained by Griffith [361] by postulating that a solid contains preexisting flaws or cracks. A solid object always has some preexisting flaws or cracks. Thus flaws in the form of minute scratches get introduced in a piece of glass by the handling of the glass or even by merely storing it for an extended length of time, due to environmental effects. A ceramic body, made by compaction of powder followed by sintering, nearly always has some remnant porosity or surface cracks due to machining or handling damage. Even if the porosity is absent and there is no machining damage, the fabrication process introduces some defects *e.g.* the high temperature sintering causes gain boundary grooving due to preferential sublimation from the grain boundaries. The length of these flaws scales as the grain size. The largest flaw size can in general be taken equal to the grain size. The fracture initiates from the largest such flaw.

The stress experienced at the tip of a crack is much higher than the uniform stress in the bulk due to the stress concentration effect of the crack tip. A quantity called K_{IC}, or *the mode I critical stress intensity factor*, is defined to quantify the resistance of a material to brittle fracture as follows. The stress intensity at a point (r, θ) in the vicinity of the crack tip of an infinitely sharp crack of length $2a$ can be calculated from the theory of elasticity and has the form

$$\sigma\,(\mathrm{r},\theta) \approx \sigma_{\mathrm{applied}}\,(\frac{a}{2r})^{1/2} f(\theta) \;= \sigma_{\mathrm{applied}} \frac{K_I}{\sqrt{2\pi r}} f(\theta) \qquad \text{(A2.16)}$$

where $2a$ is the crack length $f(\theta)$ is a function of θ (Fig. A2.13) and $K_I \equiv \sigma\, \sqrt{}(\pi a)$.The quantity K_I is called the stress intensity factor. The subscript I in it stands for mode I which is the crack opening mode in which the stress is applied perpendicular to the crack (there can be other modes of loading, called the crack sliding mode (mode II) and the tearing

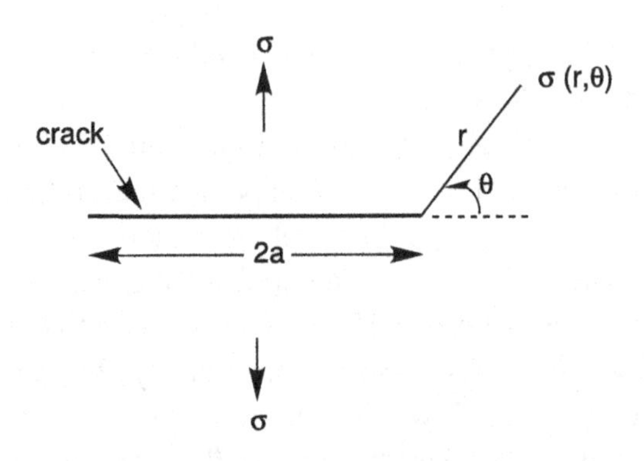

Fig. A2.13. The stress intensity at a point (r, θ) in the vicinity of the tip of an infinitely sharp crack of length $2a$, remotely loaded to a nominal stress σ.

mode (mode III)). The value of K_I at which the fracture occurs is the *critical stress intensity factor*, K_{IC}. It can be measured experimentally and is a material property which is a measure of the fracture toughness or the resistance to fracture of a material and is given by

$$K_{IC} = Y \sigma_f \sqrt{(\pi a)} \tag{A2.17}$$

where Y is a geometrical factor depending on the sample and the loading geometry and σ_f is the nominal fracture stress.

Example A2.5. *A sample of a brittle solid has five cracks of sizes 0.56 μm, 0.89 μm, 1.8 μm, 2.1 μm and 2.5 μm, all loaded in mode I. The fracture toughness, K_{IC} of the material is 10 $MPam^{1/2}$. What should be the strength of the sample? Assume the geometrical factor to be 1.1.*

Solution. It is the largest crack which results in failure. In this case $2a = 2.5$ μm or $a = 1.25$ μm.
Substituting in (Eq. A 2.17) we get

$$10 \times 10^6 = 1.1\, \sigma_f\, (\pi \times 1.25 \times 10^{-6})^{1/2}$$

or

$$\sigma_f = 4588 \text{ MPa}$$

□

It was mentioned above that the fracture initiates from the largest crack present in a component. In carbon nanotubes, flaws are introduced during their preparation. As the size of the largest flaw in ten different samples of an otherwise identical material will be different, the fracture stress of these samples will all be different. If c_0 is the half crack length of the largest crack in a material, then its fracture stress would be given by, from Eq. A2.17.

$$\sigma_f = \frac{K_{IC}}{Y\sqrt{\pi c_0}} \quad \text{(A2.18)}$$

where K_{IC} is the fracture toughness of the material.

The strength data of brittle solids, like carbon nanotubes, therefore has a lot of scatter and it is of not much use to give an average value. Instead the data has to be analyzed statistically. A statistical distribution function for the extreme values (in this case the extreme value of the crack length), called the Weibull distribution is used to analyze the fracture data.

When the samples being tested are all of the same volume then the following two–parameter form of the Weibull distribution is most commonly used

$$P(\sigma_f) = 1 - \exp\left[-\left(\frac{\sigma_f}{\sigma_0}\right)^m\right] \quad \text{(A2.19)}$$

To get the values of the parameters, the above expression is reduced to the following straight line by taking repeated logarithms

$$\ln[-\ln(1-P(\sigma_f))] = m \ln \sigma_f - m \ln \sigma_0$$

A plot of $\ln[-\ln 1-P(\sigma_f)\}$ *vs.* $\ln\sigma_f$ is a straight line whose slope is m and the intercept is $-m\ln\sigma_0$.

The Weibull modulus m is an important parameter. Figure A2.14 compares the shape of the probability plots for two values of m. A high value of m signifies a narrow distribution of strength and a more reliable material. The value of m can be as high as 100 for metals while for ceramics and glass it is usually 1 to 5.

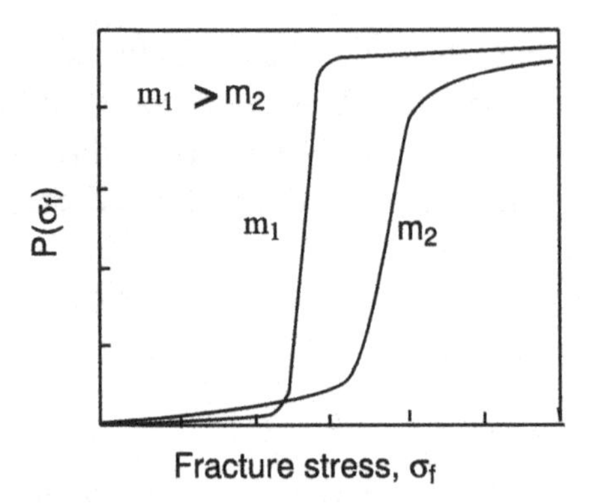

Fig. A2.14. Effect of the Weibull modulus m on the strength distribution – a larger m implies a narrower distribution and so a more reliable material.

The parameter σ_0 is called the scaling parameter and signifies the stress below which ~ 63 % of the samples fail.

Example A2.6. *Ten samples of carbon nanotubes were tested in tension and were found to have the following breaking strengths (GPa): 80.7, 38.2, 123.2, 100.2, 55.4, 130.5, 100.2, 66.1, 85.9 and 73.5.*
Plot the data on a cumulative probability $P(\sigma_f)$ vs σ_f plot. Assuming that the data follows a two parameter Weibull distribution, find the parameters m and σ_0 of the distribution.

Solution. Arrange the data in ascending order

Sample No.	1	2	3	4	5	6	7	8	9	10
σ_f (GPa)	38.2	55.4	66.1	73.5	80.7	85.9	100.2	100.2	123.2	130.5
$P(\sigma_f)$	1/11	2/11	3/11	4/11	5/11	6/11	8/11	8/11	9/11	10/11

In calculating $P(\sigma_f)$, the denominator is taken as n+1, where n is the total number of data points, to account for the limited data. A plot of σ_f vs. $P(\sigma_f)$ is shown in Fig. A2.15.

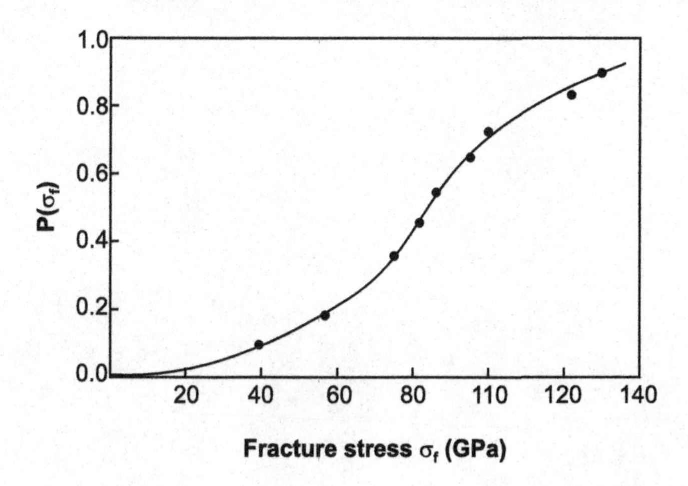

Fig. A2.15. The plot of σ_f vs. $P(\sigma_f)$ for the given data.

To find m and σ_0, fit the $ln[-ln\{1-P(\sigma_f)\}$ *vs.* σ_f data to a straight line. This gives $m = 2.68$ and $\sigma_0 = 96.6$ GPa. □

The reader should attempt the problems given at the end of Chapter 11 to test the understanding of the material covered in this Appendix.

References

1. www.whitehouse.gov/files/documents/ostp/NSTC%20Reports/NNIStrategic Plan2004.pdf.
2. www.zyvex.com/nanotech/feynman.html.
3. Moore, G.E., Cramming more components onto integrated circuits. Electronics, 1965. 38(8).
4. Moore, G.E. Progress in digital integrated electronics. in Electron Devices Meeting, 1975 International. 1975.
5. Wulff, G., Z. Krist., 1901. 34: p. 449.
6. Adamson, A.W. and A.P. Gast, *Physical Chemistry of Surfaces*. Sixth ed. 1997, New York: John Wiley.
7. Benson, G.C. and D. Patterson, J. Chem. Phys., 1955. 23: p. 670.
8. Ino, S., J. Phys. Soc. Jpn., 1969. 27: p. 941.
9. Marks, L.D., J. Cryst. Growth 1984. 61: p. 556.
10. Bonacic´-Koutecký, V., J. Burda, R. Mitric´, M. Ge, G. Zampella, and P. Fantucci, J. Chem. Phys., 2002. 117: p. 3120.
11. Baski, A.A., S.C. Erwin, and L.J. Whitman, *The structure of silicon surfaces from (001) to (111)*. Surface Science, 1997. 392: p. 69-85.
12. Behm, R.J., K. Christmann, G. Ertl, M.A. Van Hove, P.A. Thiel, and W.H. Weinberg, *The structure of CO adsorbed on Pd(100): A leed and hreels analysis*. Surface Science, 1979. 88(2-3): p. L59-L66.
13. Rousset, S., B. Croset, Y. Girard, G. Prévot, V. Repain, and S. Rohart, *Self-organized epitaxial growth on spontaneously nano-patterned templates*. C. R. Physique 2005. 6: p. 33-46.
14. Hauschild, J., H.J. Elmers, and U. Gradmann, Dipolar superferromagnetism in monolayer nanostripes of Fe(110) on vicinal W(110) surfaces. Physical Review B, 1998. 57(2): p. R677.
15. Butt, H.-J., K. Graf, and M. Kappl, *Physics and Chemistry of Surfaces*. First ed. 2003, Weinheim: Wiley-VCH.
16. Rahman, M.N., *Ceramic Processing and Sintering*. Second ed. 2003, New York Mercel Dekker.
17. German, R.M., *Liquid Phase Sintering*. 1985, New York: Plenum Press.
18. Finsy, R., *On the Critical Radius in Ostwald Ripening*. Langmuir, 2004. 20(7): p. 2975-2976.

19. Orr, F.M., L.E. Scriven, and A.P. Rivas, *Pendular rings between solids: meniscus properties and capillary force.* J. Fluid Mech., 1975. 67(4): p. 723-742.
20. Young, T., Phil.Trans.Roy Soc. London, 1805. 62: p. 567.
21. Morrison, I.D. and S. Ross, *Colloidal Dispersions: Suspensions Emulsions and Foams.* First ed. 2002, New York: John Wiley.
22. Lau, K.K.S., J. Bico, K.B.K. Teo, M. Chhowalla, G.A.J. Amaratunga, W.I. Milne, G.H. McKinley, and K.K. Gleason, Nano Letters, 2003. 3(12): p. 1701-1705.
23. Chhowalla, M., K.B.K. Teo, C. Ducati, N.L. Rupesinghe, G.A.J. Amaratunga, A.C. Ferrari, D. Roy, J. Robertson, and W.I. Milne, *Growth process conditions of vertically aligned carbon nanotubes using plasma enhanced chemical vapor deposition.* J. Appl. Phys., 2001. 90: p. 5308.
24. Koch, K., H.F. Bohn, and W. Barthlott, *Hierarchically Sculptured Plant Surfaces and Superhydrophobicity.* Langmuir, 2009. 25(24): p. 14116–14120.
25. Wenzel, R.N., Industrial and Engineering Chemistry, 1936. 28: p. 988.
26. Cassie, A.B.D. and S. Baxter, Trans. Faraday Soc., 1944. 40: p. 0546–0550.
27. Bico, J., C. Tordeux, and D. Qu´er´e, *Rough wetting.* Europhys. Lett, 2001. 55(2): p. 214–220.
28. Bico, J., C. Tordeux, and D. Quere, *Rough wetting.* Europhysics Lett, 2001. 55(2): p. 214-220.
29. Spori, D.M., T. Drobek, S. Zürcher, M. Ochsner, C. Sprecher, A. Mühlebach, and N.D. Spencer, *Beyond the Lotus Effect: Roughness Influences on Wetting over a Wide Surface-Energy Range.* Langmuir, 2008. 24(10): p. 5411-5417.
30. Bormashenko, E., *Why does the Cassie–Baxter equation apply?* Colloids and Surfaces A: Physicochem. Eng. Aspects, 2008. 324: p. 47–50.
31. Sajadinia, S.H. and F. Sharif, Thermodynamic analysis of the wetting behavior of dual scale patterned hydrophobic surfaces. J. Coll. and Interf. Sc., 2010. 344: p. 575–583.
32. Liu, X. and J. He, Langmuir, 2009. 25(19): p. 11822-11826.
33. Wang, R., K. Hashimoto, A. Fujishima, M. Chikuni, E. Kojima, A. Kitamura, M. Shimohigoshi, and W. Toshiya, *Light Induced amphiphilic surfaces.* Nature, 1997. 388: p. 431-432.
34. Buffat, P. and J.P. Borel, *Size effect on the melting temperature of gold particles.* Physical Review A, 1976. 13(6): p. 2287.
35. Dupas, C., P. Houdy, and M. Lahmani, eds. *Nanoscience - Nanotechnologies and Nanophysics.* 2007, Springer - Verlag: Heidelberg.
36. Patankar, N.A., Mimicking the Lotus Effect: Influence of Double Roughness Structures and Slender Pillars. Langmuir, 2004. 20(19): p. 8209-8213.
37. Knight, W.D., W.A. De Heer, W.A. Saunders, K. Clemenger, M.Y. Chou, and M.L. Cohen, *Alkali Metal Clusters and The Jellium Model.* Chem. Phys. Let., 1986. 109: p. 1-5.

38. Murray, C.B., D.J. Norris, and M.G. Bawendi, Synthesis and Characterization of Nearly Monodisperse CdE (E = S, Se, Te) Semiconductor Nanocrystallites. J. Am. Chem. Soc., 1993. 115: p. 8706-8715.
39. Muller, H., *Clusters of Atoms and Molecules*, H. Haberland, Editor. 1994, Springer: Berlin.
40. Echt, O., K. Sattler, and E. Recknagel, *Magic Numbers for Sphere Packings: Experimental Verification in Free Xenon Clusters.* Physical Review Letters, 1981. 47(16): p. 1121-1124.
41. Schaefer, H.-E., Nanoscince - The Science of Small in Physics, Engineering, Chemistry, Biology and Medicine. 2010, Heidelberg: Springer.
42. Martinez-Duart, J.M., R.J. Martin-Palma, and F. Agullo-Rueda, *Nanotechnology for Microelectronics and Optoelectronics.* 2006, Amsterdam: Elsevier B.V.
43. Jaros, M., Physics and Applications of Semiconductor Nanostructures. 1989, Oxford: Oxford University Press.
44. Tsurumi, T., N. Hirayama, M. Vacha, and T. Taniyama, *Nanoscale Physics for Materials Science.* 2010, Boca Raton, Fl: Taylor and Francis.
45. Cullity, B.D., S.R. Stock, and S. Stock, *Elements of X-Ray Diffraction.* Third ed. 2001, Upper Saddle River, N.J.: Prentice Hall.
46. Cohen, M.L. and T.K. Bergstresser, Band structures and pseudopotential form factors for fourteen semiconductors of the diamond and zinc-blende structures. Phys. Rev., 1966. 141: p. 789-796.
47. Pankove, J.I., *Optical Processes in Semiconductors.* 1971, New Jersey: Prentice Hall.
48. Lide, D.R., ed. *CRC Handbook of Chemistry and Physics.* 90th ed. 2009, CRC Press: Boca Raton, Fl.
49. Sapoval, B. and C. Hermann, *Physics of Semiconductors.* 1995, New York: Springer.
50. Zong, F., H. Ma, W. Du, J. Maa, X. Zhang, H. Xiao, F. Ji, and C. Xue, Optical band gap of zinc nitride films prepared on quartz substrates from a zinc nitride target by reactive rf magnetron sputtering. Applied Surface Science 2006. 252: p. 7983–7986.
51. Fehrenbach, G.W., W. Schafer, and R.G. Ulbrich, *Excitonic versus plasma screening in highly excited gallium arsenide.* Journal of Luminescence 1985. 30: p. 154-161.
52. Narayan, J., J. App. Phys., 2006. 100: p. 034309.
53. Fox, A.M., Contemp. Phys., 1996. 37: p. 111.
54. Rubio, G., N. Agrait, and S. Vieira, Atomic-Sized Metallic Contacts: Mechanical Properties and Electronic Transport. Phys. Rev. Lett., 1996. 76: p. 2302.
55. Schmid, G., ed. *Nanoparticles: From Theory to Application.* First ed. 2004, Wiley-VCH: Weinheim.
56. Legrand, B., B. Grandidier, J.P. Nys, D. Stievenard, G.J. M., and V. Thierry-Mieg, App. Phys. Lett., 1998. 73: p. 96.

57. Bruchez Jr., M., M. Moronne, P. Gin, S. Weiss, and A.P. Alivisatos, *Semiconductor Nanocrystals as Fluorescent Biological Labels* Science, 1998. 281(5385): p. 2013 – 2016.
58. Rogach, A.L., D.V. Talapin, and H. Weller, Semiconductor Nanoparticles, in Colloids and Colloid Assemblies: Synthesis, Modification, Organization and Utilization of Colloid Particles, F. Caruso, Editor. 2004, Wiley-VCH: Weinheim.
59. Perez-Centeno, A., V.-H. Mendez-Garcıa, L. Zamora-Peredo, N. Saucedo-Zeni, and M. Lopez-Lopez, *Study of the GaAs growth on pseudomorphic Si layers for the formation of self-assembled quantum dots.* J. Crystal Growth 2003. 251: p. 236-242.
60. Hellwege, K.H., ed. *Numerical Data and Functional Relationship in Science and Technology,.* Landolt-Bornstein Semiconductors III(17). Vol. III(17). 1982, Springer Verlag: Berlin.
61. Kolluri, K., L.A. Zepeda-Ruiz, C.S. Murthy, and D. Maroudasa, App. Phy. Lett., 2006. 88: p. 021904.
62. Srolovitz, D.J., Acta Metall., 1989. 37(2): p. 621-625.
63. Priester, C., Modified two-dimensional to three-dimensional growth transition process in multistacked self-organized quantum dots. Physical Review B, 2001. 63(15): p. 153303.
64. Tomlinson, I.D., M.R. Warnement, and S.J. Rosenthal, *Quantum Dots in Medicinal Chemistry and Drug Development*, in *Quantum Dots: Research, Technology and Applications*, R.W. Knoss, Editor. 2008, Nova Science Publishers: New York.
65. Ngo, C.Y., S.F. Yoon, and S.J. Chua, *Quantum dot technology for semiconductor broadband light sources*, in *Quantum Dots: Research, Technology and Applications*, R.W. Knoss, Editor. 2008, Nova Science Publishers: New York. p. 203-242.
66. Stier, O., M. Grundmann, and D. Bimberg, Phys. Rev B 1999. 59: p. 5688.
67. Luryi, S. and E. Suhir, Appl. Phys. Lett. , 1986. 49: p. 140.
68. Feynman, R.P., R.B. Leighton, and M. Sands, *The Feynman Lectures on Physics.* Vol. 2. 1964, Reading, MA: Addison Wesley.
69. Marder, M.P., *Condensed Matter Physics.* 2000, New York: Wiley.
70. Khlebtsov, N.G. and L.A. Dykman, *Optical properties and biomedical applications of plasmonic nanoparticles.* J. Quant. Spect, & Radiative Transfer 2010. 111: p. 1-35.
71. Kelly, K.L., E. Coronado, L.L. Zhao, and G.C. Schatz, *The Optical Properties of Metal Nanoparticles: The Influence of Size, Shape, and Dielectric Environment.* The Journal of Physical Chemistry B, 2002. 107(3): p. 668-677.
72. Kittel, C., *Introduction to Solid State Physics.* 8th ed. 2005, New York: John Wiley.
73. Link, S. and M.A. El-Sayed, Size and temperature dependence of the plasmon absorption of colloidal gold nanoparticles. J Phys Chem B 1999. 103(21): p. 4212–7.

74. Klabunde, K.J., *Nanoscale Materials in Chemistry*. First ed. 2001, New York: John Wiley.
75. Gans, R., Ann. Phys., 1912. 37: p. 881.
76. P´erez-Juste, J., I. Pastoriza-Santos, L.M. Liz-Marz´an, and P. Mulvaney, *Gold nanorods: Synthesis, characterization and applications.* Coordination Chemistry Reviews 249 (2005) 1870–1901, 2005. 249: p. 1870-1901.
77. Huang, X. and M.A. El-Sayed, Gold nanoparticles: Optical properties and implementations in cancer diagnosis and photothermal therapy. Journal of Advanced Research. 1(1): p. 13-28.
78. Penn, S.G., L. Hey, and M.J. Natanz, *Nanoparticles for bioanalysis.* Current Opinion in Chemical Biology 2003. 7: p. 609-615.
79. Mayer, K.M., F. Hao, S. Lee, P. Nordlander, and J.H. Hafner, *A single molecule immunoassay buy localized surface plasmon resonance.* Nanotechnology, 2010. 21: p. 1-8.
80. Vlckova, B., I. Pavel, M. Sladkova, K. Siskova, and M. Slouf, *Single molecule SERS: Perspectives of analytical applications.* Journal of Molecular Structure 2007. 834-836: p. 42–47.
81. Peng, C., Y. Song, G. Wei, W. Zhang, Z. Li, and W.-F. Dong, *Self-assembly of λ-DNA networks/Ag nanoparticles: Hybrid architecture and active-SERS substrate.* J. Coll. and Interf. Sc., 2008. 317: p. 183-190.
82. Honda, K. and S. Kaya, *On the magnetization of single crystals of iron.* Sci. Reports Tohoku Univ, 1926. 15: p. 721-753.
83. Kaya, S., *On the magnetization of single crystals of nickel.* Sci. Reports Tohoku Univ, 1928. 17: p. 639-663.
84. Honda, K. and S. Kaya, *On the magnetization of single crystals of cobalt.* Sci. Reports Tohoku Univ, 1928. 17: p. 1157-1177.
85. Iwasaki, S., K. Ouchi, and N. Honda, IEEE Trans. Magn. MAG-16, 1980: p. 1111.
86. Nave, M., S. Hasunuma, Y. Hoshi, and S. Yamanaka, IEEE Trans. Magn. MAG-17, 1981: p. 3184.
87. den Broeder, F.J.A., W. Hoving, and P.J.H. Bloemen, *Magnetic anisotropy of multilayers.* Journal of Magnetism and Magnetic Materials, 1991. 93(0): p. 562-570.
88. Bean, C.P. and J.M. Livingston, J. App. Phy., 1959. 30: p. S120.
89. Jeong, J.R., S.L. Lee, and S.C. Shin, Phys. Stat.Sol., 2004. 241: p. 1593-1596.
90. Tang, Z.X., C.M. Sorensen, K.J. Klabunde, and G.C. Hadjipanayis, *Size-dependent Curie temperature in nanoscale $MnFe_2O_4$ particles.* Phys. Rev. Lett., 1991. 67(25): p. 3602-3605.
91. Chou, S.Y., P.R. Krauss, and L. Kong, Nanolithographically defined magnetic structures and quantum magnetic disk (invited). J. Appl. Phys., 1996. 79(8): p. 6101.

92. Shi, C.X. and H.T. Cong, Tuning the coercivity of Fe-filled carbon-nanotube arrays by changing the shape anisotropy of the encapsulated Fe nanoparticles. J. Appl. Phys., 2008. 104: p. 034307.
93. Baibich, M.N., J.M. Broto, A. Fert, F.N. Van Dau, F. Petroff, P. Etienne, G. Creuzet, A. Friederich, and J. Chazelas, *Giant Magnetoresistance of (001)Fe/(001)Cr Magnetic Superlattices.* Physical Review Letters, 1988. 61(21): p. 2472.
94. Coey, J.M.D., *Magnetism and Magnetic Materials.* 2010, Cambridge: Cambridge University Press.
95. Lifshitz, E.M., The theory of molecular attractive forces between solids, Sov. Phys. Sov. Phys. - JETP, 1956. 2: p. 73-83.
96. Hamakar, H.C., The London van der Waal's attraction between spherical particles. Physica (Utrecht), 1937. 4: p. 1058-1072.
97. Hiemenz, P.C. and R. Rajagopalan, *Principles of Colloid and Surface Chemistry.* Third ed. 1997, New York: Mercel Dekker.
98. Israelachvili, J.N., *Intermolecular and Surface Forces.* 1991, New York: Academic Press. pp186-187.
99. Autumn, K., Y.A. Liang, S.T. Hsieh, W. Zesch, W.P. Chan, T.W. Kenny, R. Fearing, and R.J. Full, *Adhesive force of a single gecko foot-hair.* Nature, 2000. 405(6787): p. 681-685.
100. Morrison, I.D. and S. Ross, *Colloidal Dispersions: Suspensions Emulsions and Foams.* First ed. 2002, New York: John Wiley.
101. Mulvaney, P., Metal Nanopartilcles: Double Layers, Optical Properties, and Electrochemisttry, in Nanoscale Materials in Chemistry, K.J. Klabunde, Editor. 2001, John Wiley: New York.
102. Hunter, R.J., *Foundations of Colloid Science.* Vol. 1. 1989, New York: Oxford University Press.
103. Schartl, W., in *Nanoscale Materials*, L. Liz-Marzan and P.V. Kamat, Editors. 2003, Kluwer: Norwell, MA.
104. Jiang, P., J.F. Bertone, K.S. Hwang, and V.L. Colvin, Chem. Mater. , 1999. 11: p. 2132-2140.
105. Xia, Y., H. Fudouzi, Y. Lu, and Y. Yin, Colloidal Crystals: Recent Developments and Niche Applications, in Colloids and Colloid Assemblies: Synthesis, Modification, Organization and Utilization of Colloid Particles, F. Caruso, Editor. 2004, Wiley-VCH: Weinheim. p. 284-316.
106. Gast, A.P. and W.B. Russel, *SIMPLE ORDERING IN COMPLEX FLUIDS.* Physics Today, December 1998: p. 24-30.
107. Silva, J.M. and B.J. Mokross, Phys. Rev. B, 1980. 21: p. 2972.
108. Robbins, M.O., K. Kremer, and G.S. Grest, J. Chem. Phys., 1988. 88: p. 3286.
109. Park, S.H., D. Qin, and Y. Xia, *Crystallization of Mesoscale Particles over Large Areas.* Advanced Materials, 1998. 10(13): p. 1028-1032.

110. Park, S.H. and Y. Xia, Assembly of Mesoscale Particles over Large Areas and Its Application in Fabricating Tunable Optical Filters. Langmuir, 1998. 15(1): p. 266-273.
111. Gates, B., D. Qin, and Y. Xia, *Assembly of Nanoparticles into Opaline Structures over Large Areas.* Advanced Materials, 1999. 11(6): p. 466-469.
112. Kroto, H.W., J.R. Heath, S.C. O'brien, R.F. Curl, and R.E. Smalley Nature, 1985. 318: p. 162.
113. Iijima, S., *Helical microtubules of graphitic carbon.* Nature, 1991. 354: p. 56.
114. Bethune, D.S., C.H. Kiang, M.S. De Vries, G. Gorman, R. Savoy, J. Vasquez, and R. Beyers, *Cobalt-catalyzed growth of carbon nanotubes with single atomic layer walls.* Nature, 1993. 363: p. 605.
115. Novoselov, K.S., A.K. Geim, S.V. Morozov, D. Jiang, Y. Zhang, S.V. Dubonos, I.V. Grigorieva, and A.A. Firsov, *Electric Field Effect in Atomically Thin Carbon Films.* Science, 2004. 306(5696): p. 666-669.
116. Thomsen, C. and S. Reich, *Double resonant Raman scattering in graphite.* Phys. Rev. Lett., 2000. 85: p. 5214-5217.
117. Meyer, J.C., C. Kisielowski, R. Erni, M.D. Rossell, M.F. Crommie, and A. Zettl, *Direct Imaging of Lattice Atoms and Topological Defects in Graphene Membranes.* Nano Lett., 2008. 8(11): p. 3582-3586.
118. Ferrari, A.C., Raman spectroscopy of graphene and graphite: Disorder, electron-phonon coupling, doping and nonadiabatic effects. Solid State Communications, 2007. 143(1-2): p. 47-57.
119. Wang, Y.y., Z. Ni, T. Yu, Z. Shen, Y. Wang, Y.h. Wu, W. Chen, and A.T.S. Wee, *Raman Studies of Monolayer Graphene: The Substrate Effect.* J. Phys. Chem. C, 2008. 112(29): p. 10637-10640.
120. Iijima, S., *Growth of carbon nanotubes.* Mat. Sc. and Eng. , 1993. B19: p. 172-180.
121. Kratschmer, W., L.D. Lamb, K. Fostiropoulos, and D.R. Huffman, *Solid C_{60} : a new form of carbon.* Nature, 1990. 347: p. 354.
122. Ebbesen, T.W. and P.M. Ajayan, *Large scale synthesis of carbon nanotubes.* Nature, 1992. 358: p. 220.
123. Journet, C., W.K. Maser, P. Bernier, A. Loiseau, M. Lamy de la Chapelle, S. Lefrant, P. Deniard, R. Lee, and J.E. Fischer, *Large scale production of the single walled tubes by the electric arc technique,.* Nature, 1997. 388: p. 756.
124. Guo, T., N. Pavel, A.G. Rinzler, D. Tomanek, D.T. Colbert, and R.E. Smalley, *Self-Assembly of Tubular Fullerenes.* J. Phys. Chem., 1995. 99: p. 10694 -10697.
125. Kong, J., A.M. Cassell, and H. Dai, *Chemical vapor deposition of methane for single-walled carbon nanotubes.* Chem. Phys. Lett., 1998. 292: p. 567–574.
126. Fujita, M., R. Saito, G. Dresselhaus, and M.S. Dresselhaus, *Formation of general fullerenes by their projection on a honeycombe lattice.* Phys. Rev. B, 1992. 45: p. 13834.

127. Fana, S., W. Lianga, H. Danga, N. Franklinb, T. Tomblerb, M. Chaplineb, and H. Daib, *Carbon nanotube arrays on silicon substrates and their possible application.* Physica E 2000. 8: p. 179-183.
128. Tibbetts, G.G., *Why are carbon filaments tubular?* J. Cryst. Growth, 1984. 66: p. 632.
129. Saito, Y., *Nanoparticles and filled nanocapsules.* Carbon, 1995. 33: p. 979.
130. Soldano, C., A. Mahmood, and E. Dujardin, *Production, properties and potential of graphene.* Carbon 2010. 48: p. 2127-2150.
131. Reina, A., X. Jia, J. Ho, D. Nezich, H. Son, V. Bulovic, M.S. Dresselhaus, and J. Kong, *Large Area, Few-Layer Graphene Films on Arbitrary Substrates by Chemical Vapor Deposition.* Nano Letters, 2008. 9(1): p. 30-35.
132. Shivaraman, S., R.A. Barton, X. Yu, J. Alden, L. Herman, M.V.S. Chandrashekhar, J. Park, P.L. McEuen, J.M. Parpia, H.G. Craighead, and M.G. Spencer, *Free-Standing Epitaxial Graphene.* Nano Letters, 2009. 9(9): p. 3100-3105.
133. Cho, J., J.J. Luo, and I.M. Daniel, *Mechanical characterization of graphite/epoxy nanocomposites by multi-scale analysis.* Composites Sc. & Tech., 2007. 67: p. 2399-2407.
134. Zaeri, M.M., S. Ziaei-Rad, A. Vahedi, and F. Karimzadeh, *Mechanical modelling of carbon nanomaterials from nanotubes to buckypaper.* Carbon, 2010. 48: p. 3916-3930.
135. Liew, K.M., X.Q. He, and C.H. Wong, On the study of elastic and plastic properties of multi-walled carbon nanotubes under axial tension using molecular dynamics simulation. Acta Materialia, 2004. 52(9): p. 2521-2527.
136. Zhao, Q.Z., M.B. Nardelli, and J. Bernhole, *Ultimate strength of carbon nanotubes: a theoretical study.* Phys. Rev. B, 2002. 65: p. 144105.
137. Van Lier, G., C. Van Alsenoy, V. Van Doren, and P. Geerlings, *Ab initio study of the elastic properties of single-walled carbon nanotubes and graphene.* Chemical Physics Letters, 2000. 326(1-2): p. 181-185.
138. Zhou, G., W. Duan, and B. Gu, First-principles study on morphology and mechanical properties of single-walled carbon nanotube Chemical Physics Letters, 2001. 333: p. 344-349.
139. Yao, Z., C.-C. Zhu, M. Cheng, and J. Liu, *Mechanical properties of carbon nanotube by molecular dynamics simulation.* Computational Materials Science, 2001. 22(3-4): p. 180-184.
140. Jin, Y. and F.G. Yuan, *Simulation of elastic properties of single-walled carbon nanotubes.* Composites Science and Technology, 2003. 63(11): p. 1507-1515.
141. Li, C. and T.-W. Chou, *Elastic moduli of multi-walled carbon nanotubes and the effect of van der Waals forces.* Composites Science and Technology, 2003. 63(11): p. 1517-1524.
142. Lu, J.P., Elastic properties of carbon nanotubes and nanoropes. Phys. Rev. Lett., 1997. 79: p. 1297.

143. Lee, C., X. Wei, J.W. Kysar, and J. Hone, *Measurement of the Elastic Properties and Intrinsic Strength of Monolayer Graphene.* Science, 2008. 321(5887): p. 385-388.
144. Gomez-Navarro, C., M. Burghard, and K. Kern, *Elastic Properties of Chemically Derived Single Graphene Sheets.* Nano Letters, 2008. 8(7): p. 2045-2049.
145. Pereira, V.M. and A.H. Castro Neto, Phys. Rev. Lett., 2009. 103: p. 046801.
146. Treacy, M.M., T.W. Ebbesen, and J.M. Gibson, Nature, 1996. 38: p. 678.
147. Wong, E.W., P.E. Sheehan, and C.M. Lieber, *Nanobeam Mechanics: Elasticity, Strength, and Toughness of Nanorods and Nanotubes.* Science 1997. 277(5334): p. 1971-1975.
148. Salvetat, J.-P., J.-M. Bonard, N.H. Thomson, A.J. Kulik, L. Forro, W. Benoit, and L. Zuppiroli, Appl. Phys., 1999. A 69 p. 255.
149. Demczyk, B.G., Y.M. Wang, J. Cumings, M. Hetman, W. Han, A. Zettl, and R.O. Ritchie, *Direct mechanical measurement of the tensile strength and elastic modulus of multiwalled carbon nanotubes.* Materials Science and Engineering A, 2002. 334(1-2): p. 173-178.
150. Li, C. and T.-W. Chou, Failure of carbon nanotube/polymer composites and the effect of nanotube waviness. Composites: Part A 2009. 40: p. 1580-1586.
151. Barber, A.H., R. Andrews, L.S. Linda S. Schadler, and H.D. H. Daniel Wagnera, *On the tensile strength distribution of multiwalled carbon nanotubes.* App. Phys. Lett., 2005. 87: p. 203106.
152. Cao, A., P.L. Dickrell, W.G. Sawyer, and Ajayan, *Super compressible foamlike carbon nanotube films.* Science, 2005. 310: p. 1307.
153. Peres, N.M.R., The electronic properties of graphene and its bilayer. Vacuum 2008. 83: p. 1248-1252.
154. Bostwick, A., J. McChesney, T. Ohta, E. Rotenberg, T. Seyller, and K. Horn, *Experimental studies of the electronic structure of graphene.* Progress in Surface Science, 2009. 84 p. 380-413.
155. Ohta, T., A. Bostwick, T. Seyller, K. Horn, and E. Rotenberg, *Controlling the Electronic Structure of Bilayer Graphene.* Science, 2006. 313(5789): p. 951-954.
156. Li, X., X. Wang, L. Zhang, S. Lee, and H. Dai, Science, 2008. 319: p. 12292.
157. Deshpande, V.V., B. Chandra, R. Caldwell, D.S. Novikov, J. Hone, and M. Bockrath, *Mott Insulating State in Ultraclean Carbon Nanotubes.* Science, 2009. 323(5910): p. 106-110.
158. Sohn, J.I., S. Lee, Y.-H. Song, S.-Y. Choi, K.-I. Cho, and K.-S. Nam, *Large field emission current density from well-aligned carbon nanotube field emitter arrays.* Curr. Appl. Phys., 2001. 1(1): p. 61-65.
159. Bauhofer, W. and J.Z. Kovacs, *A review and analysis of electrical percolation in carbon nanotube polymer composites.* Composites Science and Technology, 2009. 69(10): p. 1486-1498.

160. Celzard, A., E. McRae, C. Deleuze, M. Dufort, G. Furdin, and J. F. Mareche, *Critical concentration in percolating systems containing a high-aspect-ratio filler.* Physical Review B, 1996. 53(10): p. 6209.
161. Logakis, E., P. Pissis, D. Pospiech, A. Korwitz, B. Krause, U. Reuter, and P. Pötschke, *Low electrical percolation threshold in poly(ethylene terephthalate)/ multi-walled carbon nanotube nanocomposites.* European Polymer Journal, 2010. 46(5): p. 928-936.
162. Geng, H.-Z., T.H. Kim, S.C. Lim, H.-K. Jeong, M.H. Jin, Y.W. Jo, and Y.H. Lee, *Hydrogen storage in microwave-treated multi-walled carbon nanotubes.* International Journal of Hydrogen Energy. 35(5): p. 2073-2082.
163. Wang, N., Y. Cai, and R.Q. Zhang, *Growth of nanowires.* Materials Science and Engineering R, 2008. 60: p. 1-51.
164. Rao, C.N.R., F.L. Deepak, G. Gundiah, and A. Govindaraj, *Inorganic nanowires.* Progress in Solid State Chemistry 2003. 31: p. 5-147.
165. Heoa, Y.W., D.P. Nortona, L.C. Tiena, Y. Kwona, B.S. Kangb, R. Renb, S.J. Peartona, and J.R. LaRoche, *ZnO nanowire growth and devices.* Materials Science and Engineering R 2004. 47: p. 1-47.
166. Kolmakov, A. and M. Moskovits, Ann. Rev. of Mat. Res., 2004. 35: p. 151.
167. Wang, Z.L., *Nanostructures of zinc oxide.* Materials Today, 2004. 7(6): p. 26-33.
168. Tenne, R., Inorganic Nanotubes and Fullerene-Like Materials of Metal Dichalcogenide and Related Layered Compounds, in Nanotubes and nanofibers, Y. Gogotsi, Editor. 2006, CRC Press, Taylor and Francis: Boca Raton, FL. p. 135-155.
169. Wagner, R.S. and W.C. Ellis, Appl. Phys. Lett., 1964. 4: p. 89.
170. Kwak, D.W., H.Y. Cho, and W.-C. Yang, Dimensional evolution of silicon nanowires synthesized by Au–Si island-catalyzed chemical vapor deposition. Physica E 2007. 37: p. 153-157.
171. Givargizov, E.I., *Fundamental aspects of VLS growth.* J. Crystal Growth, 1975. 31: p. 20-30.
172. Schmidt, V., S. Senz, and U. G¨osele, The shape of epitaxially grown silicon nanowires and the influence of line tension. Appl. Phys. A 2005. 80: p. 445-450.
173. Li, N., T.Y. Tan, and U. G¨osele, Chemical tension and global equilibrium in VLS nanostructure growth process: from nanohillocks to nanowires. Appl. Phys. A 2007. 86: p. 433–440.
174. Morales, A.M. and C.M. Lieber, A Laser Ablation Method for the Synthesis of Crystalline Semiconductor Nanowires. Science, 1998. 279(5348): p. 208-211.
175. Givargizov, E.I., J. Cryst. Growth, 1975. 31: p. 20.
176. Holmes, J.D., K.P. Johnston, R.C. Doty, and B.A. Korgel, *Control of Thickness and Orientation of Solution-Grown Silicon Nanowires.* Science, 2000. 287(5457): p. 1471-1473.

177. Zheng, M., L. Zhang, X. Zhang, J. Zhang, and G. Li, Fabrication and optical absorption of ordered indium oxide nanowire arrays embedded in anodic alumina membranes. Chemical Physics Letters 2001. 334: p. 298-302.
178. Panga, Y.T., G.W. Menga, L.D. Zhanga, W.J. Shana, C. Zhangb, X.Y. Gaoa, and A.W. Zhaoa, *Synthesis of ordered Al nanowire arrays* Solid State Sciences, 2003. 5: p. 1063-1067
179. Penn, R.L. and J.F. Banfield, Geochim. Cosmochim. Acta, 1999. 63: p. 1549.
180. Bunimovich, Y.L., Y.S. Shin, W.-S. Yeo, M. Amori, G. Kwong, and J.R. Heath, *Quantitative Real-Time Measurements of DNA Hybridization with Alkylated Nonoxidized Silicon Nanowires in Electrolyte Solution.* Journal of the American Chemical Society, 2006. 128(50): p. 16323-16331.
181. Iwai, H., *Roadmap for 22 nm and beyond.* Microelectronic Engineering 2009. 86: p. 1520-1528.
182. Thelander, C., P. Agarwal, S. Brongersma, J. Eymery, L.F. Feiner, A. Forchel, M. Scheffler, W. Riess, B.J. Ohlsson, U. Gösele, and L. Samuelson, *Nanowire based one dimensional electronics.* Materials Today, 2006. 9(10): p. 28-35.
183. Chen, Z.H., H. Tang, X. Fan, J.S. Jie, C.S. Lee, and S.T. Lee, *Epitaxial ZnS/Si core-shell nanowires and single-crystal silicon tube field-effect transistors.* Journal of Crystal Growth, 2008. 310(1): p. 165-170.
184. Bryllert, T., L.-E. Wernersson, T. L¨owgren, and L. Samuelson, *Vertical wrap-gated nanowire transistors.* Nanotechnology 2006. 17: p. S227–S230.
185. ITRS, *www.itrs.net/reports.html*, International Technology Roadmap for Semiconductors.
186. Leonhardt, U. and T.Å. Tyc, *Broadband Invisibility by Non-Euclidean Cloaking.* Science, 2009. 323(5910): p. 110-112.
187. Smith, D.L., *Thin-Film Deposition: Principles and Practice* 1995, New York: McGraw Hill.
188. Mahan, J.E., *Physical Vapor Deposition of Thin Films* 2000, New York: John Wiley & Sons.
189. Choy, K.L., Progress in Materials Science 2003. 48: p. 57-170.
190. Cao, G., *Nanostructures and Nanomaterials.* First ed. 2004, London: Imperial College Press.
191. Suntola, T., Thin Solid Films, 1992. 216: p. 84-89.
192. Brukh, R. and S. Mitra, Chem. Phys. Let., 2006. 424: p. 126-132.
193. Clearfield, A. and P.A. Vaughan, Acta Cryst., 1956. 9: p. 555.
194. Mak, T.C.W., Can. J. Chem., 1968. 46: p. 3491.
195. Brinker, C.J. and G.W. Scherer, *Sol - Gel Science.* 1990, San Diego: Academic Press.
196. Zarzycki, J., M. Prassas, and J. Phallipou, J. Mater. Sc., 1982. 17: p. 3371-3379.
197. Hench, L.L., in *Science of Ceramic Chemical Processing*, L.L. Hench and D.R. Ulrich, Editors. 1986, Wiley: New York. p. 52-64.

198. Kresge, C.T., M.E. Leonowicz, W.J. Roth, J.C. Vartuli, and J.S. Beck, *Ordered mesoporous molecular sieves synthesized by a liquidcrystal template mechanism.* Nature, 1992. 359: p. 710-712.
199. Beck, J.S., J.C. Vartuli, W.J. Roth, *et. al*, *A new family of mesoporous molecular sieves prepared with liquid crystal templates.* J Am. Chem. Soc., 1992. 114: p. 10834-10843.
200. Yanagisawa, T., T. Schimizu, K. Kuroda, and C. Kato, The preparation of alkyltrimethylammonium-kanemite complexes and their conversion to mesoporous materials. Bull. Chem. Soc. Jpn., 1990. 63: p. 988-992.
201. www.icknowledge.com.
202. Sneh, O., R.B. Clark-Phelps, A.R. Londergan, J. Winkler, and T.E. Seidel, *Thin film atomic layer deposition equipment for semiconductor processing.* Thin Solid Films 2002. 402: p. 248-261.
203. Lin, J. and C. Lee, J. Alloys and Compds, 2008. 449: p. 371-374.
204. Parka, S.-J., W.-H. Kima, H.-B.-R. Leea, W.J. Maenga, and H. Kim, Microelectronic Eng., 2008. 85: p. 39-44.
205. Jakschika, S., U. Schroedera, T. Hechta, Martin Gutschea, H. Seidla, and J.W. Barthab, *Crystallization behavior of thin ALD-Al_2O_3 films.* Thin Solid Films 2003. 425: p. 216-220.
206. Klaus, J.W. and S.M. George, Atomic layer deposition of SiO_2 at room temperature using NH_3-catalyzed sequential surface reactions. Surface Science 2000. 447: p. 81-90.
207. Terasawaa, N., K. Akimotoa, Y. Mizunoa, A. Ichimiyaa, K. Sumitanib, T. Takahashib, X.W. Zhangc, H. Sugiyamac, H. Kawatac, T. Nabatamed, and A. Toriumie, *Crystallization process of high-k gate dielectrics studied by surface X-ray diffraction.* Applied Surface Science 2005. 244: p. 16-20.
208. Aarika, J., A. Aidlaa, A.-A. Kiislera, T. Uustarea, and V. Sammelselgb, *Influence of substrate temperature on atomic layer growth and properties of HfO_2 thin films.* Thin Solid Films 1999. 340: p. 110-116.
209. Kim, J., H. Hong, K. Oh, and C. Lee, Properties including step coverage of TiN thin films prepared by atomic layer deposition Appl. Surf. Sc., 2003. 210(3-4): p. 231-239.
210. Mitchell, D.R.G., D.J. Attard, and G. Triani, Characterisation of epitaxial TiO_2 thin films grown on MgO(0 0 1) using atomic layer deposition. J. Crystal Growth, 2005. 285: p. 208-214.
211. Cassir, M., F. Goubin, C. Bernay, P. Vernoux, and D. Lincot, Synthesis of ZrO2 thin films by atomic layer deposition: growth kinetics, structural and electrical properties. App. Surf. Sc., 2002. 193: p. 120-128.
212. Wang, C.-C., C.-C. Kei, Y.-W. Yu, and T.-P. Perng, Nano Lett., 2007. 7: p. 1566-1569.
213. Volger, D. and P. Doe, *ALD Special Report: Where's the metal?* Solid St. Technology, 2003: p. 35.

214. Farmer, D.B. and R.G. Gordon, Nano Lett., 2006. 6: p. 699-703.
215. Caruso, F., R.A. Caruso, and H. Mohwald, Science, 1998. 282: p. 1111.
216. Antipov, A.A., G.B. Sukhorukov, E. Donath, and H. Mohwald, J. Phys. Chem. B, 2001. 105: p. 2281-2284.
217. Chung, A.J. and M.F. Rubner, Langmuir, 2002. 18: p. 1176-1183.
218. Shim, B.S., P. Podsiadlo, D.G. Lilly, A. Agarwal, J. Lee, Z. Tang, S. Ho, P. Ingle, D. Paterson, W. Lu, and N.A. Kotov, Nano Lett., 2007. 7(11): p. 3266-3273.
219. Shiratori, S.S. and M.F. Rubner, Macromolecules, 2000. 33: p. 4213-4219.
220. Chung, A.J. and M.F. Rubner, Langmuir 2002. 18: p. 1176-1183.
221. Mamedov, A.A., A. Belov, M. Giersig, N.N. Mamedova, and N.A. Kotov, J. Am. Chem. Soc., 2001. 123: p. 7738.
222. Schlenoff, J.B., S.T. Dubas, and T. Farhat, Langmuir, 2000. 16.
223. Izquierdo, A., S.S. Ono, J.-C. Voegel, P. Schaaf, and G. Decher, Langmuir, 2005. 21: p. 7558.
224. Jiang, C., S. Markutsya, and V.V. Tsukruk, Langmuir, 2004. 20: p. 882.
225. Sun, J., M. Gao, and J.J. Feldmann, Nanosci. Nanotechnol., 2001. 1: p. 133.
226. Fujimoto, K., S. Fujita, B. Ding, and S. Shiratori, Jpn. J. Appl. Phys., 2005. 44: p. L126-L128.
227. Picart, C., J. Mutterer, L. Richert, Y. Luo, G.D. Prestwich, P. Schaaf, J.-C. Voegel, and P. Lavalle, Proc. Natl. Acad. Sci. U.S.A., 2002. 99: p. 12531.
228. Yashar, P.C. and W.D. Sproul, *Nanometer scale multilayered hard coatings.* Vacuum, 1999. 55 p. 179-190.
229. Setoyama, M., A. Nakayama, M. Tanaka, N. Kitagawa, and T. Nomura, *Nanometer scale multilayered hard coatings.* Surf. Coat. Technol., 1996. 86-87: p. 225.
230. Helmersson, U., S. Todorova, S.A. Barnett, J.E. Sundgren, L.C. Markert, and J.E. Greene, J. Appl. Phys., 1987. 62(2): p. 481.
231. Sevillano, J.G., *Strength of Metals and Alloys*, in *Proc. ICSMA 5*, P. Haasen, V. Gerold, and G. Kowtorz, Editors. 1980, Pergamon Press: Oxford. p. 819-824.
232. Chu, X. and S.A. Barnett, *Model of superlattice yield stress and hardness enhancements.* J. Appl. Phys., 1995. 77(9): p. 4403.
233. Koehler, J.S., *Attempt to Design a Strong Solid.* Phys. Rev B, 1970. 2: p. 547.
234. Dieter, G.E., *Mechanical Metallurgy.* 1986, New York: McGraw Hill.
235. Shinn, M. and S.A. Barnett, *Effect of superlattice layer elastic moduli on hardness.* Appl. Phys. Lett., 1994. 64(1): p. 61-63.
236. Voevodin, A.A., J.S. Zabinski, and C. Muratore, *Recent Advances in Hard, Tough, and Low Friction Nanocomposite Coatings.* Tsinghua Sc. and Tech., 2005. 10(6): p. 665-679.
237. Veprek, S., P. Nesladek, A. Niederhofer, and F. Glatz, *Search for superhard materials: nannocrystalline composites with hardness exceeding 50 GPa.* Nanostructured Materials 1998. 10(5): p. 679-689.
238. Hua, F., T. Cui, and Y.M. Lvov, Nano Lett., 2004. 4: p. 823.

239. Gleiter, H., *Nanocrystalline Materials.* Prog. in Mat. Sc., 1989. 33: p. 223-315.
240. Aihara, T., E. Akiyana, K. Aoki, M. El-Eskandarany, H. Habazaki, and K. Hashimoto, in *Amorphous and nanocrystalline materials*, Inoue A and Hashimoto K, Editors. 2000. p. 103.
241. Valiev, R.Z., R.K. Islamgaliev, and I.V. Alexandrov, *Bulk nanostructured materials from severe plastic deformation.* Progress in Materials Science, 2000. 45(2): p. 103-189.
242. Zhang, H., X. Cheng, L. Zhang, and B. Bai, An ultrahigh strength steel with ultrafine-grained microstructure produced through superplastic forming. Mat. Sc. and Eng. A 2010. 527: p. 5430–5434.
243. Cahoon, J.R., W.H. Broughton, and A.R. Kutzak, *The Determination of Yield Strength From Hardness Measurements.* Met. Trans., 1971. 2: p. 1979.
244. Lucca, D.A., K. Herrmann, and M.J. Klopfstein, *Nanoindentation: Measuring methods and applications.* Manufacturing Technology 2010. 59: p. 803-819.
245. Conrad, H. and J. Narayan, *On the grain size softening in nanocrystalline materials.* Scripta Materialia, 2000. 42(11): p. 1025-1030.
246. Schiotz, J., F. Di Tella, and K. Jacobson, Nature 1988. 391: p. 561.
247. van Swygenhoven, H., M. Spaczer, and A. A. Caro, Acta Mater., 1999. 47: p. 3117.
248. Hall, E.O., Proc. Phys. Soc., Lond. B, 1951. 64: p. 747.
249. Petch, N.J., J. Iron Steel Inst., Lond., 1953. 173: p. 25.
250. Conrad, H., *Electron Microscopy and Strength of Crystals*, G. Thomas and J. Washburn, Editors. 1961, Interscience: New York. p. 299.
251. Li, J.C.M., Trans. TMS-AIME, 1963. 227: p. 239.
252. Ashby, M.F., Phil. Mag., 1970. 21: p. 399.
253. Conrad, H., Effect of grain size on the lower yield and flow stress of iron and steel. Acta Met., 1963. 11: p. 75-77.
254. Qi, Z.B., P. Sun, F.P. Zhu, Z.C. Wang, D.L. Peng, and C.H. Wu, *The inverse Hall–Petch effect in nanocrystalline ZrN coatings.* Surface & Coatings Technology 2011. 205: p. 3692–3697.
255. Chen, J., L. Lu, and K. Lu, *Hardness and strain rate sensitivity of nanocrystalline Cu.* Scripta Materialia 54 (2006) 1913–1918, 2006. 54: p. 1913-1918.
256. Dao, M., L. Lu, R.J. Asaro, J.T.M. De Hosson, and E. Ma, *Toward a quantitative understanding of mechanical behavior of nanocrystalline metals.* Acta Materialia 2007. 55: p. 4041-4065.
257. Wang, Y.M. and E. Ma, Strain hardening, strain rate sensitivity, and ductility of nanostructured metals. Mat. Sc. and Eng. A, 2004. 375-377: p. 46-52.
258. Koch, C.C., Optimization of strength and ductility in nanocrystalline and ultrafine grained metals. Scripta Materialia 2003. 49: p. 657-662.
259. Wang, Y.M. and E. Ma, Strain hardening, strain rate sensitivity, and ductility of nanostructured metals. Mat. Sc. and Eng. A 2004. 375-377: p. 46-52.

260. Youssef, K.M., R.O. Scattergood, M.K. Linga, J.A. Horton, and C.C. Koch, *Ultrahigh strength and high ductility of bulk nanocrystalline copper.* App. Phys. Lett., 2005. 87: p. 091904.
261. Chokshi, A.H., A.K. Mukherjee, and T.G. Langdon, *Superplasticity in advanced materials.* Mat. Sc. and Eng., 1993. R10: p. 237-274.
262. Nieh, T.G., C.M. McNally, and J. Wadsworth, *Superplastic behavior of a 20% Al203/YTZ ceramic composite.* Scripta Metallurgica, 1989. 23: p. 457-460.
263. Prasad, M.J.N.V. and A.H. Chokshi, Extraordinary high strain rate superplasticity in electrodeposited nano-nickel and alloys. Scr. Mater. , 2010. 63: p. 136–139.
264. Wan, H., F. Delale, and L. Shen, *Effect of CNT length and CNT-matrix interphase in carbon nanotube (CNT) reinforced composites.* Mechanics Research Communications 2005. 32: p. 481-489.
265. Cronin, S.B., A.K. Swan, M.S. Unlu, B.B. Goldberg, M.S. Dresselhaus, and M. Tinkham, Phys. Rev. Lett., 2004. 93(16): p. 4.
266. Ma, W., L. Liu, Z. Zhang, R. Yang, G. Liu, T. Zhang, X. An, X. Yi, Y. Ren, Z. Niu, J. Li, H. Dong, W. Zhou, P.M. Ajayan, and S. Xie, *High-Strength Composite Fibers: Realizing True Potential of Carbon Nanotubes in Polymer Matrix through Continuous Reticulate Architecture and Molecular Level Couplings.* Nano Letters, 2009. 9(8): p. 2855-2861.
267. Coleman, J.N., U. Khan, W.J. Blau, and Y.K. Gun'ko, Small but strong: A review of the mechanical properties of carbon nanotube–polymer composites. Carbon 2006. 44: p. 1624-1652.
268. Carman, G.P. and K.L. Reifsnider, *Micromechanics of short-fiber composites.* Compos Sci Technol, 1992. 43(2): p.137–46.
269. Halpin, J.C., *Primer on Composite Materials Analysis.* 2nd ed. 1992, Lancaster, Pa.: Technomic.
270. Leblanc, J.L., *Filled Polymers: Science and Industrial Applications.* 2010, Boca Raton, FL.: CRC Press.
271. Kardos, J.L., Critical issues in achieving desirable mechancal properties for short fiber composites. Pure Aopl. Chem, 1985. 57(11): p. 1651-1657.
272. Fornes, T.D. and D.R. Paul, Modeling properties of nylon 6/clay nanocomposites using composite theories. Polymer 2003. 44: p. 4993-5013.
273. Spitalsky, Z., D. Tasis, K. Papagelis, and C. Galiotis, *Carbon nanotube–polymer composites: Chemistry, processing, mechanical and electrical properties.* Prog. in Polym. Sc., 2010. 35: p. 357-401.
274. Theodore, M., M. Hosur, J. Thomas, and S. Jeelani, *Influence of functionalization on properties of MWCNT-epoxy nanocomposites.* Materials Science and Engineering: A. 528(3): p. 1192-1200.
275. Bakshi, S.R. and A. Agarwal, An analysis of the factors affecting strengthening in carbon nanotube reinforced aluminum composites. Carbon 2011. 49: p. 533 –544.

276. Spitalskya, Z., D. Tasisb, K. Papagelisb, and C. Galiotis, *Carbon nanotube–polymer composites: Chemistry, processing, mechanical and electrical properties.* Prog. in Polym. Sc., 2010. 35: p. 357-401.
277. Kojima, Y., A. Usuki, M. Kawasumi, A. Okada, Y. Fukushima, T. Kurauchi, and O. Kamigaito, J Mater Res, 1993. 8: p. 1185–9.
278. Usuki, A., Y. Kojima, M. Kawasumi, A. Okada, Y. Fukushima, and T. Kurauchi, J Mater Res, 1993. 8: p. 1179–84.
279. Alexandre, M. and P. Dubois, Polymer-layered silicate nanocomposites: preparation, properties and uses of a new class of materials. Mat. Sc. and Eng. A, 2000. 28: p. 1-63.
280. Manias, E., A. Touny, L. Wu, K. Strawhecker, B. Lu, and T.C. Chung, *Polypropylene/montmorillonite nanocomposites. Review of the synthetic routes and materials properties.* Chem. Mater., 2001. 13: p. 3516–23.
281. Ploehn, H.J. and C. Liu, Quantitative Analysis of Montmorillonite Platelet Size by Atomic Force Microscopy. Ind. Eng. Chem. Res., 2006. 45: p. 7025-7034.
282. Paul, D.R. and L.M. Robeson., *Polymer nanotechnology: Nanocomposites.* Polymer 2008. 49: p. 3187-3204.
283. Kornmann, X., H. Lindberg, and L.A. Berglund, Synthesis of epoxy–clay nanocomposites: influence of the nature of the clay on structure. Polymer 2001. 42: p. 1303-1310.
284. Pavlidou, S. and C.D. Papaspyrides, *A review on polymer-layered silicate nanocomposites.* Progress in Polymer Science, 2008. 33(12): p. 1119-1198.
285. Powell, C.E. and G.W. Beall, *Physical properties of polymer/clay nanocomposites.* Curr. Opinion in Sol. St. and Mat. Sc., 2006. 10: p. 73-80.
286. Okada, A. and A. Usuki, Mater Sci Eng 1995. C3: p. 109–15.
287. Kojima, Y., A. Usuki, M. Kawasumi, A. Okada, T. Kurauchi, and O. Kamigaito, *Sorption of water in nylon 6–clay hybrid.* J Appl Polym Sci, 1993. 49: p. 1259–64.
288. Hasegawa, N., H. Okamoto, M. Kato, A. Usuki, and N. Sato, Polymer, 2003. 44: p. 2933–7.
289. Fornes, T.D., P.J. Yoon, H. Heskkula, and D.R. Paul, Polymer, 2001. 42: p. 9929–40.
290. Liu, X. and Q. Wu, Macromol Mater Eng, 2002. 287: p. 180–6.
291. Chavarria, F. and D.R. Paul, Polymer, 2004. 45: p. 8501–15.
292. Han, B., G. Ji, S. Wu, and J. Shen, Eur Polym J, 2003. 39: p. 1641–6.
293. Zhang, Q., K. Wang, Y. Men, and Q. Fu, Chinese J Polym Sci, 2003. 21: p. 359–67.
294. Oya, A., Y. Kurokawa, and H. Yasuda, J Mater Sci, 2000. 35: p. 1045–50.
295. Wang, D. and C. Wilkie, Polym Degrad Stabil, 2003. 80: p. 171–82.
296. Goettler, L.A., Overview of property development in layered silicate polymer nanocomposites. Ann Tech Confr Soc Plast Eng, 2005: p. 1980-82.
297. Gopakumar, T.G., J.A. Lee, M. Kontopoulou, and J.S. Parent, Polymer, 2002. 41: p. 5483–91.

298. Wang, K.H., C.M. Koo, and I.J. Chung, J Appl Polym Sci, 2003. 89: p. 2131–6.
299. Wei, L., T. Tang, and B. Huang, J Polym Sci Part A, 2004. 42: p. 941–9.
300. Hotta, S. and D.R. Paul, Polymer, 2004. 45: p. 7639–54.
301. Tjong, S.C. and Y.Z. Meng, J Polym Sci, B: Polym Phys, 2003. 41: p. 2332–41.
302. Zhao, C., H. Qin, F. Gong, M. Feng, S. Zhang, and M. Yang, *Mechanical, thermal and flammability properties of polyethylene/clay nanocomposites.* Polym Degrad Stabil, 2005. 87: p. 183–9.
303. Chang, J.-H., S.J. Kim, Y.L. Joo, and S. Im Polymer, 2004. 45: p. 919–26.
304. Pegoretti, A., J. Kolarik, C. Peroni, and C. Migliaresi, Polymer, 2004. 45: p. 2751–9.
305. Sanchez-Solis, A., A. Garcia-Rejon, and M. O., Macromol Symp., 2003. 192: p. 281-92.
306. Chang, J.-H., Y.U. An, S.C. Ryu, and E.P. Giannelis, Polym Bull, 2003. 51: p. 69–75.
307. Chisholm, B.J., R.B. Moore, G. Barber, F. Khouri, A. Hempstead, and M. Larsen, et. al. Macromolecules, 2002. 35: p. 5508–16.
308. Finnigan, B., D. Martin, P. Halley, R. Truss, and K. Campell, Morphology and properties of thermoplastic polyurethane nanocomposites incorporating hydrophilic layered silicates. Polymer, 2004. 45: p. 2249–60.
309. Wu, Y.-P., Q.-X. Jia, D.-S. Yu, and L.-Q. Zhang, *Modeling Young's modulus of rubber–clay nanocomposites using composite theories.* Polymer Testing 2004. 23: p. 903-909.
310. Brule, B. and J.J. Flat, Macromol Symp., 2006. 233: p. 210-216.
311. Kim, S.H. and S.C. Kim, J Appl Polym Sci, 2007. 103: p. 1262–1271.
312. Choi, W.J., H.J. Kim, Y. K.H., O.H. Kwon, and C.I. Hwang, J Appl Polym Sci, 2006. 100: p. 4875–4879.
313. Matayabas Jr, J.C. and S.R. Turner, *Polymer-clay nanocomposites.*, T.J. Pinnavaia and G.W. Beall, Editors. 2000, John Wiley p. 207-25.
314. Takahashi, S., H.A. Goldberg, C.A. Feeney, D.P. Karim, M. Farrell, K. O'Leary, and D.R. Paul, *Gas barrier properties of butyl rubber/vermiculite nanocomposite coatings.* Polymer 2006. 47: p. 3083-3093.
315. Nielsen, L.E., J Macromol Sci (Chem), 1967. A1: p. 929-942.
316. Lape, N.K., E.E. Nuxoll, and E.L. Cussler, J Membr Sci 2004. 236: p. 29–37.
317. Gusev, A.A. and H.R. Lusti, Adv Mater 2001. 13: p. 1641–3.
318. Fredrickson, G.H. and J. Bicerano, J Chem Phys 1999 110: p. 2181-2188.
319. Bharadwaj, K., Macromolecules 2001. 34: p. 9189-92.
320. Merkel, T.C., B.D. Freeman, R.J. Spontak, Z. He, I. Pinnau, P. Meakin, and A.J. Hill, *Ultrapermeable, Reverse-Selective Nanocomposite Membranes.* Science 2002. 296: p. 519.
321. Templin, M., A. Franck, A. DuChesne, H. Leist, Y.M. Zhang, R. Ulrich, V. Schadler, and U. Weisner, Science, 1997. 278: p. 1795.

322. Ulrich, R., A. Duchesne, M. Templin, and U. Wiesner, Adv.Mater., 1999. 11: p. 41.
323. Lahiri, J., E. Ostuni, and G.M. Whitesides, J. Am. Chem. Soc., 1999. 15: p. 2055-2060.
324. Guo, P., RNA Nanotechnology: Engineering, Assembly and Applications in Detection, Gene Delivery and Therapy. J Nanosci Nanotechnol 2005. 5(12): p. 1964-1982.
325. Win, M.N. and C.D. Smolke, Higher-Order Cellular Information Processing with Synthetic RNA Devices. Science 2008. 322(5900): p. 456-460.
326. Carcia, P.F., P.F. Meinhaldt, and A. Suna, *Perpendicular magnetic anisotropy in Pd/Co thin fUm layered structures.* Appl. Phys. Lett. , 1985. 47: p. 178.
327. Corkill, J.M., J.F. Goodman, and R.H. Ottewill, Trans. Faraday Soc. , 1961. 57: p. 1927.
328. Newkome, G.R., C.N. Moorefield, and F. Vogtle, eds. *Dendritic Molecules.* 1996, VCH Weinheim.
329. Klajnert, B. and M. Bryszewska, *Dendrimers: properties and applications.* Acta Biochimica Polonica, 2001. 48(1): p. 199-208.
330. Bosman, A.W., H.M. Janssen, and E.W. Meijer, *About Dendrimers: Structure, Physical Properties, and Applications.* Chemical Reviews, 1999. 99(7): p. 1665-1688.
331. Hsueh, H.-Y., H.-Y. Chen, M.-S. She, C.-K. Chen, R.-M. Ho, S. Gwo, H. Hasegawa, and E.L. Thomas, *Inorganic Gyroid with Exceptionally Low Refractive Index from Block Copolymer Templating.* Nano Lett., 2010. 10: p. 4994.
332. Landschulz, W.H., P.F. Johnson, and S.L. McKnight, *The leucine zipper: a hypothetical structure common to a new class of DNA binding proteins.* Science, 1988. 240(4860): p. 1759-1764.
333. Allis, D.G. and J.T. Spencer, *Nanostructural Architectures from Molecular Building Blocks*, in *Handbook of Nanoscience, Engineering and Technology*, W.A. Goddard III, et al., Editors. 2003, CRC Press: Boca Raton, Florida.
334. Huang, M.H., S. Mao, H. Feick, H.Q. Yan, W.W. Wu, H. Kind, E. Weber, R. Russo, and P.D. Yang, Science, 2001. 292: p. 1897.
335. Yang, H., N. Coombs, and G.A. Ozin, Nature, 1997. 386: p. 692.
336. Mann, S., *Biomineralization: Principles and Concepts in Bioinorganic Materials Chemistry.* First ed. Oxford Chemistry Masters, ed. R.G. Compton, S.G. Davies, and J. Evans. 2001, New York: Oxford University Press.
337. Tanford, C., The Hydrophobic Effect, the Formation of Micelles and Biological Membranes 2nd ed. 1980, New York: Wiley.
338. Hamley, I.W., *Introduction to Soft Matter.* First ed. 2000, New York: John Wiley.
339. Blodgett, K.B., Use of Interface to Extinguish Reflection of Light from Glass. . Phys. Rev., 1939. 55.
340. Petty, M.C., *Langmuir-Blodgett Films, An Introduction.* 1996, Cambridgge: Cambridge University Press.

341. Dulcey, C.S., J.H. Georger, V. Krauthamer, D.A. Stenger, T.L. Fare, and J.M. Calvert, Science, 1991. 252: p. 551.
342. Stine, K.J. and B.G. Moore, *Langmuir Monolayers: Fundamentals and Relevance to Nanotechnololgy*, in *Nano-Surface Chemistry*, M. Rosoff, Editor. 2002, Marcel Dekker: New York.
343. Schwartz, D.K., R. Viswanathan, and J. Zasadzinski, J. Phys. Chem., 1992. 96: p. 10444-10447.
344. Sastry, M., V. Patil, K.S. Mayya, D.V. Paranjape, P. Singh, and S.R. Sainkar, Thin Solid Films, 1998. 324: p. 239-244.
345. Abe, K., T. Hanada, Y. Yoshida, N. Tnaigaki, H. Takiguchi, H. Nagasawa, M. Nakamoto, T. Yamaguchi, and K. Yase, Thin Solid Films, 1998. 327-329: p. 524-527.
346. Schmid, G., Chem Rev, 1992. 92: p. 1709-1727.
347. Yi, K.C., Z. Horvolgyi, and J.H. Fendler, J Phys Chem, 1994. 98: p. 3872-3881.
348. Ravaine, S., G.E. Fanucci, C.T. Seip, J.H. Adair, and D.R. Talham, Langmuir, 1998. 14: p. 708-713.
349. Li, L.S., J. Jin, Y.Q. Tian, Y.Y. Zhao, T.J. Li, S.M. Jiang, Z.L. Du, G.H. Ma, and N. Zheng, Supramol Sci, 1998. 5: p. 475-478.
350. Torimoto, T., N. Tsumura, M. Miyake, M. Nishizawa, T. Sakata, H. Mori, and H. Yoneyama, Langmuir, 1999. 15: p. 1853-1858.
351. Dierking, I., G. Scalia, P. Morales, and D. LeClere, *Aligning and Reorienting Carbon Nanotubes with Nematic Liquid Crystals.* Advanced Materials, 2004. 16(11): p. 865-869.
352. Chang, W.S., L. S., A. Yethiraj, and P.F. Barbara, Single molecule spectroscopy of conjugated polymer chains in an electric field-aligned liquid crystal. J. Phys. Chem. B., 2008. 112(2): p. 448-53.
353. Khatua, S., W.-S. Chang, P. Swanglap, J. Olson, and S. Link, *Active Modulation of Nanorod Plasmons.* Nano Letters. 11(9): p. 3797-3802.
354. Srivastava, S., B.L. Frankamp, and V.M. Rotello, *Controlled Plasmon Resonance of Gold Nanoparticles Self-Assembled with PAMAM Dendrimers.* Chem. Mater. , 2005. 17: p. 487-490.
355. Ung, T., L.M. Liz-Marzan, and P.J. Mulvaney, Phys. Chem. B, 2001. 105: p. 3441-3452.
356. Callister, W.D., *Callister's Materials Science and Engineering*. Adapted by R, Balasubramaniam ed. 2007, New York: John Wiley.
357. Nye, J.F., Physical Properties of Crystals - Their Representation by Tensors and Matrices. 1985, Oxford: Oxford University Press.
358. Hart, E.W., Acta Mater., 1967. 15: p. 351.
359. Cottrell, A.H., *Dislocations and plastic flow in crystals*. 1965, London: Oxford University Press.

360. Murr, L.E., Strain-induced Dislocation Emission from Grain Boundaries in Stainless Steel. Mat. Sc. and Eng., 1981. 51: p. 71-79.
361. Griffith, A.A., Philos. Trans. R. Soc. London, 1920. 221A: p. 163-198.

Index